LIFE-SPAN HUMAN DEVELOPMENT

LIFE-SPAN HUMAN DEVELOPMENT

Dale Goldhaber
University of Vermont

HBJ
Harcourt Brace Jovanovich, Inc.
New York San Diego Chicago San Francisco
Atlanta London Sydney Toronto

to
Ben and Dan
and all who will share their future

ISBN: 0-15-540380-X
Library of Congress Catalog Card Number: 85-71399

Printed in the United States of America

PREFACE

This is an interesting time to write a life-span human development textbook for many reasons. We are in the midst of a time of changing roles for men and women. Each future generation can expect to live longer and enjoy better health than did previous generations. As we become more prosperous as a nation, our notions of life satisfaction and happiness continue to shift from ones associated with material advantage to ones associated with interpersonal relationships.

Perhaps because of the changes in our life patterns, the study of human development is also undergoing change. Scholars who have tended to focus on limited segments of the life-span are now debating the relative influence of events at different ages on the course of development across the entire life-span. This debate has become most focused on the impact of early experience. To what extent are our adult years primarily a reflection of earlier experiences? Are the events of the second half of life so different from those of the first half that the course of development is best understood as discontinuous rather than continuous?

These changes in the study of development are not only occurring in the work of age-based researchers. Those who approach the study of development from other perspectives are also beginning to consider how their conclusions might be influenced by research in related fields. For example, those who study the development of families are giving greater attention to the findings of those who study individual development. The reverse is equally true. What all of these changes in the study of development mean is that perhaps for the first time we are in a position to begin to understand the variety of factors that influence the course of development over as grand a scale as an entire life-span.

This text represents an attempt to bring you up to date on the study of development. It pictures the process of development as a lifelong activity, one that is as evident in the behavior of the very old as it is in the behavior of the very young. I have chosen to "package" this process in a succession of developmental stages, each reflecting a unique balancing act between the changing demands of the physical and social environments and the changing capabilities of the individual.

From this perspective, known as a cognitive-developmental view, we are each seen as actively involved in determining the course of our individual life-spans. Quite literally, the process of development is the process of constructing meaning out of the myriad life experiences we each encounter. If, after reading the text, you find this as interesting a time to be reading about development across the life-span as I have to be writing about it, my time will have been well spent.

ORGANIZATION OF THE CHAPTERS

The human life-span is a long and complex phenomenon. The impact of events occurring at one age may not be evident until another. Factors that are prominent at one stage exert little influence at another. Even the very meaning of events changes as new experiences place old ones in a new light.

Chapter 1 provides a framework for understanding this long and complex phenomenon. This framework is presented in five sections. The first examines the various ways in which the sequence of developmental events across the life-span has been organized and studied. The second examines the "human ecology" of life-span development, that is, the variety of factors at the different levels of society that exert an influence on the course of development. The third section provides a synthesis of the first two sections. It is this synthesis that presents the process of development across the life-span as a succession of balancing acts.

The fourth section concerns the actual study of life-span development. You may never be in a position to research the process of development; you may never even want to be in such a position. But you will always be in a position to be affected by the findings of such research, and for this reason alone it is important to have at least an intuitive grasp of both the power and the limitations of human development research.

The last section of Chapter 1 concerns the relationship of the study of human development to the provision of human services. Just as only some of you will go on to do research, only some of you will pursue careers in the various human service professions. Your lives, however, will be affected by what happens within these professions, and again, for this reason alone it is important to understand how the study of development translates into the provision of human services. This is not the only reason for this section, however. Whether we identify ourselves as human service providers or not, we are all nevertheless involved in the provision of human services. Directly or indirectly, intentionally or unintentionally, what each of us does in some way touches the lives of others. The second reason for this last section of Chapter 1 is to help you be more aware of this fact and of its implications.

Chapter 2 concerns prenatal development and birth. It follows the course of development from conception through birth and describes the various factors that influence development at this time; in other words, it describes the context of prenatal development. This context exists at a variety of levels — it is determined by such diverse factors as the intricate workings of the genetic code and the greater humanizing of the childbirth process. The chapter ends with a discussion of the relationship between human development and human services.

Chapters 3, 4, 5, and 6 all follow the same general outline. Chapter 3 examines the course and context of development during the first three years of life, while Chapters 4, 5, and 6 focus on early childhood, middle childhood, and adolescence, respectively. Each of the four chapters starts with an overview of the particular stage. The work of Erikson is particularly useful in this regard: he offers a good sense of the primary developmental events typical of each stage. The introduction of each chapter is followed by a review of the course of physical/motor, cognitive, and social development during the particular stage. The section on the course of development is followed by a discussion of the contexts of development during the stage. This discussion examines the developmental impact of the various settings children typically encounter. Each chapter then examines the significance of development during the stage for the course of development across the entire life-span. While development may involve a succession of stages, these stages are not unrelated. Certain themes or issues seem to appear and reappear; certain problems are encountered and reencountered. This section examines this developmental fabric. The final section of each chapter focuses on the relationship between human development and human services.

Chapter 7 serves two purposes. The first is to distinguish the nature of development through adolescence from the nature of development during adulthood. As will become apparent in this discussion, the nature of developmental events and the presumed causes of these events change following adolescence. In fact, some developmentalists even question whether the developmental events of adulthood are really developmental.

The second part of Chapter 7 reviews the course of development between the ages of eighteen and twenty-five. This is a very important stage of the life-span since the events occurring at this time, including the decision to attend college, establish the points of entry into the adult world. For most people, the transition years are a unique experience. They are as distinct from adolescence as they are from adulthood. The chapter ends with a discussion of the significance of the transition years in the life-span and the role human service providers can play in this transition.

Chapters 8, 9, and 10 each review the three phases of the adult years. Early adulthood typically consists of the age range of twenty-five to forty; middle adulthood lasts from the early forties to the early sixties; and late adulthood consists of the remaining years until death. Each of the three chapters reviews the way the three "careers" of the adult years — intimate, parent, and worker — are pursued by men and women in different segments of our society. Like earlier chapters, each chapter concludes with a discussion of the significance of the particular stage in the life-span and the role played by human service providers.

Chapter 11 could perhaps as easily be the first chapter as the last. It is an attempt to summarize and integrate, at a more general level, the various specific issues and topics presented throughout the text. It is the last chapter rather than the first because I think that learning at the introductory level is an intuitive process. The sometimes abstract notions discussed in this chapter are more likely to be appreciated after you have read the data concerning the life-span than they would be before.

Each chapter in the text is followed by three endnotes. The first endnote provides a very short summary of the key points made in the chapter. The second endnote lists the key words used in the chapter. Some of these words are of a technical nature and are unique to the discipline of human development; others are words that are in common usage but take on a more specialized meaning within the discipline. The third endnote provides a short annotated bibliography of works on some of the topics discussed in the chapter.

A WORD ABOUT WORDS

All writers find themselves facing the dilemma of style and gender. Do you use the pronoun "he" throughout the text, assuring the reader that no sexist connotations are intended? Do you use both "he" and "she" throughout the text, therefore necessitating such inelegant phrases as "when he or she takes his or her"? Do you alternate "he" and "she" chapter by chapter, so that the reader must always be conscious of whether a chapter is a "he chapter" or a "she chapter?"

As you have probably surmised, I have chosen none of these options. Rather I have chosen to write in the plural, asking myself each time I talked about infants or adolescents or adults if the statement is equally true of men and women. The answer I got often depended on what stage of the life-span I was discussing. In general, the early and later years of the life-span appear to be considerably less sex typed than the middle years (approximately early adolescence to the early forties), a pattern that has significant implications for the course of development across the entire life-span.

So, as you read, you may be assured that "he" refers to "he," that "she" refers to "she," and that anything else refers to all of us — with one exception. I have not taken the liberty of changing the wording of quotes. Rather I trust that the context of the discussion surrounding the quote will provide you a clear indication whether the quote concerns one or the other or both sexes. I also trust that you will share with me the belief that no matter what words the quoted authors used, they meant not to be sexist, especially the last author quoted.

ACKNOWLEDGMENTS

The writing of this text would not have been possible without the support and advice of a number of people. Heading this list is my colleague Professor Armin Grams. His encouragement for me to pursue the project, his careful and critical readings of early drafts of the chapters, and more generally, his support of my academic career have all helped create whatever excellence this text can claim.

To Nancy Datan of the University of Wisconsin, Green Bay, I owe a special debt of gratitude. The very fact this text exists at all has much to do with her convincing my publisher that the project had merit — a foresighted argument given the state of the first drafts of the first few chapters. I owe Datan for more than her support, however. Her careful and critical reviews of each chapter have also significantly contributed to, again, whatever excellence this text can claim. She is a very supportive, very critical reviewer in the best sense of both words.

The writing of this text began during a sabbatical leave from the University of Vermont. I would like to thank Larry Shelton and Charles Tesconi for making the sabbatical possible and George Forman of the University of Massachusetts, Amherst, for arranging a Visiting Scholar appointment for me during this period.

A number of scholars have reviewed drafts of the text. Mary Attig (Pennsylvania State University) and Jane Ellen Maddy (University of Minnesota) each reviewed drafts of the entire manuscript. Their comments and suggestions for improvements have been very useful. They will find most of them incorporated in this final version.

James Youniss (Catholic University) reviewed early drafts of the first five chapters. His recommendations, particularly those pertaining to the cognitive aspects of development, proved valuable in the reorganization of these chapters.

Mary Kay Biaggio (Indiana State University), Paul Bomba (University of Texas, Arlington), Ann Mullis (North Dakota State University), and Sandra Singer (Indiana State University) each reviewed final drafts of the text. Their comments were also very useful, particularly in terms of chapter organization and clarity of expression.

I want to thank Martha MacDonald, Betty Bourgea, and Kit Landry for their help in typing the manuscript. If you ever saw my handwriting, you would realize the extent of my debt to them. Julie Bode, Cheryl Farnum, and Nancy Littlehale played a major role in the compilation of the bibliography. Their work was consistently excellent.

Finally, I want to thank my publisher for its support of this project. Warren Abraham, Psychology Editor, Sue Miller, Developmental Editor, and Julia Morrissey, Copy Editor, could not have been more helpful.

CONTENTS

8 THE EARLY YEARS OF ADULTHOOD: AGES TWENTY-FIVE TO FORTY 381

9 THE MIDDLE YEARS OF ADULTHOOD: AGES FORTY TO SIXTY 437

LIFE-SPAN HUMAN DEVELOPMENT

1
PROLOGUE TO THE STUDY OF THE LIFE-SPAN

CHAPTER OUTLINE

The Course of Development

Developmental Stages As Sequences of Behaviors

A Biologically Based Sequence
An Environmentally Based Sequence

Developmental Stages As Constructions of Reality

Developmental Stages As Responses to Social Expectations

Developmental Tasks
Psychosocial Tasks

An Overview of Stage-Based Theories

Causes of Development
Timetable for Development
Specificity of Focus

Limitations of Stage-Based Theories

Determination of Causality
Deemphasis of Variability
Coherence
Validity
Cultural Bias

The Context of Development

The Human Ecosystem: A Systems View
The Human Ecosystem: A Family View
Learning and Development

Direct Learning
Indirect Learning

Limitations of the Ecological Approach

Determinants of Causality
Explaining Complex Behavior
Behavior versus Development

A Synthesis

The Study of Development Across the Life-Span

The Process of Research: Defining the Topic

The Process of Research: Methods of Data Collection

Longitudinal and Cross-Sectional Research Strategies
Cohort Research Strategies
Controlled Research Strategies
Naturalistic Research Strategies
Clinical, or Interview, Strategies

The Process of Research: Methods of Data Analysis

Descriptive Statistics
Patterns of Relationships
Inferential Statistics

The Process of Research: Limitations

Human Development and Human Services

A Common Standard
Identification of Services
Effective Design
A Common Frame of Reference

Summary

Key Terms and Concepts

Suggested Readings

We are born, we grow and develop, and eventually, we die. How this sequence of events is regulated and why it takes different forms for different people are questions that have been asked at least since the dawn of recorded history. It has only been within the past one hundred years, however, that the manner in which these questions have been asked has changed from the contemplative and philosophical to the analytic and the scientific. This text reports on the findings of these more recent efforts to unravel the patterns of development across the life-span. It surveys the course of human development at the various stages of the life-span, discusses ways in which development is influenced by the various contexts in which it occurs, and suggests ways in which an understanding of development across the life-span can aid the design and provision of services that are intended to facilitate the process of development. Since many of you who read this text are likely to pursue careers in the psychological, health, educational, or social service components of the human service professions, this text also serves as a foundation for more advanced study in these professions.

This chapter serves as a prologue. The dictionary defines a prologue as a preliminary act or course of action foreshadowing greater events. On this optimistic note, let's begin. This chapter introduces the various perspectives that are used to describe and relate the numerous events that constitute one's life-span. Because the events are so numerous and the causes so intertwined and embedded in each other, any attempt to unravel the process and present the components individually soon confronts the same problem that must be confronted in finding the beginning of a circle or deciding which came first—the chicken or the egg. That is, the solutions must of necessity be somewhat arbitrary. The two most common solutions have been to take either a stage-based approach or an ecological approach. In the **stage-based approach,** the process of development is studied by observing the changes that occur in individuals as they age. The **ecological approach,** on the other hand, focuses on ways in which the various institutions and people that an individual encounters in a lifetime influence the course of that person's development. Each approach provides an insight into one facet of the process of development. Both are needed to gain a complete understanding of the process.

It is also possible to organize the study of development across the life-span in terms of the different theories of development. Theories of development differ in a variety of ways. They differ in terms of method, some preferring highly controlled laboratory studies and others more naturalistic approaches. They differ in terms of what they will accept as evidence to support their views, some accepting subjective impressions and others only objective observations. They differ in terms of their focus of interest, some being concerned with the way people think and learn while others examine the interpersonal aspects of development. Finally, they differ in terms of explanation, some arguing that the process of development is determined largely by biological events, others placing great emphasis on the environment, and still others focusing on the role individuals play in directing the course of their own development.

Generally, developmentalists can be associated with one of four broad theoretical orientations—cognitive, humanistic, learning, and psychoanalytic. Each of the four reflects a unique combination of the factors listed in the preceding paragraph (see Table 1-1). As is true of the stage-based approach and the ecological approach, no one theory can adequately

TABLE 1-1 *Comparing Developmental Theories*

	COGNITIVE	HUMANISTIC	LEARNING	PSYCHOANALYTIC
COMMON METHODS	Controlled and naturalistic	Self-report	Controlled	Self-report and naturalistic
DATA SOURCE	Objective and subjective	Subjective	Objective	Subjective
PRIMARY FOCUS	Development of thought	The self	Change in behavior	Interpersonal relations
PRIMARY CAUSE	Biological, environmental, the individual	The individual	Environmental	The individual

describe and explain the process of development across an entire life-span. Each contributes an important piece to the puzzle.

It is possible to present the material in a life-span text using either a stage-based/ecological orientation or a theoretical orientation. I have chosen the former because, at an introductory level, it offers the broadest and most comprehensible frame of reference.

THE COURSE OF DEVELOPMENT

Expectations of development based on age are common in our everyday speech. We explain the behavior of a child by pointing out that two-year-olds are very negative. They always say "no" to every parental request. A nine-year-old is told by his father that he is too old to be afraid of the dark. The parents of a twenty-eight-year-old frequently inform her of the marriages of various friends and "inquire" about her plans for the future. Worried parents ask their pediatrician if there is reason to be concerned that their one-year-old isn't walking. A middle-aged couple is told by the landlord that the new couple moving into the apartment across the hall isn't married. The next day, the unmarried couple moves into the apartment. The middle-aged couple is very surprised to find that their new neighbors aren't in their twenties but rather are a retired couple in their late sixties.

Although age in no way explains why people behave as they do, it is a generally accepted standard against which the course of development can be observed and evaluated. For the purpose of analysis, ages are grouped together to form **stages of development.** These stages of development are then analyzed from a particular theoretical perspective. In general, the type of theoretical perspective employed determines what aspects of development will be examined. The majority of stage-based theories of human development can be assigned to one of three broad classifications. One set of theories views development as a sequence of behaviors. Among theories in this category are Gesell's maturational theory of child development (Gesell, Halverson, Thompson, Castner, & Ames, 1940; Gesell & Ilg,

1949), Levinson's (1978) stage theory of adult male development, and Frieze's (1978) stage theory of adult female development. A second set of theories deals with the cognitive aspects of development. Piaget's (1952, 1980) theory of cognitive development is representative of this type of theory. A third set of theories views development as a response to social expectations. Havighurst's (1972) developmental task theory and Erikson's (1950, 1968) theory of psychosocial stages are examples of this type of theory.

Developmental Stages As Sequences of Behavior

Theorists who view development as sequences of behaviors attribute these sequences to biological and environmental factors. While these factors work together to influence behavior, theorists often concentrate on one of the broad categories.

A Biologically Based Sequence

Arnold Gesell was a pediatrician teaching at Yale during the 1940s. Through his work with children, he became very impressed with the regularity of their development. To Gesell, this regularity was evident in the fact that all children reach various developmental milestones (for example, sitting upright unsupported, taking the first steps, drawing a circle or a square, throwing a ball, hopping, skipping, and so forth) at about the same age. Gesell believed that this uniformity of development was a reflection of the role that heredity plays in human development, specifically the regulation of the process of **maturation**—the process through which biological structures achieve their mature state. This belief led him to develop the **maturational theory of child development.** Much of Gesell's work was devoted to documenting the changes in behavior associated with the maturation process, and he has left as a fitting legacy many tables of **developmental norms,** which are in use today to evaluate the normality of a child's rate of development. Examples of these developmental sequences are provided in Table 1-2.

Gesell's contributions highlight the genetic component of growth and development. They remind us that part of who we are and what we do and what we can do is defined by our species-specific characteristics—that is, those aspects of our development that are common and unique to all of us irrespective of culture.

An Environmentally Based Sequence

The species component is an integral part of the developmental process, but it is not the only part. Development is also a reflection of the environment in which it occurs. Even the relatively universal aspects of development that Gesell studied show a considerable degree of variability. Not all children of the same age perform the same tasks equally well. Further, Gesell's focus on the developmental milestones of childhood obscures the fact that the various components of the developmental process exert their influence in varying degrees at different points in the life-span. Maturation is a much more significant regulator of the course of development during

TABLE 1-2 *Gesell's Developmental Norms*

AGE IN MONTHS	EATING: SELF-FEEDING	EATING: GENERAL RESPONSE TO MEALS	SLEEPING	DRESSING
15	Holds cup with digital grasp	Interested in participating in eating	Put to bed easily	Cooperates in dressing by extending arm or leg
18	Lifts cup to mouth and drinks well	Hands empty dishes to mother	Lying down next to child or sitting next to crib usually induces sleep	Can take off mittens, hat, and socks; can unzip zippers
21	Handles cup well: lifting, drinking, and replacing		Keeps demanding things such as a drink, food, or the toilet before going off to sleep	
24	Holds small glass in one hand while drinking	Is apt to dawdle and play with food, especially stirring it	Demands to take toys to bed with him	Can remove shoes if laces are untied
30			Prolongs process of going to bed by setting up complicated ritual to be adhered to	
36	Pours well from a pitcher	Frequent getting up from table		Is able to unbutton all side and front buttons by pushing button through buttonhole
48		Likes to serve self		Is able to dress and undress with little assistance

Source: Adapted from Gesell, 1940.

childhood than it is during adulthood. During adulthood, the culture assumes an increasingly directive role. Levinson's (1978) stage theory of adult male development and Frieze's (1978) stage theory of adult female development each demonstrate how the biological clock of childhood is replaced by the social clock of adulthood (Neugarten, 1968).

Levinson's research on adult men led him to develop the **stage theory of adult male development.** Like Gesell, he was impressed with the similarity in the life patterns of the subjects he studied. To Levinson and his colleagues, the similarity of the patterns clearly indicated that cultural factors can produce the same age-related changes that biological factors can produce.

Through his research, Levinson differentiated three eras, or stages, in the male life cycle. The first stage, *early adulthood,* lasts from approximately

Even though he has achieved his goals of an executive position, a private secretary, and a corner office with a view, like many of the middle-aged men in Levinson's sample, this man may feel unfulfilled. Now that his push for the top has slowed, he has time to contemplate the cost of occupational success in terms of personal happiness.

age eighteen to age forty-five. It is a time when men leave home and come to think of themselves as adults, when they make an initial and then a more permanent commitment to a vocation, and, for almost all men in our culture, when they enter marriage and start a family. Levinson found *middle adulthood,* from approximately the early forties to the middle sixties, to be initially a period of taking stock (the fabled mid-life crisis) and then a period of reordering priorities. Typically, this reordering places greater emphasis on the family and relationships and less emphasis on occupation, a reversal of the pattern typical of early adulthood. The transition to *late adulthood* in the early sixties again calls for taking stock and reordering priorities.

Levinson's work makes evident the often conflicting demands that are placed on adults in our culture. The men in his study continually faced the dilemma of attending to the instrumental side of their lives in the form of their careers and the expressive side in the forms of their marriage, family, and personal friendships.

In her **stage theory of adult female development,** Frieze (1978) notes a similar but often more complicated dilemma in the lives of adult women. The added complications are due to the fact that women are the primary care givers in most families and as a result face a more difficult task in attempting to integrate the instrumental and expressive components of their lives. Many women find it necessary either to temporarily "retire" from their careers to care for their young children or to manage somehow to simultaneously meet the demands of a job and the demands of a family. Later, having managed to get the youngest child off to school, these women are often eager to return to work. Ironically, they may form this commitment to work at the very point at which their husbands, like Levinson's men, are beginning to realize that their career goals will not be fulfilled and, therefore, to ask if their commitment to work should continue to be as great as it has been. These changes usually take place during the middle or late thirties.

Levinson's and Frieze's views of adult development suggest that there are probably a number of social clocks during adulthood and, as is true of the ones we each wear on our wrist, that they are usually out of sync with each other. Like Gesell's more precise biological clock, however, these social clocks also emphasize the fact that behavior often correlates with age and that one way individuals often evaluate their own development is by determining the degree to which it is "on schedule."

Developmental Stages As Constructions of Reality

Piaget's (1952, 1980) **theory of cognitive development** is an example of another way of using stage theory as an organizer of development over the life-span. Rather than relating age to specific behaviors, Piaget delineated four stages of development that are defined in terms of the types of mental processes, or operations, individuals have available at different ages (see Table 1-3). These mental operations serve as the filters through which individuals understand the events that occur in their environments. Each stage of **cognitive development** is based on a unique construction of reality that is reflected in individual behavior.

Piaget was a wonderful observer of children. He became intrigued by the fact that the child's view of the world isn't simply an incomplete or incorrect version of the adult's view but rather that the child's view has a logic all its own. In other words, there seem to be qualitative differences in the way children and adults view the world. Piaget devoted most of his professional life to trying to understand the nature of these differences.

Why is it that six-month-olds will not search under a blanket for a favorite rattle that they have just seen placed there but eleven-month-olds will? Why is it that four-year-olds believe that when you pour milk from a short, wide glass into a tall, thin one, it somehow becomes more milk? What makes it possible for the five-year-old to discover that no matter how you arrange five marbles, every time you count them, you will still get five? Or, what is it that makes the fourteen-year-old so idealistic and critical and the ten-year-old so pragmatic and (usually) cooperative? Piaget said that each of these instances reflects the child's unique construction of reality—a construction that results from both the influence of the environment and the continued maturation of our biological system. Out of this continuing interaction between environment and biological maturation develop the four unique sets of mental structures that are associated with Piaget's stages of development. The first set of structures emerges during the *sensorimotor stage.* This stage occurs during approximately the first two years of life. The second stage, the *preoperational stage,* occurs approximately between the ages of two and seven, while the third stage, the *concrete operational stage,* occurs approximately between the ages of seven and twelve. The *formal operational stage* is the last stage. It begins about age twelve and continues through the remainder of the person's life.

Piaget saw this developmental process as a never-ending attempt to adapt to the ever-changing demands of the environment. This adaptation process has two components. The first is a process of **assimilation.** We understand events by integrating them with previous experiences. But the process isn't merely one of matching new experiences to old ones: our construction of reality isn't a carbon copy of reality. We act upon information in a manner analogous to the way we digest food. Information, like food, must be transformed before it becomes useful. Food is transformed by being acted upon at various sites in the digestive system. In a like manner, information is acted upon by or assimilated into existing cognitive structures. The end product of this assimilation reflects the nature of the information and the cognitive structures. We are unable to digest certain food and assimilate certain types of information at different stages in our lives. Three-year-old John always wants his mother's attention. He is un-

TABLE 1-3 *Piaget's Four Stages of Cognitive Development*

STAGE	APPROXIMATE AGE RANGE	REPRESENTATIVE SKILLS
SENSORIMOTOR	Birth to 24 months	Coordination of simple behaviors
		Simple means-ends separation
		Awareness of permanence of objects in the physical world
PREOPERATIONAL	2 to 7 years	Early representational ability
		Use of language and pretend play
		Rudimentary notions of cause, number, and spatial relationships
CONCRETE OPERATIONAL	7 to 12 years	Conventional notions of time, cause, number, spatial relationships
		Ability to apply logical thinking to real events
		Sensitive to viewpoint of others
FORMAL OPERATIONAL	Age 12 onward	Ability to use hypothetical-deductive logic
		Introspective thought
		Appreciation of abstract concepts and metaphor

able to realize that his mother has many roles in addition to the one of mother. The preoperational nature of his mental structures prevents him from viewing people and events from more than one perspective. His mother is only too aware of her many, sometimes conflicting, roles. Eight-year-old Anne reasons from past experience. Her father asks her not to ride her bike in the street because she might be hit by a car. She replies that she rode her bike in the street yesterday and "not even one car even came close." The nature of her concrete operational reasoning prevents her from realizing that her past experience, not being hit by a car, is only one form of possible experience, which includes being hit by a car.

Assimilation is only one half of the adaptation process. **Accommodation** is the other half. We change as a result of new experience and continuous biological maturation. We adapt to new situations that can't be assimilated into existing mental structures by modifying those structures, that is, by accommodating to the new experience. Although the process is gradual, there do seem to be intervals when the continuing impact of successive accommodations produces a relatively abrupt and radical reorganization of mental structures. These transitions mark the passage between

One of the achievements of the sensorimotor stage of development is the realization that objects continue to exist when they are not in sight. It is this sense of object permanence that may be motivating this child to await the return of her mother.

the four stages. Behavior during transitions appears somewhat more disorganized and tentative than previous behavior. Five-year-old Kevin used to be certain that when you pour milk from a short, wide glass to a tall, thin one you get more milk because "it's taller." Now he is not so sure. Sometimes he says it's more; sometimes he says it's the same. In explaining his answers, he mentions both the height and the width of the two glasses but doesn't seem able to relate the two dimensions. He seems confused and dissatisfied with his answers. He feels fairly sure that his old answer is wrong but isn't sure why. Two years later Kevin, now seven, is presented the problem again. His answer is now quick and sure: "There has to be the same because you didn't add any and you didn't take any away." Kevin's transition experience is typical of most of the stage transitions we experience throughout the life-span.

Some events or series of events produce greater disruption than others. When transitions involve changes in well-established relationships or long-standing patterns of behavior, a period of relatively disorganized, often ineffective, and frequently contradictory behavior ensues. Old patterns are no longer effective, but new ones have not yet been established to replace them. Gradually, however, new patterns of organized, relatively effective, and coherent behaviors emerge.

This process of reorganization is evident in the following example. Jack and Helen have been married for thirty years. At two on a Sunday afternoon, Helen receives a call informing her that Jack has died of a massive heart attack while playing golf. The next few days are a blur as arrangements for the funeral are made and friends and relatives pay their condolences. Jack and Helen had been a close couple. They enjoyed many of the same interests and activities. In the weeks and months that follow the funeral, Helen encounters situation after situation that reminds her how intertwined their lives were. She turns down social invitations from friends because she feels odd about going places alone. Household routines such as eating and cleaning become increasingly irregular as there seems to be less reason to do them anymore. Every now and then, for no apparent reason, she starts crying. Gradually, over the course of the next year, the crying episodes become less frequent, and she feels less tired and irritable. She finally accepts one of many invitations to spend a week with her sister's family. She decides to move to a smaller apartment, one with "fewer memories." She takes a part-time job at a local clothing store, joins an amateur theater group, and starts dining again with old friends. She has managed to establish a new life-style for herself. She wishes Jack were still alive and that her life could still be as it was. But she knows that that is not possible, and in learning to cope without him, she has discovered new strengths in herself. Although Helen's circumstances differ from Kevin's in many ways, both demonstrate the same general transition pattern that is evident throughout the life-span.

The complementary processes of assimilation and accommodation are the means by which the maturing biological organism and the changing social environment interact to form the developing psychological person. Psychological development involves growing in awareness, learning to see oneself in relation to others, learning to understand the events that occur in the spatial and temporal domains of experience, and acquiring the ability to adapt to change. Through this process, the individual develops a sense of mastery, that is, a sense of being an active participant in and director of the course of his or her development. For Piaget, the psychological level is distinct from the biological and the social. Although the latter both contribute to psychological development, the interaction process creates a level of functioning that is no more reducible to biology or social environment than the "wetness" of water is reducible to either hydrogen or oxygen. Just as wetness is a characteristic of the interaction of hydrogen and oxygen, psychological awareness is a characteristic of the interaction of the biological and the social. The specifics of Piaget's stages will be presented in subsequent chapters. For now it is sufficient to understand Piaget's theory as a sequence of four successive psychological stages in the way we construct our understanding of the people, objects, and events in our environment.

Developmental Stages As Responses to Social Expectations

A third way of using developmental stages to explain the process of development emphasizes the role that culture plays in determining age-related behaviors. The two examples of this kind of explanation are Havighurst's (1972) developmental tasks and Erikson's (1950, 1968) psychosocial stages.

TABLE 1-4 Havighurst's Developmental Tasks

INFANCY AND EARLY CHILDHOOD	MIDDLE CHILDHOOD	ADOLESCENCE
Learning to walk	Learning physical skills necessary for ordinary games	Achieving new and more mature relations with age mates of both sexes
Learning to take solid food	Building wholesome attitudes toward oneself as a growing organism	Achieving a masculine or feminine role
Learning to talk	Learning to get along with age-mates	Accepting one's physique and using the body effectively
Learning to control the elimination of body wastes	Learning an appropriate masculine or feminine social role	Achieving emotional independence of parents and other adults
Learning sex differences and sexual modesty	Developing fundamental skills in reading, writing, and calculating	Preparing for marriage and family life
Forming concepts and learning language to describe social and physical reality	Developing concepts necessary for everyday life	Preparing for an economic career
Getting ready to read	Developing conscience, morality, and a scale of values	Acquiring a set of values and an ethical system as a guide to behavior—developing an ideology
Learning to distinguish right and wrong and beginning to develop a conscience	Achieving personal independence	Desiring and achieving socially responsible behavior
	Developing attitudes toward social groups and institutions	

Developmental Tasks

In Havighurst's **developmental tasks theory,** development is viewed as a continuous process of learning new tasks:

> Living is learning, and growing is learning. One learns to walk, talk and throw a ball; to read, bake a cake, and get along with age-mates of the opposite sex; to hold down a job, to raise children; to retire gracefully when too old to work effectively, and to get along without a husband or wife who has been at one's side for forty years. These are all learning tasks. To understand human development, one must understand learning. The human individual learns his way through life. (Havighurst, 1972, p. 1)

Havighurst has provided a list (see Table 1-4) of developmental tasks appropriate to each stage of the life-span. These tasks are defined by the particular society and reflect the society's need to maintain itself. The or-

EARLY ADULTHOOD	*MIDDLE AGE*	*LATE MATURITY*
Selecting a mate	Assisting teenage children to become responsible and happy adults	Adjusting to decreasing physical strength and health
Learning to live with a marriage partner	Achieving adult social and civic responsibility	Adjusting to retirement and reduced income
Starting a family	Reaching and maintaining satisfactory performance in one's occupational career	Adjusting to death of spouse
Rearing children	Developing adult leisure-time activities	Establishing an explicit affiliation with one's age group
Managing a home	Relating oneself to one's spouse as a person	Adopting and adapting social roles in a flexible way
Getting started in an occupation	Accepting and adjusting to the physiological changes of middle age	Establishing satisfactory physical living arrangements
Taking on civic responsibility	Adjusting to aging parents	
Finding a congenial social group		

dering of the tasks reflects the society's awareness of the biological basis of development. All of the various components of a society, such as family, media, schools, economy, and government, are involved in the learning of these tasks. Some, like the school, are involved during limited segments of the life-span; others, like the family, exert influence throughout an individual's life. Some, like the economy, play an indirect role, for example by influencing job opportunities; others, like government, can even legislate developmental tasks.

Havighurst believed that the developmental tasks approach has value for two reasons. First, it defines the basic expectations a society holds for each of its members. Second, and perhaps more important, it defines the services a society must provide its members. If a society expects its school-age children to develop fundamental skills in reading, writing, and calculating, then it must provide schools. If a society expects young adults to start a family and manage a home, then it must provide the necessary social and economic services to facilitate these tasks. And if a society expects its senior members to establish satisfactory physical living arrangements, then it must ensure that suitable housing is available for the elderly. In general, the developmental tasks approach defines the support services that must be available to individuals at each stage of the life-span.

Psychosocial Tasks

According to Erikson's (1950, 1968, 1980) **theory of psychosocial stages,** each stage presents a psychosocial task that must be adequately resolved if development is to continue (see Table 1-5). Each stage serves as a necessary building block for the next. Incomplete resolution of the psychosocial task of one stage adversely affects resolution of the next. For example, the psychosocial task of the first stage is developing a sense of trust. The environment for the newborn is strange and unpredictable. During the first year and a half, the infant must discover that the environment is predictable—that the sight of the breast or bottle is associated with food, that the parent's touch is gentle and loving, that cats sometimes scratch when their tails are pulled. Through these early encounters with a predictable environment, infants develop a sense of mutuality, or trust, with their care givers. Through this sense of mutuality, infants begin to learn the patterns of giving and getting that are necessary for their continuing socialization. Specifically, this sense of trust in self and in the environment is a necessary prerequisite for the second stage of psychosocial development: developing a sense of autonomy. Infants who have little trust in themselves or their environment are not very likely to venture out into a seemingly unpredictable environment to explore its many potential wonders.

Although the major focus of Erikson's theory is the resolution of the particular task associated with each stage, he also maintains that a secondary task of each new stage is the incorporation of previous tasks into the new level of functioning (Erikson, 1980). For example, developing a sense of trust is not restricted to infancy. Each new stage exposes us to new relationships and situations. To deal successfully with new encounters, we must be able to extend our trust in ourselves and in others to include these new experiences. Similarly, adolescents develop a sense of identity for the first time, but they must continually extend and modify that initial sense of identity as they encounter the various tasks of adulthood, particularly mar-

Erikson would probably consider this child's wary expression a typical reaction to being confronted by a stranger. Infants who fail to develop a sense of trust in their general environment often have difficulty mastering subsequent stages of psychosocial development.

TABLE 1-5 *Erikson's Stage Theory of Psychosocial Development*

STAGE	TYPICAL AGE RANGE	POSITIVE RESOLUTION	NEGATIVE RESOLUTION
1	Birth to 24 months	A sense of trust	A sense of mistrust
2	2 to 3 years	A sense of autonomy	A sense of shame and doubt
3	3 to 6 years	A sense of initiative	A sense of guilt
4	6 to 12 years	A sense of industry	A sense of inferiority
5	12 to 18 years	A sense of identity	A sense of role confusion
6	18 to 35 years	A sense of intimacy	A sense of isolation
7	35 to 60 years	A sense of generativity	A sense of stagnation
8	60 years to death	A sense of ego integrity	A sense of despair

Source: From Erikson, 1968.

riage and parenthood. Like Piaget's theory, Erikson's theory illustrates the fact that one common thread or pattern of development across the entire life-span is the repeated encounter of developmental events at successively more sophisticated or mature levels of functioning.

Also like Piaget, Erikson views development in a relative sense. Piaget talked about the individual establishing a degree of equilibrium with the environment; Erikson talks about optimum balance. Too much trust in the environment is as counterproductive as too little trust. Too little industry (the task of the fourth stage) interferes with the school-age child's need to acquire the basic educational skills that our society demands of its members; too much industry, however, results in people who believe that others value them only to the extent that they can produce something. The pattern is also true for the other six stages.

An Overview of Stage-Based Theories

While all stage-based theories use chronological age as the index of level of development, it is evident from the previous discussion that they differ widely in all other aspects of their focus. This diversity of focus is not the result of correct or incorrect perceptions, but rather it is the result of the multifaceted nature of the developmental process itself. Because development is a multifaceted process, it is difficult (if not impossible) for one theory to examine every contributing factor. Thus individual theories focus on specific aspects of development. As a result, it is often difficult to get a complete view of development over the life-span. In an effort to provide a broader focus, this section will examine how the various theories differ in three key areas: the causes of development, the timetable for development, and the specificity of their focus.

BOX 1-1
FREUD'S LEGACY

Sigmund Freud and his psychoanalytic theory occupy a unique place in the discipline of human development. From a broad historical perspective, no other single theorist, not even Piaget, has exerted such a profound influence on our study and understanding of the process of development. Yet Freud's theory today generates little developmental research. Taken as a literal statement of the process of development, the theory is considered either too vague to be systematically evaluated or too reflective of the pre-World War I Victorian era in which it evolved. This does not mean that Freud's ideas are no longer of importance. On the contrary, the questions that Freud asked continue to be raised by others, and the answers he derived continue to be debated and reevaluated. Freud thought that behavior and development were motivated by a psychic energy which he called libido. In his view, as children develop, this energy tends to focus or concentrate in different parts of the body. His sequence of developmental stages (oral, anal, phallic, latency, and genital) reflects the migration of the libido.

According to Freud's theory, for the infant in the oral stage, most behavior is motivated by the libido's concentration around the mouth. Chewing, eating, sucking, biting, and spitting are all common behaviors during infancy and are all focused around the mouth. Freud described similar patterns for the other stages. His theory suggests that if the normal expression of these behaviors is blocked, perhaps by inappropriate parenting, then frustration develops. If the frustration becomes sufficiently intense for a prolonged period, the individual fixates at that stage. Fixation is an important notion since Freud used it to explain much of the adult's behavior. The adult whose oral experiences were sufficiently frustrated so as to become fixated might, literally, eat constantly as a way to compensate for these early frustrations or, more figuratively, might be obsessively motivated to acquire a large personal fortune. If the fixation involved more expressive rather than incorporative as-

Causes of Development

Theories such as Gesell's emphasize the relatively universal aspects of human development, those aspects that are evident in all societies. Univer-

pects of oral behavior (such as severe punishment for biting), then the fixation might take an expressive form such as excessive sarcasm. The specific characteristics of the adult's inappropriate behavior reflect the nature and stage of the fixation.

Over the course of development, the location of the libido continues to be the primary determinant of behavior. As children gain more experience in dealing with others, however, their mode of reducing tension or frustration changes. For the young infant, behavior is motivated by an instinctual need for immediate gratification and tension reduction. The structure regulating this behavior is called the id. Gradually, infants and preschoolers gain an elementary understanding of the relationships between their behaviors and those of others. As this discovery becomes further understood, a second structure, the ego, also serves to regulate gratification and tension reduction. The ego is a very efficient structure. That is, it allows the preschooler to realize that, for instance, throwing yourself on the floor of the supermarket and screaming and yelling and threatening to hold your breath until you die or until you get a candy bar (whichever comes first) may not be the most effective way to get what you want. Finally, toward middle childhood, children begin to appreciate the fact that others have needs just as they and that there may well be times when the needs of others may have a more legitimate claim to gratification. Such realizations mark the emergence of the superego, or conscience. According to Freud, most of our adult behavior is a reflection of the unmet needs of childhood and the dynamic interplay between the impulsive id, the efficient ego, and the sometimes guilt-ridden superego.

Freud's legacy is most clearly evident in the work of Erik Erikson. Erikson took Freud's notion of stages defined in terms of body zones and reformulated them in terms of relationships. Erikson's work is discussed throughout this text. Freud's ideas are also present in many of the topics that developmentalists of all theoretical orientations pursue. They are evident in the continuing debate over the importance of early childhood experiences in determining the course of adult development. They are evident in our appreciation of infants as active (perhaps Freud would have said lustful) participants in directing the course of their development. They are evident in the debate concerning the degree to which parents should behave in an indulgent manner toward their infants. They are evident in our appreciation of the fact that we are rarely fully aware of the motivations for our behavior. Finally, they are evident in our continuing investigation of the process of sex-role identification.

sal aspects are most evident in the physical and motor components of development (for example, the creeping, crawling, and walking sequence; the coordinated movement patterns involved in throwing a ball, hopping, skipping, or jumping; and the fine muscle control involved in copying a design) and in the systematic expansion of language patterns. These aspects of development are very evident during the early part of the life-span and no doubt are strongly associated with the biological maturation of the individual. The theories of Levinson and Frieze follow an opposite pattern. Their focus is on patterns of development that are unique to a culture rather than to the species. Radical shifts in a society (for example, a reversal in male and female roles) would be expected to produce equally radical shifts in age-related behaviors. Men would then experience greater ambiguity in defining the adult life course than would women.

The fact that Gesell emphasizes the biological aspects of development while Levinson and Frieze emphasize the societal aspects should not be taken to mean that the opposing aspects are not important for each theorist. All of these researchers recognize that biology and society are both important to the overall developmental process. To say that children learn to walk and talk in all societies does not mean that the particular culture of a society is unimportant. Rather, it means that all societies seem to possess the ingredients necessary to ensure the development of movement and speech. For these aspects of development, all societies seem equivalent. In other aspects of development, such as the particular language spoken or the age at which young adults leave home, the society's unique influence becomes evident. The issue isn't the degree to which environment or heredity influences development: each influences all aspects of development. The real question is the extent to which each contributes to the variability or range of expression of a particular aspect of development.

Timetable for Development

In the theories of Levinson and Frieze, the timetables for development are socially defined, while Gesell's timetable is biologically determined. As a result, Gesell's timetable is more specific than Levinson's or Frieze's. Development for Gesell is measured in weeks or at most months. On the other hand, years are the measure of adult societal-based development. In general, the intervals become increasingly longer as the life-span continues. Events during the prenatal period are measured in hours, days, or weeks. Weeks and months measure the events of early childhood. As adulthood approaches, the measure is years; by middle age, even decades. The shifts in measurement reflect the changes both in the rate of development and in the components undergoing change. Learning to throw a ball accurately may not be any easier than learning to maintain an intimate relationship, but it does seem to take less time to accomplish.

Specificity of Focus

Stage theories differ as to the specificity of their focus. Gesell and Havighurst are each primarily interested in the development of relatively specific behaviors. Gesell, for example, reports that at thirty months, many children prolong the process of going to bed by setting up a complicated ritual that must be rigidly adhered to. At age two, children often demand to take toys or stuffed animals to bed with them. By age four, these requests are rare. Havighurst talks about the learning of specific skills—walking, talking, and taking solid foods during infancy; getting along with playmates and adopting appropriate sex roles during middle childhood; selecting a mate and managing a home during early adulthood; adjusting to the death of a spouse and establishing satisfactory living arrangements

This five-year-old Montessori student is using a trinomial cube as a sensorial exercise. If this exercise were part of a study by Piaget, interest would focus on what the girl's ability to master the exercise said about her level of cognitive functioning, rather than on the exercise itself.

during late maturity. What distinguishes these various examples is their specificity.

Erikson and Piaget, on the other hand, define stages very broadly, not in terms of specific behaviors but rather in terms of level of functioning. Specific behaviors merely serve as indicators of the presence or absence of trust or industry or concrete operations or formal operations. In their role as indicators, a wide variety of behaviors could serve the same purpose.

Another way to think about the matter of specificity is to view it in terms of performance and competence. Gesell and Havighurst are interested in what people actually do at different ages; Piaget and Erikson, in what they are potentially able to do. For example, in Piaget's work, the various behaviors children demonstrated in the situations Piaget devised (for example, searching for a hidden object, arranging sets of blocks or pictures of animals, predicting which objects will float and which will sink, determining which combinations of solutions produce a particular color) are of less interest in themselves than they are as an index of level of development. This use of specific behaviors stands in sharp contrast to their use by Havighurst and Gesell.

Limitations of Stage-Based Theories

As useful as stage-based theories are, they are not without their limits. These limits do not invalidate stage-based theories, however. Rather, they point out topics in need of further research, and they qualify the breadth and weight of the generalizations possible from the perspective.

Determination of Causality

Perhaps the most frequent criticism of stage-based theories is that they are all better at describing behavior than at explaining it. Since the mere passage of time cannot be a causal factor, something else must be happening to cause behavior to develop as it does. What is this something else? The process of development involves both heredity and environment as causal factors, but how each exerts its influence at each stage of the life-span remains poorly understood.

Deemphasis of Variability

A second problem particularly common in performance-based stage theories is that they tend to deemphasize the variability of behavior. To say that the average child copies a cross correctly at forty-eight months or a diamond at seventy-two months tells us little about the normal range of behaviors at those ages. Should parents whose children copy a diamond at sixty months consider early admission to college? Should parents of children who don't copy the diamond until eighty months consider institutionalization? These examples are clearly exaggerations, and in fact most performance-focused stage theories do provide a range as well as an average. But whenever behaviors are linked to ages, the inherent variability of behavior tends to be deemphasized. At a practical level, this deemphasis of variability can lead to incorrect evaluation of individual ability.

Competence-based stage theories are less troubled by the problem of variability. They focus on the sequence rather than the timing of development. Although both Erikson and Piaget associate their stages with age, they only provide approximations. For them, development is measured "in sequence" rather than on a specific timetable.

Coherence

Although spared the problem of timetables, competence theories are vulnerable to the problems of coherence and validity. If adolescence is the time of identity formation, should it not be evident in all aspects of the adolescent's life? If the formal operational stage of cognitive development is characterized by the ability to deal with abstract thought, how do we explain the inability of the forty-year-old to understand the adage "Make hay while the sun shines" as anything other than a prescription for good farming? Such problems are usually explained by noting that, for example, the timetable for identity formation in the realm of occupation may differ from those in the realms of politics or religion or interpersonal relations. The forty-year-old's problem suggests that there may be factors that interfere with the expression of competence. Lack of coherence does not prove the theory wrong. Rather it suggests that additional factors other than those presently offered by the theory may be necessary to provide a complete explanation of development.

Validity

In its broadest usage, validity concerns correctness. For performance theories, the issue is relatively straightforward. If the theory says that 90 percent of all three-year-olds are able to build a block tower seven blocks high, we can try to duplicate its results by getting a group of three-year-olds together and seeing how many in fact can build a block tower seven blocks high. Our results may not be exactly the same as those of the original study, but they should be close enough to give us confidence in its accuracy, or validity.

Validity is a more significant issue for competence theories. In Piaget's theory, for example, infants who search under a blanket for an object they have just seen hidden there are said to have acquired the concept of object permanence. Infants who do not search are said to lack the concept—in this case, out of sight, quite literally, does seem to mean out of mind. Presumably, the differing behaviors are caused by the presence or absence of the concept. A potential weakness exists in this argument, however, as the following questions indicate. How do we know the concept is there in the successful infants? Because they lifted the blanket to search for the rattle. What caused them to search for the rattle? The object concept. The circularity of the argument is evident. The behavior that is caused by the concept is in turn used to validate the concept's very existence. The way out of this dilemma is to demonstrate that a variety of logically related behaviors can be linked to the same concept. Such relationships have been demonstrated at each of Piaget's four stages (Goldhaber, 1981; Uzgiris, 1977; Wohlwill, 1973). In the case of object permanence, infants also begin to protest the withdrawal of a parent at about the same age as that at which they begin searching for hidden objects. Both examples reflect the ability to mentally represent a formerly present and now absent object. In general, the greater the cross-situational coherence of a theory, the greater is its validity.

Cultural Bias

A final problem concerns bias. Cultural-based theories are particularly vulnerable to bias. There is no problem with Gesell's saying that most infants walk by fifteen months. Walking is a universal characteristic of humans. It is present in all societies, and with the exception of the harried parents of a fifteen-month-old, it's hard to imagine anyone arguing that infants shouldn't be allowed to walk at fifteen months. But what about Havighurst's statement that selecting a mate is one of the developmental tasks of young adulthood? What if someone doesn't want to select a mate? Can't single people be adults? Similarly, Levinson associates the early forties with self-reflection. He sees it as a time when many men realize that their commitment to their professional development has often come at the expense of their families.The reordering of priorities is seen as a necessary component of the resolution of the mid-life transition. But what if someone doesn't want to reorder his priorities? What if he would rather wine and dine prospective clients at posh restaurants than watch his seven-year-old child's theatrical debut as one of Santa's elves in the school Christmas play? He may be a creep but can he not resolve mid-life questions?

The problem arises when what most people do (for example, marry) is equated with what people should do or when what is typical and appropriate for one segment of society (for example, a particular social, racial, or ethnic group) is considered equally so for all segments of society. The idea of taking stock at mid-life suggests the possibility of change. It implies that priorities can be reordered, that a man can reduce his work week or at least rearrange his schedule. What about the person who isn't so fortunate? Assembly-line workers don't enjoy such privileges. Neither do office clerks or secretaries. The point is that when we extend a cultural event appropriate for one segment of the population to all segments (Levinson's sample consisted of well-educated men), we inadvertently introduce our own biases and in the process create untenable situations for others.

In light of all these problems, do stage theories have any value? The answer is a little like deciding if the six-gallon bucket filled with three gallons of water is half empty or half full. Stage theorists think it's half full; those who advocate an ecological approach think it's half empty.

THE CONTEXT OF DEVELOPMENT

Age-related changes in development are not an automatic or inevitable process. They occur within an environment and reflect the impact of that environment. The environment is a major determinant of the knowledge, values, beliefs, and attitudes we hold. It is an equally important determinant of the level of our development. An environment lacking the supports and stimulation necessary for continued development is as detrimental to full development as one lacking in adequate nourishment. This section reviews three ways of examining the environment. The first two suggest ways of structuring the environment, while the third concerns some of the processes that regulate human interaction.

The Human Ecosystem: A Systems View

The environment is neither unitary nor stable. It exists at many levels, and like the developing individual, it changes over time. In his **human ecology model,** Bronfenbrenner (1979) has demonstrated that there are really four interconnected levels of the environment: the microsystem, mesosystem, exosystem, and macrosystem. Each has an influence on the course of development, but the manner in which these influences are expressed differs.

The settings and experiences that constitute day-to-day reality reflect our immediate environment, or **microsystem.** The home, school, workplace, neighborhood, and church are examples of the settings of the microsystem. These are the places where we spend our time, where we play, work, love, and learn. Others also spend time in these places, and these family members, coworkers, teachers, friends, and neighbors represent the people in our microsystem (Garbarino, 1982).

Each of us lives in a variety of microsystems. The interplay among these various microsystems defines a second level of the environment—the **mesosystem.** Mesosystems can differ in terms of their degree of complexity and integration.

Highly complex mesosystems are ones that involve a variety of distinct microsystems. The mesosystem of a thirty-five-year-old adult who has been divorced and remarried, who has one full-time and one part-time job, who has two children through the first marriage and one through the current marriage, and who is totally responsible for the care of an aged

In this Hare Krishna commune in West Virginia, the classroom doubles as the dormitory. Children raised in this type of microsystem will have a different world view than children raised in traditional communities.

parent is, to say the least, very complex. It is certainly a more complex mesosystem than the one experienced by a seventy-year-old adult living in a retirement community that offers a coordinated program of educational and recreational activities.

Integration refers to coordination and is therefore a measure of stress in the ecosystem. Children growing up in homogeneous mesosystems experience a world that is very different from the one experienced by children whose mesosystems are composed of a variety of conflicting microsystems. For example, children living in suburban or small-town environments usually experience a higher degree of community integration than do children living in larger cities. Since housing tends to be similar in size and cost, families living in suburban communities and small towns are likely to exhibit many of the same characteristics, for example, in income and family composition. This fact, in turn, results in a community that possesses a fairly uniform set of values. Uniformity of values reduces community stress by lessening conflict among the various microsystems. Schools tend to reflect the values of the community, thereby gaining the support of parents. Likewise, the community is viewed as a safe environment, and thus children are encouraged to participate in community activities and explore their surroundings.

Larger cities, on the other hand, represent a wider range of variability and thus present a less uniform system of values. Children growing up in this kind of environment are likely to experience a higher level of conflict among the various microsystems. This higher level of conflict does not necessarily mean that these children are living in a less desirable setting than their suburban counterparts. It does mean, however, that they will possess a different worldview.

We play no part in making many of the decisions that affect our development. The city planning board decides that a house must be demolished to make way for a new highway. An employer decides to move the corporate office from New York to Florida. The city decides to improve the local parks and playgrounds. These are events in the **exosystem.** Each has a dramatic impact on the people affected. Each significantly alters both the

BOX 1-2
THE CONTEXT OF CHILD ABUSE AND NEGLECT

The study of development across the life-span is a complex process. There are few instances where causal statements are absolute, that is, where behaviors are determined by only one or two factors. More typically, a variety of factors, each affecting the impact of the others, cumulate to increase or decrease the probability of a behavior or a change in behavior. The complex pattern is nowhere more evident than in the study of child abuse and neglect.

Over one million children per year suffer physical or emotional abuse or neglect or both, and this is almost certainly a conservative estimate. Many cases remain either undetected or unreported. There is no single cause. There is no special type of adult who commits these offenses. Rather, the evidence suggests that given a particular set of circumstances, most of us could conceivably abuse or neglect our children.

Imagine yourself in the following set of circumstances. You are the parent of a premature infant who seems particularly unresponsive to your parenting efforts, you consider your marriage unsatisfying, you have had three children in the past five years, the first when you were seventeen, your spouse has been layed off from work and is unable to find another job that will provide

Research has shown that housing conditions are one of the factors that influence the incidence of child abuse. Abuse is more prevalent when individuals live in large, overcrowded housing projects in deteriorating neighborhoods.

microsystem and the mesosystem. For example, the well-integrated mesosystem of the small town probably reflects the local zoning board's actions concerning the size of lots, the size of housing, and the materials that can be used in the construction of housing in the community. These decisions define the types of people that are likely to live in the community.

Among the most significant influences on the development of children are the ones over which they have no control. These factors constitute their exosystem. Primary among these factors are the conditions of parental employment. Companies that adopt family-oriented policies, such as

sufficient income to meet minimal family expenses, you live in substandard housing in a community that has a high crime rate and a large transient population. You know few people in the neighborhood, and those that you do know are as overwhelmed with their circumstances as you are. You have very negative feelings about your own childhood experiences, in particular the fact that your parents never seemed to have much interest in your development. In sum, you are an individual having to deal with a variety of highly stressful life situations with few material resources or interpersonal coping skills, living in a neighborhood that offers few social or material supports. You are at risk for child abuse or neglect. Each of these circumstances correlates with the incidence of abuse and neglect; cumulatively they represent a disproportionately higher probability that abuse or neglect will occur.

Any serious attempt to reduce the incidence of abuse and neglect in this country must first recognize the many factors that contribute to its likelihood. Second, it must recognize that many of these factors are indirect. For example, we live in a society that condones moderate levels of physical punishment as a way to discipline children. Other industrialized countries that do not tolerate any degree of physical punishment have considerably lower incidences of abuse and neglect. We are generally not very supportive of dispersing low-income housing across a wide variety of neighborhoods, yet the incidence of abuse and neglect is higher when low-income housing projects are large and concentrated in deteriorating neighborhoods than when the units are few in number and dispersed. In the latter case, the positive benefits of the neighborhood transfer to the new residents. The better able we are to reduce the burdens on families, the better able will those families be to cope effectively with the expected stresses of life. The more effectively they deal with the stresses of life, the less likely they are to deal with their children in abusive and neglectful ways. Ultimately, the incidence of child abuse and neglect is a social indicator of the degree to which society's social and economic policies support development across the entire life-span.

Sources: Belsky, 1980; Garbarino & Gilliam, 1980; Goldberg & DiVitto, 1983.

subsidized day care, flexible work hours, and the coordination of transfers with the school calendar, facilitate the parents' jobs and, in the process, foster good human development.

The events of the microsystem, mesosystem, and exosystem have specific effects on people's lives. The events of the fourth level of the environment, the **macrosystem,** are more on the order of blueprints (Garbarino, 1982). They reflect people's shared assumptions about the way things should be done. They reflect the values of a society. The 1954 Supreme Court decision that ruled the "Separate but Equal" law unconstitutional expressed a value that any form of separation was discrimination and that discrimination was inconsistent with the law of the land. Local communities were ordered to develop and implement school desegregation plans. The desegregation plans that were developed, or in some instances the refusal to develop such plans, became elements of the mesosystem and eventually of the school-age child's microsystem. What must the school environment come to mean to children who are accompanied by armed National Guard soldiers as they go from class to class?

Decisions at the macrosystem level often determine the role of government in our lives or reach some consensus among conflicting viewpoints. For example, they may decide what role the government should play in protecting the consumer, and they may reach the consensus of opinion necessary to establish policies governing income tax deduction. It is important to bear in mind that macrosystems decisions define direction, not specific actions. Most children did not need an armed escort to enter a desegregated classroom.

The Human Ecosystem: A Family View

Of the various components at the different levels of the environment, the family is the single most significant context for development. Garbarino (1982) refers to the family as the "headquarters for human development." We spend our entire lives as part of one family and usually much of our lives as part of two families. We are born into a family (the family of origin), and we are always a part of its nuclear (siblings and parents) and extended (cousins, aunts, uncles, nephews, and nieces) components. Most adults also form new families (the family of procreation). Due to death and the increasing number of divorces, many adults are forming third, or blended, families. The pivotal role of the family is evident when we consider that much of the direction of our lives is a reflection of the relationships of members within the family, of relationships between families, or of the relationship between the family and other microsystem components, typically the school and the workplace.

Families differ greatly. Each has a unique internal structure. Families differ in terms of size, spacing between members, sex of members, and age of parents (Rodgers, 1973). The meaning of the family to the youngest child is certainly different from its meaning to the oldest. The ten-year-old whose parents are forty-five and fifty and the ten-year-old whose parents are thirty and thirty-five have different relationships with their parents. Only children and children with two or five or ten siblings come to know their parents in different ways.

Families differ not only in terms of structure. They differ in terms of internal organization, in terms of goals and values, and in terms of the relationships between members of the family and between the family and the society (Kantor & Lehr, 1975). Some families involve their children in the decision-making process; others reserve such power for adult members. Some families share responsibility for family tasks. Both husband and wife work, and everyone takes a turn doing the dishes, raking leaves, and vacuuming the rugs. Other families have a very clear division of labor. The father is responsible for earning a living; the mother, for raising the children. The girls do the dishes and the boys take out the garbage. Some families always seem to be doing things together; others seem more like strangers to each other.

Families differ in terms of their perception of the world and as a result in the way they prepare their children for entry into that world. Kohn (1977) notes that the major difference between families of higher and lower social class involves the expectation that what someone does makes a difference.

> The essence of the higher class position is the expectation that one's decisions and actions can be consequential; the essence of the lower class position is the belief that one is at the mercy of forces and people beyond one's control, often beyond one's understanding. (p. 189)

Given such expectations, it is not surprising that higher class child rearing often emphasizes self-direction, while lower class child rearing is more likely to emphasize conformity.

Neither the role the family plays in an individual's development nor the meaning it has to different members is always the same. As individuals change, so too must families. The family has a life cycle (see Figure 1-1) that parallels that of its members (Duvall, 1977; Rodgers 1973). Each of the stages of the family life cycle corresponds to a unique family configuration. Each configuration influences the structural constraints on the family (such as size, allocation of resources, division of labor, decision making). For example, marital satisfaction differs across the family life cycle (Rollins & Feldman, 1970). In general, satisfaction is at its highest during the early and late stages of the family life cycle and lowest during the middle. This "curvilinear" pattern is more evident in the reports of wives than in those of husbands.

> These data suggest that experiences of childbearing and childrearing have a rather profound and negative effect on marital satisfaction for wives, even in their basic feelings of self-worth in relation to their marriage. Perhaps this is partly a consequence of the great reduction in positive companionship experiences with their husbands instigated by the pressures of childrearing responsibilities. (p. 27)

The context of development, especially the family, defines the settings in which development occurs. Within these settings, education and socialization occur. The nature of the mechanisms involved in these processes has been the concern of learning theorists and social learning theorists. The next section discusses their contribution to the understanding of development across the life-span.

Learning and Development

Learning theorists are interested in the way individuals acquire the information and behaviors necessary to function in a society. They have tended to emphasize the roles of direct learning, or conditioning, and the role of indirect learning, or modeling.

Direct Learning

Both instrumental and classical conditioning involve learning an association. In the case of **classical conditioning,** the association is between a

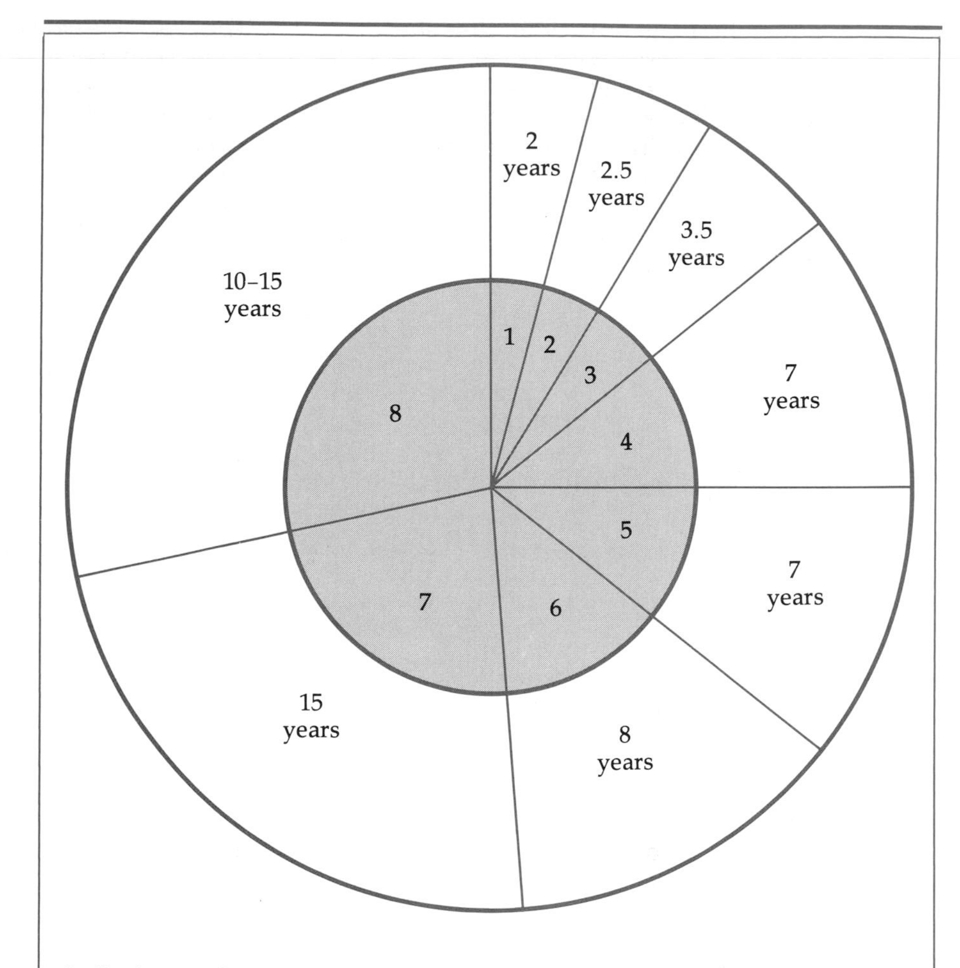

1. Beginning Families (married couple without children)
2. Childbearing Families (oldest child birth to 30 months)
3. Families with Preschool Children (oldest child 30 months to six years)
4. Families with School Children (oldest child 6 to 13 years)
5. Families with Teenagers (oldest child 13 to 20 years)
6. Families as Launching Centers (first child gone to last child leaving home)
7. Families in the Middle Years (empty nest to retirement)
8. Aging Families (retirement to death of both spouses)

FIGURE 1-1
Each family has a life cycle that corresponds to specific family configurations. The numbered segments of the shaded inner circle correspond to Duvall's eight stages (from Duvall, 1977).

stimulus and a response. Infants, for example, will begin to suck when a nipple is placed in their mouths. No learning is involved in this association; it is one of the reflex responses that infants have at birth. Within a matter of weeks, however, parents often note that just the sight of the breast or bottle or in some cases even the preparations that precede the presentation of the breast or bottle come to elicit the sucking reflex. A new association has been formed between the sucking reflex and the sight of the breast or bottle. In more formal terms, the breast or bottle is an *unconditioned stimulus.* The frequent pairing of the sight of the breast or bottle with its placement in the infant's mouth transfers the response-producing potential of the breast or bottle to its image (now becoming the *conditioned stimulus*). The sucking-reflex response to the image becomes a *conditioned response* in that it has become associated with a stimulus that was initially neutral. In a like manner the sight of the dentist's drill or the doctor's needle is often enough to elicit strong conditioned responses. In fact many of our emotional responses probably have a conditioned response as one of their key elements. Classical conditioning is also known as *Pavlovian conditioning.* Ivan Pavlov was the Russian physiologist who first formally demonstrated classical conditioning.

Instrumental conditioning, or Skinnerian conditioning (after B. F. Skinner, its best-known researcher and advocate), is a second means of changing the patterns of behavior. Such a change occurs as a result of the association between a response and its consequences. Behavior that meets with favorable consequences, or positive reinforcement, tends to increase; behavior that meets with negative consequences, or punishment, tends to decrease. Further, behavior that has previously been positively reinforced will begin to decrease, or extinguish, when the positive reinforcement is removed. Behavior can change in a number of ways through reinforcement. It may become more frequent, the likely consequence of a high school girl's telling her classmate that she really appreciated his calling the other night. Such reinforcement will almost certainly result in another call. It may increase in amplitude or strength, as may the studying of a student whose teacher says she has real potential in a particular area. It also may increase in duration, as may the jogging time of the person succeeding in losing weight.

In addition to the nature of the reinforcement, two other equally important determinants of the success of instrumental conditioning are the schedule of reinforcement and the interval between the response and its consequences. The most effective way to establish a new behavior is through *continuous reinforcement,* that is, to reinforce the behavior every time it occurs. Once the behavior has been established, the most effective means of maintaining it is through *partial reinforcement.* One way of thinking about the employment history of an individual is to consider the typically frequent and relatively small promotions and raises that occur early in a career as examples of continuous reinforcement and the increasingly less frequent promotions that occur later on as examples of partial reinforcement.

There is also a very good reason why letting daddy punish young children for the day's misdeeds after he gets home in the evening is usually not a very effective deterrent (besides the fact that daddy is often too tired to give the punishment his best effort). The amount of time separating the act and its consequences is too long for an effective association to be formed. In general, the longer the interval is, the less effective the association. The frequent efforts to reform the judicial system so that trial and sentencing will more closely follow arrest are an implicit recognition of this fact.

Over the course of a lifetime, individuals acquire, maintain, and extinguish a wide variety of classical and instrumental responses. Given the variety of events that are occurring simultaneously at any one time, it is understandable why people sometimes find it difficult to determine what

While it merely looks like play, by imitating his favorite superhero, this young boy is developing important social and cognitive skills.

behaviors go with what consequences. One of the key elements of most counseling and therapy techniques is helping individuals unravel the interwoven components of their lives. This is particularly important when longstanding relationships such as marriages begin to fail. So many associations have been formed over so many years that trying to pinpoint the causes of the problem can be a very difficult task. Further, the instrumental value of behavior is subject to different interpretations under different circumstances. Two children will respond very differently to the same parental behavior if one child views the parent's behavior as motivated by a sincere concern for the child's welfare and the other child does not.

Indirect Learning

As powerful an influence as direct learning is, it has one major limitation. For an association to be formed, a response must first be made. To social learning theorists (Ahammer, 1973; Bandura, 1971), the concept of direct learning therefore seems to provide a very uncertain basis for accounting for behavioral change. They believe that the more significant components of behavior are not acquired through direct learning but rather through indirect, or vicarious, learning. Vicarious learning emphasizes the role of imitation in behavioral change.

Like direct learning, the study of **social learning** or **imitation** has involved an examination of the ways in which the various parameters of the modeling situation influence the imitator's behavior. Frequency of contact is an important factor. We tend to imitate those with whom we have frequent contact. It is no wonder that married couples are said to begin to act and sometimes even to look like each other as the years go by. We are also more likely to imitate those whose behavior appears distinct or powerful. Most children imitate their favorite superheroes. They want to grow up to be just like their favorite sports hero or to look like their favorite television or movie personality. Fortunately for most parents, relevance is also an important determinant.

We not only observe what others do, we also observe the consequences of their actions. (The advertising industry has invested billions of dollars on this simple notion.) Children find it much easier to adopt comfortable sex roles if their same-sex parent's behavior is appreciated by the other parent. As the four-year-old who proposes to his mother quickly learns, however, behavior that is reinforced when performed by the model doesn't necessarily meet the same fate when performed by the imitator. The four-year-old is not yet able to appreciate the fact that an important element of the modeling process is learning the conditions under which the imitated behavior will be reinforced. In circumstances like these, modeling is a matter of delayed imitation, a component of the socialization process that is especially relevant in taking on the roles of parent and intimate. We learn these roles, in part, through observation of our parents. The modeling, however, takes place many years later in a different setting with different people. The fact that many adults are amazed at how similar their parenting styles are to those of their own parents suggests that much delayed imitation occurs unintentionally.

Limitations of the Ecological Approach

Ecological theories, whether they focus on events in the broad environment or in the most immediate, highlight the fact that much of what we do is a reflection of the circumstances of our individual lives. By focusing on the structure of environments and the processes each of us uses to influence others, the ecological perspective stresses the interdependent nature of development across the life-span. Like stage-based perspectives, the ecological view adds another piece to the puzzle, and also like stage-based approaches, it has unique limits.

Determinants of Causality

The ecological perspective places great importance on the settings in which people find themselves and on the things that happen to them within these settings. There is no question that setting is an important determinant of behavior; however, it is not the only one. Advocates of this perspective tend to disregard the individual within the setting. They tend to portray the individual as a passive being whose behavior is primarily a response to the nature of the surrounding events. Not all developmentalists share this view of the organism. Some place much more emphasis on the role the individual plays in interpreting the events encountered within the environment. Because they tend to underestimate the role of the individual, however, adherents of the ecological perspective may overestimate the importance of the setting and the circumstances of the ecology.

Explaining Complex Behavior

Most of the research within the learning theory tradition has examined the variety of association and reinforcement patterns and the way each influences behavior. For the most part, the behaviors studied are fairly simple in nature. This approach has been much less successful in explaining more complex behaviors such as language or reading or interpersonal attraction. This inability to explain more complex behaviors may reflect little more than the amount of time these complex behaviors have been studied. However, it may also reflect the fact that there are limits to the degree to which learning-based theories can adequately explain the complex character of human behavior.

Behavior versus Development

The ecological perspective does a much better job of explaining why we act as we do than it does of explaining why we change over the course of time. To argue that people change in response to the changes occurring in other people is consistent with the model but merely begs the question. Why did the other people change? There appears to be a systematic and orderly quality to at least some parts of human development, and the ecological model has not yet been very successful in explaining its nature.

A SYNTHESIS

Our discussions of the course of development and the context of development highlighted the complementary components of the developmental process.The discussions of the course of development focused on the aspects of development that correlate with age; the discussions of the context of development focused on the aspects of development that correlate with social structure and interpersonal relationships. Both age-related developmental processes and social structure influence all aspects of human development across the entire life-span. The process is multidetermined, interdependent, and cumulative. It involves the successive products of a biological organism maturing within a sociohistorical context. The successive products are levels of psychological awareness.

Three aspects of the process of development need to be emphasized (Werner, 1957). First, the process of development is characterized by an ever-increasing competence over increasingly wider perspectives. The adult is better able than the child to deal with any one situation and better able to deal with more situations. Second, the process of development is characterized by an increasing harmony between the individual and the environment. Adults are better able than children to respond effectively to the demands of the environment while at the same time they are also better able to modify the environment to meet their particular needs. Third, the process of development is characterized by an increasing sense of individual unity. Once again, adults are better able than children to coordinate the various components of their lives and the various components of their individual identities. It is the adult who, in Piaget's words, is capable of reason. An understanding of the ways in which these progressions occur across the life-span requires an understanding of the multidetermined, interdependent, and cumulative nature of the developmental process and of the way development occurs between individuals, such as a parent and child, across an individual life-span, and across successive generations of people. Much of this text is devoted to providing this understanding, but for the moment a few examples will have to suffice.

Watch parents with their newborn child. Watch how the relationship develops between parents and child during the first year or two of the infant's life. What you will observe is that a type of synchrony develops. Both parents and child are learning to respond to each other. Both are learning ways to influence each other. Both are learning how to use information provided by the other to make their behavior more responsive and effective. When the process works, watching it is like watching a pair of figure skaters gliding gracefully across the ice; when it doesn't work,

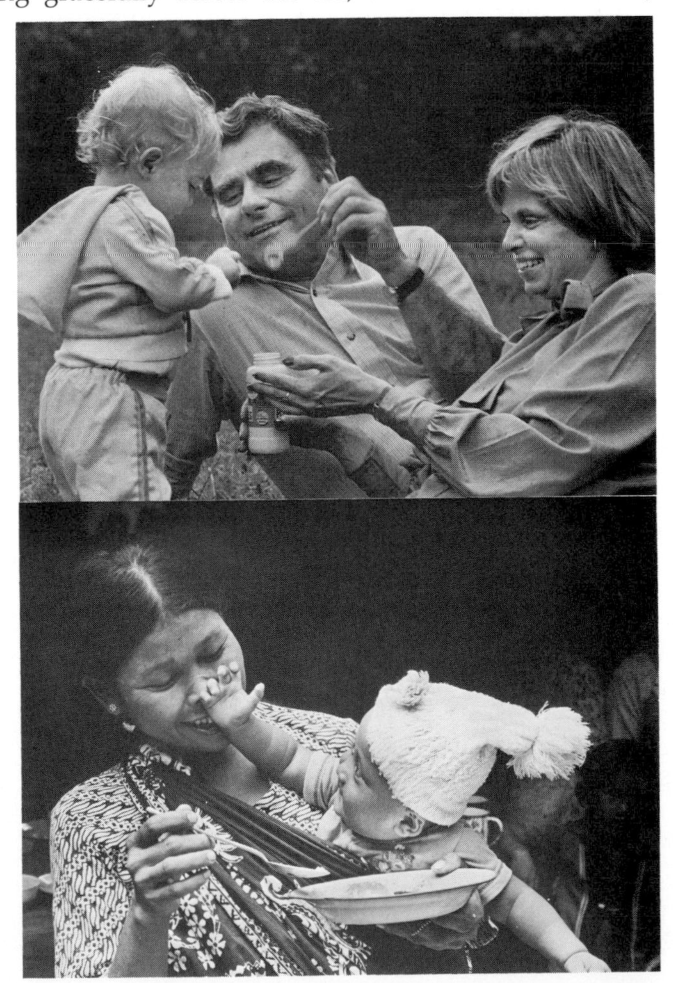

When parents and infants develop a strong and loving relationship, it is a joy to see.

watching it is like watching young adolescents negotiating their first goodnight kiss. In both instances the behaviors are interdependent: each influences and, at the same time, is influenced by the other. The difference is that in one instance the process of development is facilitated; in the other, the process is hindered.

Much of our society is influenced by the developmental level of the individual. The structure of the family, especially its emphasis on continuity, in part reflects the immaturity of the newborn and the relatively slow rate at which humans mature. The age-graded structure of our school systems and the increasing segmentation of the curriculum into specific subjects reflect the learning rate and learning style of children and adolescents. The juvenile justice system, child labor laws, child custody laws, and mandatory school attendance laws all reflect the unique developmental status of minors. The privileges and responsibilities given adults reflect the same awareness of developmental level. We assume adults have reached a level of development at which they can understand the consequences of their actions and therefore be held responsible for them.

Although development across the life-span is cumulative, the process is not inevitable; that is, earlier events influence but do not determine later events. In some instances the significance of early experience can only be defined when later events become known. For example, it is only possible to determine the developmental consequences of some complications associated with labor and delivery when the socioeconomic conditions of the family are known. The initial disadvantage is only maintained when the perinatal complication is combined with and supported by poor environmental circumstances (Sameroff & Chandler, 1975).

More generally, society can only prepare its members for the tasks of the future to the extent that those future tasks are knowable. If there are significant shifts in the fabric of a society, people may find themselves having to learn new skills because the ones society provided are no longer effective. One of the most obvious examples of such a shift (see Table 1-6)

TABLE 1-6 Pattern of Family Changes, 1970 to 1980

NATURE OF CHANGE	1970	1980	% CHANGE
Marriages performed	2,159,000	2,317,000	+ 7.3
Divorces granted	708,000	1,170,000	+65.3
Married couples	44,728,000	47,662,000	+ 6.6
Unmarried cohabiting couples	523,000	1,346,000	+157.4
People living alone	10,851,000	17,202,000	+58.5
Married couples with children	25,541,000	24,625,000	− 3.6
Children living with two parents	58,926,000	48,295,000	−18.0
Children living with one parent	8,230,000	11,528,000	+40.1
Families with both husband and wife working	20,327,000	24,253,000	+19.3

Source: From James Garbarino and Associates, *Children and Families in the Social Environment* (New York: Aldine Publishing Company). Copyright © 1982 by James Garbarino. Used by permission.

and its resulting consequence has been the change in family structure over the past generation. Melson (1980) finds that within this generation the family has decreased in size, that it has become more isolated from its extended members, and that more frequently than before either both parents work or only one parent is present. The family has also assumed greater responsibility for decision making as society has become increasingly tolerant of a variety of adult life-styles, and it has learned to cope with government and social services agencies for services previously provided by religious groups or extended family. It is no wonder that so many adults find the knowledge of family gained from their childhoods of limited value in starting and maintaining their own families.

The restructuring of one societal component such as the family has consequences for other components as well. The decrease in family size, for example, not only has consequences for family relations but, in the long term, for many facets of society. The decrease in family size results from a declining birth rate, which also changes the age distribution of our population. If the number of people born equals the number that die, then the number of people of any given age will be eventually the same as those of any other age. In fact, this is the present and probably the future pattern of population growth within the United States. The effects of this change in birth rate are already evident in the debates over the future of the social security system, in revisions in the laws governing the age of retirement, in the frequency of age discrimination cases in the courts, in the provision and distribution of medical care, in the platforms of political parties, and more generally in the debate about the role of the elderly in society.

One of the consequences of trends such as shifts in the population or changes in family structure is that each generation is faced with its own unique timetable for accomplishing the major tasks of its lifetime. Consider how different the lives of women must have been in 1890 from what they were in 1966 given the timetables in Table 1-7. As Elder (1974) puts it, each generation is bounded by its peculiar biography. This biography regulates features of life such as entry into the labor market, the timing of marriage and parenthood, the number and spacing of children, the number and

TABLE 1-7 Changes in the Timing of Life Events[1]

MEDIAN AGE AT	*1890*[2]	*1966*[2]	*1983*
Leaving school	14	18	19
Marriage	22	20	22
Birth of first child	24	21	23
Birth of last child	32	26	29
Death of husband	53	64	68
Marriage of last child	55	48	52
Death	68	72	74

[1]For women only
[2]*Sources:* 1890 column from Glick, P.C., Heer, D.M., & Beresford, J.C. (1963). Family formation and family composition: Trends and prospects. In M.B. Sussman (Ed.), *Sourcebook in marriage and the family*. New York: Houghton Mifflin. 1966 column from U.S. Department of Labor 1967 Manpower Reports. 1983 column from U.S. Census data.

types of job changes, and the scheduling of acquisitions. In sum, an understanding of development across an individual life-span requires an understanding of the various biological and ecological processes that influence that development. Since these biological and ecological processes do not remain constant over successive generations, an understanding of development at any particular point in history requires an understanding of the particular character of these processes at that historical moment or interval.

THE STUDY OF DEVELOPMENT ACROSS THE LIFE-SPAN

We study the process of human development for two reasons. The first is to determine the course of development across the life-span. The second is to determine why the course of development occurs as it does. The scope of human development research may be grand. It may involve scores of researchers, cost millions of dollars, and have a potential impact on a large percentage of the population. On the other hand, it may be as limited in scope as a study to determine which of two basic reading series is more effective in teaching children how to read. Few professionals are ever involved in a systematic program of research. But almost everyone, professional and nonprofessional alike, is influenced by research. In short, we are the consumers of research. As a result, it is important to have at least an elementary understanding of the ways in which information about the process of development is collected, interpreted, and applied. This section is a step in that direction.

The Process of Research: Defining the Topic

Research is a process of systematically sorting out events and the relationship between events and then making inferences as to the cause of the observed patterns. The actual procedures involved reflect the nature of the problem. Research often involves extremely complex statistical analyses. But no matter how complex or simple the research strategy, the same kinds of questions are asked, and the procedures for answering them follow the same general pattern.

The first step in designing a research project is to ask a question in such a way that it can be answered. This may appear to be a fairly simple task, but in fact it probably causes more problems than any other element of the research process. For example, it is certainly important to ask about the effects of poverty, racism, and sexism on individual development; but unless these questions are posed in a focused way, they are unanswerable. Asking, for example, how poverty influences a child's educational aspirations won't explain all the effects of poverty, but it will give us a clearer understanding of one of them. In other words, the question must ask about the nature of a relationship (poverty and educational aspiration, smoking and cancer, racism and employment, and so forth).

Once the question is posed in an answerable fashion, the next step is to define a set of procedures for collecting the relevant data. How do you define poverty? Would it be sufficient to use government guidelines that define the poverty level? What about the person whose income is above the poverty level but who feels poor? How do you define educational aspirations? How do you determine that a person's educational aspirations reflect what he or she wants to be and not simply a limited knowledge of educational options? These are not the kinds of questions that have absolute answers. In some cases, as in the measurement of intelligence, for example, standardized tests are generally accepted as a valid procedure. In other cases, the researcher must develop a procedure. In either case, the researcher is always subject to the criticism that the procedure isn't an accurate and valid measure of the focus of study. As a result, the researcher must be able to argue effectively why the procedure is an appropriate **operational definition** of the variable. For example, educational aspiration could be operationally defined as the number of years of schooling adolescents expect to complete. In other words, the operational definition is the actual procedure used to collect the information about the topic.

Armed with an answerable question and a method for answering it, the next task is to find someone to ask. Since it is usually impossible to obtain information from all the people in a population (for example, all four-year-olds attending preschool or all divorced couples or all retired teachers), the researcher must obtain a **sample** of individuals from the particular population. Since the ultimate interest is the entire population and not simply a sample from the population, it is crucial that the sample be of sufficient size and representativeness to allow for generalizations about the population. If, for example, 65 percent of the retired teachers in this country are women, then that should also be the approximate percentage of women in the sample. If 20 percent of the four-year-olds attending preschool come from families with an income below ten thousand dollars, then the sample should reflect that fact. Adequate sample size is harder to define. Often it depends on the types of statistical analyses that will be performed on the data. But in general, the larger the sample, the more valid the generalization.

Once the sample is obtained, the questions asked, and the answers given, the analysis of the data begins. Data analysis is almost always a quantitative endeavor. Its purpose is first to obtain an accurate description of the findings and then to determine what patterns of relationships exist between the pieces of information obtained.

The Process of Research: Methods of Data Collection

The choice of research strategy is dictated by the nature of the question asked. Questions that ask for description and questions that concern explanation are different and require different strategies. Questions that concern changes in behavior over long periods of time and questions that focus on the interrelationships of behavior at any one time are different and require different strategies. Questions that concern the individual's construction of reality and questions that concern the influence of various developmental contexts also require different strategies. In essence, each strategy focuses on the influence of one particular cause of the pattern of development.

Ideally, a research strategy ought to examine the interdependent, cumulative influence of every cause on every pattern of development, but the technical and conceptual problems of such an ideal solution make it impossible. Sometimes people who study the process of development across the life-span inadvertently come to resemble the blind men who were asked to describe an elephant. One felt the tail and concluded that the elephant resembled a rope; another felt the leg and with equal confidence concluded that a tree provided the more appropriate comparison; a third, feeling the ear, concluded that the elephant most closely resembled a palm leaf. Although debates among researchers do, on occasion, take on this flavor (especially when the argument concerns the relative merits of one *or* the other influence on development), more often they focus on the relationship between variables and therefore provide the lifeblood for the continued study of life-span human development.

Longitudinal and Cross-Sectional Research Strategies

The classic research strategies of human development are the longitudinal and cross-sectional research designs. Both are designed to provide information across a wide age range. The **longitudinal design** strategy identifies a group of people and at various points in their lives obtains relevant information from them. The information can be about anything that has a developmental course. It might be as simple and direct as measures of physical development such as height and weight, or the questions might concern attitudes or values or use of leisure time or learning ability. The particular questions asked, the frequency with which they were asked, and the characteristics of the people who were asked to answer them indicate the focus of the study. Irrespective of the particulars, the researcher wants to be able to show how some aspect of development changes as individuals age.

The **cross-sectional design** asks the same type of questions the longitudinal design asks but does so with one significant difference. Instead of identifying a population and testing it repeatedly over some extended period of time, the cross-sectional design tests different groups of people at the same time. The cross-sectional design is certainly more economical than the longitudinal design, but in spite of its practical advantages, it has one serious limitation—it uses different groups of people at different ages. That is, it isn't possible to conclude from a cross-sectional design whether the observed differences between the different age groups (for example, their views on premarital sexual behavior or their ability to learn computer programming) are due to their different ages or to the fact that they are groups who have experienced different historical events at comparable periods of their life-spans. For example, if a group of twenty-year-olds seemed more tolerant of premarital sexual behavior than a sample of sixty-year-olds, I could not determine whether the reason was the forty-year age difference (which would lead me to predict that in forty years the twenty-year-olds of the original study would become equally conservative in their attitudes) or the fact that the world in which the sixty-year-olds formed their early attitudes about premarital sexual activity forty years ago was a very different world from the one the twenty-year-olds are experiencing today (which would lead me to predict that in forty years the twenty-year-olds of the original study would still profess more liberal attitudes). The longitudinal design solves this problem by having all of the research subjects of the same generation or cohort (a group of people born at the same time). Any changes in their attitudes concerning premarital sexual behavior are most likely due to factors that correlate with age since they all experienced the same historical events at the same age. (Figure 1-2 shows how research findings on the same topics can vary depending on whether longitudinal or cross-sectional designs are employed.)

The longitudinal solution is only a solution, however, if the question

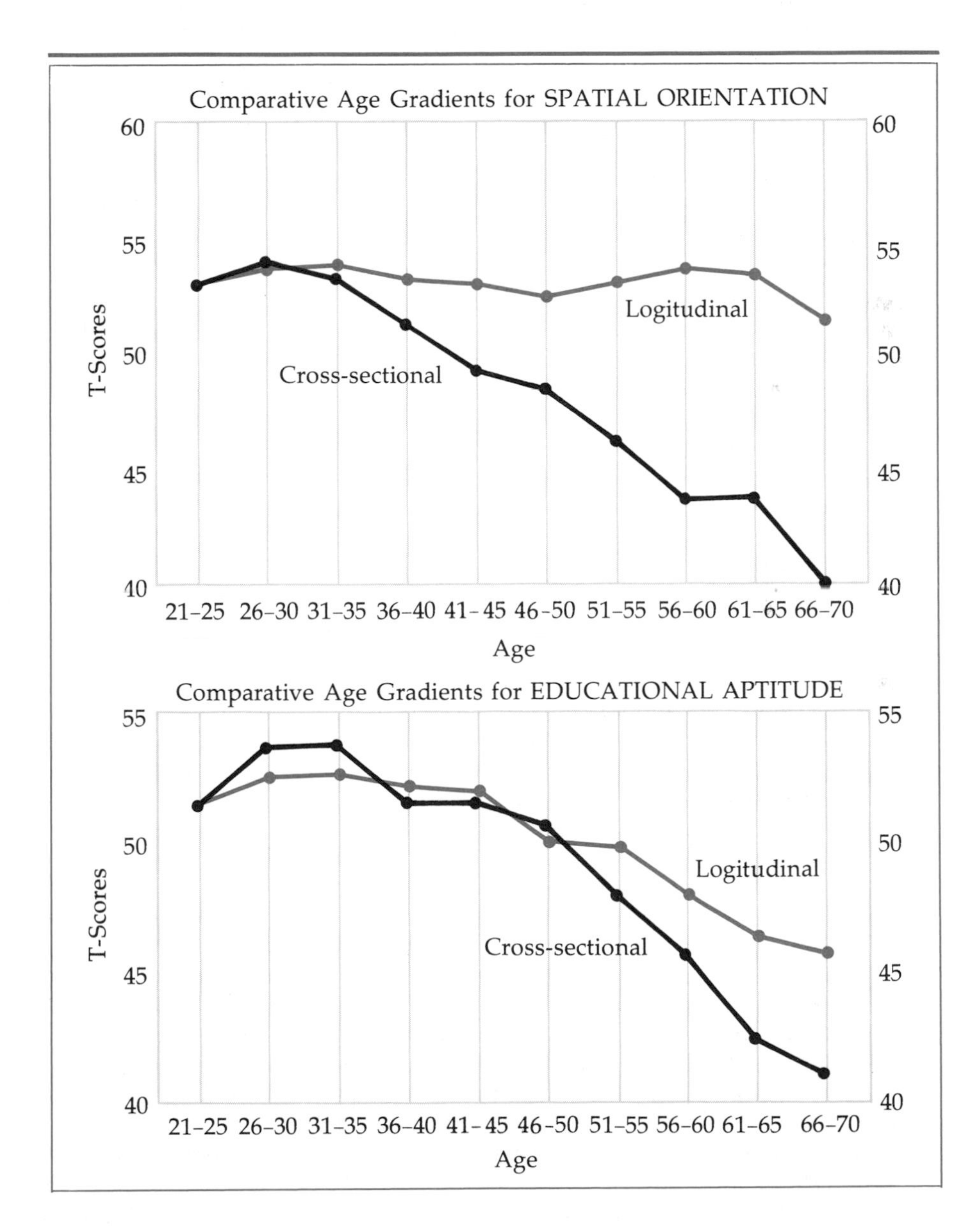

FIGURE 1-2
Research on the same topic will produce slightly different findings depending on whether a cross-sectional or a longitudinal design strategy is employed (from Schaie and Strother, 1968).

concerns a subject like changing attitudes on premarital sexual behavior during the past sixty years. If the question is a more basic one—for example, what factors influence changes in attitudes concerning topics such as premarital sexual relations—then the longitudinal design encounters the same limitation that the cross-sectional design encounters. That is, there is no way of determining if the patterns observed in this particular longitudinal cohort are the same as the ones that would be observed if the study were repeated with a second and a third longitudinal cohort, each cohort, for example, born ten years apart. Longitudinal studies involving successive cohorts are called cohort designs. They are the most involved of all research strategies, but they are also the most comprehensive and the most revealing about the course and the context of life-span human development.

Cohort Research Strategies

The **cohort design** strategy is designed to resolve the ambiguities of interpretation found in the cross-sectional design and also in the longitudinal design. Unfortunately the cost of resolving the theoretical problems is to create some very large practical ones. In essence, the cohort strategy is simply a series of longitudinal studies. Look at Figure 1-3. It represents a simulated cohort design involving five successive longitudinal samples. Each longitudinal sample is a separate cohort (that is, a group of individuals born at the same time). Any one of the five diagonals illustrates a longitudinal research design. For example, the 1920 cohort was born in 1920; each member of the cohort was age ten in 1930, age twenty in 1940, and so forth.

Each of the three dashed vertical lines on the graph illustrates a cross-sectional research strategy. For example, in the year 1960, a cross-sectional study would involve five different groups of individuals, a group of infants born in 1960, a group of ten-years-olds born in 1950, a group of twenty-year-olds born in 1940, and so forth. The results obtained from this cross-sectional sample would be subject to the same problems of interpretation as the cross-sectional study of attitudes concerning premarital sexual behavior. Namely it would be impossible to determine the extent to which the attitudes reflected age-related changes in development rather than the changing values of society.

On a broad scale, this problem is also true of the longitudinal design. The longitudinal design does unambiguously show how these values change with time, but it does not show the extent to which changing cultural patterns can influence this age-related transition. If the sixty-year-olds were born in 1920, then their changing attitudes would have been influenced by the particular slice of history they experienced between 1920 and 1980. What if the study were repeated using a cohort born in 1930? This second cohort would become sixty in 1990. Both groups when interviewed at age sixty would have lived a total of sixty years and, since their life-spans overlap, would have been exposed to similar historical events. But they would have experienced the same events at different points in their life-spans. The value of the cohort design is that it allows these various influences to be separated. It emphasizes the fact that the process of development involves the cumulative interaction between a maturing biological organism and an evolving sociohistorical context.

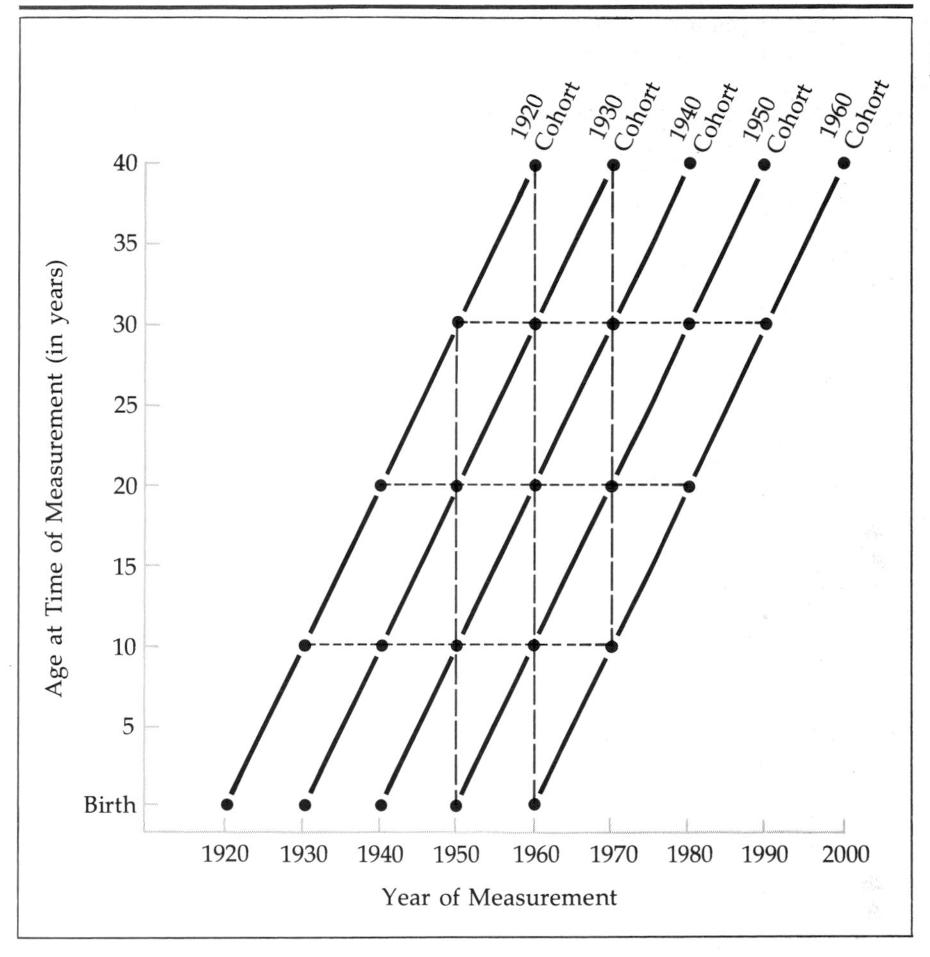

FIGURE 1-3
This simulated cohort design consists of five successive longitudinal samples, each of which is a separate cohort.

Controlled Research Strategies

The **controlled research experiment** is the most rigorous of all research strategies. It represents a method by which the influence of one variable on a particular behavior is examined independently of all other possible influences. To accomplish this separation of variables, the researcher arranges the experimental conditions in such a way that if a difference between groups is found, the only logical explanation is the variable in question. For example, it is frequently noted that older people often have more difficulty learning new tasks than younger people do. If this is in fact true, it has some very important implications for curriculum planning for different age groups, for the design of job retraining programs, and, in the extreme, for the justification for excluding older adults from certain types of tasks or occupations. One way to determine if this observation is correct and, if so, why, is to set up a series of tasks and, by systematically manipulating the conditions under which the tasks are learned, to see if older adults have more difficulty learning the tasks and maintaining a predefined level of performance than a group of younger adults.

The variables chosen to be studied would be determined by the researcher's knowledge of possible factors influencing learning ability across the life-span. For example, the researcher might want to study the effects of different teaching methods on different age groups and therefore might vary such conditions as the number of discrete steps provided in the instructions. Similarly, the researcher might want to study the way in which different features of the background environment affect learning in different age groups and therefore might vary conditions such as the amount of light or background noise present during instruction. Yet another variable to be studied might be the means by which members of different age groups are able to maintain an acceptable level of performance, and experimental conditions might therefore be designed to measure the effects of different learning procedures such as self-paced and timed work sessions.

There might be more relevant variables than could practically and adequately be studied in one experiment. Perhaps a series of experiments would be required. In either case, the research design would ensure that all possible combinations of participants and variables were included and that the resulting measures of learning and retention were recorded for each. Then, using a process analogous to the cohort strategy, the researcher would determine the relative influence of each variable on learning and retention. Since all of the variables were presented in a systematic and controlled fashion, the researcher would be able to show with a great deal of confidence which, if any, factors influence learning ability and performance in older adults.

The main advantage of the controlled research experiment—the high degree of control possible—is also its main disadvantage. Situations that are highly controlled, that systematically vary conditions in some predetermined fashion, are quite different from the "relative chaos" that characterizes most life-styles. For this reason, the highly controlled laboratory environment runs the risk of being too artificial, that is, of producing behavior that would not be typical in less well controlled environments. The significance of this problem depends on the nature of the question. If the experiment concerns learning ability, as in the example, then it focuses on a basic

These women are undergoing a manual dexterity test as part of a larger controlled research experiment on the effects of age on task performance. The fact that the test is timed will play a large role in determining the scores of the three women. Speed of performance is more affected by age than is accuracy of performance.

process, and the problem of artificiality is probably not too significant. If, however, the experiment concerns patterns of parent-child interaction, then the problem of artificiality may be very significant. For example, an experimenter wishing to determine how children react to different parenting styles might ask parents to behave toward their children in a number of different ways, such as by being very intrusive in their play or being quite removed from it. After observing the children's behavior, however, the experimenter might still be unable to determine whether the children's behavior was reflecting the specific parent behavior or whether it was a more general reaction to the fact that the parent was behaving in a strange and unexplained fashion. A naturalistic research strategy would be more appropriate to the study of parent–child interaction.

Naturalistic Research Strategies

In a **naturalistic research strategy,** researchers assume the role of observers rather than experimenters. They might, for example, observe parents with their children at various times of the day and in various settings. Since it is impossible to observe everything, the observer, like the experimenter, has to decide what behaviors will be observed and in what fashion the observations will be made. Typically, the observer focuses on the frequency of a particular type or types of behavior, for example, how often the parent praises or reprimands the child. This strategy is called **event sampling.** The other approach is **time sampling.** In this case, at predetermined intervals, whatever behaviors are taking place are recorded. The choice of procedure again reflects the nature of the question. Event sampling may be appropriate if the focus is on a particular type of behavior such as aggression or requests for help. Time sampling may be more appropriate for behaviors that occur more frequently such as language use. In either case, the observer makes the observations and then, like the experimenter, attempts to draw conclusions about the behavior in question on the basis of the particular patterns found.

Although the observer may be more confident than the experimenter in believing that what was observed is in fact what really happens in everyday life, the observer's interpretations are made with less confidence because the situation was less controlled. That is, in the study of parent-child interaction in the laboratory, all possible interaction patterns were created, and the behavior resulting from each was noted. In the natural setting, not all combinations occur. Some parents use a great deal of praise in dealing with their children; others, a great deal of punishment. The observer notes the result from each pairing and makes an inference as to the value of praise and punishment on behavior but can never be completely sure if the behavior was actually due to the praise or punishment or to some other factor also unique to the two types of parenting styles. For example, parenting styles often correlate with education. From the observation, it might be impossible to determine whether the differences were due to the specific technique used or to some more general quality acquired through years of education. Again, the issue is not whether one approach leads to good research and the other to bad research. Rather, the issue is which kind of research is appropriate for any particular question. To use either method effectively, the researcher must be aware of its limitations, in interpreting the results and making generalizations from the interpretation.

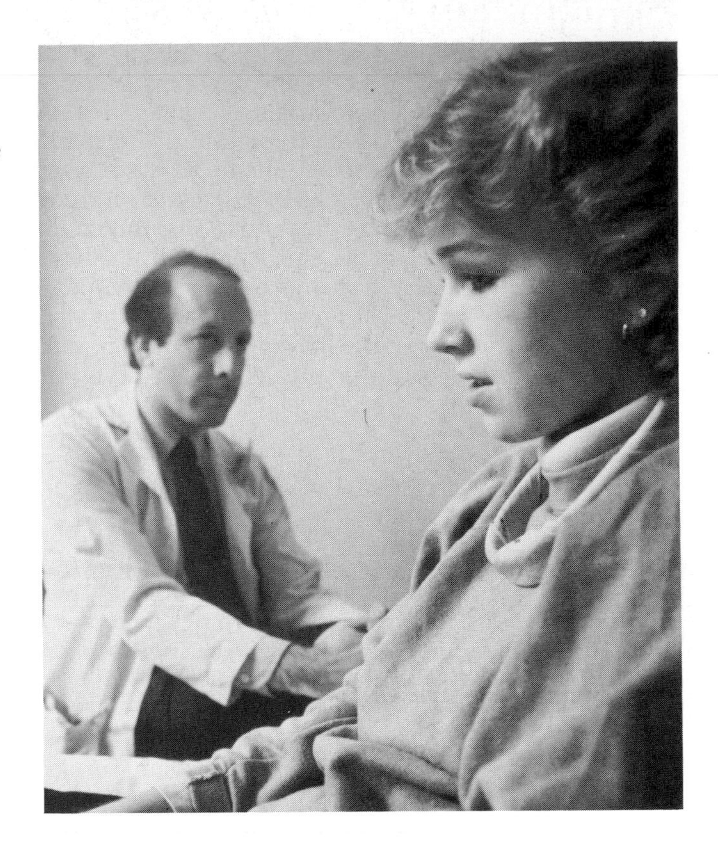

Clinical strategies allow researchers to go beyond recording behaviors and delve into the motives that give rise to these behaviors.

Clinical, or Interview, Strategies

The **clinical, or interview, strategy** is designed to gain insight into the motives or reasons for an individual's behavior or to determine the extent to which individuals understand various aspects of their lives. In the case of the parent-child example, the interviewer might ask parents why they frequently use praise or punishment as their primary means of dealing with their child. Questions might also be asked to determine the degree to which parents are aware of the effect of their parenting technique on their child's behavior. The observer must have answers to these questions if some attempt is to be made to change a parent's behavior. For example, the parent who uses punishment because of ignorance of other techniques would require one intervention strategy; the parent who prefers the technique because of anger and hostility directed toward the child would require another. The behavior in both cases would be the same; the motive, however, quite different.

The Process of Research: Methods of Data Analysis

Once the sample is obtained, the questions asked, and the answers given, the analysis of the data begins. Data analysis is almost always a quantitative endeavor. Its purpose is first to obtain an accurate description of the findings and then to determine what patterns of relationships exist among the pieces of information obtained.

Descriptive Statistics

Descriptive statistics give information about central tendency (the most common responses) and variability (the range of responses). **Measures of central tendency** tell us what is most typical, average, or common about the population. This information takes one of three forms. The most common is the **mean,** or arithmetic average. It is obtained by summing all the scores and then dividing by the number of scores summed. The mean score is the most representative value of the behavior being measured for the population being studied. In other words, if you had to tell someone about a population and you could only tell them one thing, the mean would be the thing to tell.

The other measures of central tendency are the median and the mode. The **median** represents the middle point of the distribution. The median separates the upper half of the cases from the lower half. The **mode** is the single most frequent score or category in the distribution.

Of the three measures of central tendency, the mean is both the most useful and the most frequently used. It is the statistic that is most often used in more complicated statistical procedures.

As useful as the mean is in describing a population, it can be very misleading when used alone. Although it tells us about central tendency, it tells us nothing about an equally important characteristic of the sample, its variability. Specifically, the mean tells us nothing about the range of scores or how frequently each possible score in the distribution is represented. The greater the variability in the distribution, the less representative is the mean of any individual score. Conversely, the less variability there is, the more representative the mean.

Degree of variability is expressed in two ways. The simplest of the **measures of variability** is the range. The **range** is the difference between the highest and the lowest score in the distribution. A more useful measure is the standard deviation. The **standard deviation** tells, on the average, how much each individual score differs, or deviates, from the sample

While height, weight, and age are correlated in children, there is a fairly wide range of variability—a fact that is evident in this photograph of nine-year-old girls.

mean. Like the mean, the standard deviation is a component of almost all complex statistical procedures.

It is essential to know both the mean and the standard deviation before reaching conclusions about the population from which the sample is drawn. If, for instance, you are responsible for making program or policy decisions concerning a particular population, knowledge of both the mean and standard deviation allows you to accurately define not only what level or type of service is most needed, but the full range of services needed as well. Two third-grade teachers whose average class reading level is 9 years, 2 months will find themselves developing very different lesson plans if the range in one class is 8 years, 7 months to 9 years, 8 months and in the other 6 years, 3 months to 12 years, 4 months.

Patterns of Relationships

Once we have an accurate description of the responses given by the sample, the next step is to see what relations exist among the responses. Measures of correlation are usually used to determine patterns of relationship. There is a wide variety of measures of correlation (the nature of the data determining which are appropriate to use), but they are all used to answer the same question—what is the degree of relationship between two or more variables?

Specifically, the **correlation coefficient** tells us the degree to which uncertainty about one variable is reduced by knowledge about another. If, for example, you tell me a person is six feet tall, you are also telling me something about his weight. There is a correlation between height and weight. The frequent correlations between and among various aspects of family, school, work setting, and neighborhood attest to the interdependent nature of development across the life-span.

The correlation coefficient is usually expressed as a number ranging from −1.0 to +1.0 (the "+" is inferred if no sign is given). The sign indicates the direction of the correlation. A minus sign indicates a negative correlation; a plus or no sign, a positive correlation. A correlation of zero indicates absolutely no relationship; a correlation of 1.0, a perfect relation. Direction of the correlation is a separate piece of information from the magnitude. A correlation of −.75 expresses a stronger relationship than one of .50. For example, there is a negative correlation (that is, as the magnitude of one variable increases, the magnitude of the other decreases) between competitiveness in young boys and maternal protectiveness. The more protective the parent is, the less competitive the son. There is a positive connection between years of school and income: the more years of school, the higher the income.

Correlations are very useful bits of information. When they are misinterpreted, however, they can be extremely misleading. The most common misinterpretation occurs when a measure of correlation is taken as a measure of causality. Two variables that are correlated may exist in a causal relationship. That is, variations in one may be caused by variations in the other. However, there is nothing in the correlation coefficient that indicates such a relationship. In fact, patterns of correlation may reflect the influence of another variable. For example, height and weight are correlated, but one doesn't cause the other. Rather, both reflect the biological and nutritional status of the individual. There is a correlation between grade point average and standardized achievement test scores, but one doesn't cause the other. Both may reflect the influence of a third factor such as the quality of the school system or family setting. The misinterpretation becomes most significant when a decision is made to change one element of the relation in order to change or "cure" the other. Not only is the result frequently a waste of time and money, but it often leads to the erroneous conclusion that the particular condition is not changeable.

A second common misinterpretation occurs when a measure of correlation is taken as a measure of absolute rather than relative standing (McCall, 1977). If I found, for example, a strong correlation between class standing in fifth grade and class standing in tenth grade, I would be noting that the more able fifth graders tend to be the more able tenth graders, the less able fifth graders, the less able tenth graders, and so forth. However, my correlation tells me very little about actual ability level. The more able fifth graders know more than their less able classmates but may not know nearly as much as the least able tenth grader.

Inferential Statistics

The ultimate goal of any statistical analysis is the understanding of relationships. **Inferential statistics** are methods that can document the presence of causal relationships. Does a work incentive program significantly reduce the numbers of adults needing financial assistance? Does a marriage counseling program improve the quality of family life? Do low-income children do poorly in school because of inadequate diet?

In the strictest sense, statistical analysis does not absolutely determine cause. Rather, it allows you to make a probability statement. In the social and behavioral sciences, cause is usually inferred if the researcher concludes, through statistical analysis, that the probability of a particular relationship's occurring by chance is less than five in a hundred. In the physical and biological sciences, the probability may be one in a thousand or less. The differences reflect the degree to which the social and physical sciences are each able to use rigorous experimental controls when testing for cause as well as the greater variability in events studied in the social and behavioral sciences. Through various statistical procedures involving **analysis of variance,** it is possible to determine the degree to which many different variables jointly cause some events to occur.

Statistical significance, that is, the probability that two events are not related merely by chance, is not necessarily the same as practical significance. As a result, a finding of statistical significance is not always a call for action. Rather, all of the different factors that contribute to the cause of the phenomenon should be found and their relative importance (in other words, the percentage of variance each accounts for) determined. Those factors accounting for the most variance have the most practical significance.

The Process of Research: Limitations

Probably the single greatest problem in studying the process of development is that people have an unfortunate tendency to act like people. That is, they show a high degree of variability in their behavior, which makes generalizations sometimes unreliable. Furthermore, because behavior is modifiable across the entire life-span, it becomes extremely difficult to pinpoint the antecedents or causes of development. These "idiosyncrasies" of the species are of course what make people so interesting, but for the researcher who is trying to explain and predict behavior and for the human service professional who must take the findings of the researcher and translate them into some sort of policy or program, they make the task more difficult.

Not all the problems of research are due to the nature of the subjects. Some problems are caused by the procedures of the research design, and some are caused by decisions about the way to use the findings. Research findings are not the only determinants of policy and practice in the human services; questions of economics, of ethics, of politics, and of values are equally relevant. These are not the kinds of questions that are addressed through research, but they are the ones that determine the application of research. In other words, research in human development doesn't provide the answers as to what we should do with our lives or the lives of others. Rather, it provides information relevant to determining the answers. For example, research has shown how we can prolong the lives of terminally ill individuals. Through the use of life support systems that assume the functions of diseased organs and through the use of treatments that fight disease, we can extend lives. But the use of life support systems creates a dependency that some people find unacceptable, and the side effects of some therapies are sometimes viewed as a fate worse than death. The researcher has provided us with these tools; the decision to use them must consider factors other than their availability.

HUMAN DEVELOPMENT AND HUMAN SERVICES

The last section of each chapter of this text discusses the relationship between human development at that particular stage of the life-span and the provision of human services. For those of you who are planning to pursue careers in helping professions such as social work, counseling and guidance, family relations, and psychotherapy, this should seem an obvious and logical pairing. Certainly an understanding of the way we develop across the life-span is useful if we are in the business of fostering that development.

But what about the rest of you? What about the person interested in the health professions? What about the person who is studying psychology but wants to do basic research? What about the person who wants to be a demographer and simply chart the movements of people? What about the business major or the economics major or the prelaw student? Of what relevance are these last sections to these people?

Simply put, each of us is involved in the provision of human services. What each of us does affects the lives of others. The number affected may be great or small, the consequences of short or long duration, the scope broad or narrow, the intent direct or indirect, the timing before or after the fact; but in every case there are consequences, and if the reading of this text provides nothing else but a greater awareness of this fact, then my efforts in writing it have been well spent.

Ironically, those who are in the position to exert the greatest influence on the lives of others are sometimes the ones who are least likely to think of their actions in terms of human services. None of us has any problem seeing that the social worker, the marriage counselor, and the school psychologist provide human services. Each of them makes an effort to help an individual, a couple, or a family deal with some specific problem or crisis. But what about the urban planner designing a housing project or the judge deciding a child custody case or the personnel manager of a large corporation rearranging work schedules or the government economist proposing changes in the tax structure? The actions of these people have consequences for others and therefore can be considered human services like those of the social worker, marriage counselor, and school psychologist. The consequences may be indirect, they may be qualified by a variety of other factors, they may be difficult to evaluate separately from other factors; but they nevertheless are consequences and each of us, in all of our endeavors, both public and private, needs to be aware of that fact.

What role does an understanding of life-span human development play in all this? It plays four roles. It provides a standard against which to evaluate the human consequences of individual and group effort. It aids in the identification of needed human services. It aids in the design of more effective human services, and it provides a shared frame of reference to aid in the coordination of individual efforts.

A Common Standard

The study of human development provides a description of the course of development across the life-span. This description specifies the behaviors that are most typical at each age or stage of the life-span, and, equally important, it indicates the range or diversity of behavior that is typical at each stage. Furthermore, the study of human development provides information about the antecedents or causes of behaviors and the extent to which behaviors are linked together. In essence, it provides a basis for making judgments about people and for making predictions about what people are likely to do in the future. Consider the following situations.

(1) You are in charge of a program to provide educational services for children from low-income families. You know that these children don't do as well on standardized achievement tests as children from more economically advantaged backgrounds. But you don't know if the differences in test scores reflect a genuine deprivation or a cultural difference. That is, you don't know if the parenting patterns of these low-income families in some way actually interfere with normal developmental events or if the situation is better understood as two distinct cultural patterns each producing developmentally equivalent but distinct behaviors.

(2) You are the personnel director of a local school district. There is an opening in one of the elementary schools, and you have found a very qualified candidate. The candidate, however, is homosexual. If the candidate were "straight," you would hire him immediately: in effect you realize that competence as a teacher is unrelated to the sex-related behavior of heterosexuals. But you don't know if the same is true about homosexuals. Homosexuals form a small percentage of the population. In this sense they are atypical. But what, if any, is the relationship between being atypical

and being abnormal; that is, how extensive are the range and diversity of human behavior?

(3) You are responsible for developing a set of recommendations to be considered by a subcommittee of the U.S. Senate. The subcommittee's task is to recommend changes in federal laws regulating mandatory retirement. How do you determine if the aging process is sufficiently uniform to justify a federal policy requiring mandatory retirement at some specific age? Or how do you demonstrate that the variability is so great that some more individualized approach is necessary?

In all three of these examples, a knowledge of the developmental patterns across the life-span would be extremely valuable. In particular, this knowledge would tell you that social class deficits in children's learning abilities are almost always reversible (Ginsburg, 1972; Lazar & Darlington, 1982), that homosexual teachers are no different from heterosexual teachers in their relationships with students (Masters, Johnson, & Kolodny, 1985), and that there is virtually no relationship between age and work performance for the vast majority of work roles (Foner & Schwab, 1983).

Identification of Services

The design of human service programs intended to prevent undesirable outcomes involves predicting the consequences of historical and cultural events on the course of development across the life-span. Such prediction, in part, requires an understanding of the multidetermined, interdependent, cumulative nature of human development. For example, the past decade has witnessed a startling increase in the divorce rate. The immediate impact of this increase is evident in terms of its effect on the divorced husband, the divorced wife, their children, and the subsequent relationship of the children with each of their divorced parents (Hetherington, Cox, & Cox, 1976). A variety of services have been developed and implemented to help cope with these immediate consequences. What about less immediate consequences? What about the possibility that some of the effects of divorce on children may not be evident until those children are old enough to marry and parent? We can wait and see if any patterns develop, but then the services become intervention rather than prevention, a certainly more costly and often less effective approach. What do we know about the development of social relations across the life-span that might lead us to believe that some long-term consequences are possible? How accurately can we predict the nature of these consequences, and how effectively can we design appropriate long-term prevention programs?

Our society is currently experiencing an increase in the rate of technological growth, an evening of the population distribution due to a declining birth rate, and an increase in the average life expectancy due to medical accomplishments. Taken together, these facts then suggest that the educational, recreational, and resource needs of future generations will be quite different from those of the current generation. Lifelong education will need to be more common, and postretirement living will occupy a larger percentage of the life-span (Woodruff & Birren, 1975). The ability to anticipate the influence of these trends on development across the life-span and to plan the services that will therefore be necessary is a reflection of our understanding of the contextual nature of development.

As the number of people over the age of 65 increases in the United States, more attention will have to be paid to providing services that enable the elderly to live active and productive lives.

Effective Design

Life events do not occur in isolation. Each life event can be viewed as having a history and a future. Human services programs that treat life events as isolated units will probably have a limited impact. An understanding of the contextual nature of events across the life-span is one good way to ensure that human service programs have maximum influence and long-lasting consequences.

Compensatory education programs for preschool children are a good example to show the importance of understanding the contextual nature of life events. These programs, of which Project Head Start is the best known, began in the 1960s and were intended to provide children a variety of educational experiences and skills that were assumed to be lacking from their environments (Evans, 1975). What started as a six-week summer program grew to half-day and then to full-day programs as it became increasingly apparent that these skills were more difficult to teach and to maintain than they were initially believed to be (Goldhaber, 1979). Although some (e.g., Jensen, 1969) interpreted this pattern as indicating that the nature of the problem was genetic rather than environmental, others (for example, Bronfenbrenner, 1975; Shipman, 1976) felt that the impact of these programs would not improve until the family became a more integral component of the educational experience.

> Of primary importance was the finding that a warm, supportive home atmosphere combined with a warm, supportive school setting creates an upward spiraling in a child's ability to achieve. In other words, cognitive gains are likely to be greatest when the total ecology of the child is supported at home and in school. A corollary to that observation is that the quality of either parent–child or teacher–child interaction alone is generally not sufficient for modifying the deleterious effects of poverty that act upon the developing child. Rather sustained intellectual growth depends on the quality of the relationships established between parents, teachers, and child. (Shipman, 1976, p. 4–5)

The history of compensatory education programs demonstrates the importance of understanding all the components that influence a particular facet of development. In this case, it pointed out that learning is not done exclusively in the school and that instruction is not provided solely by the teacher. For the educator, understanding the nature of the home environment enhances the impact of the classroom experience, and helping the parent appreciate the value of the classroom experience increases the likelihood that the child's gains will be maintained even after the program is completed.

Life events are not only embedded in the context of other life events; they are also part of an individual's unique construction of the meaning of events. A failure to appreciate the active role individuals take in defining the course of their own development will also reduce the impact of human service programs. This omission sometimes is evident in programs designed to deter substance abuse among adolescents as well as in programs providing services for the aged. In the case of adolescents, programs designed to deter smoking, excessive use of alcohol and drugs, and indiscriminate sexual behavior are often designed to provide the adolescent objective information about the negative effects of substance abuse and the serious burdens encountered in adolescent pregnancy, parenting, and marriage. Most of the information is accurate, but it often misses the mark because it fails to consider the adolescent's perspective. To many adolescents, these activities are indicators of adulthood and are not likely to be stopped until other equally potent indicators come to replace them. In the same vein, programs that fail to consider the egocentrism of many adolescents ("it won't happen to me") may do relatively little to reduce the incidence of substance abuse, teenage pregnancy, or reckless driving.

A Common Frame of Reference

Designing effective human service programs or simply trying to determine the effect of some policy or action on the course of development is a difficult task. It is difficult because the process of development is long and complex. It is also difficult, however, because by law, custom, and philosophy, our society places great value on individual autonomy. We shun broad national planning, preferring instead to decide many matters locally. We prefer specific categorical services designed to remedy specific problems to broad comprehensive programs intended to meet a diversity of needs. The categorical training of human service professionals reflects this national value. It is equally reflected in the irony of the fact that people who are in positions to influence development the most are often the ones least likely to think in such terms. A categorical approach is consistent with our macrosystem value of "checks and balances," but it can prove inefficient and even insensitive to individual and family needs. Families with the least material and personal resources often find themselves having to go from one social service agency to another, each time providing the same information that they provided to the previous agency. Further, because each agency evaluates the provision of services solely in terms of its specific category of service, families may find themselves judged ineligible for services because they do not meet the categorical requirements—that is, they may not be able to demonstrate enough need in any one category to be eligible for aid, even though the cumulative effects of their needs in

By making it necessary to deal with numerous agencies, categorical approaches to the provision of social services place the burden of obtaining help on those individuals least able to handle it—the low-income recipients.

several categories show that they do need help (Garbarino, 1982). Given the fact that the categorical system is more consistent with our macrosystem value structure than a comprehensive service system would be, the problem of inefficiency is perhaps best resolved through coordination. Coordination, however, requires a common denominator.

The study of human development provides a common denominator by creating a common frame of reference. It provides a countermeasure to the increasingly specialized nature of categorical services in the health, educational, and social welfare sectors of the human services. This common frame of reference fosters an awareness that factors external to a category of service or area of endeavor influence the provision of services and the quality of effort. It also makes us more aware that our efforts have side effects beyond our sphere of interest or influence.

SUMMARY

THE COURSE OF DEVELOPMENT

1. Both biological and social clocks are useful rulers to measure the course of development. Although they provide no explanation for the course of development, they do suggest that the process proceeds at a fairly orderly rate and in a fairly orderly sequence.
2. The meaning of the events in our lives is as much a reflection of the meaning we construct for these events as it is a function of the nature of the event itself. The two are usually similar but rarely identical.
3. Stage theories differ in a variety of ways. Some focus on actual behavior, others on the potential for behavior; some concern those aspects of development common to the species, others only those common to a culture; some emphasize the social aspects of development, others the cognitive aspects.

THE CONTEXT OF DEVELOPMENT

1. We function within a human ecosystem. Events that occur at each of the four levels of the ecosystem, even though we may not be involved directly in the making of those events, influence the course of our development.
2. Both direct and indirect learning mechanisms play a significant role in creating the associations that form the data for our constructions of meaning across our individual life-spans.

A SYNTHESIS

1. The process of development is multidetermined, interdependent, and cumulative. It involves the successive products of a biological organism maturing within a sociohistorical context. The successive products of this interaction are levels of psychological awareness.

2. Over the course of a life-span, individuals acquire competence over increasingly wider domains, become better integrated within the environment, and develop a greater sense of individual unity.

THE STUDY OF DEVELOPMENT ACROSS THE LIFE-SPAN

1. Research is a process of systematically sorting out events and the relationships between events and then making inferences as to the cause of the observed patterns.
2. Cohort research strategies, by incorporating the elements of both longitudinal and cross-sectional research designs, offer the most comprehensive means of studying the course of development across the life-span.
3. Controlled research designs provide the most precise means of making statements about the relationships between developmental events. Because they are controlled, however, they may place artificial constraints on the observed relationships. Naturalistic research designs may not provide enough control.
4. An accurate description of behavior requires both a measure of central tendency and a measure of variability.
5. Correlation coefficients express the degree of relationship that exists between two or more measures.
6. Inferential statistics provide a measure of the degree of certainty that can be attributed to a hypothesized causal relationship between two or more factors.

HUMAN DEVELOPMENT AND HUMAN SERVICES

1. For the provision of human services to be efficient and effective, it should have a solid foundation in the study of development across the entire life-span.

KEY TERMS AND CONCEPTS

Stage-Based Approach
Ecological Approach

THE CONCEPT OF DEVELOPMENT

Stages of Development
Maturation
Gesell's Maturational Theory of Child Development
Developmental Norms
Levinson's Stage Theory of Adult Male Development
Frieze's Stage Theory of Adult Female Development
Piaget's Theory of Cognitive Development
Cognitive Development
Assimilation
Accommodation
Havighurst's Developmental Tasks Theory
Erikson's Theory of Psychosocial Stages
Bronfenbrenner's Human Ecology Model
- Microsystem
- Mesosystem
- Exosystem
- Macrosystem

Classical Conditioning
Instrumental Conditioning
Social Learning or Imitation

THE STUDY OF DEVELOPMENT

Operational Definition
Sample
Longitudinal Design
Cross-Sectional Design
Cohort Design
Controlled Research Experiment
Naturalistic Research Strategy
Event Sampling
Time Sampling
Clinical, or Interview, Strategy
Descriptive Statistics
Measures of Central Tendency
- Mean
- Median
- Mode

Measures of Variability
Range
Standard Deviation
Correlation Coefficient
Inferential Statistics
Analysis of Variance
Statistical Significance

SUGGESTED READINGS

Much of this chapter has focused on the various contexts in which we grow and develop and on the way in which each influences the course of our development. Any of the following readings should help you gain a better understanding of this relationship.

Bronfenbrenner, U. (1979). *The ecology of human development.* Cambridge: Harvard University Press.

Elder, G. H. (1974). *Children of the great depression.* Chicago: University of Chicago Press.

Erikson, E. (1950). *Childhood and society.* New York: Norton.

Garbarino, J. (1982). *Children and families in the social environment.* New York: Aldine.

Kohn, M. L. (1977). *Class and conformity: A study in values* (2nd ed.). Chicago: University of Chicago Press.

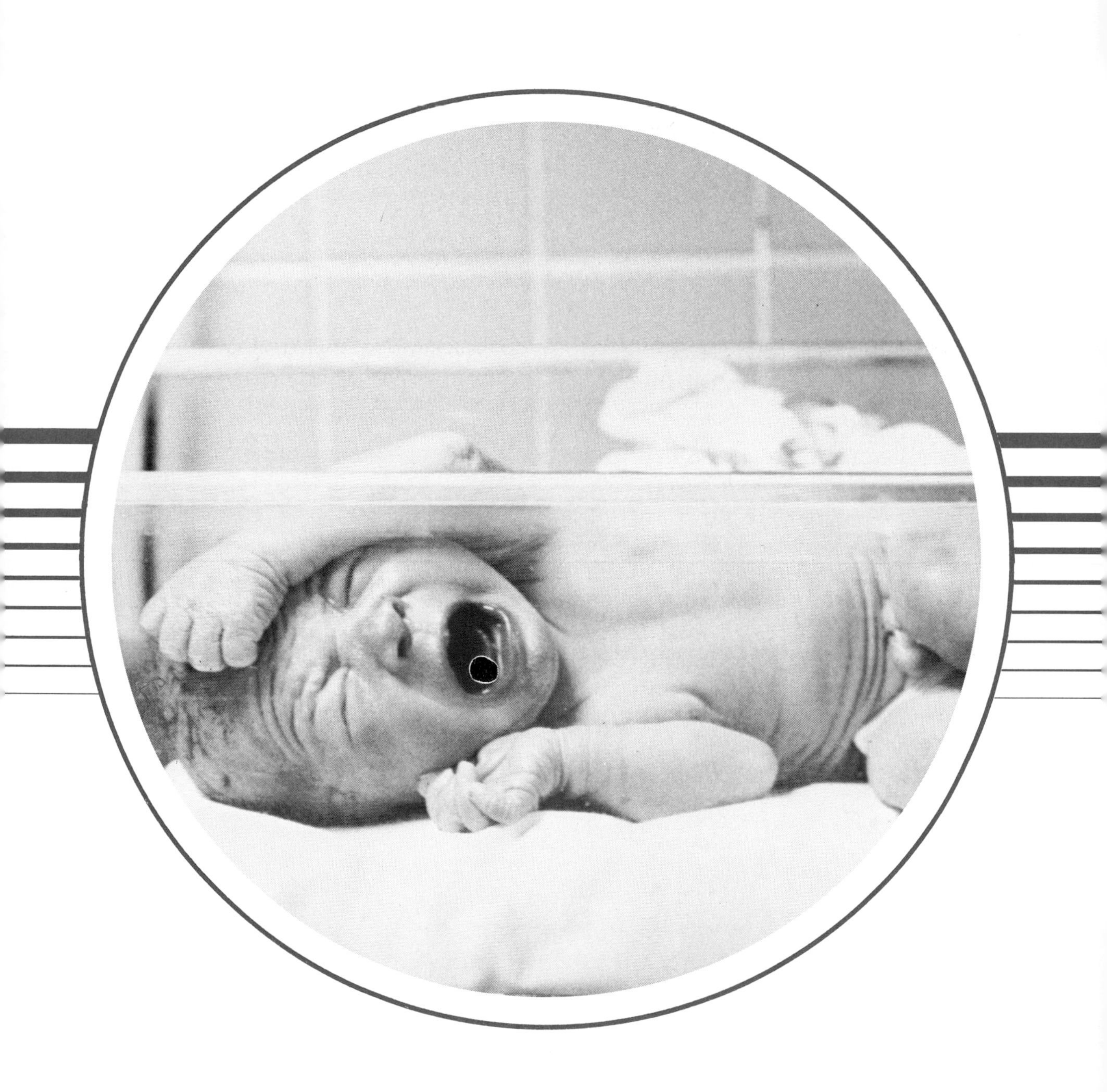

2
PRENATAL DEVELOPMENT AND BIRTH

CHAPTER OUTLINE

The prenatal period is unmatched by any other period in terms of both the rate of growth and the degree of change that occur in it, a short span of nine months. Perhaps with the exception of death, the end of life, there is no other time in the life-span that is as likely to be described in poetic or romantic terms or as likely to be used to demonstrate the mystery of life as the prenatal period, the beginning of life. Surely there is no other stage of the life-span that is bounded by such extraordinary events as conception and birth.

From a more mundane perspective, development during the prenatal period concerns the differentiation and maturation of the human biological organism. The course of this development, like all aspects of development across the life-span, involves the cumulative interaction of the biological organism with its sociohistorical environment. What is unique about the study of prenatal development is that it focuses on the biological organism itself rather than, as will be true in subsequent chapters, on the psychological functioning of that biological organism. Since our psychological functioning at any point in the life-span is, in part, a reflection of our biological status, and since the converse is equally true, an understanding of one must entail an understanding of the other.

This chapter is organized around four aspects of prenatal development. These four aspects can be thought of as the what, why, how, and so what of prenatal development. The what aspect concerns the normative course of development from conception through birth. The why aspect examines the various genetic and environmental influences on prenatal development. The how aspect studies the mechanism through which these genetic and environmental influences interdependently exert their influence on the course of development. The so what aspect concerns the relevance of our knowledge of normative prenatal development and the factors that regulate it for the provision of services and the development of public policy concerning children and families.

THE COURSE OF PRENATAL DEVELOPMENT

The prenatal period is divided into three distinct phases. The first, the period of the ovum, encompasses the ten days immediately following conception and is characterized by the initial cell division of the fertilized ovum and its implantation in the uterus. The second period, the period of the embryo, continues through the seventh week of prenatal life. During this period, most of the organ systems are formed. The remainder of prenatal life is called the period of the fetus. During this last period, the fetus develops greatly in both size and weight.

Period of the Ovum

The **period of the ovum,** lasting approximately ten days, begins at conception. Figure 2-1 illustrates the sequence of events that occurs during this first phase of prenatal life. Following ovulation, the ovum is released into the Fallopian tube where it is fertilized by a sperm cell. The fertilization of the ovum by the sperm restores the complete number of chromosomes (forty-six), which was previously halved when the sex cells were formed. The fertilized ovum is now known as a **zygote** (Moore, 1974). During the next several days, the zygote continues to pass through the Fallopian tube and into the uterus. During this time, the zygote undergoes a series of cell divisions that by the third day produces an approximately sixteen-cell structure known as a **morula.** When the morula enters the uterus, it undergoes a rather drastic change in shape. Because fluid in the uterine cavity enters the morula, it begins to hollow out. It now is called a **blastocyst.** The change from morula to blastocyst involves more than simply a change in shape: it also involves a differentiation, or specialization, of function. The outer wall of the blastocyst, known as the **trophoblast,** attaches itself to the uterine wall and begins to implant itself. The trophoblast will evolve into the **placenta,** the organ through which the developing fetus will receive oxygen and nourishment and excrete waste products.

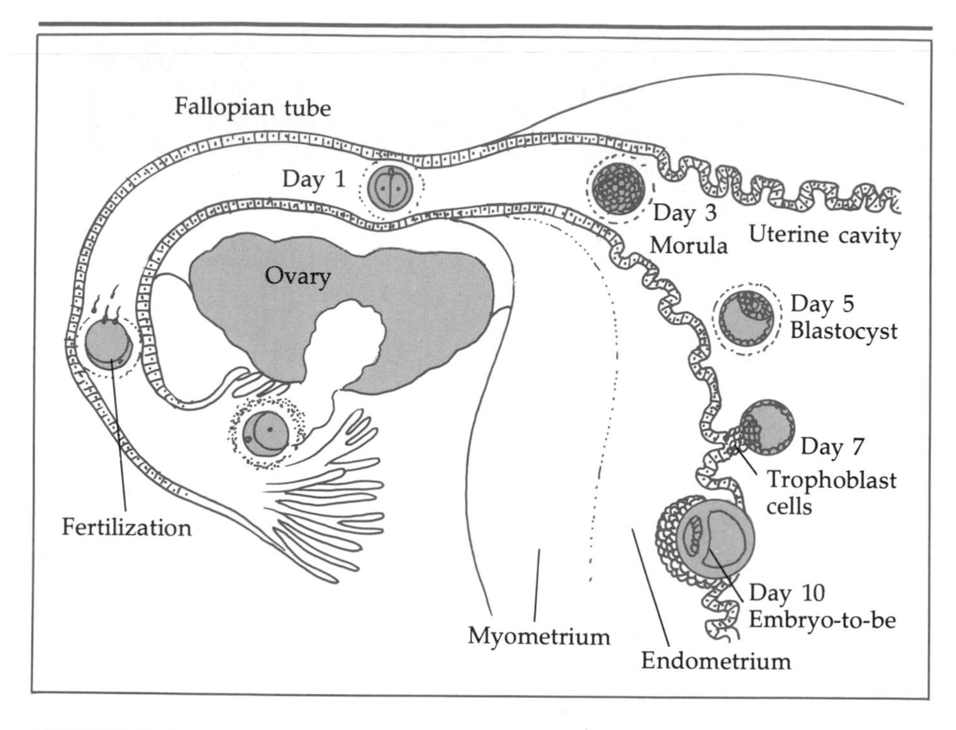

FIGURE 2-1
During the period of the ovum, the ovum moves out of the Fallopian tube and into the uterus, finally implanting itself in the uterine wall.

Period of the Embryo

The **period of the embryo** is probably the most dramatic of the three phases. During this approximately seven-week period, all of the major organ systems are formed, limbs become evident, and facial features begin to emerge.

The transition from the period of the ovum to the period of the embryo is marked by the formation of the first two **embryonic cell layers.** While the trophoblast is attaching itself to the uterine wall, the ball of cells within it aligns itself into two layers, the **embryonic ectoderm** and the **embryonic endoderm.** By the middle of the second week following fertilization, a third cell layer, the **embryonic mesoderm,** is formed when an inner layer of cells separates from the trophoblast. What then follow are the folding and hollowing out of the three cell layers that produce the specific sites from which the various organ systems develop (see Figure 2-2).

From the endoderm develop the liver, pancreas, urinary bladder, and parts of the trachea, lungs, gastrointestinal tract, tonsils, and thyroid

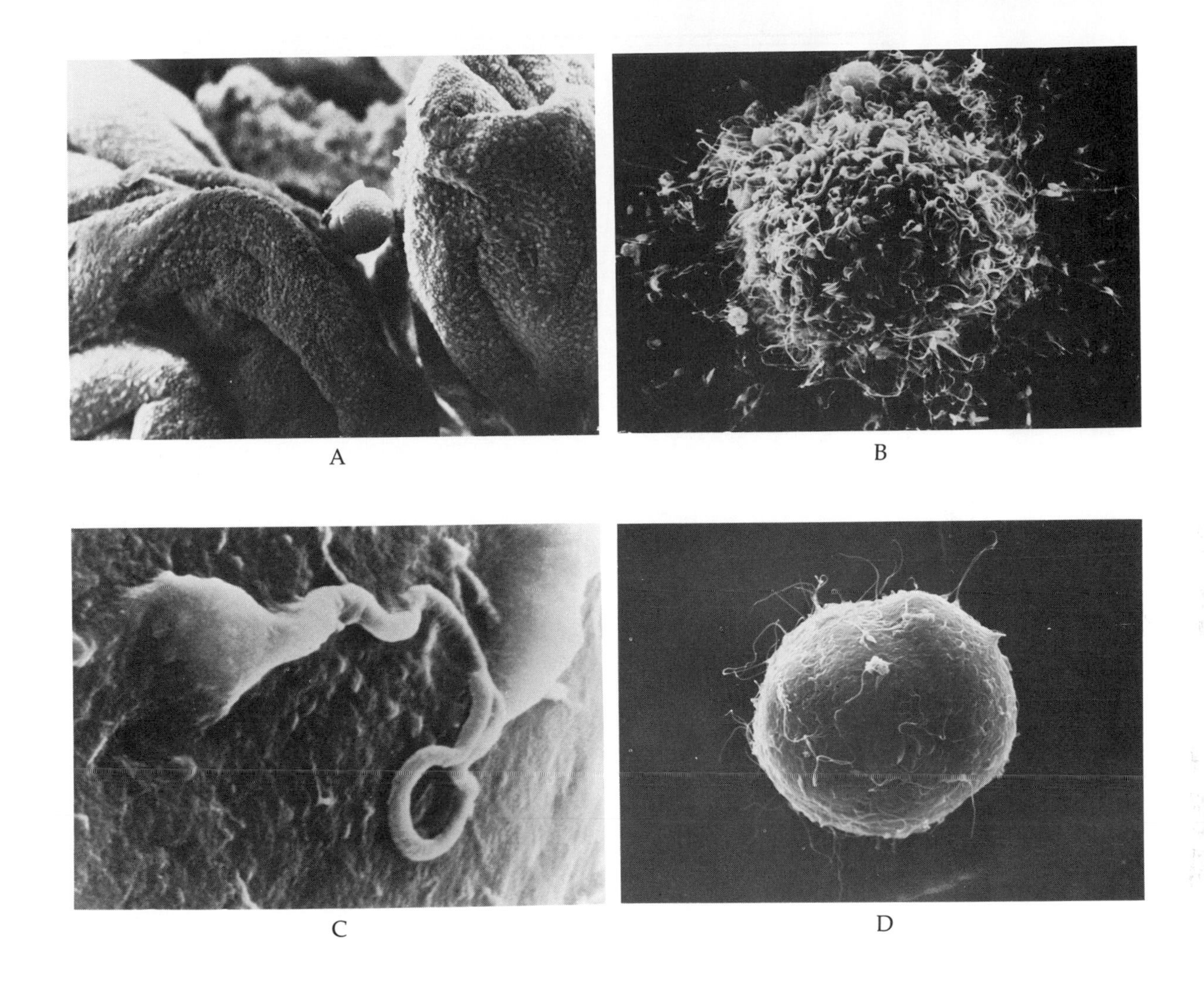

These electron microscope photographs show several of the early events illustrated in Figure 2-1: (a) the egg nestled in the folds of the Fallopian tube; (b) several hundred sperm attacking the egg; (c) a sperm as it begins to penetrate the egg; and (d) the fertilized egg beginning to divide and multiply.

gland. From the mesoderm develop the muscles, skeleton, urogenital system, spleen, blood and lymph cells, and the cardiovascular and lymphatic systems. From the ectoderm develop the hair, nails, mammary glands, eyes, ears, nervous system, and pituitary gland (Moore, 1974). By the seventh week, the embryo begins to acquire a distinctly human appearance. The head is clearly outlined, the neck is distinct, and even the eyelids are evident. Once development has reached this point, the embryo becomes known as a fetus and enters the fetal period of prenatal development. The fetal period continues until birth.

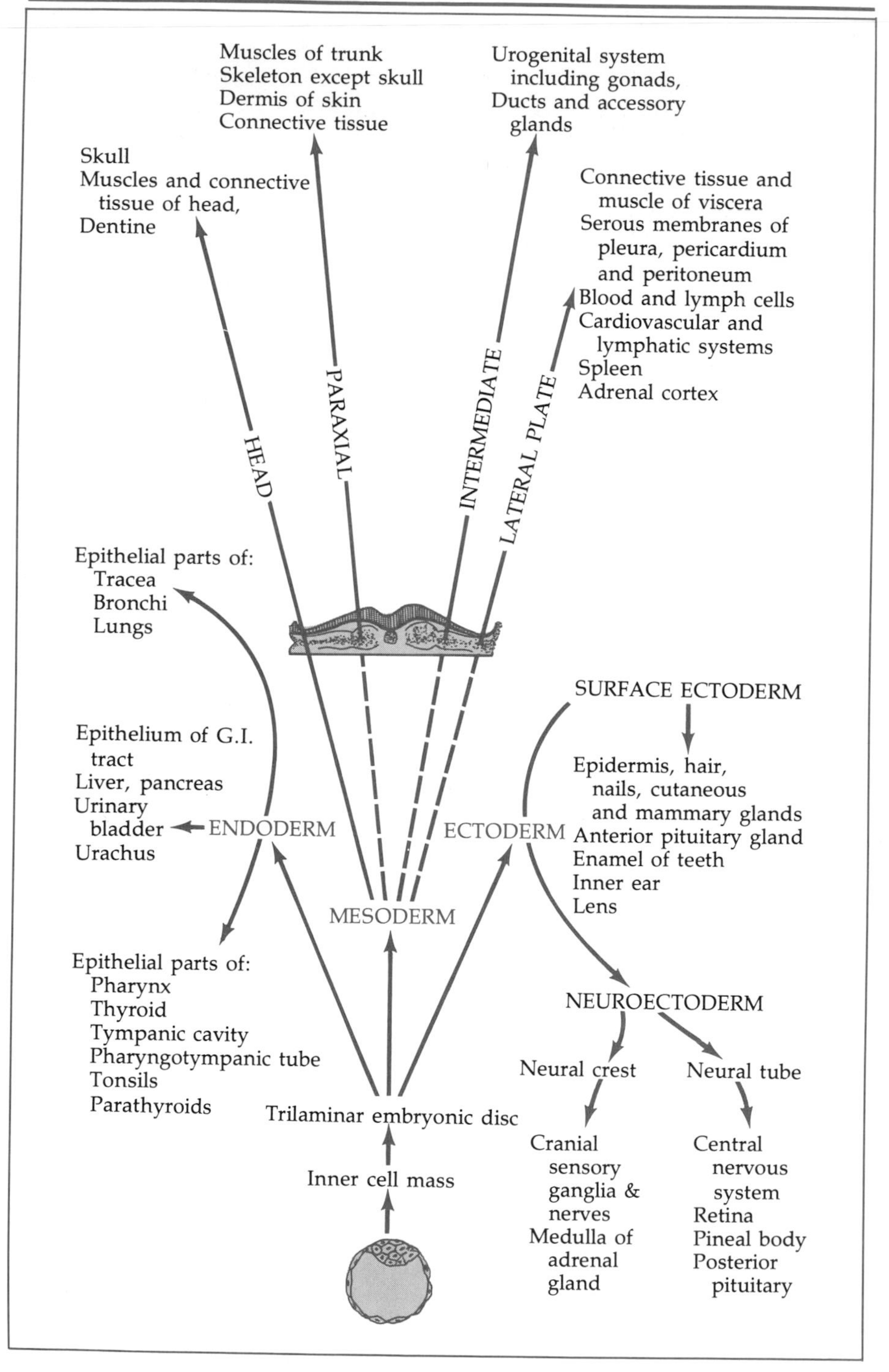

FIGURE 2-2
During the period of the embryo, the three primary cell layers develop. Out of these three layers come the various organ systems of the body (from Moore, 1974).

Period of the Fetus

The **period of the fetus** encompasses the balance of the prenatal period. By the eighth week, organ **differentiation** is relatively complete, and from this point until birth, change takes the form of growth in size and weight and further maturation of the various systems and parts of the body. The amount of growth is incredible. (See Figure 2-3.) Growth in length is from approximately 40 to 360 millimeters. Growth in weight is from 5 grams at two months to 3410 grams at birth.

By the end of the third month, the external genitalia are finally differentiated. By the end of the fourth month, the limbs have attained their at-birth proportions. Sometime during the fifth month, fetal movements (quickening) are clearly felt by the mother. By the seventh month, the fetus has a high probability of surviving a premature birth. When premature birth does result in death, it is usually due to respiratory problems. Birth usually occurs approximately 280 days from the onset of the last menstrual period. As most parents will attest, however, this average figure can be and often is highly variable.

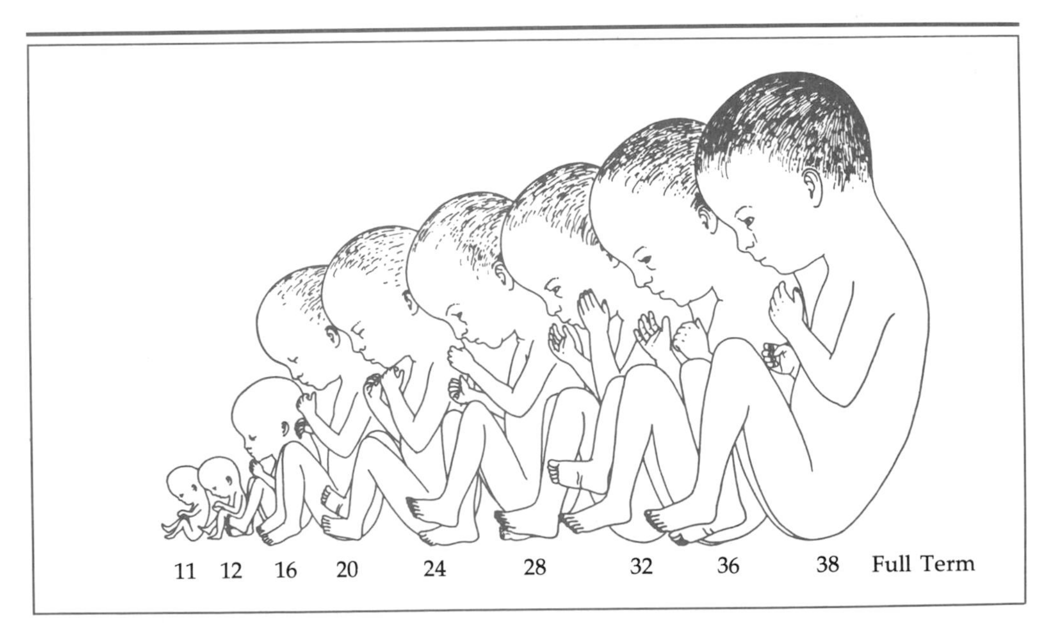

FIGURE 2-3
During the period of the fetus, the fetus grows from a length of approximately 40 millimeters to 360 millimeters and increases in weight from approximately 5 grams to 3410 grams.

THE CONTEXT OF PRENATAL DEVELOPMENT

This text focuses on the way people behave at various points in their lives. It also considers how their behavior is influenced by their continually changing understanding of their environment. An understanding of the way this process occurs requires an understanding of the cumulative interactions of the biological organism with its environment. The study of prenatal development provides an initial understanding of the nature of the biological organism, both in terms of its course of development and in terms of the factors that regulate that development. The previous section provided a short review of the course of prenatal development and this section examines its regulation.

The "Why"

All aspects of development, postnatal as well as prenatal, are influenced by heredity. Likewise, all aspects of development are influenced by environment. However, neither heredity nor environment alone determines any aspect of development across the life-span. Rather, it is the interaction of heredity and environment across the life-span that determines how individuals behave and how their behavior reflects their understanding of their environment. Although it is possible to statistically separate the relative contributions of heredity and environment for large populations of individuals, the meaning of such estimates is highly debatable (see Hirsch, 1963, and Jensen, 1969, for two contrasting views), and at the level of a single individual, such partitions have about the same usefulness as debating the relative importance of hydrogen and oxygen in the formation of water.

The Genetic Code

At conception, a person inherits twenty-three **chromosomes** from the father and an equal number from the mother. The chromosomes align themselves to form twenty-three pairs. On each chromosome are strands of **deoxyribonucleic acid (DNA),** which arrange themselves in a double helix, or twisting ladder, fashion (see Figure 2-4). DNA is composed of four molecules (adenine, guanine, thymine, and cytosine), and these four molecules always arrange themselves so that adenine pairs with thymine and cytosine pairs with guanine. The sequence of these two pairs on each chromosome forms the **genetic code.** Specifically, each set of three pairs (a codon) specifies the formation of a particular amino acid, which in turn initiates a series of biochemical reactions, which again in turn regulates the formation of tissues and then organs and then organ systems and eventually behavior (Thiessen, 1972).

Each of the twenty-three chromosome pairs regulates a unique set of biological reactions. The eventual expression of a particular pair of chromosomes depends on what has actually been inherited from each parent and the degree of similarity between each half of each pair. The composition of the gene—the genetic makeup of the person—is called the **genotype.** Since different genotypes can produce the same outcome, it is also necessary to

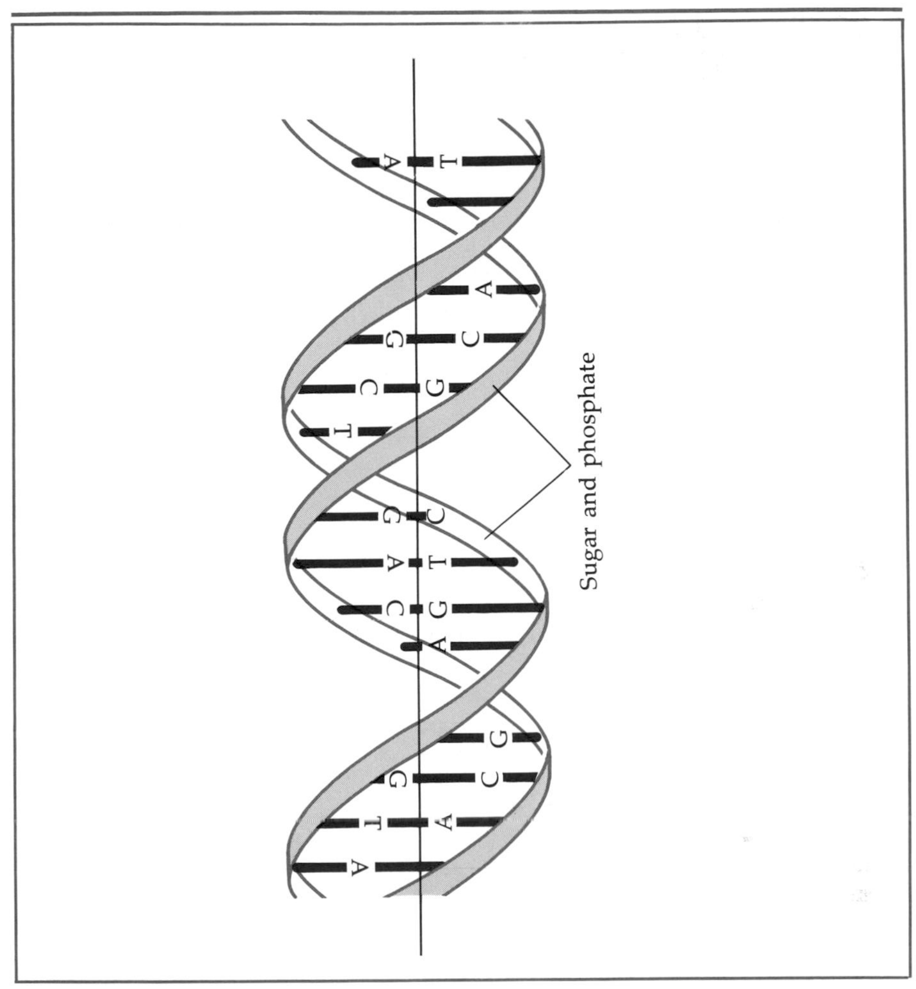

FIGURE 2-4
The molecular structure of the double helix DNA is composed of adenine (A), guanine (G), thymine (T), and cytosine (C).

distinguish an individual's genotype from his or her **phenotype.** The phenotype is the ultimate expression of the individual's unique genotype. Two individuals with identical phenotypes for a particular characteristic may nevertheless have different genotypes for that characteristic. When the genes on the two chromosomes of the pair are identical, they are said to be **homozygous** for that particular trait. If they have different forms, they are said to be **heterozygous** for the trait. The significance of the degree of correspondence between the two was first studied by Gregor Mendel, an Austrian priest living about the time of the American Civil War. Mendel worked with plants, particularly peas, but the principles he discovered are equally relevant for certain aspects of animal inheritance, including the genetic inheritance of humans.

Individuals who are homozygous for a trait such as eye color or hair color will demonstrate the trait in the form in which it is present in the parent. If the parents have blond hair, the child will have blond hair; if the parents have blue eyes, so will the child. If the genotype for a particular trait is heterozygous, one of three possible phenotypes will emerge. The heterozygous genes may average themselves, in which case the phenotype will be a blend of the two genes; they may express themselves in one or the other form, in which case the expressed gene is considered the **dominant gene** and the nonexpressed gene is considered the **recessive gene;** or they may express themselves in a form closer to but not identical with that of one or the other gene. In this last instance, the more influential gene is said to exert partial dominance over the other gene. If the trait in question is hair color and if one parent has brown hair and the other blond, then the first phenotype may be light brown hair, the second may be either color depending on dominance, and the third closer to the color expressed by the dominant gene than the color expressed by the recessive gene.

It is important to bear in mind that even though a particular gene is not expressed in a person's phenotype, the person still carries it and may pass it on to the next generation. Since some recessive genes express themselves in undesirable forms such as mental retardation, physical defects, and diseases such as hemophilia, an individual carrying this kind of recessive gene is faced with the dilemma of whether or not to conceive a child. Genetic counseling may be able to inform a couple how likely it is that the recessive gene will be expressed in the offspring. Similarly, prenatal diagnostic techniques such as amniocentesis can tell the parents whether or not the gene is present and, if so, whether or not it is homozygous (and therefore expressed). Even so, the parents still face the decision to conceive a child, and if the child is found to express the gene, the question of abortion.

Sickle-cell anemia and **phenylketonuria (PKU)** are two examples of major recessive gene effects. Sickle-cell anemia results from the mutation, or distortion, of one of the codons, or sets of DNA pairs, that regulate the formation of red blood cells. The consequence of this mutation is depicted in Figure 2-5. The chart clearly demonstrates that a single mutation can have pervasive effects on development. PKU is caused by a recessive gene that prevents the formation of an enzyme needed to metabolize, or break down, one of the amino acids formed by DNA. The amino acid builds to toxic levels in the fetal bloodstream and seriously interferes with nervous system formation. Untreated, infants with PKU develop severe mental retardation. Fortunately, there is a relatively simple test to detect PKU in infants. The test is a routine part of all newborns' examinations, and if PKU is detected, the newborn is placed on a diet low in phenylalanine, the amino acid the body is unable to metabolize. Newborns placed on a restricted diet have a good chance of normal intellectual development (Birren, Kinney, Schaie, & Woodruff, 1981) and are often able to resume a normal diet by their teens.

Chromosomes influence development not only in terms of the message they carry on their genes. There are also a number of adverse genetic conditions caused by defects in the number or completeness of the chromosomes themselves. Down's Syndrome, a particular form of mental retardation is the most common, but as Table 2-1 shows, it is not the only form of chromosomal aberration. People with Down's Syndrome carry an extra chromosome on the 21st pair (hence the other name of Down's Syndrome, trisomy 21). Most chromosomal aberrations appear on the sex chromosomes and are either due to the absence of one member of the pair as in

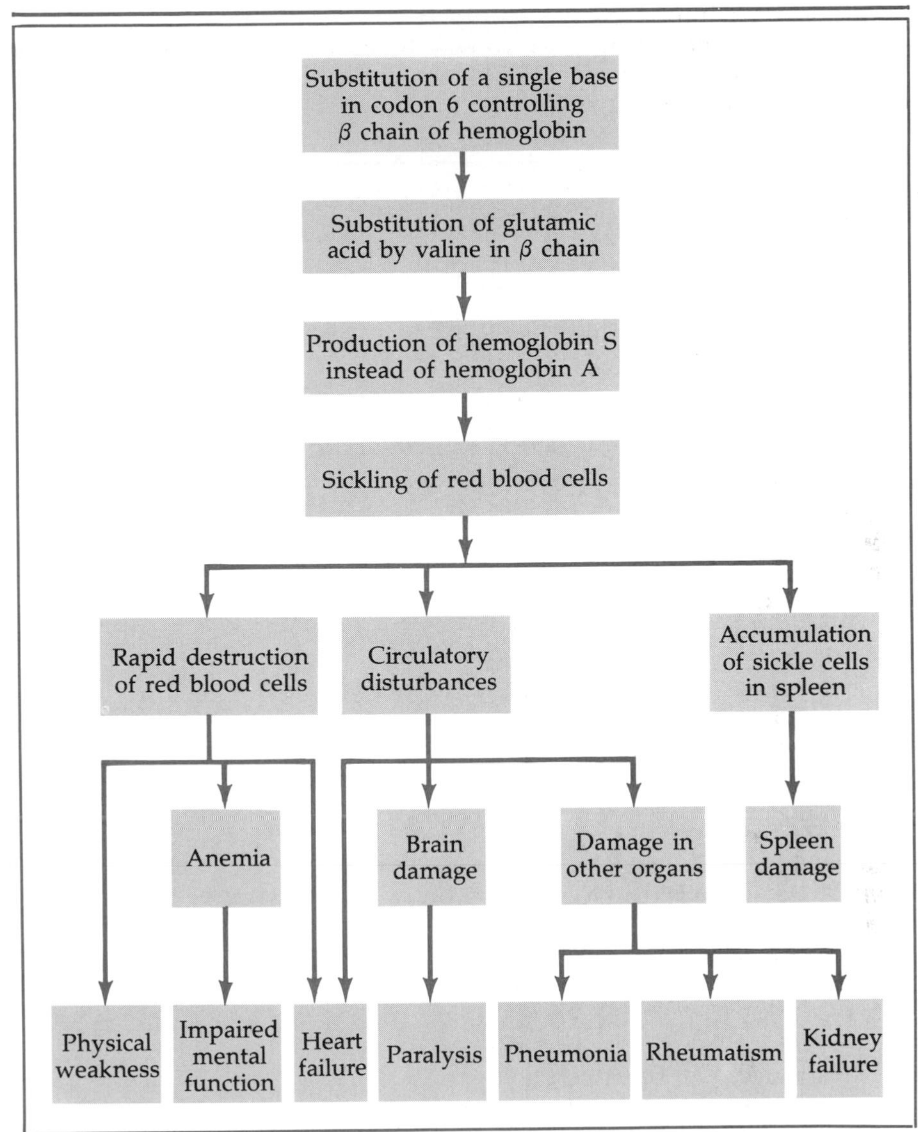

FIGURE 2-5
Sickle cell anemia results from the mutation of one of the DNA sets. This single mutation eventually leads to a number of serious health consequences. (From *Heredity, Environment, and Society,* second edition, I. Michael Lerner and William J. Libby, New York: W. H. Freeman and Company, 1976, Figure 3.6, p. 97, reprinted by permission.)

Turner's Syndrome or to an excess of either X or Y chromosomes. Aberrations of the nonsex chromosomes (the autosomes) with the exception of the 21st pair are apparently lethal. It is believed that such aberrations may account for as many as one hundred thousand miscarriages a year, fully one-fifth the total number per year (Birren, Kinney, Schaie, & Woodruff, 1981).

TABLE 2-1 Summary of Chromosomal Anomalies

	TYPE OF ANOMALY	INCIDENCE PER LIVE BIRTHS	SYMPTOMS
Autosomal anomalies:			
Edward's syndrome	Trisomy-18	1 in 5,000	Early death; numerous congenital problems
D-trisomy syndrome	Trisomy-13	1 in 6,000	Early death; numerous congenital problems
Cri du chat	Deletion of part of short arm of chromosome 4 or 5	1 in 10,000	High-pitched monotonous cry; severe retardation
Down's syndrome	Trisomy-21; 5 percent involve 15/21 translocation	1 in 700	Congenital problems; retardation
Sex chromosomal anomalies:			
Turner's syndrome	XO or XX-XO mosaics	1 in 2,500	Some physical stigmata and hormonal problems; specific spatial deficit
Females with extra X chromosomes	XXX XXXX XXXXX	1 in 1,000	For trisomy X, no distinctive physical stigmata; possibly some retardation
Klinefelter's males	XXY XXXY XXXXY XXYY XXXYY	2 in 1,000	For XXY, sexual development problems; tall; possibly some retardation
Males with extra Y chromosomes	XYY XYYY XYYYY	1 in 1,000	Tall; possibly some retardation

Note: Reprinted, by permission of the publisher, from R. Plomin, J. C. DeFries, & G. E. McClearn, *Behavioral Genetics: A Primer* (San Francisco: Freeman, 1980), p. 173.

Environmental Influences

Certain conditions existing within the environment can exert as direct and as devastating an influence on prenatal development as the genetic factors can. These factors are called **teratogens,** which literally means monster forming. Drugs, viruses, and radiation have each been shown capable of teratogenic influences. Both the degree and nature of the teratogenic influence is determined by the type of teratogen as well as by the time at which it is present during the pregnancy.

The developing organism exhibits three phases of susceptibility to teratogens (Fishbein, 1976). The first phase lasts for the first few weeks following fertilization. Depending upon the agent, the result is either the death of the zygote or embryo or minimal damage. In the latter case, the cells must be sufficiently undifferentiated to allow for almost complete regeneration. The second phase, occurring toward the end of the first trimester, produces the most marked malformations. During this time, the cells are undergoing the most rapid period of differentiation. (Figure 2-6 details this three-phase pattern.)

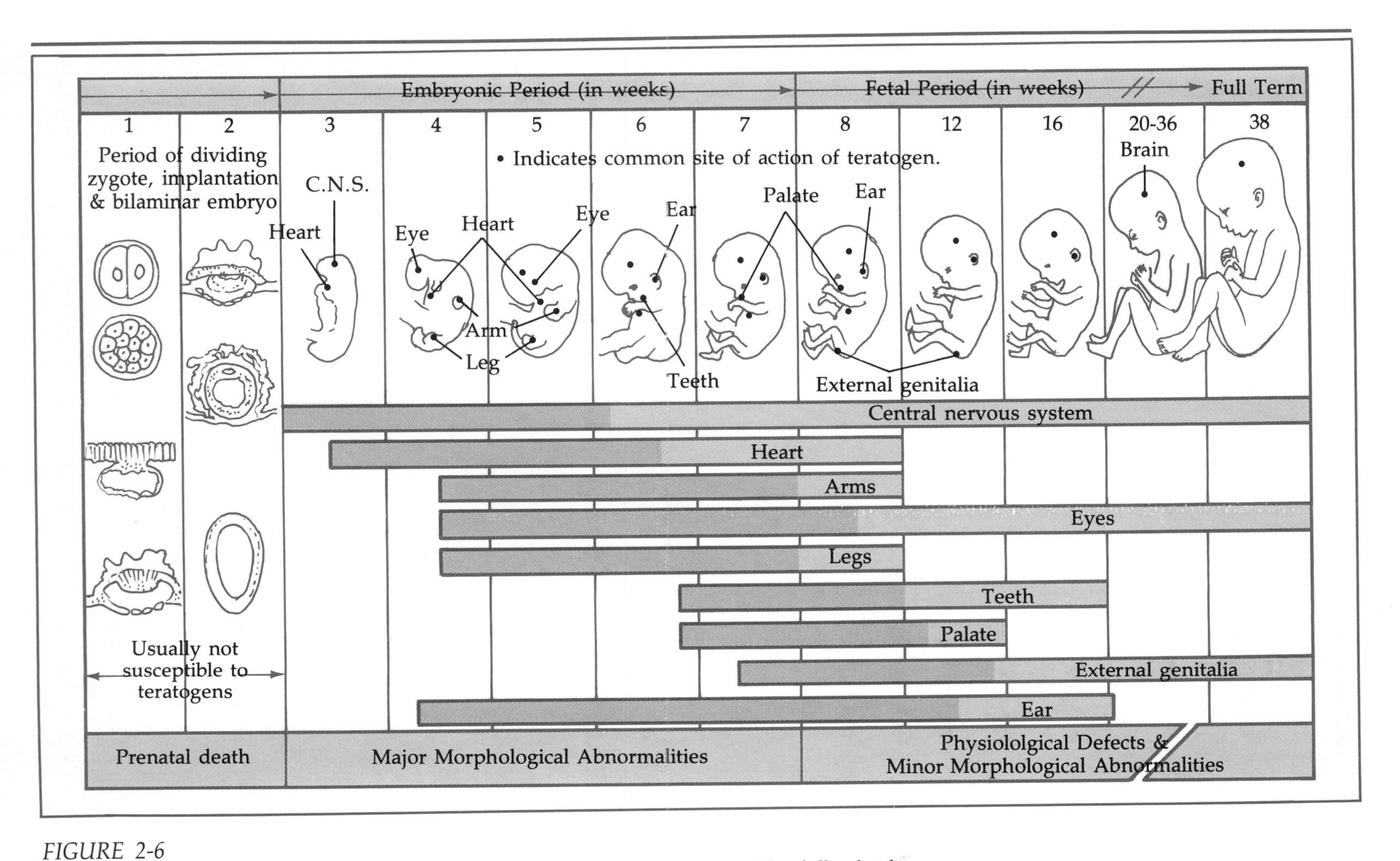

FIGURE 2-6
Susceptibility to teratogens varies depending on the stage of development. The full-color bars represent highly susceptible periods, while shaded bars represent periods that are less sensitive to teratogens (from Moore, 1974).

Environmentally induced factors (teratogens) exert maximal influence toward the end of the first trimester because of the manner in which they exert their influence. They interfere with the instructions the genetic code gives to the cell. Normally, a cell would differentiate into an appropriate structure in accordance with the genetic instructions provided. However, the teratogens alter the coding process so that the instructions to the cell are in error. The cell nevertheless follows the instructions and differentiates in an abnormal fashion. Since subsequent development is in part a function of previous development, development either ceases or takes an abnormal course. These dysfunctions often take the form of diminishing the size of the anatomical structure or of creating two structures when only one would have normally developed. During the second and third trimester, teratogens usually do not produce serious congenital effects (except on the nervous system).

The tragic consequences of the drug Thalidomide provide a good example of the influence of timing. Until the early 1960s, the drug was used extensively in Europe to aid sleep and to relieve the symptoms of morning sickness during early pregnancy. It then became apparent that an increase in multiply deformed infants was directly traceable to the mother's use of the drug during her pregnancy. Typically, these children were born without limbs, with digits developing directly out of the trunk, with flipperlike limbs, or with deformities of the ear or hip. What proved crucial in determining the nature of the malformation was the time during the pregnancy at which the mother took the drug. As you can see from Figure 2-6, for each type of deformity, there was a period of little or no impact. The periods of damage corresponded to the periods of initial differentiation of the particular structures. Before a structure began to differentiate, and once differentiation was well established, there were virtually no negative consequences associated with the drug. Some other drugs that have been associated with prenatal and perinatal complications are listed in Table 2-2.

The use of alcohol during pregnancy provides additional evidence of the adverse effects of specific substances on prenatal development. Women who use alcohol excessively during their pregnancy are significantly more likely to produce in their offspring any one of a number of conditions, which taken together are now referred to as the **fetal alcohol syndrome (FAS)** (Abel, 1984; Sokol, 1981). Infants suffering from the syndrome have a low birth weight, exhibit a slower rate of growth through childhood irrespective of diet, have a smaller head circumference, score below average on intelligence tests, and usually have a variety of congenital anomalies, especially of the cardiovascular and urogenital systems. In addition to alcohol and drugs, smoking, radiation, and certain viral infections such as rubella and venereal diseases have all been shown to adversely influence prenatal development.

Prenatal factors such as recessive genes, chromosomal aberrations, radiation, and substance abuse point out the impact that a single factor can have on the course of development. However, these factors are extreme in their consequence and relatively rare in occurrence. More commonly, a number of factors, each in itself perhaps of little lasting consequence, may jointly contribute to determining the developmental status of the individual. One of the most striking examples of the way such influences seem to determine developmental outcome is provided in the research of Birch and Gussow (1970) on the **cycle of poverty.** Their findings show how the cycle of poverty is passed from generation to generation (see Figure 2-7).

TABLE 2-2 *Drugs and Substances That May Affect the Fetus If Taken during Pregnancy*

DRUG OR SUBSTANCE	EFFECT ON FETUS OR NEONATE
Alcohol	Cardiac anomalies, growth retardation
Antibiotics	
Chloramphenicol	Vomiting, respiratory problems, circulatory collapse in newborns
Streptomycin	Deafness
Tetracycline	Stained teeth, cataracts (rare)
Anticancer drugs (Aminopterin, Busulfan, Chlorambucil, Cyclophosphamide, Mercaptopurine, Methotrexate)	Abortion, various malformations
Anticoagulant drugs	
Dicumarol	Fetal bleeding or death
Warfarin	Facial deformities, mental retardation, fetal death or bleeding
Anticonvulsant drugs	
Dilantin	Cleft palate, heart defects
Paramethadione	Growth retardation
Phenobarbital	Bleeding
Trimethadione	Multiple defects
Aspirin	Bleeding
Hormones	
Birth control pills	Limb defects, genital malformations, possible heart or windpipe defects
Diethylstilbestrol (DES)	Delayed effects: vaginal cancer in adolescent and young adult women, reproductive tract abnormalities in men
Estrogen	Genital defects in male fetus
Progestins	Masculinization of female fetus
Testosterone	Masculinization of female fetus
Lithium	Heart defects
Organic mercury	Cerebral palsy
Poliomyelitis immunization	Death or neurologic damage
Polychlorinated biphenyls (PCBs)	"Cola"-colored neonates with developmental defects
Quinine	Deafness
Sedatives, hypnotics, tranquilizers	
Meprobamate	Developmental retardation
Phenobarbital (excessive)	Neonatal bleeding
Phenothiazines	Excessive amount of bilirubin in blood
Smallpox vaccination	Death or fetal vaccinia
Thalidomide	Fetal death; limb defects; deafness; cardiovascular, gastrointestinal, or genitourinary anomalies
Thiazides	Abnormal number of blood platelets
Tobacco	Low birth weight and size

Source: Data primarily taken from Benson, 1978; 1983; and Wilson, 1977.

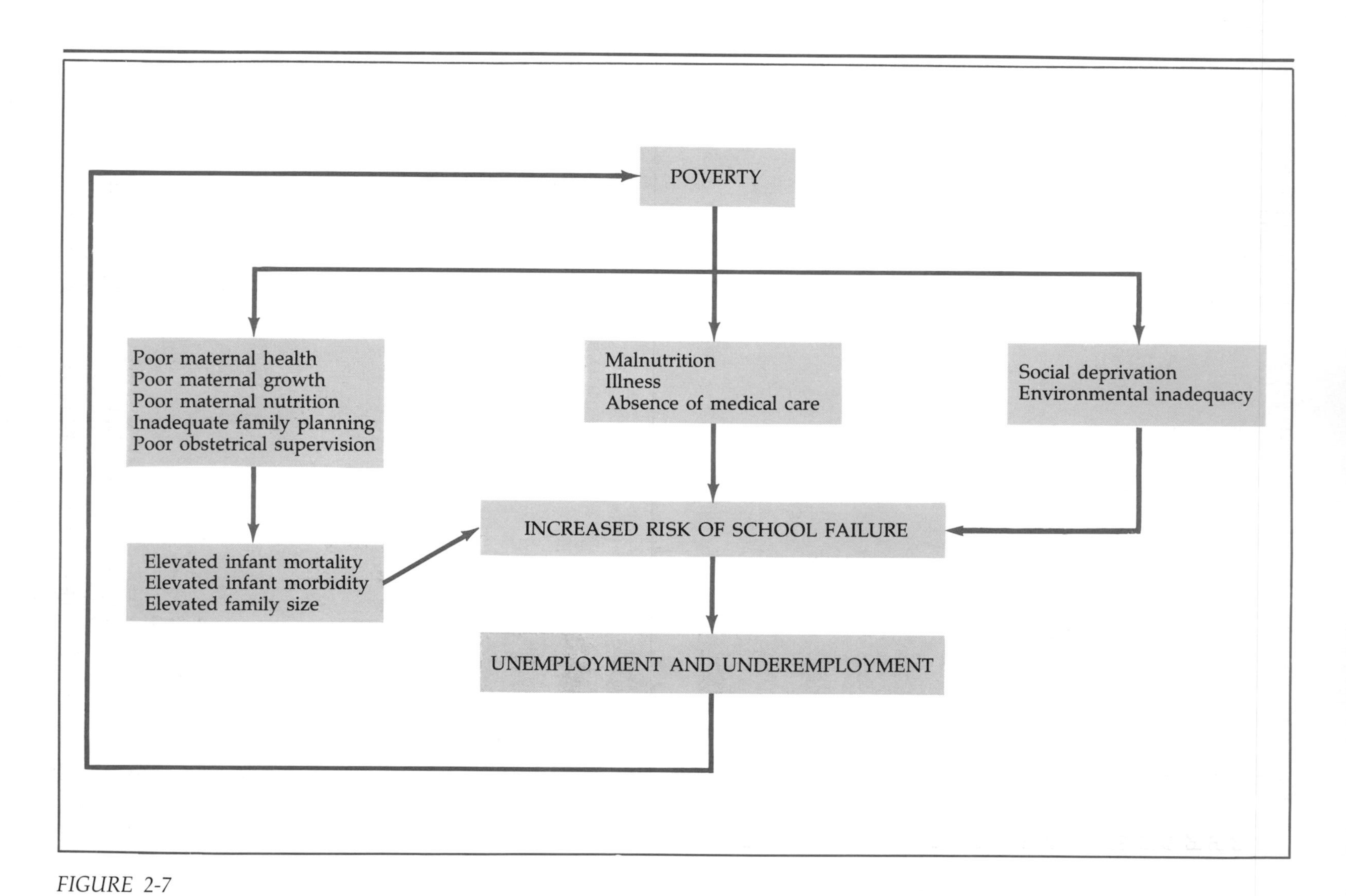

FIGURE 2-7
Based on their research findings, Birch and Gussow (1970) feel that poverty gives rise to a set of conditions that results in the cycle of poverty being transmitted from generation to generation (from Birch

Birch and Gussow found that the effects of poverty begin long before conception. They begin with the developmental history of the mother. Because she grows up in poverty, her height potential is usually not reached. Since growth in height is an excellent correlate of health and physical well-being, it is also a good predictor of the mother's adequacy as a prenatal environment for her child. Women whose height potential is not realized because of poor health and nutrition are almost twice as likely as full-grown, healthy women to encounter a stillbirth, have difficulty in delivery, or give birth to a premature infant.

The situation is further compounded by the fact that the poor are significantly more likely than more economically advantaged women to have a first child early, give birth to many children, and have the last one late. All three of these related circumstances predict complications in pregnancy, labor, and delivery.

The disadvantages continue to cumulate during the pregnancy itself. Diet is a large part of the problem. Although pregnant women usually meet calorie requirements, their diets are often deficient in specific nutriments. This problem is especially true in the dairy, fresh vegetable, citrus, and fruit categories. Birch and Gussow report that evidence clearly demonstrates a relationship between diet and a variety of prenatal and perinatal conditions influencing the viability of the child.

Diet, however, is only part of the problem. Availability and quality of prenatal care are also important. Most prenatal medical care is protective; that is, sustained supervision serves to forestall preventable abnormalities of the reproductive process and to minimize the effects of those that are not preventable. The absence of such care does not necessarily have negative consequences for every woman and her infant. Unfortunately, the women who are most likely to have negative consequences from lack of prenatal care are also the ones least likely to receive it. Birch and Gussow found that one-third to one-half of all women (irrespective of social class) delivering at city hospitals in major cities had had no prenatal care at all. When data is available by social class, it is staggering. Fully three-quarters of the lower-class women surveyed did not receive any form of prenatal care until the third trimester.

All of these prenatal conditions continue to exert their influence after the birth of the child. The developmental and educational history of the parents leaves them less well qualified to compete for well-paying jobs. The age of the mother coupled with both the number and spacing of the children creates a situation that makes effective parenting difficult. The prenatal history of the children may leave them more susceptible to illness and prolonged recovery. These circumstances contribute to poor school performance, and the cycle starts all over again.

More recent research on the poverty cycle, especially the influence of prenatal malnutrition (Lester, 1979; Zeskind & Ramey, 1981), has begun to identify some of the specific links that maintain the cycle. Zeskind and Ramey believe that one of the consequences of prenatal malnutrition is that it reduces the general responsiveness of the infant. The parents, whose own developmental histories as well as their limited resources leave them with limited coping abilities, may be less able to provide appropriate stimulation to their unresponsive infant. Zeskind and Ramey also note that malnourished infants have a higher pitched, more irritating cry than the healthy newborn has. An irritating cry from a relatively unresponsive infant may be more than some parents can tolerate. In extreme cases the

BOX 2-1
HAVING A HEALTHY BABY

There are a variety of things a pregnant woman can do to increase the likelihood of giving birth to a healthy infant. Those that are commonly known include proper diet and exercise, avoidance of smoking, alcohol, and unnecessary medications, and avoidance of situations that present the possibility of physical injury or pollution. Perhaps less commonly known are the variety of procedures that are now available to help prospective parents determine the probability of conceiving a child carrying a particular genetic defect and the prenatal procedures that are available to inform parents whether or not their child does carry a particular defect.

Genetic counseling involves determining if a prospective parent carries an abnormal gene and how likely it is to be both passed on to and expressed in the offspring. Simple blood tests are available to determine the presence of a variety of recessive genes such as those for Tay-Sachs disease and sickle-cell anemia. In some instances, a family genealogy will help provide the same information. Not all inherited conditions are yet detectable through genetic counseling. In the

In amniocentesis, a needle is inserted into the amniotic sac and a sample of amniotic fluid is withdrawn. This fluid is then used to prepare cultures that are used for chromosome and biochemical analyses (from Francoeur, 1982).

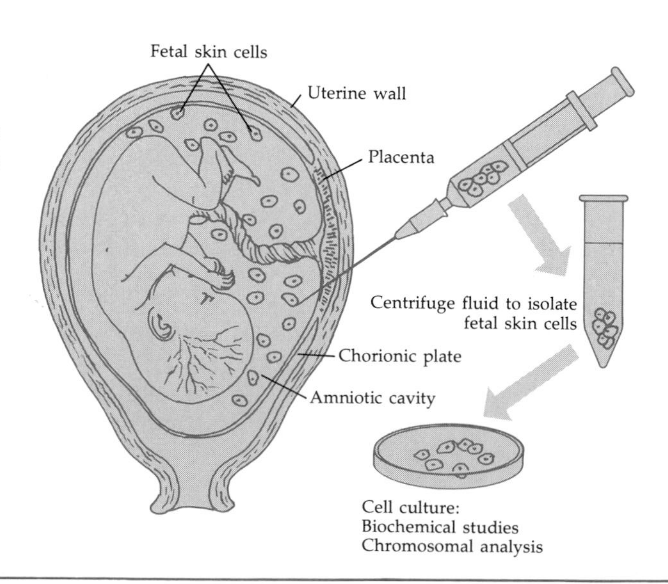

case of muscular dystrophy, for example, the only basis for making predictions about future children would be to have had one child with the condition. Genetic counseling provides information as to probability. Prospective parents are then faced with the agonizing dilemma of conceiving a child and hoping for the best, adopting a child, or remaining childless.

For parents who take the chance as well as for parents who represent high-risk groups or for those who have reason to be concerned about the prenatal status of their child, there are techniques for prenatal diagnosis. Amniocentesis is one common technique. In this procedure, a needle is inserted through the mother's abdomen and uterus into the amniotic fluid surrounding the embryo or fetus. This fluid contains fetal skin cells, which can be analyzed for the presence of genetic defects. Amniocentesis is frequently used to determine the presence of Down's syndrome, a condition most common in mothers over the age of forty. The procedure may also be used to detect Tay-Sachs disease and hemophilia, among other life-threatening or seriously handicapping conditions.

A more recent procedure providing the same type of information is called chorionic villa sampling (CVS). In this procedure, a catheter is inserted through the vagina and cervix into the uterus. Small tissue samples are taken from the villa (thin threadlike projections) that surround the fetus. A major advantage of CVS is that it can be performed earlier in the pregnancy than amniocentesis can. Ultrasound is a third prenatal diagnostic procedure. High frequency sound waves are reflected off the fetus providing a "picture." Although ultrasound does not provide genetic information, it can detect gross physical anomalies. Given the rapid development of prenatal, in utero, surgical procedures, these diagnostic procedures are becoming both more common and more important.

Once parents have information about the prenatal status of their offspring, they face the difficult question of continuing the pregnancy and raising a child who may have a significantly shortened lifespan, a decision involving many medical complications and a great deal of physical and emotional trauma for all, or, if possible, of aborting the pregnancy, an equally traumatic decision. The decision is individual and not easily made under any circumstances. The fields of genetic counseling and prenatal diagnosis make clear the fact that more information does not always lead to easier decision making.

Sources: Francoeur, 1982; Masters, Johnson, & Kolodny, 1985.

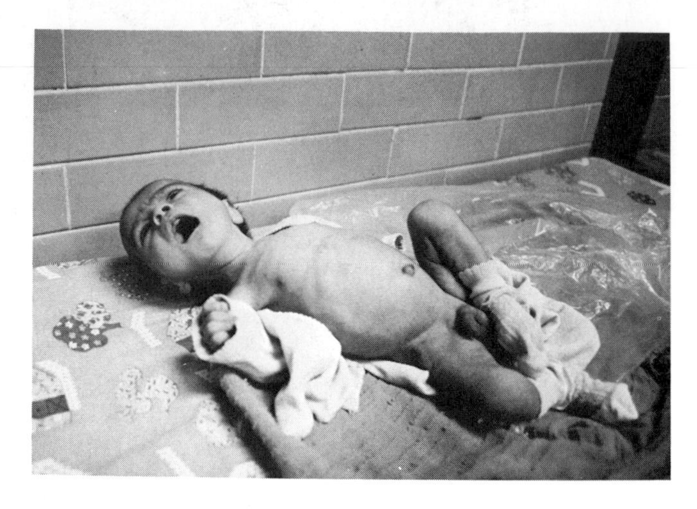

Caught in the cycle of poverty, this six-month-old infant has not progressed past his birth weight. Inadequate prenatal care and an impoverished postnatal environment have resulted in malnutrition and developmental delay.

result may be abuse or neglect, but more generally the result is a child requiring more than the typical amount of intervention, developing in an environment that is able to provide less intervention than is typical.

The "How"

How do heredity and environment interact over the life-span? How is it possible that even though all the cells of the body have virtually the same genetic material, they function in many different ways? Further, how is it possible that any one cell can change its mode of functioning across the life-span? The answer involves more than references to changing environmental circumstances. The genetic code not only regulates the initial differentiation of the organ systems during prenatal development but also continues to regulate all aspects of physiological functioning across the entire life-span. On the other hand, the numerous ways environmental events have been shown to counter the expression of a gene (for example, diet to correct PKU or, in a more everyday situation, putting on a pair of eyeglasses to correct an inherited defect in vision) suggest that genotype is no guarantee of good or bad fortune.

The first part of the answer to the "how" question concerns the many steps that are involved between the direct action of the gene and its ultimate expression at the behavioral level. The gene itself determines a biochemical reaction which, during the embryonic phase, eventually results in organ differentiation and, after that time, either biological **homeostasis** (that is, the maintenance of regular function), or at periods such as puberty or advanced age, further organ differentiation. At any step in the sequence, at any point in the life-span, environmental events interact with the biological structure to determine its further differentiation or its continued maintenance. The environmental event may be as specific as a particular drug or some nutritional component, or it may be as general as the nature and number of stresses in an individual's life. The environmental event may have a general effect on the biological structure as, for example, radiation would, or it may have a specific effect such as the staining of the infant's first teeth when the antibiotic tetracycline is taken by the mother during her pregnancy. The event need not occur early in life. The middle-aged adult who is overworked, overweight, underexercised, and overstressed is exerting as much of an influence on the expression of genetically regulated biochemical reactions as are the parents who are unable to provide their infant an adequate diet.

The actual interaction mechanism is still poorly understood, but it is likely that the endocrine system serves as the mediator between the gene and the environment. In particular, researchers think that changes in endocrine levels turn genes on and off, in effect, regulating their impact (Plomin, DeFries, & McClearn, 1980; Vale, 1980).

The interaction mechanism was first described in 1961 by two French scientists, Jacob and Monad, in a study for which they eventually received the Nobel Prize. Although the **Jacob-Monad model** was developed through research with microorganisms, it is probably a close approximation to what occurs in higher organisms such as humans (Gottlieb, 1976; Plomin, DeFries, & McClearn, 1980). The model (see Figure 2-8) describes most genes as structural. That is, they serve to regulate particular biochemical reactions. However, since not all genes are functional at any one time, the model suggests that there are also two other types of genes, whose purposes are to regulate the activity of the structural genes. One of these, the operator gene, serves as the switch that activates a particular structural gene. The operator gene is in turn regulated by a regulator gene. The regulator gene produces a substance that deactivates the operator gene, an occurrence that results in the deactivation of the structural gene. The structural gene again becomes active when a new substance binds with the repressor produced by the regulator gene. Once bound, the regulator gene no longer inhibits the functioning of the operator gene. The end result is

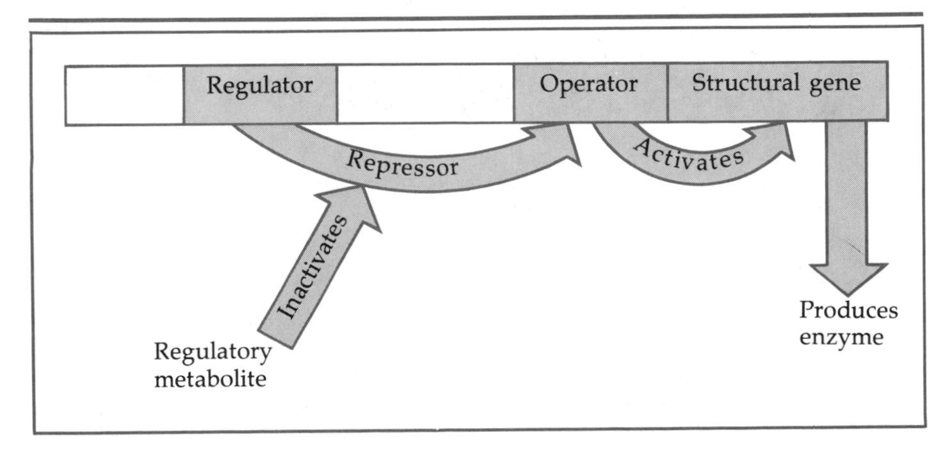

FIGURE 2-8
According to the Jacob-Monad model, regulator genes control operator genes, which in turn control structural genes. Structural genes are responsible for the regulation of particular biochemical reactions (adapted from Vale, 1980).

that the structural gene again becomes functional (Plomin, DeFries, & McClearn, 1980).

Since the endocrine system is considered the likely source of the regulatory metabolites that originate outside the cell, any research that helps us understand the relationship between behavior and the endocrine system provides potential insight into the relationship between behavior and the regulation of gene activity.

The Jacob-Monad model of gene-environment interaction helps explain how genes influence the course of development across the entire life-span and how environmental events can interact with genetic information to determine developmental outcomes. However, the Jacob-Monad model provides little insight into the mechanisms regulating the direction of human development. For this insight we turn to a second mechanism, called by its originator, C. H. Waddington (1962, 1966), **canalization.**

Although we differ in a variety of ways at all points of the life-span, in other ways we remain remarkably alike. We all share the same anatomical features, and the basic mechanisms that regulate our physiological functioning are common to us all. There are even many developmental patterns that appear to exist across cultures and historical periods. These common aspects of our biological and psychological structure and function are said to be canalized (Fishbein, 1976). To say that these characteristics are canalized means that they are quite impervious to the variety of factors that are

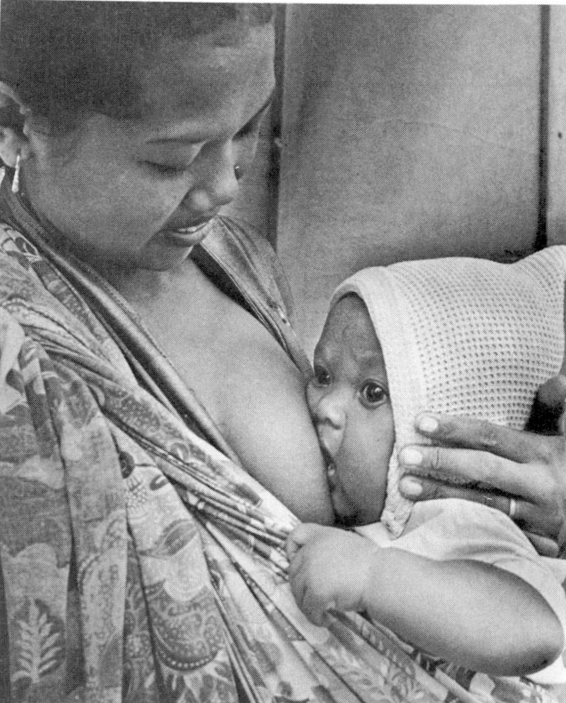

Although its form may vary from culture to culture and generation to generation, the bonding process between a parent and infant is probably highly canalized.

capable of altering the expression of the various related genotypes. In other words, canalized characteristics have a high probability of developing a particular phenotype.

Canalization mechanisms can be thought of as the backup systems of human development. They help explain why, for example, children who recover from a prolonged illness or who have been subjected to severe child-rearing conditions such as might be found during a war usually show a growth spurt once these conditions end and, when this growth spurt ends, are about the same height they would have been if the rate of growth before the illness or adversity had continued unchanged. The canalization mechanism also helps explain why even though there are sixty-four unique DNA triplets, only twenty unique amino acids are formed. Of the twenty that are formed, there are at least two unique triplets that can form any one of the twenty and some have as many as six unique sets. Those that have more alternative pathways are considered more protected, and research has shown that these more protected amino acids usually play a more prominent role in the regulation of biochemical activity (Levine, 1973).

Canalized aspects of development are basic aspects of development. They are not only fundamental to individual development but also to the survival of the species. As such, they are not likely to be aspects of development unique to a particular generation or a particular culture. For example, the bonding process between infant and parent is probably highly canalized, but the particular form the bond may take in different cultures is not. The need of the growing child for novel stimulation might be canalized, but whether the stimulation comes from contact with objects typically found in most American homes or from those found in a mountain village in Peru is probably of little significance in terms of the survival of the species.

Canalized aspects of development do not develop automatically, and their development is not inevitable. Rather, the mechanism ensures that certain aspects of development are more resiliant than others. However, as the research of Birch and Gussow (1970) on the cycle of poverty and Lester (1979) on the effects of prenatal malnutrition clearly indicates, while individual factors such as prenatal maternal stress or a complicated delivery or poor responsiveness in the newborn may each have limited influence on the developmental outcome, their cumulative influence may be considerable, even for those aspects of development considered highly canalized.

BIRTH

The birth of a child is a truly amazing process. It's one thing to know about the events surrounding labor and delivery; it's another thing to experience them. I was fortunate to be present at the birth of our youngest child and value the experience as one of the high points of my life. It was clear from the expression of the doctors and nurses present in the delivery room that it's not the type of experience that fades with exposure. They seemed almost as excited and emotionally involved as my wife and I were.

The Course of Labor and Delivery

Labor begins approximately 280 days after conception. This figure is of course highly variable. First, it is hard to determine the actual date of conception. Usually the physician will define it as two weeks after the last menstrual period. The accuracy of this estimate depends on the regularity of the woman's menstrual cycle and on the time during the cycle when conception actually occurred. Second, it is not uncommon for the onset of labor to be as much as two weeks early or late. In fact, only 4 percent of all women actually deliver on their due date, and 25 percent are more than two weeks off (Smart & Smart, 1977).

Our inability to accurately predict the onset of labor is related to our limited understanding of the factors that regulate its onset. At least three different hormones are known to be involved in the process: progesterone, prostaglandins, and oxytocin. Progesterone serves as an inhibitor of uterine activity. Its production decreases as the pregnancy approaches term. Conversely, there is an increase in prostaglandins, which serve as a stimulator of uterine muscle activity. Oxytocin exerts its influence during the later stages of labor.It increases the force of the uterine contractions that are necessary to pass the child through the birth canal (Katchadourian & Lunde, 1975). It is still uncertain what regulates the action of these hormones.

The **stages of labor and delivery** consist of three fairly distinct phases (Figure 2-9). The first is the longest. It may last twelve to fifteen hours for first deliveries (primigravidas) and six to eight hours for subsequent deliveries (multigravida). The first phase involves the **effacement and dilatation of the cervix** (Figure 2-10). These changes in the cervix occur through a series of muscular contractions that are initially spaced as much as twenty minutes apart and become more frequent as labor progresses. One of the consequences of these contractions is the rupturing of the membranes surrounding the fetus. When these membranes are ruptured, the woman experiences a rush of amniotic fluids being expelled through the vagina. By the completion of this first phase of labor, the opening of the cervix has reached a diameter of approximately four inches (10 cm).

The second phase of labor is shorter and more intense. It begins when the cervix is completely dilated and ends with the actual **expulsion of the fetus.** For first deliveries, this phase takes about ninety minutes; for second and subsequent births, it is about half as long. Whereas in the first phase, the contractions were involuntary, in this second phase the mother's active involvement in the pushing of the child through the birth canal can provide as much as 50 percent of the necessary force (Smith & Bierman, 1973).

The third phase is the **expulsion of the afterbirth,** which consists of the placenta and fetal membranes. The mother is often so engrossed in her newborn child that she is unaware of the expulsion of the afterbirth. At this time, the physician stitches any tears to the perineum (the skin and deep tissues between the vagina and anus) or the episiotomy incision if one has been made. The episiotomy is an incision made in the perineum. It is believed that this incision facilitates passage of the baby's head and enables the perineum to heal faster and better than it would if a tear occurred.

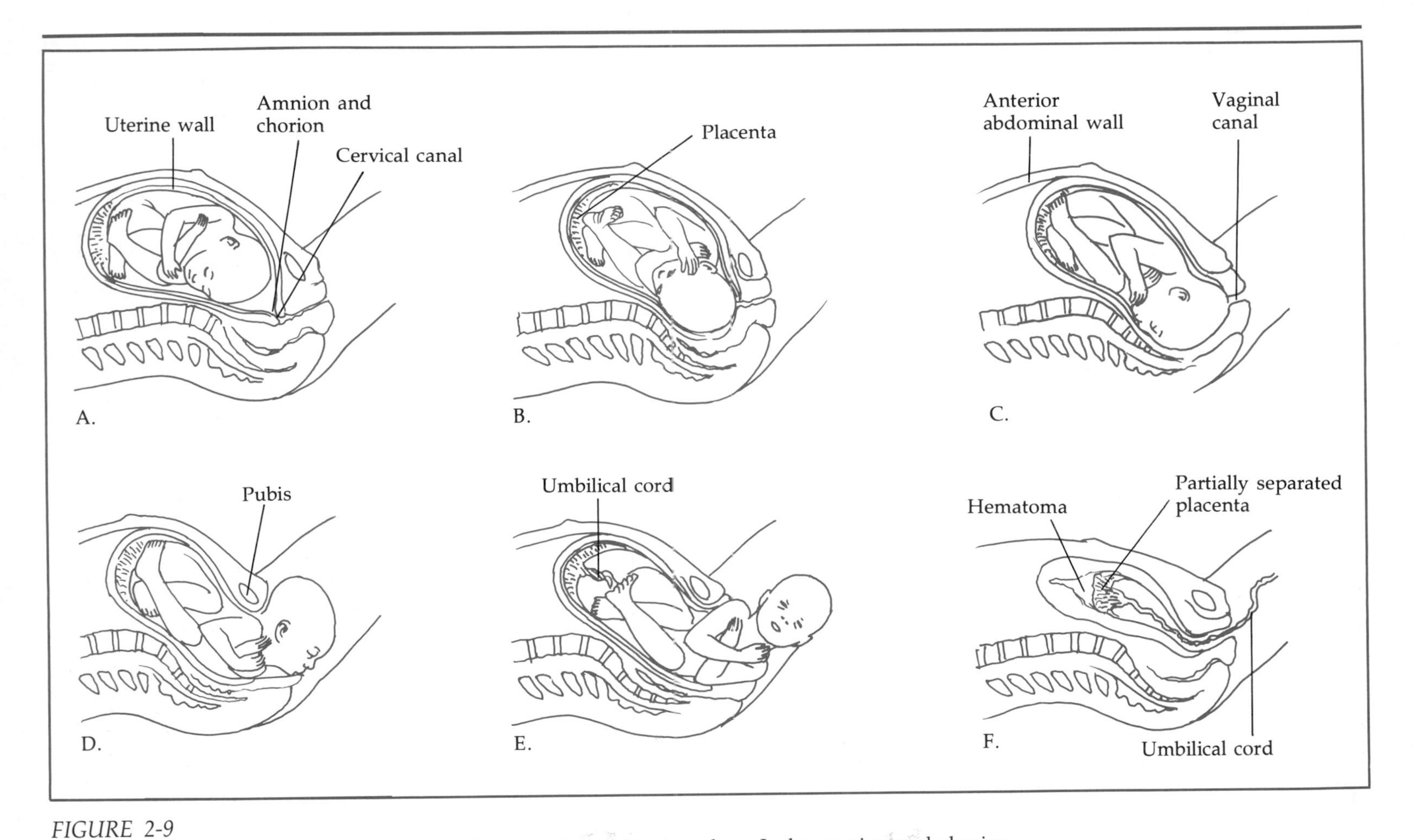

FIGURE 2-9
The stages of delivery consist of three fairly distinct phases. During phase I, the amnion and chorion are forced into the cervical canal causing dilatation to begin (A and B). In phase II, the fetus passes through the cervix and vagina (C through E). Finally, in phase III, the placenta folds up and pulls away from the uterine wall, resulting in the formation of a large hematoma (F). Later, the placenta and its membranes are expelled (not shown).

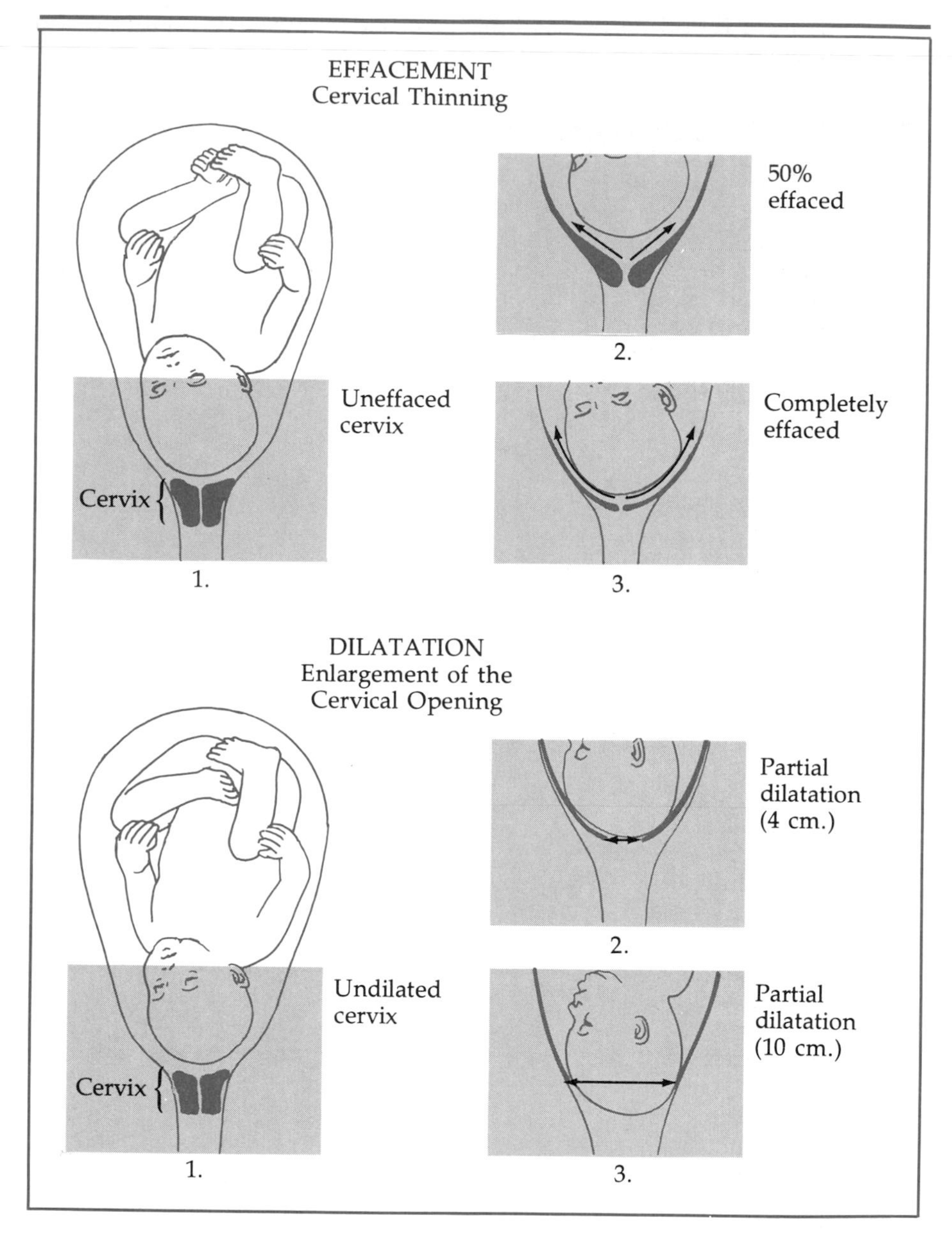

FIGURE 2-10
Effacement and dilatation of the cervix occurs during the first phase of labor.

The Context of Labor and Delivery

Even though the duration of labor and delivery is measured in hours, its potential impact can be a matter of months or even years. Like the impact of all transitions in the life-span, the impact of the events of labor and delivery is best understood from a broad perspective—both in terms of the newborn child and in terms of the parents.

The Neonate's Perspective

Although scholars have always differed as to the psychological significance of birth for the **neonate** (literally, new born), there is little question as to its physical significance. The birth process is hard on the neonate. Since the width of the birth canal is almost always less than the width of the neonate, bruising is common. Further, the head often assumes a somewhat pointy and "squished" appearance since the plates of the skull are compressed and overlap each other as the neonate passes through the birth canal. This is almost always a temporary condition lasting at most a few days, but during prolonged labor or when the birth canal is very narrow relative to the size of the neonate's head, the potential for long-term trauma is present. Often, such circumstances prompt a decision to deliver the fetus by a cesarean section. A cesarean section is an operation whereby the fetus is delivered through an incision in the walls of the abdomen and uterus. The incidence of cesarean sections has tripled over the last few years (from 5 to 15 percent of all deliveries), prompting many in the health community to investigate whether the increase reflects a change in the ability of women to deliver vaginally or a change in medical preference among physicians (Masters, Johnson, & Kolodny, 1985).

The likelihood of physical trauma is further increased if the physician is required to use forceps (a flat-bladed instrument used to "pull" the neonate through the birth canal) during delivery. The use of forceps is infrequent today but was more common when high doses of anesthesia were routinely administered to women in labor. The nature, degree, and duration of physical injuries sustained during birth depend on the site of the injuries and the degree of trauma. An injury may involve some sort of motor damage, (as, cerebral palsy does), mental retardation, or sensory system dysfunction.

Perinatal anoxia (literally, without oxygen) is another common birth-related problem. Anoxia is also a major cause of mental retardation and cerebral palsy. It may result from insufficient oxygenation of the mother's blood (due to low blood pressure, reduced respiration, or protracted contraction of the uterus), problems in the transfer of oxygenated blood from the mother to the placenta (due to premature separation of the placenta) or from the placenta to the fetus (due to a collapsed umbilical cord), a knot in the cord, or the cord's wrapping itself around the neck of the fetus (Smith, Bierman, & Robinson, 1978). Minimal incidents of anoxia are common during labor, especially during contractions. There is no evidence to suggest that these incidents have a prolonged influence on the child. More pronounced incidents of anoxia, however, can have a significant and long-lasting impact on the child (Lyle, 1970).

Medication administered during labor and delivery is also a potential source of birth trauma. The use of medication has decreased significantly over the past few decades, but the necessity of its use at all, as well as its consequence for the neonate, continues to be the focus of a major controversy (Brackbill, 1979).

Some medication is used in virtually all hospital deliveries, even those in which the parents have participated in prepared childbirth classes. It is administered at various times during labor and delivery and serves to reduce pain (demoral, novocain, lidocaine), relieve anxiety (tranquilizers), and facilitate labor contractions (oxytocin). The concern about medication reflects the fact that the neonate is less able to deal with potentially toxic agents than the adult is, and since dosage is often based on adult body weight, an appropriate dosage for the mother can constitute an overdose for the fetus. As Brackbill (1979) notes,

> The newborn human being is an organism poorly positioned to deal with toxic agents. Drugs enter the central nervous system readily because of incomplete blood/brain barriers; they lodge in brain structures that are still developing and therefore at high risk for injury; they cannot be readily transformed to nontoxic compounds since the necessary liver functions are immature and they cannot readily be excreted because of inefficient kidney function. (p. 78)

In a normal full-term delivery, the use of medication may influence the newborn during the first few minutes or hours after birth. The most noticeable initial effects may be a slight delay in crying, depressed breathing or the depressed functioning of a reflex such as the sucking reflex. These effects are short-lived; there is no indication that they have any long-range consequences. The liability must be weighed against whatever assets the parents and the physician believe are provided by the use of limited medication. When the use of medication is extreme or when a host of other factors, such as those Lester (1979) described, are present, then there may well be both direct and indirect long-term consequences.

Concerns about complications during labor and delivery are further heightened when the birth is premature. Prematurity is defined in terms of weight at birth (less than 2500 grams) or length of gestation (less than 36 weeks). Although premature infants seem to catch up with their full-term cohorts by six or seven years of age (Sameroff & Chandler, 1975), the fact that prematurity rates are as much as four times greater (16 percent versus 4 percent) in economically disadvantaged populations than in more economically advantaged populations and the fact that prematurity is often associated with a variety of other high-risk factors, such as maternal health, family stress, and difficult pregnancy, are further indications that factors having little or no lasting impact on development when taken individually can have significant adverse cumulative effects when taken jointly (Smith, Bierman, & Robinson, 1978).

The vast majority of neonates cope very well with the ordeal of birth. In fact, birth-related trauma occurs in less than 1 percent of all live births. There is therefore little reason for most parents to be concerned about the effects of typical labor and delivery practices on their newborn child. Concern should exist, however, about differences in the quality of life in different segments of our population and in different segments of the world population and about the influence these differences have on pregnancy, labor, delivery, and development. Concern should also exist over the fact that even when the incidence of birth-related trauma is less than 1 percent, approximately ten thousand children suffer some degree of trauma from birth for every one million children born. Both in terms of human suffering and in terms of the resources necessary to meet the special needs of these children, this percentage is still much too high.

The Parents' Perspective

Before World War II, most births in this country occurred at home. In many parts of the industrial and nonindustrial world, births at home are still common. In some instances, a trained midwife is present; in others, an "experienced" layperson. Since World War II, most births in this country have occurred in the hospital. The trend to have children in a hospital has accompanied the increasing professionalization of the field of obstetrics, as well as significant advances in medical science. It has certainly reduced the risk of death from childbirth for both the mother and the child and has vastly improved the chances that a prematurely born infant or an infant born with prenatal or perinatal complications will receive the interventions necessary to ensure a full and productive life. But this trend has also had some negative consequences. It has excluded the father from the birth process; it has fostered the view of childbirth as a "medical condition" rather than a normal biological function; and it has reduced the role of the mother in the birth of her child to that, in the extreme case, of a medicated bystander.

The past decade has witnessed a revision of this trend. The aim of this revision is to reemphasize the interpersonal aspects of the childbirth experience and to increase the participation and decision-making roles of both the mother and the father while at the same time retaining the medical advantages of modern obstetric and pediatric practice. This revision is evident in the increasingly common practice of allowing fathers to be present in both the labor and delivery rooms. It is also evident in the use of the labor room for delivery when medically warranted, the development of homelike birthing centers within hospitals, the small but significant increase in the number of home births, the revitalization of the profession of midwifery in this country, the reduced use of instruments and medication during labor and delivery, and the increased frequency of rooming-in procedures (letting the newborn spend most of the day in the mother's room). Many of these changes have occurred as parents have become more concerned about the way their children enter the world. This concern has been apparent in the considerable increase in the number of parents involved in **prepared childbirth training classes.** These classes are based on the ideas of the English physician Grantly Dick-Read (1959) and the French physician Fernand Lamaze (1958).

Prepared childbirth classes are attended by both the father and the mother. Their primary purpose is to provide the mother a variety of psychological and physical techniques to help cope with the emotional and physical stresses that are part of labor and delivery. The role of the father during labor and delivery is to provide both emotional support and direct assistance to his partner. A sense of personal control, effective coping techniques, and knowledge of the events of labor and delivery, as well as the emotional support provided through a shared experience, are all intended to reduce the mother's fear and anxiety and thus to reduce tension and therefore pain. Less pain means less need of medication, and less medication means both a greater involvement on the part of the mother in delivering her own child and a reduced risk of trauma associated with medication to the newborn. Compared to women not participating in prepared childbirth classes, those that do have shorter labors (Tanzer & Block, 1972), experience less pain, and report the experience as more positive (Entwisle & Doering, 1981).

Increasing numbers of parents and physicians are also being influenced by the work of another French physician, Frederick Leboyer (1975). Leboyer believes that the typical hospital delivery is hard on the child not only physically but emotionally as well. To make the transition into the world less abrupt and more humane, Leboyer advocates softly lit delivery rooms and skin-to-skin contact with the mother immediately after birth followed by immersion in warm water. There is as yet little evidence evaluating the technique, but it seems certain that labor and delivery practices will continue to emphasize the health and welfare of the child, the active participation of both parents, and the humanness of the entire childbirth experience.

The greater involvement of parents in defining the course of labor and delivery as well as their increasing involvement in prepared childbirth classes has influence far beyond the birth experience itself. These activities help the couple make the transition from a two-person to a three-person family.

Although the birth of a child is a unique and special experience, its full impact, both in terms of the child's future development and in terms of the family's future as a unit, is best understood when it is viewed as one experience in a series of experiences that starts with the decision to conceive a child, continues through the pregnancy and preparation for childbirth, and then is followed by the events that define the early bonding of the child to the parents and the beginnings of the family as a new unit. As is true of the series of events that leads from the genetic code to behavior, an event at any point in this sequence can have repercussions far beyond its direct and immediate impact. For example, parents who participate in prepared childbirth classes may gain not only a greater sense of personal control over the birth of their child, but also, more generally, the belief that they can be a potent influence in directing the course of their child's development. One general benefit for men involved in prepared childbirth classes seems to be greater involvement in the care of their children. They are more likely to become involved in the various caretaking aspects of rearing an infant such as bathing, feeding, and changing a diaper (Entwisle & Doering, 1981).

An event at another point in the sequence, such as a premature birth, can have as potent and potentially long-lasting an influence but in a different direction. The unexpectedness of the birth and the relative immaturity of the neonate, both in terms of size and level of functioning, present barriers to early parent-child bonding. The isolation of premature infants in an intensive care nursery, their prolonged stay in the hospital, and the various monitors and tubes that may be attached to them all hinder contact. Premature infants usually sleep more and are less responsive to stimulation than full-term infants. The parents' normal reaction, which is to increase their efforts to arouse such a neonate, may have the opposite effect to the intended one. The relationship becomes stressed, and if the pattern continues, the parents begin to feel a sense of helplessness in dealing with the neonate (Goldberg, 1979).

Participation in prepared childbirth classes does not guarantee that a child's development will be ideal, nor is prematurity a sentence on the child to be always at a disadvantage (most children do catch up). Rather each possibility represents one event that, when coupled with other events, increases the likelihood of a particular developmental outcome. There are certainly circumstances, such as the direct genetic effects discussed earlier, that have a prolonged and pervasive influence on the child. But the more typical pattern is a set of circumstances reflecting the psychological status of the parents, the social and economic circumstances of the family, and the developmental status of the child that together define the pattern of parent-child interactions and the resulting course of individual development.

HUMAN DEVELOPMENT AND HUMAN SERVICES

Our understanding of the process of development from conception through birth has influenced the definition and delivery of human services in a variety of ways. This closing section examines four of these influences. The first is the influence that our growing understanding of the interaction of heredity and environment has had on the direction of human service delivery. The second concerns the ways in which the various prenatal diagnostic and screening measures that have evolved from basic research are forcing us to realize that decisions concerning the direction of human development must be based on considerations that extend beyond the boundaries of the field of human development. The third is the influence our understanding of human development has had on social policy, and the fourth concerns the influence of the human services on the cumulative nature of human development.

The Interaction of Heredity and Environment

Historically, heredity has been viewed as the major determinant of individual ability and circumstance. The distribution of wealth and power that was passed from generation to generation was justified not on the grounds that the older generation could teach the new but rather on the grounds that one generation passed along, through the gene, the wisdom to rule or the ability to lead or, for those less fortunate, a life of misfortune. From such a perspective, the role of the human services was essentially conservative (Goldhaber, 1982; Lomax, 1978). Since people believed that one's circumstances could not be altered, those who were more fortunate assumed the responsibility of caring for those who were less fortunate (Rothman, 1973).

By the turn of the century, people became increasingly aware of the influence of life circumstances, especially those fostered by the industrial revolution, on the course of development. A new, infinitely more optimistic, pattern of human services emerged. The fields of child welfare, early childhood education, pediatrics, public health, and social work as well as child labor laws and compulsory school attendance laws all emerged during this period (Takanishi, 1978). A strong environmental view began to replace the strong genetic view of development. The course of development came to be seen as a reflection of the conditions in which it occurred. People began to think that if conditions were changed, the course of development would change too.

BOX 2-2
FAMILY-ORIENTED CHILDBIRTH

Who controls the labor and delivery experience: the mother, the parents, or the physician and other health care professionals? During the past ten to twenty years, this question has been asked with greater frequency. Why? A variety of contributing factors include the "back to basics" movement, the women's movement, the increasing importance given to early bonding experiences, the desire of couples to share "peak experiences," the increasing concern over the effects of standard obstetric practices on the viability of the neonate, and evidence from other countries, both industrial and nonindustrial, that more "natural" childbirth practices are associated with less pain, less need for medication, shorter labors, and no less healthy neonates. In general, unless there is clear medical evidence to the contrary, the mother and her partner should be given the necessary training and support to allow the greatest possible latitude in defining the course of labor and delivery.

In many hospitals in this country, partners are allowed to stay during labor and delivery, women are generally allowed to nurse their newborns on the delivery table, and rooming-in procedures are increasingly common. These changes have not come about easily. The medical profession has accepted them sometimes only reluctantly.

Other changes in labor and delivery practice, common in most other countries, are still uncommon here. Most women in this country deliver while lying on their backs with their legs raised and often supported by stirrups. Surveys in other countries, however, show that women in labor find a kneeling or squatting position more comfortable, especially when they are able to rock during contractions. These positions also facilitate the actual delivery process since they make maximum use of gravity and they provide the optimum stretch of the perineal muscles. In countries such as Holland or Sweden where these labor and delivery positions are common, the episiotomy is quite rare. In this country, it is still common practice. The use of pain medication is also quite rare in other countries, again a circumstance different from ours. In some hospitals

The mass social action programs of the 1960s are examples of such a strong environmental view. Many of these programs have made significant improvements in people's lives (Project Head Start and the Maternal/Infant Nutrition programs being two examples); however, still others have provided little benefit. Certainly part of the problem results from inadequate funding levels, resistance by various political pressure groups, and even inadequate administration and implementation. Whatever the reasons for the success or failure of these programs, their results show that an unbridled faith in the environment to direct alone the course of development is incompatible with our current understanding of the interaction of heredity and environment.

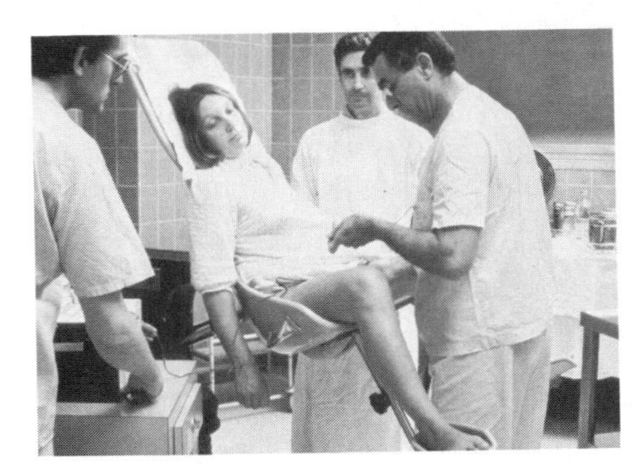
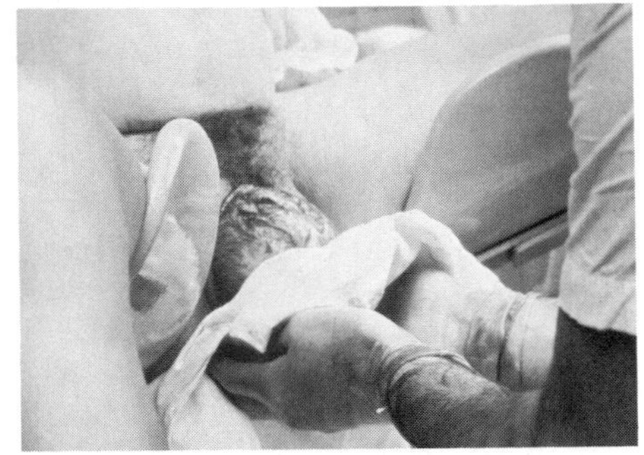

The modern birthing chair employs the force of gravity to ease the birth process.

in France, contractions during the first stage of labor are enhanced not through the use of drugs but through a warm bath or shower.

When parents are provided greater latitude in directing the labor and delivery experience, they gain a sense of mastery and control over the experience. Most researchers believe such a sense enhances the early parent-infant bonding process. The presence of the partner not only provides psychological support for the mother, it also seems to have positive effects on the actual labor process. The presence of a partner during labor has even been associated with increased uterine blood flow and effective uterine contractions.

All of these changes in the labor and delivery and postpartum experience are serving to make the birth of a child not only a safe procedure but a humane one as well. It is important that all prospective parents make sure that their physician or midwife and the hospital they are likely to use not only allow but fully support and encourage such practices.

Sources: Fogel, 1984; Garbarino, 1980; Klaus, 1982; MacFarlane, 1977.

Although actions at the level of the exosystem, such as a commitment to end poverty, must be broad in nature and therefore prone to stress the influence of environment, actions at the level of the mesosystem and the microsystem must interpret and tailor these broad guidelines to ensure the optimum fit for each individual. To do less is no more likely to result in success than having everyone wear the same size shoe is likely to result in equal comfort for all.

The Element of Choice

As discussed in the section on research methods in Chapter 1, research provides information, not decisions. To reach a decision, people must take the information gained from the research and in some fashion integrate it with information obtained from sources such as economics, politics, religion, and personal moral values. Perhaps nowhere in life-span research is this more true than in research on the prenatal period. Our understanding of the process of prenatal development has provided incredible tools to mark and assess its progress. It has provided tools to prevent conception or increase its likelihood. It has also provided tools to terminate pregnancy. All of these tools make choices possible. With the use of safe and effective contraceptives, a couple can choose whether to conceive a child and, if so, when to conceive it. Similarly, a couple can choose whether to conceive a child once they know, as a result of genetic counseling, the likelihood of an inherited genetic defect's being passed on to their child, or through procedures such as amniocentesis and fetal monitoring, they can determine the necessity of terminating an abnormal pregnancy through abortion. Once their child is conceived, a couple can choose whether to participate in prepared childbirth classes.

The availability of choice makes planning possible, but it also forces one to make a choice and then to live with the consequences of that choice. An appreciation of this fact leads to two observations. The first is that for individuals to make a truly informed decision, society must ensure that all the information needed to make such decisions is readily available and comprehensible. The second is that since the individual or the couple must live with the consequences of that decision, society must make every effort to ensure that those most affected by the decision are free to make it.

The Nature of Social Policy

Social policy should have two goals for the period from conception through birth. The first is to ensure that the biological development of the organism proceeds optimally. The second is to ensure that the family is prepared to receive a new member and that the "introductions" occur under the best possible circumstances.

Our understanding of human development indicates that events occurring at any level of the ecosystem can influence the course of development. At the broadest level of social policy, this awareness should serve as a guideline in evaluating the potential influence of any course of action. In a sense, we need to make an environmental impact statement for the family at the same time that we make one for the environment. The most obvious influences on the course of prenatal development would be policies specifying proper nutrition, adequate medical care, avoidance of toxic substances, and protection from physical injury. However, since not all influences on prenatal development are as direct and as evident as diet or toxic substances, all policy decisions should be evaluated as to their potential impact on the developing child.

Children are born into families. Some of these families already have other children; others do not. Some extend across generations; others are nuclear and relatively isolated. Some have two adults; others, one. Whatever the nature of the family constellation, it should be provided every opportunity to learn about the impact the new member will have on the family, to prepare to be an active participant in its arrival, to ease the transition to a new family form, and to integrate the new family constellation into the other aspects of the family's life.

These goals for the family translate into a number of specific human services. For most families, these include parent education programs and prepared childbirth training classes. For some, they might also include counseling to help the family understand and cope with the transition to parenthood. Some of these services are provided by employers in the form of maternity and, in some instances, paternity leaves. Giving the parents the opportunity to adjust to their new child without the added strains of coping with a job, even for just a few days or weeks, may make the transition easier and, from the perspective of the employer, may produce a more loyal and motivated employee. The most immediate services are provided in the hospital. Every effort should be made to ensure the active involvement of the parents in all aspects of labor and delivery. Equally important, every effort should be made to ensure contact between the parent and newborn from the moment of birth onwards. There is a growing body of evidence (Entwisle & Doering, 1980; Garbarino, 1980; Macfarlane, 1977) that under all but the most extreme medical circumstances, there need be little if any conflict between policies governing optimal medical practice and those governing optimal parent-child relations.

The Synergistic Nature of Human Development

A synergistic effect is one in which the whole is greater than the sum of the parts. With respect to the study of human development, it might be defined as the manner in which individual elements cumulatively interact to influence the course of development. For example, Rutter (1979), in studying families' reactions to stress, found that the type of stress encountered was considerably less important in predicting family reaction than the total number of stresses. He considered six sources of stress in a family: severe marital discord, poverty, overcrowding, paternal criminality, maternal psychiatric disorder, and the prolonged psychiatric hospitalization of a parent. The presence of any one of these factors was not in itself sufficient to adversely affect family functioning. The family certainly had to cope with the stress, but almost all families were able to make the necessary adjustments. When the stresses began to cumulate, however, then their influence changed dramatically, or, in this case, synergistically. The additional second source of stress resulted in children being four times more likely to develop some sort of problem than if only one source of stress had been present. When there were three sources of stress, the rate increased to six times the base rate, and with the addition of a fourth, the rate ballooned to twenty times the base rate. Our earlier discussions of the poverty cycle and the effects of prenatal malnutrition showed essentially the same pattern. And, although there is little research on the topic, our everyday experiences suggest that the same pattern can also function in a positive manner.

The significance of the synergistic nature of human development is that it is not necessary to remove all of the problems in an individual's life. It is certainly not realistic to attempt such a goal and in fact may actually be detrimental—we develop in part because we are able to overcome adversity. Rather, what is necessary is to reduce the number of problems to a point where the individual or the family can successfully cope with those that remain. In some cases, it may require little more than providing a babysitter so that a low-income single parent can have some time away from the child. In more extreme situations, it may involve more services. But in either case, the goal is to reduce the number of problems to a point where the rest are manageable with the resources of the family or the individual.

SUMMARY

THE COURSE OF PRENATAL DEVELOPMENT

1. The prenatal period is divided into three phases: the period of the ovum (conception through the first ten days), the period of the embryo (ten days through the seventh week), and the period of the fetus (seventh week till birth).
2. Organ differentiation occurs during the period of the embryo, each structure evolving from one of the three embryonic cell layers: ectoderm, endoderm, and mesoderm.

THE CONTEXT OF PRENATAL DEVELOPMENT

1. The influences of heredity and environment are continuous across the entire life-span. Efforts to distinguish their individual influences are at best limited to large population studies. At the level of the individual, such efforts are misleading and inappropriate.
2. Teratogens have their greatest negative impact when organ systems are undergoing their maximum rate of differentiation. This is true because teratogens alter the biochemical signals cells respond to.
3. The cycle of poverty demonstrates how relatively insignificant individual factors can cumulate to significantly influence the course of development.
4. The endocrine system seems to be the link between environmental events and the functioning of genetic regulatory mechanisms within the individual.

BIRTH

1. Labor is divided into three phases. The first and longest involves the thinning and enlargement of the cervix; the second, the actual delivery of the neonate; and the third, the delivery of the placenta and other fetal membranes.
2. Prepared childbirth practices provide parents a more deliberate role in the delivery process and reduce the possibility of drug-related injuries to the infant.

HUMAN DEVELOPMENT AND HUMAN SERVICES

1. Our increasing understanding of the complex process of development and the active role that individuals play in that process continues to make the provision of human services more efficient and more effective.

KEY TERMS AND CONCEPTS

THE COURSE OF PRENATAL DEVELOPMENT

The Period of the Ovum

Zygote
Morula
Blastocyst
Trophoblast

Placenta
The Period of the Embryo
Embryonic Cell Layers
- Embryonic Ectoderm
- Embryonic Endoderm
- Embryonic Mesoderm

The Period of the Fetus
Differentiation

THE CONTEXT OF PRENATAL DEVELOPMENT

Chromosomes
Deoxyribonucleic Acid (DNA)
Genetic Code
Genotype
Phenotype
Homozygous
Heterozygous
Dominant Gene
Recessive Gene
Sickle-Cell Anemia
Phenylketonuria (PKU)
Teratogens
Fetal Alcohol Syndrome (FAS)
Cycle of Poverty
Homeostasis
Jacob-Monad Model
Canalization

BIRTH

Stages of Labor and Delivery
- Effacement and Dilatation of the Cervix
- Expulsion of the Fetus
- Expulsion of the Afterbirth

Neonate
Perinatal Anoxia
Prepared Childbirth Training Classes

SUGGESTED READINGS

If you are interested in a more detailed discussion of the course of prenatal development, I would suggest

Moore, K. L. (1974). *Before we are born*. Philadelphia: W. B. Saunders.

There are two books that provide a good discussion of the psychological and interpersonal aspects of pregnancy.

Entwisle, D. R., & Doering, S.G. (1981). *The first birth: A family turning point*. Baltimore: Johns Hopkins University Press.

MacFarlane, A. (1977). *The psychology of childbirth*. Cambridge: Harvard University Press.

Childbirth practices have become increasingly varied over the past years. Three books provide a good discussion of the advantages of the more "people-oriented" approaches.

Leboyer, F. (1975). *Birth without violence*. New York: Knopf.

Lamaze, F. (1958). *Painless childbirth: Psychoprophylactic method*. London: Burke.

Tanzer, D. T., & Block, J. L. (1972). *Why natural childbirth?* New York: Doubleday.

Two sources provide insight into the way the pregnancy and labor experience interacts with other life events to influence the course of early development.

Birch, H. G., & Gussow, J. D. (1970). *Disadvantaged children: Health, nutrition and school failure*. New York: Harcourt, Brace & World.

Lester, B. M. (1979). A synergistic approach to the study of prenatal malnutrition. *International Journal of Behavioral Development*, 2, 377–395.

3
INFANTS AND TODDLERS

CHAPTER OUTLINE

This chapter surveys development during the first three years of postnatal life. During the first one and a half to two years, the child is called an infant. For the remainder of the period, the child is called a toddler (a label that aptly describes how the two-year-old walks).

Perhaps the best way to appreciate the amount of change that occurs during this interval is to consider a question asked by Flavell (1977). If you had a newborn infant, a two-year-old, and an adult, which two would you pair together as being most alike? Flavell suggests that most people would pair the two-year-old and the adult. What the question notes is that by the end of the second year the infant demonstrates a basic competence in all those skills that characterize us as human.

The most obvious skill is language. By age three, most children have a vocabulary of approximately one thousand words and a grammar equivalent to that of adult everyday speech (Lenneberg, 1967). Perhaps less obvious but equally important are the facts that by age three, children demonstrate coordinated motor skills, rudimentary notions of space, time, cause, and number, a well-developed attachment to their parents, the beginnings of peer-relation skills, and a very active interest in exploration and manipulation of their environment. What makes the presence of these abilities so significant is that none is present at birth.

The acquisition of these abilities takes place within a social context. One of the best ways of understanding this social context is to review the two psychosocial tasks that Erikson (1950) ascribes to these first three years. The task associated with infancy is the development of a sense of basic trust. The task for toddlers is the development of a sense of autonomy.

Erikson sees the newborn as entering a strange and unpredictable world. For development to proceed optimally, infants must develop a sense of mutuality, or responsiveness, between themselves and their primary caretakers, usually the parents. Through this sense of mutuality, infants begin to learn the patterns of giving and getting that are necessary for their continuing interaction with objects and people in their environment.

The major determinant of a sense of trust is the quality of the parent-infant relationship. If parents provide their infants an environment that is predictable and sensitive to their needs, then the infants will develop a basic trust in the environment. A sense of basic trust is a necessary prerequisite to the continued exploration of the environment. It is also through this responsiveness that infants develop a basic trust in themselves. A basic trust in self is a necessary prerequisite to the development of self-control.

It is important to note that Erikson considers too much trust as undesirable as too little trust. Therefore, a healthy resolution of this first stage is a balance between enough trust in yourself and others to venture into the world and sufficient caution to protect yourself from the world's uncertainties. The parent who caters to every whim and want of the infant is as much a hindrance to the development of a healthy sense of trust as the neglectful parent is.

The age range of two to three years is sometimes described as the "terrible twos." Although the description is clearly an overstatement (two-year-olds can be the most delightful of creatures), it does reflect the fact that the two-year-old is very much involved in developing what Erikson calls a sense of autonomy. Autonomy builds on trust. The ability of infants to trust in themselves and the predictability of their environments gives them the confidence to exercise choices, that is, to be autonomous. Care givers play a particularly crucial role in this process. They must manage a fine balance between providing sufficient support to enable toddlers to act on their own and at the same time protecting them from what Erikson calls "the potential anarchy of their as yet untrained sense of discrimination" (Erikson, 1980, p. 71).

Erikson's recommendations for a predictable and consistent environment for the infant and firmness and tolerance for the toddler set the tone and the context in which development during these first three years occurs. To one degree or another, all of the physical, cognitive, and social developments occurring during this period influence the children's sense of trust in themselves and their environments. This trust, in turn, defines how eagerly children will explore and comprehend their environments.

The content of this chapter is arranged in three sections. The first offers a normative description of the course of physical-motor, cognitive, and social development during the first three years. Although the focus is

BOX 3-1
NEOTANY

Have you ever wondered why most people find Mickey Mouse so lovable and adorable? What is it about Mickey that Goofy doesn't have? What is it that seems to make Donald Duck so jealous? Stephen Gould (1980), a very well-respected naturalist at Harvard, thinks he has the answer. Mickey Mouse has the facial characteristics and body proportions of an infant. When compared to adults, infants and Mickey Mouse (who is now quite middle-aged) have proportionally larger heads and eyes, smaller jaws, and smaller and pudgier legs and feet.

Gould's interest in the celebrated rodent comes from a more general interest in the process of evolution and in the way our evolutionary heritage influences the formation of human behavior patterns. He and others have noted that primates, including humans, as well as many other species treat their young differently from the way they treat adult members of the species. This preferential treatment is essential if the young are to survive since they are rarely able to provide for their own needs. In our case, this preferential treatment is more than merely a case of survival. It also involves the element of socialization. Gould suggests that this preferential treatment might in part be triggered by the distinct physical appearance of the infant. Specifically, the facial characteristics and body proportions shared by Mickey Mouse and infants might trigger an "innate releasing mechanism" for affection and nurturance in humans. The pervasiveness of this mechanism is reflected in the diagram created by the Nobel Laureate Konrad Lorenz (1950). Human infants are not alone in having the power to change the behavior of adults; the young of many species elicit the same response. The study of this evolutionary process is the study of neotany.

Most of us seem to prefer to play down our very distant pasts. Although we generally acknowledge the validity of the process of evolution, we prefer to think that we have transcended biological constraints, that we are creatures of culture rather than creatures of evolution. To a large extent this is true: we are the species most able to modify the environment to meet our particular needs. But as Gould and others continue to point out, we are creatures that have evolved through a process that has probably left distinct impressions on us. In particular, it may well have significantly increased the likelihood that we will care for our young, that we will socialize them, and that in the process, we will help maintain the species.

Sources: Gould, 1980; Lorenz, 1950.

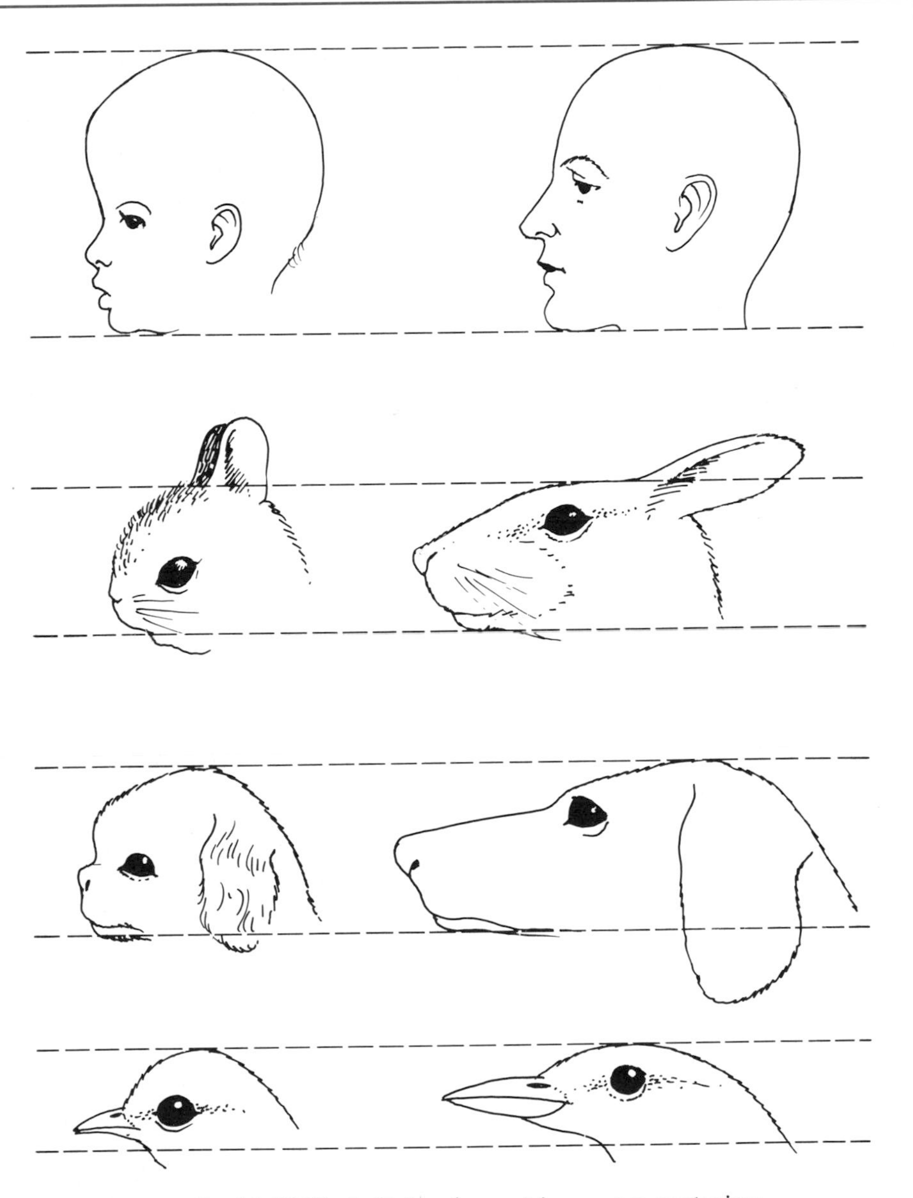

According to Gould (1980), individuals react in a more nurturing manner to animals that have proportionally larger heads and eyes, smaller jaws, and pudgier legs and feet. For example, the animals on the left (child, jerboa, Pekinese dog, and robin) would trigger this protective response, while those on the right (man, hare, hound, and golden oriole) would not (after Lorenz, 1950).

In his effort to develop a sense of autonomy, this two-year-old is beginning to actively explore and test his surroundings.

on normative development, it is important to bear in mind that variability is the rule in each of these domains. Therefore, all references to chronological age should be viewed as rough approximations rather than exact specifications.

The second section offers a discussion of the context of development. This section emphasizes the interactive nature of development. An interactive view stresses the fact that infants both influence and are influenced by the events and people in their environment.

The third section concerns the implications of the statements of the first two sections. Specifically, this section discusses the influence of the first three years of life on the rest of the life-span and the application of this knowledge to the provision of human services to infants and their families.

THE COURSE OF DEVELOPMENT DURING THE FIRST THREE YEARS

Physical development refers both to the physical stature of children and adults and to the functioning of their various biological systems. Motor development refers to their ability to control their own bodies. In all aspects of physical and motor development, neonates are quite immature at birth in comparison with most other mammals. Their biological systems are less resistant than those of adults, and less regular in functioning. Neonates have very little control over movement or posture. One of the most dramatic aspects of early development is the increasing degree of control and stability infants come to exercise over their physical and motor systems.

Physical and Motor Development

Full-term infants may weigh anywhere from 5½ to 9½ pounds—averaging 7½ pounds—and range in length from 19 to 21 inches. Their skin is wrinkled and has a thin, waxlike protective layer called the vernix. After a few days, the vernix dries and peels away. Their noses are small and flat, and their heads may be elongated or lopsided from the pressure exerted on the skull during delivery. The head is comparatively quite large. Whereas an adult's head constitutes one-eighth of total body length, a newborn's constitutes one-fourth. The eyes of a newborn are even more prominent than the head. They are already one-half of their future adult size. The neck is barely distinguishable; the head, quite wobbly. The abdomen is large and rounded; the legs, comparatively short, bowed, and usually bent at the knees (Kaluger & Kaluger, 1974).

The **Apgar score** (Apgar, 1953) is the most common evaluation of newborn status. As you can see from Table 3-1, the Apgar is a simple index

TABLE 3-1 Apgar Scale

SIGN	RATING		
	0	1	2
Heart rate	Absent	Less than 100/min	More than 100/min
Respiratory effort	Absent	Weak cry; Hypoventilation	Good strong cry
Muscle tone	Limp	Some flexion of extremities	Well flexed
Reflex irritability (Response to stimulation of feet)	No Response	Some motion	Cry
Color	Blue; pale	Body pink; extremities blue	Completely pink

Note: Reprinted with permission from the International Anesthesia Research Society from "A Proposal for a New Method of Evaluation of the Newborn Infant," by V. Apgar, *Anesthesia and Analgesia*, 32, 264.

of the status of the major biological systems of the newborn. A perfect Apgar score is 10; any score over 7 is considered healthy and normal. If there is reason to be concerned about the status of the newborn, a more complete neonatal evaluation is made.

The **Brazelton neonatal behavioral assessment scale** (Brazelton, 1973) is probably the most frequently used scale. The Brazelton is a more sophisticated means of evaluating the developmental status of the neonate. It attempts to assess the full range of the infant's response to the environment. The items on the test fall into two categories. The first measures the various reflexes that should be present at birth; the second, behavioral responses to various forms of simple stimulation (see Table 3-2). While conducting the test, the examiner also makes a general evaluation of the infant's states. In addition to providing a measure of developmental status, the Brazelton has also been used as a way of helping new parents appreciate the remarkable competency of newborn children.

Physical Growth

Gains in height and weight are rapid during the first three years of life. Birth weight typically doubles by four to six months of age and triples by one year of age. By two years of age, infants often weigh as much as four times their birth weight.

During the first six months, the infant may grow as much as six inches in length. During the second six months, an additional three inches may be added. During the second year, rate of growth begins to slow, and infants begin to assume the leaner, more slender build that is more typical of preschoolers. By four years of age, children are approximately twice their birth length. (See Figure 3-1.)

It is important to remember that all of these indices of physical growth have a wide range of individual differences. Boys tend to be slightly larger during the first two years than girls. To the extent that size and weight differences reflect nutritional intake, socioeconomic differences are also evident.

Establishment of Physiological Stability

The physiological functioning of the neonate is more irregular than that of the adult. Only gradually during the first few years do the biological systems assume their more regular adultlike patterns. Infants and young children, for example, are more sensitive to external temperature changes than older children and adults are (the trend will begin to reverse itself with advanced age). They are less able to maintain a constant body temperature and during illness are likely to show higher temperatures than they would show during a similar illness later in life.

Changes in respiration are also marked during the first two years. The young infant's breathing is shallow, rapid, and irregular. It often sounds harsh. For parents who don't expect this pattern, the infant's respiration can be a frequent source of anxiety. By the end of the second year, breathing becomes slower, more regular, and deeper. As the chest cavity continues to grow, the breathing sounds become more muffled.

Sleep-wake cycles undergo major changes during the first few years. Although children do not establish a stable adult sleep cycle until six years of age, by the latter part of the first year, they do have one eight-hour sleep

TABLE 3-2 Neonatal Behavioral Assessment

	BEHAVIORAL ITEMS
1.	Response decrement to repeated visual stimuli
2.	Response decrement to rattle
3.	Response decrement to bell
4.	Response decrement to pinprick
5.	Orienting response to inanimate visual stimuli
6.	Orienting response to inanimate auditory stimuli
7.	Orienting response to animate visual—examiner's face
8.	Orienting response to animate auditory—examiner's voice
9.	Orienting response to animate visual and auditory stimuli
10.	Quality and duration of alert periods
11.	General muscle tone in resting and in response to being handled (passive and active)
12.	Motor maturity
13.	Traction responses as he is pulled to sit
14.	Cuddliness—responses to being cuddled by the examiner
15.	Defensive movements—reactions to a cloth over his face
16.	Consolability with intervention by examiner
17.	Peak of excitement and his capacity to control himself
18.	Rapidity of buildup to crying state
19.	Irritability during the exam
20.	General assessment of kind and degree of activity
21.	Tremulousness
22.	Amount of startling
23.	Lability of skin color (measuring autonomic lability)
24.	Lability of states during entire exam
25.	Self-quieting activity—attempts to console self and control state
26.	Hand-to-mouth activity

Note: Reprinted by permission, from T. Berry Brazelton, "Behavioral Competence of the Newborn Infant," in P.M. Taylor (Ed.), *Parent-Infant Relationships* (New York: Grune & Stratton, 1980), 75.

cycle per day plus other shorter cycles (Healy, 1972). Newborns sleep almost sixteen hours per day. By six months, infants sleep more regularly, stay awake for longer periods, and do most of their sleeping at night. Children continue to take a morning and afternoon nap until sometime during the second year. At that time the morning nap is given up. By age four or five, children also give up the afternoon nap. The giving up of naps seems to reflect decreasing sleep needs since it is not accompanied by an increase in nighttime sleep.

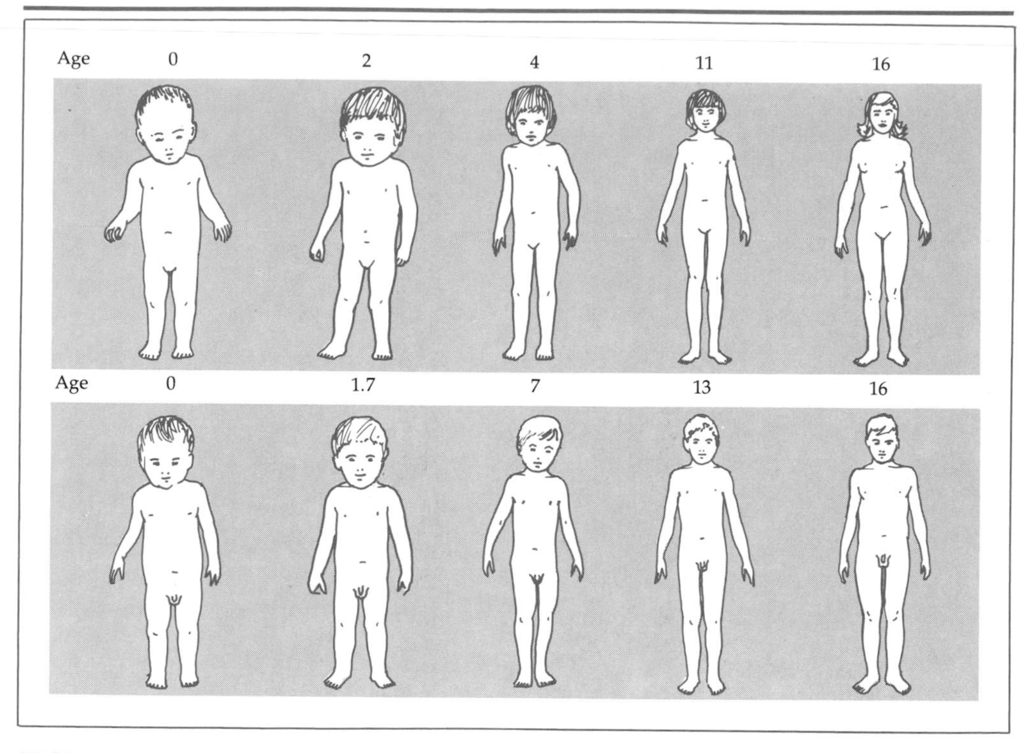

FIGURE 3-1
Body proportions shift as children age. In young children, the head represents a large proportion of total body length. From the preschool years through adolescence, the head occupies an increasingly smaller percentage of body length, the legs increase in length and proportion, and the trunk becomes slimmer (from The Diagram Group, 1977).

Temperament in Infants and Toddlers

One way to describe the stability of functioning of infants and toddlers is to describe their temperament (Thomas & Chess, 1977). As defined by Thomas, Chess, and Birch (1968), temperament describes the "how" of behavior. **Infant temperament characteristics** describe the tempo or rhythm of infants. They are a measure of the way infants go about a task rather than an answer telling why they pursue one task over another. Consequently, two infants pursuing the same task—even for what appear to be the same reasons—may display energy expenditures, moods, degrees of adaptability, and other aspects of temperament that are greatly different.

Table 3-3 defines and gives examples of the nine elements that make up an infant's temperament. Thomas, Chess, and Birch found that individual temperament traits tended to cluster together to form temperament patterns. Some children were described as "easy" children. They displayed a pattern consisting of positiveness of mood, regularity in bodily functions, low or moderate intensity of reaction, adaptability, and a positive approach to, rather than withdrawal from, new situations. Thomas

TABLE 3-3 *Categories of Temperamental Organization*

ACTIVITY LEVEL	*RHYTHMICITY*	*APPROACH OR WITHDRAWAL*
This category describes the level, tempo, and frequency of movement. Highly active children move about when sleeping, squirm during feeding, kick when being diapered, and show a great deal of body movement during play.	This category reflects the degree of regularity of repetitive biological functions. Highly rhythmic infants usually go to sleep and awake about the same time each day, show comparatively little fluctuation in appetite or behavior, and have regular and predictable bowel function.	This category describes the child's initial reaction to new situations. A child who withdraws might take a long time to warm up to a new babysitter or a new food.
ADAPTABILITY	*INTENSITY OF REACTION*	*THRESHOLD OF RESPONSIVENESS*
This category concerns the extent to which initial responses to situations or individuals can be modified in the direction desired by the parents or others. For instance, a very adaptive infant would easily adjust to a new sleeping place or a change in feeding schedule.	This category describes the child's intensity of response. Intensity of response can be in a negative direction (e.g., crying loudly and constantly or strongly pushing food away when full) or a positive direction (e.g., showing great delight in a new toy or great pleasure in a regular activity such as a bath).	This category refers to the level of extrinsic stimulation that is necessary to evoke a discernible response. A child with a high threshold might be hard to wake, not seem annoyed at having a very full diaper, or react rather calmly to a loud noise such as a door slamming.
QUALITY OF MOOD	*DISTRACTABILITY*	*ATTENTION SPAN AND PERSISTENCE*
This category describes the amount of pleasant, joyful, friendly behavior as contrasted with unpleasant, crying, unfriendly behavior. It assesses the way the infant "usually" behaves.	This category refers to the effectiveness of extraneous environmental stimuli in interfering with, or in altering the direction of, the ongoing behavior. Distractibility can work both ways. It is a blessing when a crying child is easily calmed by an attentive adult; it is less than a blessing when it seems impossible to ever finish a feeding.	Attention span measures the length of time a particular activity is pursued. Persistence refers to the extent to which a child will maintain an activity when faced with obstacles. As with distractibility these qualities can at times be desirable (e.g., when the child is trying to learn to walk) and at other times undesirable (e.g., when someone is telling the child to quit spilling water out of the tub).

Source: Adapted from Thomas, Chess, & Birch, 1968.

and Chess (1977) found that these infants quickly developed regular sleep and feeding schedules, took to most new foods easily, smiled at strangers, and accepted most frustrations with minimal fuss. Difficult children displayed almost the opposite pattern. "As infants, they are often irregular in feeding and sleeping, are slow to accept new foods, take a long time to adjust to new routines or activities and tend to cry a lot. Their crying and their laughter are characteristically loud. Frustration usually sends them into a violent tantrum" (Thomas & Chess, p. 23). A third group was described as slow-to-warm-up children. These infants tended to have a low activity level, to withdraw on first exposure to new situations, to be slow in adapting to new situations, and to be somewhat negative in mood. Approximately 40 percent of the infants Thomas, Chess, and Birch studied were classified as easy, 10 percent as difficult, and 15 percent as slow to warm up. The remaining 35 percent of the infants seemed to show some blend of the three patterns. The behavior of these infants was more situation specific. That is, some situations might produce behavior typical of the easy child; other situations might produce behavior typical of the difficult child.

Temperament seems to be a relatively enduring quality of the infant and child. Although these characteristics do undergo modification during childhood and adolescence, they do not disappear. Rather they remain an obvious component of the child's personality.

Temperament is a significant factor in determining the quality of the parent-child relationship. Much of the character and effectiveness of early parent-infant interactions is determined by the parents' ability to correctly read and interpret the infant's behaviors. The more readable and predictable the signals are, the more likely the parent is to make an appropriate response. Further, the more responsive the infant is to the parents' efforts (for example, ceasing to cry when held), the more confidence parents have in their ability to parent, and consequently, the more rewarding the parenting experience becomes. Although it would be incorrect to say that easy children have better relations with their parents than other children have, it is nevertheless true that difficult children present more obstacles to the establishment of a quality relationship.

Early Perceptual Development

At birth, the neonate's sensory systems are already remarkably accurate. By age two, the infant has sensory abilities that are virtually equivalent to the adult's.

Vision. Visual acuity develops rapidly during the first year. By four months of age, the infant can focus almost as well as the adult. By the end of the first year, the infant is able to focus on small objects and to visually track or follow distant objects.

Infants have at birth or soon thereafter a well-developed sense of depth perception. This point has been illustrated by Gibson and Walk (1960). Using the visual cliff, an apparatus designed to test for depth perception (see Figure 3-2), they found that even though it was safe to crawl anywhere on the apparatus, infants would not crawl on the side that appeared unsafe (that is, the side that had the checkerboard pattern some distance below the plexiglass surface). The avoidance was so strong that even the mother's attempt to get the child to crawl to her across the deep side usually proved unsuccessful.

FIGURE 3-2
The fact that infants have a well-developed sense of depth perception has been demonstrated by experiments using the visual cliff. Even though it is safe to crawl anywhere on the surface of the apparatus, infants will not crawl across the area that appears to be recessed.

In addition to having depth perception, infants show decided pattern perception. They will look at lights of different brightness for different amounts of time. They will often stop what they are doing to watch a light move across their visual field. In one of the more intriguing studies in the research literature, Fantz and Nevis (1967) have shown that very young infants decidedly prefer to view a human face rather than one in which the facial features are scrambled or one that has no features but the same percentage of light and dark areas.

Infant pattern preferences are not constant. They change as infants develop. Specifically, infants prefer looking at sights that are different from and moderately discrepant with previous objects and sights. This behavior

BOX 3-2
INTERVIEWING INFANTS

Studying the process of development involves the use of a variety of methods with a variety of age groups. In all cases, the developmentalist strives to ensure that the data obtained from the subjects are an accurate reflection of the process in question. Sometimes this task is more easily accomplished than others. Consider the unique case of infants.

In the chapter, I discuss the fact that infants from birth have pattern preferences. That is, they prefer to look at some sights more than at others. How is it possible to obtain such information from very young infants? Since they have very little motor control, even as simple a response as head turning proves unreliable. However, one of the few motor acts neonates can control is eye movement. Robert Fantz (1967) developed a procedure capitalizing on this fact. Neonates were placed in a "looking chamber." The chamber was an enclosed criblike structure where infants, lying on their backs, could view pairs of objects projected on a screen above their heads. The experimenter stood over the chamber looking at the neonate through a peephole. Because of the lighting in the chamber, it was possible to observe the object that the infant was looking at by seeing the object reflected in the pupil of the infant's eye. If infants have no pattern preferences and if each object is shown equally often to the left and to the right of the other, then each object should be reflected in the pupil about as often as the other. In fact there is a preference. Neonates will turn their eyes to look at some objects over others, and this preference is reflected in the eyes of the child.

Is it possible for very young infants to learn? That is, is it possible to demonstrate that infants will change their behavior in order to obtain something they want? It isn't too hard to demonstrate that nine-month-old infants have the ability to learn, but what about nine-day-old infants? What aspects of their behavior can they control? The answer is their ability to suck, and a number of studies have demonstrated that young infants' sucking rate can be modified by the presence or absence of preferred visual patterns. In one particularly ingenious procedure, a very young infant's pacifier was wired to a slide projector in such a way that the rate of infant sucking regulated the

suggests that infants are soon able to relate new visual experiences to old ones. Infants seem to use these comparisons to "test" very simple hypotheses about the way the world around them works (Kagan, 1971).

Audition. Infants hear rather well. They are able to discriminate sounds of different pitch as well as sounds of different loudness. The lower threshold for loudness in an infant (that is, the minimum sound they can hear) is about the same as it is in an adult. Infants are even able to discrimi-

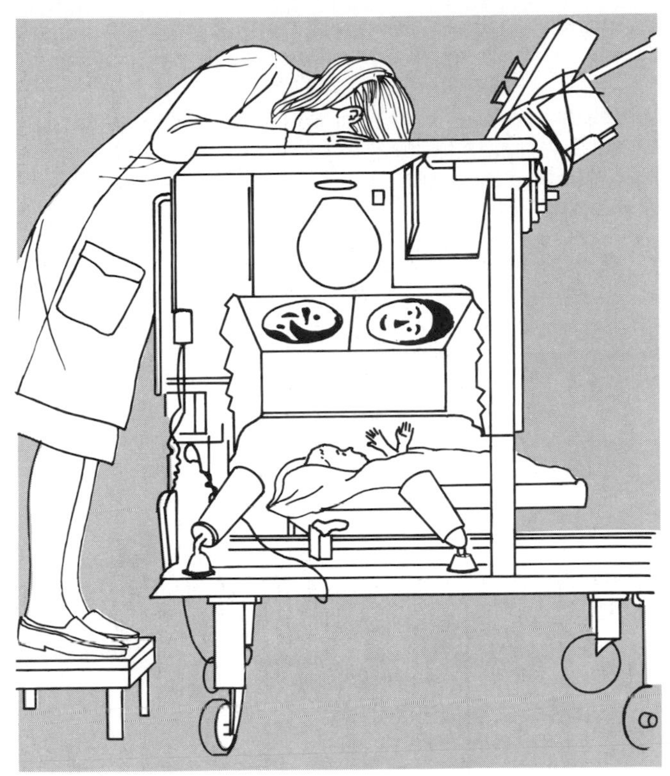

Robert Fantz (1967) used this "looking chamber" to study pattern preferences in infants (from Fantz, 1967).

degree of focus of the pattern shown through the projector. The researchers were able to clearly demonstrate that whenever the slide was taken out of focus, there was a corresponding change in sucking rate to bring it back in focus again (Siqueland & DeLucia, 1969).

Studies such as these provide a much greater insight into both the capacities of the young infant and the creativeness of those who study them.

Sources: Fantz & Nevis, 1967; Kaye, 1982.

nate between sounds as similar as "p" and "b" (Eimas, Siqueland, Jusczyk, & Vigorito, 1971).

Infants also have adaptive auditory responses. When viewed from an evolutionary perspective, these responses, like those for vision, enhance the organism's chance for survival. High-frequency sounds such as a scream make the infant freeze. Low-frequency sounds tend to quiet and calm infants who are crying and showing signs of distress. These same low-frequency sounds cause an increase in motor behavior in infants who are alert but already calm. Sudden sounds such as a car backfire cause infants to close their eyes. In contrast, sounds with a slow onset such as speech prompt infants to open their eyes and attempt to localize the source of the sound. Finally, infants tend to be more calmed by rhythmic than arhythmic sounds. Probably one of the reasons infant lullabies such as "Rock-a-Bye Baby" and Brahms' Lullaby are so soothing is that their sound patterns are regular.

Olfaction, taste, and tactile senses. Like vision and audition, taste and smell are well developed by birth. Infants even seem to have a sweet "tooth" (gum actually) already, since they will increase their sucking rate in direct proportion to the concentration of sugar in water (Lipsitt, Kaye, & Bosack, 1960).

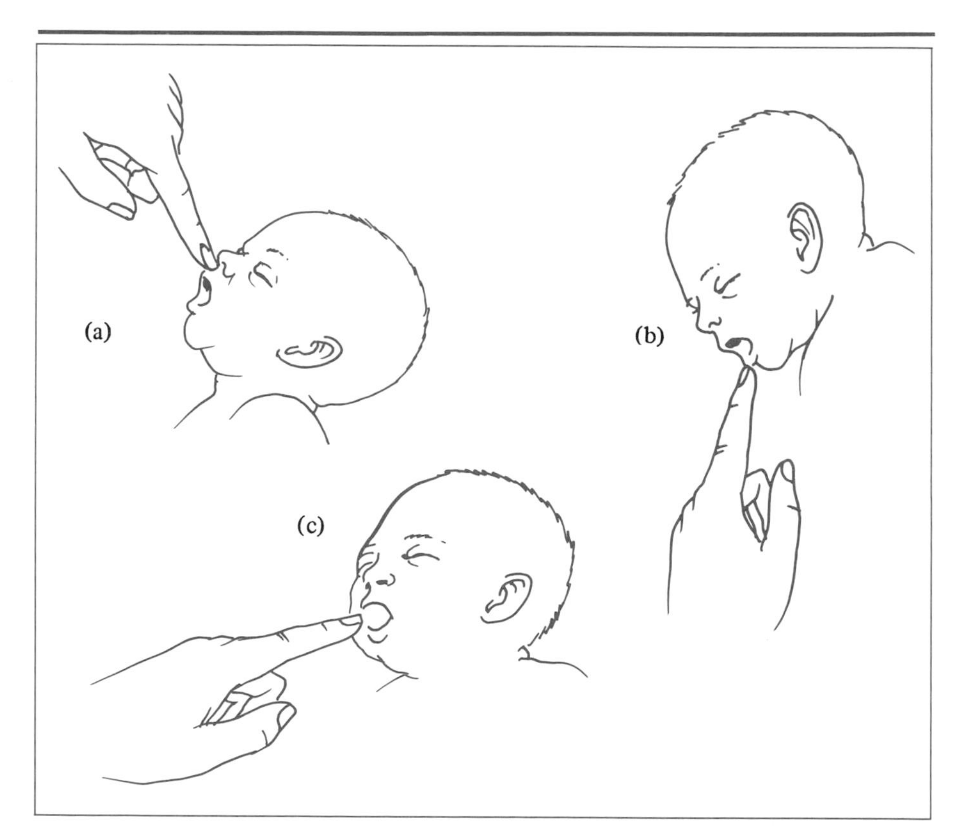

FIGURE 3-3
The rooting reflex enables infants to seek food. (a) When the middle of the upper lip is touched, the upper lip and tongue are elevated; as the finger is moved toward the nose, the head extends. (b) When the lower lip is touched in the direction of the chin, the tongue and lower lip move downward and the head is flexed. (c) When the corner of the mouth is stroked and the finger is moved toward the cheek, the head, mouth, and tongue move toward the stimulated side.

At birth, infants are almost as able to discriminate smells as adults are. They also show the same ability adults have to habituate to odors. The tactile senses, including the skin senses of pressure, pain, heat, and cold are also well functioning by birth.

Motor Development

The term motor development refers to the movements of the individual and the coordination of those movements. Movements that involve relatively large muscle masses are referred to as **gross motor skills.** The movements necessary for sports activities are good examples. Activities such as writing, weaving, block building, and piano playing are examples of **fine motor skills.** These skills require good prehensile skills and good eye-hand coordination.

At birth, neonates demonstrate two very distinct patterns of motor ability. The first is a set of well-developed reflexes that respond to specific forms of stimulation. These reflexes serve three purposes. First, they have a survival value for the neonate. They enable the neonate, for example, to seek nourishment (the rooting reflex, Figure 3-3), to suck (the sucking reflex), to cling (the grasping reflex, Figure 3-4), and to stop rolling (the tonic

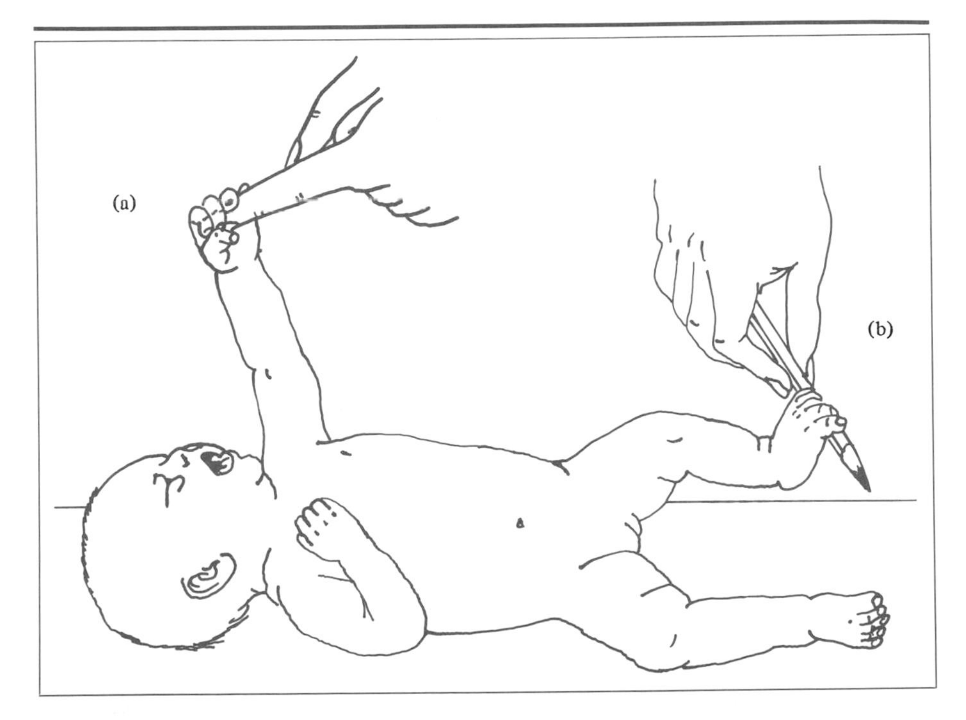

FIGURE 3-4
(a) When finger pressure is placed on an infant's palm, the infant will grasp the finger. This grasp, termed the palmar reflex, is usually strong enough to allow the infant to be lifted. (b) When an object is placed against the sole of an infant's foot, just behind the toes, the toes will curl around the object. This grasp is termed the plantar reflex.

neck reflex, Figure 3-5). Although the ability to cling and the ability to stop rolling are not highly significant for human survival, their presence probably reflects our common primate heritage.

Second, these reflexes provide information about the maturational level of the nervous system. For example, the absence of the Moro reflex (arms extended and then brought in as a response to a sudden loud noise or jarring) at birth indicates a serious disturbance of the central nervous system, and the absence of the grasp reflex in the toe suggests damage to the lower spinal cord. Further, since most of these reflexes fade during the first year, their disappearance provides an index of the level of development of the central nervous system.

Third, the reflexes are one of the first avenues through which infants come to know the world. As such, they provide the base from which more advanced behavior develops.

The second pattern of motor ability at birth is best described as a lack of pattern. Stott (1974) describes it as an amorphous mass of activity. There is no coordination of movement; there is no specificity of movement. There

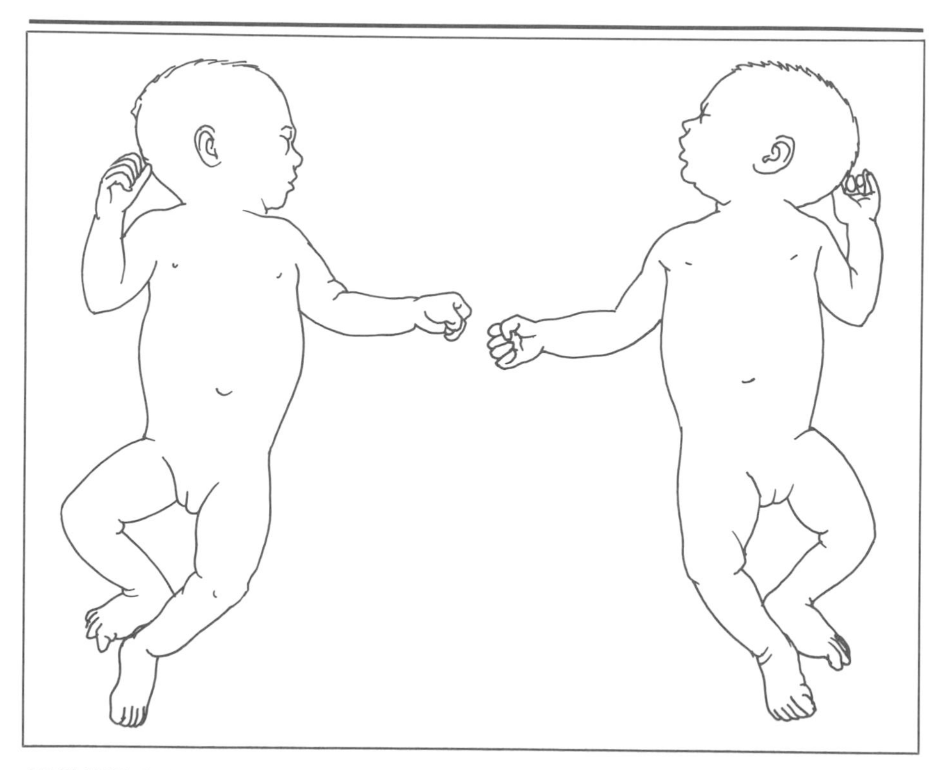

FIGURE 3-5
When an infant's head is turned toward one side, the infant will extend the arm and leg on the side to which the head is turned and flex the limbs on the opposite side. This reflex is termed the tonic neck reflex.

is no goal direction or purpose to the movement. There is merely movement. The development of this mass of activity into the relatively well-coordinated, voluntary, intentional movement of the three-year-old follows a well-defined course. In general, voluntary motor control develops progressively from the head and upper torso to the lower trunk and legs in a sequence termed the **cephalocaudal progression;** from the muscle groups of the midline of the body to those of the extremities in a sequence termed the **proximal-distal progression;** and from large motor groups to muscles involved in increasingly finer movement in a sequence termed the **mass-to-specific progression.**

Within each of these three progressions, motor activity proceeds from a global response (the newborn's cry, for example, involves movement of the entire body) to a specific limited response (for example, the infant comes to use only one hand in reaching for an object) to a coordinated response (such as the integration of arm, leg, and trunk movement involved in throwing a ball). This pattern is a postnatal example of the undifferentiated, differentiated, integrated sequence also typical of prenatal development. The process of learning how to walk is a good example of the sequence. Although a child's first steps may seem to appear suddenly, they are actually one development in a sequence of developments that began at birth (Shirley, 1931).

The sequence first involves gaining gradual postural control of the head and upper trunk (see Figure 3-6). Mastery of this control evolves over the first twenty weeks of life. Infants first become able to lift up their chin and then their chest from the mat. Soon thereafter they become able to sit with support at the lower ribs. At that time they are able to control their head movements.

From twenty to thirty weeks of age, they continue to gain postural control through the lower trunk and the limbs. They are able to sit by themselves, to stand firmly when held erect, and to roll over.

By thirty to forty weeks of age, infants are trying hard to move. They balance themselves on their hands and legs and try to move by rocking back and forth. They usually can't get anywhere because the two sides of the body are moving in unison. Sometimes they do manage to push themselves backward but usually they just manage to get very frustrated.

By ten months or so, they finally seem to realize that to get anywhere one arm and one leg must go in front of the other, first one side then the other. They are now able to crawl. They also have gained enough control of their arms to pull themselves to a standing position and enough control of their legs to maintain a standing posture. They are often able to walk when led. Sometime around the first birthday, the first true steps are taken.

During the second and third year, motor skills continue to be refined and extended into new areas. By age two, toddlers walk more smoothly and more continuously. They no longer need to pause between steps or to extend their arms for balance as a tightrope walker does. Their hands and arms can now be used for other tasks such as pulling a toy. Two-year-olds have enough fine motor control to turn the pages of a book and enough gross motor coordination to attempt walking up and down stairs unaided. By age three, toddlers have developed enough motor skill to perform such tasks as partially dressing and undressing themselves, riding a tricycle, pouring milk from a small pitcher, and drawing a circle.

FIGURE 3-6
Postural control and locomotion occur in a series of steps over approximately the first fifteen months of life.

Cognitive Development

During the first three years of life, infants acquire a very rudimentary understanding of the way the world works. They acquire a basic understanding of the nature of the objects and the people they encounter. In a sense, this early period of cognitive development is similar to learning how to put a puzzle together, but with one important difference. The task is not simply one of discovering which pieces to connect. Rather it involves first realizing that the pieces go together at all—that is, that there are relationships between actions, objects, and people; that there are reasons why things happen; that things can stand in relationship to one another; that one event can cause another event to happen. These discoveries are not all made at once, and they are certainly not completed by the age of three, but during these first three years all of the elements of mature, adult cognition make their initial appearance.

It's hard for adults to appreciate what the world must look like to an infant. It seems obvious and natural for us to look at a group of objects and think of them in terms of their number or their physical or spatial arrangement. We notice a parent and child playing together and observe how the

Unlike adults, young infants do not have the capacity to view the world as an ordered and predictable place. Parents who fail to appreciate this fact may at times be unnecessarily frustrated in their attempts to deal effectively with their infant.

actions of one cause the actions of the other, how the behavior of one is modified by the behavior of the other. We read a passage about a beautiful countryside and immediately create a mental picture of this scene. We know that behavior appropriate in one setting is inappropriate in another, that we treat parents differently from the way we treat peers. We do all these things and have all this knowledge without having to expend much deliberate thought. It's as if we see things in relationship because that's the way they really are. But to the infant things are not that way. Rather these basic fundamental relationships of cause and effect, of spatial and temporal patterns, of number, and of physical properties only gradually begin to be understood during these first three years. As a result of the repetitive behavior of early infancy and the trial-and-error behavior more typical of older infants, infants learn about the attributes of objects as well as people and in the process begin to understand their place in the scheme of things. That is, what starts as an interest in the properties of objects gradually broadens into an interest in the ability to do different things with different objects (McCall, 1976).

A number of researchers have studied the course of cognitive development during the first three years. The most notable was Piaget (1952), but significant contributions have also been made by Bower (1979), Emde (Emde, Gaensbauer, & Harmon, 1976), Kagan (1971; Kagan, Kearsley, & Zelazo, 1978), McCall (1976; 1979; McCall, Eichorn, & Hogarty, 1977), Sroufe (1979), and Uzgiris (1973; 1976; 1977). Although the specific interests of these researchers were somewhat different, their findings, when considered together, suggest five relatively distinct phases of cognitive development during the first three years of life. These first three years constitute Piaget's **stage of sensorimotor development** (birth to approximately age twenty-four months) and the beginnings of his second stage of cognitive development—the preoperational stage.

Newborns enter the world with a remarkably well-functioning sensory system and a set of well-defined motor reflexes. Piaget calls these reflexes "schemes." The label **scheme** refers not only to the actual set of behaviors typical of a particular reflex but also to the cognitive capacity necessary for the reflexive behaviors to be elicited by the appropriate stimulus. Therefore, Piaget would say, for example, that infants demonstrate the sucking scheme when they make organized sucking movements in response to a nipple's being placed in their mouth. Development during the sensorimotor stage is primarily a process of these elementary schemes being modified, elaborated, and coordinated with each other and with larger units made up of combinations of more elementary schemes. "As elementary schemes gradually become differentiated, generalized, and above all, intercoordinated and integrated with one another in diverse and complex ways, the infants' behavior begins to look more and more unambiguously 'intelligent' and 'cognitive' " (Flavell, 1977, p. 17).

Phase I: Birth to Approximately Ten to Twelve Weeks

During the first few months of life, the behavior of infants initially appears to be a response more to internal states than to external stimulation. The presence of the various reflexes certainly indicates that infants are responsive to environmental stimulation, but they don't yet appear to be very interested in people or objects in the environment. When an interesting sight or sound does elicit a response, it is usually passive—that is, the infant may look at the object but will not attempt to reach for it. More generally, the behavior of infants appears to be a reflection of hunger, fatigue, discomfort, or level of arousal.

As this first phase progresses, infants begin to show systematic modification of their behavior. They show a very elementary form of anticipation when, for example, the sight of the bottle or breast becomes sufficient to elicit the sucking response. Infants do not yet demonstrate new behaviors, but the stimuli that prompt their reflexes broaden. These anticipatory responses are the first indication that infants are becoming responsive to or at least increasingly interested in environmental events. This change is reflected in Piaget's observation that by the second month infants begin to demonstrate a behavior pattern that he calls a **primary circular reaction.** The reaction involves an infant's attempt to replicate a particular event: thumb sucking is the most typical at this age. Probably unintentionally, an infant manages to place his or her thumb in the mouth. The thumb in the mouth elicits a sucking response that appears to be pleasurable to the infant. Since the infant has little motor control, however, the thumb soon falls out of the mouth. Previous to this time, that would have ended the sequence of events, but now the infant attempts to reproduce the pleasurable event, that is, to get the thumb back in the mouth. The effort is not very efficient. The infant has limited motor control, and it's even hard to judge from the infant's behavior the extent to which the infant is aware that the pleasure was due to the thumb sucking. But in any case, the thumb eventually finds its way back into the mouth, the sucking is recommenced (hence a circular reaction), and the infant seems to acquire an early lesson in contingent relationships.

Toward the end of the third month, a new behavior pattern emerges. It involves the infant's initial attempts at coordinated actions. The passive responses of turning the head to locate a sound or visually tracking a light or object are now joined by the more active behavior of reaching and holding. An object looked at is now also reached for; an object placed in the infant's hand is now brought to the mouth and sucked.

Phase II: Approximately Three Months to Seven or Eight Months

By the third month, the infant's behavior changes dramatically. Most parents note this transition by commenting that the infant seems more fun to be with, or that they now are sure that their child knows them, or more generally, that the child somehow seems more "human." What these parents are noting is that infants, around three months of age, begin to become increasingly interested in the people and events in the environment. And, equally important, infants begin to smile and vocalize as ways of demonstrating this new interest in and pleasure from interactions with the environment. This is a time when the infant seems to derive great delight from games. For example, a parent may touch the infant's nose when the infant smiles or tickle the infant when the arms are raised. All of these games between the infant and the care giver are examples of what Piaget called **secondary circular reactions.** Again the infant is attempting to make

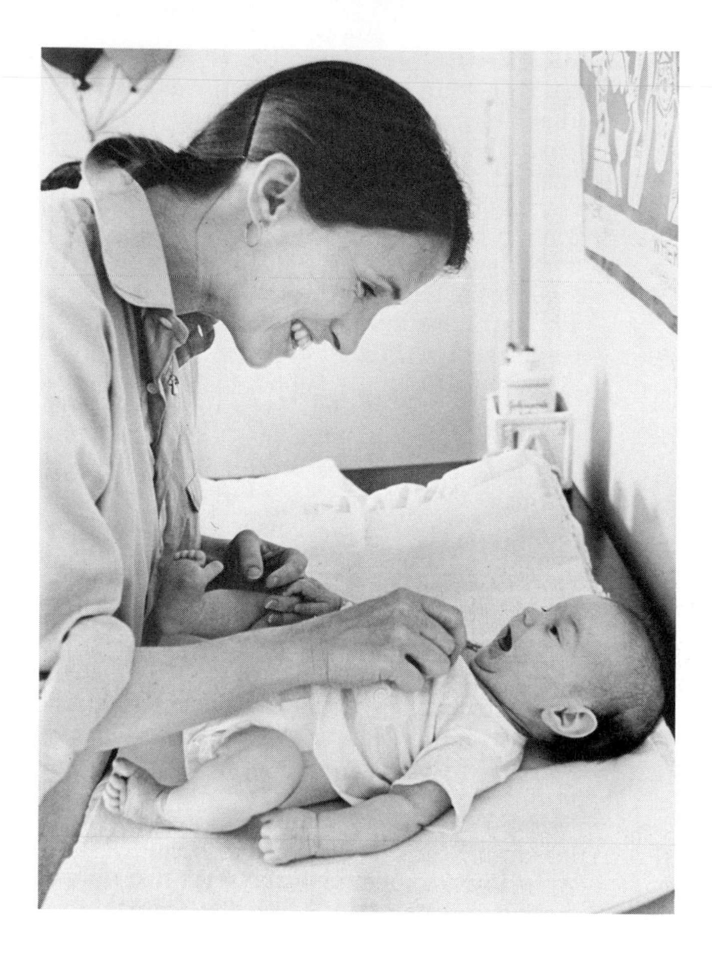

This mother is eliciting a secondary circular reaction from her young infant. By laughing at the mother's touch and facial expression, the infant is attempting to keep the game going.

an interesting event reappear, but this time the focus is no longer limited to the infant's own body, as was true in the case of thumb sucking. Now the horizon is broadened to include events in the environment as well.

Infants show delight in response to repeating an interesting event not only because of the pleasure derived from the event itself (making the mobile move, having the parent touch the infant) but also because of their growing awareness of the more basic fact that there are relationships between actions and events. At this age, infants are beginning to realize that there is indeed a puzzle to assemble.

This first puzzle is a very subjective one, however. The pieces exist only to the extent that infants deal with them. That is, infants attribute meaning to actions and objects and people only in terms of their relationship to them. Things that can be held all come to have the same meaning for the infant; so too do objects that can be sucked, moved, pushed, banged, or dropped. The fact that two objects that, for instance, can be banged, such as a rattle and a cup, are different in other ways (one can be used to drink from and the other cannot), or that they stand in relationship to each other in some ways (one is larger than the other or they are different colors), or that they even exist when they aren't being banged doesn't yet seem apparent to the infant. The object only has meaning in terms of the infant's immediate relationship to it.

Phase III: Approximately Eight Months to the First Birthday

The emergence of this third phase is an important milestone in the cognitive development of infants. For the first time, they clearly demonstrate intentional goal-directed behavior. For Piaget, the demonstration of intentionality is the first indication of true intelligence. Piaget (1952) describes an episode with his son Laurent that demonstrates this newly developing skill:

> I present a box of matches above my hand, but behind it, so that he cannot reach it without setting the obstacle aside. But Laurent after taking no notice of it suddenly tries to hit my hand as though to lower or remove it; I let him do it to me and he grasps the box. I recommence to ban his passage, but using a sufficiently supple cushion to help impression the child's gestures. Laurent tries to reach the box, and bothered by the obstacle, he at once strikes it, definitely lowering it until the way is clear. (p. 217)

When Piaget had attempted similar experiments previously, Laurent had protested the presence of the obstacle but had not made an attempt to displace it. This time, however, Laurent's behavior provided for the first time a clear indication of intentionality. We see a clear separation between means (removing the obstacle) and ends (gaining the match box).

This feat clearly indicates that at this age for the first time, infants are able to sequence schemes (that is, separate means from ends) to obtain some desired goal. From this simple sequence of two behavior patterns, more complex problem-solving sequences will eventually evolve.

The **means-ends separation** makes possible the emergence of new behavior patterns. The first is the concept of **object permanence.** For the first time, infants actively search for objects hidden from view such as a favorite toy or rattle that has been placed under a blanket. The search requires the means-ends separation, but even more important it indicates that for the first time infants are forming a mental representation of an object. Since the toy can't be seen under the blanket, the removal of the blanket and grasping of the toy must be guided by a mental representation of the object.

The emergence of the concept of object permanence also signals a fundamental shift in infants' relationships to their environments. They become extremely curious about their environment, spending a great deal of time searching and exploring. They now loudly protest when a parent leaves the room and become increasingly wary of unfamiliar people. Kagan (Kagan, Kearsley, & Zelazo, 1978) believes this wariness may result from infants' failure to resolve the inconsistency between an event and their limited representation of that event. When infants do resolve their uncertainty about a situation, a quite different response—laughter—results.

The wariness of strangers and the protests over the parent's leaving were not present at earlier ages because there was no discrepancy between infants' understanding of the situation and their ability to deal with it. Since their understanding was defined in terms of their ability to deal with a situation, there was a perfect correspondence between the two and therefore no source of anxiety. Now, however, with the infants' awareness that objects and people can exist independently of their actions, there is a discrepancy between the ability to understand and the ability to cope. Some of this imbalance is removed during the next phase as infants acquire increasingly more competent and varied means of acting upon their environment.

Phase IV: Approximately Twelve to Eighteen or Twenty Months

Phase IV sees the beginning of a third pattern of circular reaction: **tertiary circular reactions.** Unlike the first two kinds of circular reactions, tertiary reactions are neither conservative nor unintentional. Rather they show the infant actively and intentionally trying to discover new means and new ends. No longer is the environment viewed as a place for merely reproducing interesting past events. Now, the environment is seen as a potential source of new and interesting events. It is as if infants are now spending much of their day asking, "I wonder what would happen if." The increase in active exploration makes the infant a more accurate imitator of others and a decidedly more playful individual.

Infants' seemingly endless "experiments" further help them differentiate objects from the actions performed on them. As infants perform a variety of actions such as shaking, banging, throwing, dropping, and mouthing on the same objects, they come to realize that actions and objects are two distinct categories. That is, they come to realize that they can do the same thing to different objects and different things to the same object. The active experimentation typical of this age appears to be an attempt to try out all of the possible combinations of actions and objects that the infant can discover.

Since infants' efforts now appear intentional, they are better able to use the outcomes of their explorations to modify or eliminate particular behavior patterns. Feedback comes not only from the consequences of their actions but also from the reactions of significant people in the infants'

Between the ages of one and one and a half, infants begin to exhibit tertiary circular reactions. This infant's act of striking the triangle is not accidental; he is intentionally seeking to produce a sound. Tertiary reactions represent the infant's attempt to find new sources of interesting events.

environment to their behavior. In particular, Uzgiris (1976) notes that even though infants begin to play with objects by doing all sorts of things to them, gradually only the cars are pushed, the dolls and soft animals cuddled, the blocks stacked, the hats worn, and so forth.

The increase in infants' ability to use feedback to modify and maintain appropriate actions on objects coupled with the increasing motor competence of one-and-a-half-year-olds (most of whom are walking fairly well) probably explains why fear of being separated from parents and anxiety toward strangers peak early during this phase and then gradually begin to decline in intensity. The infant is simply acquiring more effective ways of coping with situations.

The infant's increasing competence is not limited to direct physical actions on objects. Infants of this age are very good imitators of simple actions performed by others. Understandably, parents derive great delight from seeing their child mimic their own behaviors.

Phase V: Approximately Twenty Months to the Third Birthday

Toward the end of the second year, infants' behavior undergoes another significant transformation. Up to this point, infants' contacts with the environment have been action oriented. The nature of these actions certainly evolved as infants gradually acquired a more mature understanding of the nature of objects and the things that could be done with them. But this knowledge was only able to be expressed in terms of direct action on objects. Now this situation changes. There now emerges a clear indication of the infant's ability to form **mental representations** of actions and objects as well as to have one thing stand for, or symbolize, another.

One-year-olds solve problems by trial and error. They try one solution, then another, and perhaps still another until they accomplish the task or lose interest in the problem. By the time infants are two years of age, however, their behavior follows a different pattern. In a typical problem-solving situation, an infant has to retrieve a favored toy placed beyond arm's reach of the playpen and is given a prop such as a stick. Rather than immediately trying some means of retrieving the toy, the infant now first pauses and then takes the stick, places it through the slats of the playpen, and using the stick, draws the toy to within arm's reach. The toy is then picked up and in all probability, an especially satisfied smile appears all over the infant's face. What is so significant about the infant's insight into the solution of this problem is that it is mental. The infant is able to form a mental representation of the elements involved in the problem, act upon them in such a way that one image (the distant toy) can be placed in relationship to another (the stick), and then finally, transfer these representations into the appropriate solution. These initial mental actions are limited to those events that the infant has had direct contact with (it is unlikely that infants will solve the problem if they have never used a stick before), but they nevertheless mark a great advance over previous means of dealing with the environment.

In the insight problem, the infant's mental representations are copies of real objects and actions. But the advances of this phase are not limited to the ability to mentally copy an action or object. Infants are also able to have one thing stand for, or symbolize, another. Kagan (1978) provides an excellent example. It involves a twenty-six-month-old child who is playing with two small dolls, a small doll bed, and a large doll bed.

> After she placed one of the small dolls in the small bed, she scanned the rest of the toys, noting aloud that she needed another bed. She looked directly at the large bed, touched it, but did not pick it up. After almost two minutes of quiet and careful study of all the available toys she finally selected a small wooden sink, about four inches long, placed the second small doll in it and placed this arrangement next to the other small bed which already had a doll. Apparently satisfied, she declared "Now mummy and daddy are sleeping." (p. 119)

The significance of this episode is that the child did not put the second doll to sleep in a sink but rather in a bed that was the same size as the first and was represented in the substance of a sink. The distinction is not one of mere semantics. The child's need to have a second bed the same size as the first coupled with her inability to find such an object forced her to create one, that is, to pretend that the sink was a bed. In the span of less than three years, infants have gone from dealing with the environment as it presents itself to them to now, for the first time, being able to transform the environment to meet their particular need. These first mental representations and symbolic transformations are very elementary, but they nevertheless elevate the infant from what Piaget called the plane of action to the plane of representations. Infants are no longer restricted to direct action on objects; they are now also able to act upon representations and transformations of those objects.

The new competencies that emerge during phase V are reflected in three types of activity that are very characteristic of infants and toddlers: **deferred imitation, pretend play,** and language. Imitation and play are complementary ways in which infants and children learn about their environments and how to influence them. Imitation corresponds to the accommodative side of the interaction process: the child attempts to make his or her behavior comply with another person's. Play represents the opposite or assimilative aspect of the interaction process. The child's actions are not an attempt to comply with another person's but rather to make the environment comply (as in the case of the sink becoming a bed) with some internal representation or transformation.

The emergence of deferred imitation during phase V reflects the infant's ability to create a mental representation of some action, person, or object, store it in memory for some period of time, and then retrieve it from memory and reproduce it. The pretend or make-believe aspect of play reflects the infant's ability to mentally transform actions or objects, to have one thing stand for, or represent, another. Pretend play occupies a large part of the toddler's day and seems to be an essential element in learning how to effectively deal with the objects, people, and events in the environment.

The Development of Language

The development of language skills is a special event because language is the primary means that we use to learn about our environment, communicate with others, and reflect on our own thoughts and actions.

During phase V development, play becomes more make believe. Possibly this young girl is taking her children for a ride in the family car.

Language development during the first three years passes through three relatively distinct periods. During the first, which covers the first year of life, a number of precursors of language develop. During the second, which covers the second year, children begin to use single words, and during the third, which covers the third year, they combine single words into sentences.

Infants communicate from birth. They cry. But the cry is merely a response to some internal state. As such, it implies no representation of that state or intention to communicate about the state. Although the cry continues to be a potent form of communication during the first year and after, it is not the only form. Infants make eye contact; they smile, point, gesture, and imitate. All of these actions are forms of communication that become intentional as the infant's cognitive development proceeds through the sensorimotor substages.

Cognitive development during the first year concerns the discovery of the regularities in the environment. Out of this search for regularity come object permanence, the separation of means and ends, and the ability to form mental representations of objects, relations between objects, and relations between objects and actions. These skills are the precursors of early language development. The use of an object label, for example, requires a memory of that object. Memory of an object presupposes object permanence. In other words, language development is not distinct from cognitive development but is a reflection of the same basic processes.

The close correspondence between cognitive development and language development is evident in the types of words that make up the infant's initial vocabulary. These words are listed in Table 3-4 (Nelson, 1979). The words that appear most commonly in early vocabularies are either the names of objects that the infant uses or acts upon (such as juice, cookie, ball, key, bottle) or the names of objects that are themselves active (such as dog, cat). Equally common items such as a shirt or dress are low on the list. Perhaps the most common item in the infant's wardrobe, diapers, doesn't even appear. In other words, the determining feature of early vocabulary is not frequency of contact but rather the nature of the contact. Placing a label on an object is another way in which infants attempt to act upon and in so doing understand their environments. Although they use words individually, they may pair them with gestures, for example by raising their arms when saying "up." Parents come to understand their infant's meaning by seeing the word paired with a gesture and by understanding the context in which the word is spoken. The nature of the parents' response to these early uses of language as well as to all other forms of interaction becomes a major determinant of the rate, direction, and adequacy of further development.

TABLE 3-4 Words Acquired in the First 50-Word Vocabularies by Semantic Categories

CATEGORY AND WORD[a]	FREQUENCY[b]
Food and drink:	
Juice	12
Milk	10
Cookie	10
Water	8
Toast	7
Apple	5
Cake	5
Banana	3
Drink	3
Bread	2
Butter	2
Cheese	2
Egg	2
Pea(s)	2
(Lolli) pop	2
Candy	1
Clackers	1
Coffee	1
Cracker	1
Food	1
Gum	1
Meat	1
Melon	1
Noodles	1
Nut	1
Peach	1
Pickle	1
Pizza	1
Soda	1
Spaghetti	1
Total	90
Animals:	
Dog (variants)	16
Cat (variants)	14
Duck	8
Horse	5
Bear	4
Bird	4
Cow (variants)	4
Bee	1
Bug	1
Donkey	1
Frog	1
Goose	1
Monkey	1
Moose	1
Pig	1
Puppy	1
Tiger	1

CATEGORY AND WORD[a]	FREQUENCY[b]
Turkey	1
Turtle	1
Total	67
Clothes:	
Shoes	11
Hat	5
Socks	4
Boots	2
Belt	2
Coat	2
Tights	1
Slippers	1
Shirt	1
Dress	1
Bib	1
Total	31
Toys and play equipment:	
Ball	13
Blocks	7
Doll	4
Teddy bear	2
Bike	2
Walker	1
Swing	1
Total	30
Vehicles:	
Car	13
Boat	6
Truck	6
Bus	2
Plane	1
Choo choo	1
Total	29
Furniture and household items:	
Clock	7
Light	6
Blanket	4
Chair	3
Door	3
Bed	1
Crib	1
Pillow	1
Telephone	1
Washing machine	1
Drawer	1
Total	29
Personal items:	
Key	6
Book	5
Watch	3
Tissue	1
Chalk	1
Pen	1
Paper	1
Scissors	1
Pocketbook	1
Money	1
Total	21
Eating and drinking utensils:	
Bottle	8
Cup	4
Spoon	2
Glass	1
Knife	1
Fork	1
Dish	1
Tray	1
Total	19
Outdoor objects:	
Snow	4
Flower	2
House	2
Moon	2
Rock	2
Flag	1
Tree	1
Map	1
Total	15
Places:	
Pool	3
Beach	1
School	1
Porch	1
Total	6

[a] Adult form of word used. Many words had several variant forms, in particular the animal words.
[b] Number of children in the sample who used the word in the 50-word acquisition sequence.
Note: Reprinted, by permission of the publisher, from K. Nelson, "Structure and Strategy in Learning to Talk," *Monographs for the Society of Research in Child Development, 38* (1973), Serial No. 149, No. 1–2, 32–33.

By the second birthday, most infants, now toddlers, have a vocabulary of about fifty words. During the third year, vocabulary growth continues at an extremely rapid rate. By age three, the toddler has a vocabulary of one thousand words. Even more important than this incredible increase in vocabulary is the fact that toddlers have begun to put words together to form sentences (Bloom & Lahey, 1978).

The first word combinations are two-word sentences. Because these early sentences are so curt, they are usually referred to as **telegraphic speech.** They are not grammatically correct, and they often use one word to modify another. Examples of these early two-word sentences are "more cookie," "ride bike," "dog go," and "Mommy read."

Toward the end of the third year, the toddler's sentences become longer as other words become embedded in the action-object relation. "Get big ball" is a sentence of this kind. Finally, words are combined in grammatically correct sentences. For the first time, sentences have subject-predicate agreement as well as agent-action-object relations. At this time, a child might say "John threw the ball." The evolution of these early sentences during the third year corresponds with the more general effort by toddlers to represent the experiences in their environment, to categorize the elements of this experience, and to better understand their relationship to those experiences.

Although toddlers make great strides in the use of language, their language still has an idiosyncratic quality. Toddlers may use or even invent words that only have meaning to them. Further, one need only talk to two- or three-year-olds to realize that they may use the same words adults use but do not always attribute the same meaning to those words. Toddlers frequently overgeneralize the meaning of words. They may, for example, refer to all four-legged animals as "doggie." On the other hand, they may unnecessarily restrict the use of a word. Piaget provided an example of this latter case in an excerpt of a dialogue between his wife and his daughter Jacqueline (age two years, seven months):

> . . . seeing Lucienne (a younger sister) in a new bathing suit, with a cap, Jacqueline asked: "What's the baby's name?" Her mother explained that it was a bathing costume, but Jacqueline pointed to Lucienne herself and said: "But what's the name of that?" (indicating Lucienne's face) and repeated the question several times. But as soon as Lucienne had her dress on again, Jacqueline exclaimed very seriously: "It's Lucienne again," as if her sister had changed her identity in changing her clothes. (Piaget, 1952, p. 224)

The type of trial-and-error learning that occurs at the level of sensorimotor activity also seems to occur at the level of representation. In both cases, toddlers are trying to make an accommodation between themselves and their environment.

Social Development

Infants use the same developing cognitive skills to learn about their social world that they use to learn about their physical world. But the understanding they acquire of the social world is distinct from their understanding of the physical world. In the first place, people behave differently from physical objects. They are more variable, and they can be more responsive.

Second, the relationships infants develop with people are different from their relationships with physical objects. Infants develop a unique bond or attachment to the few special people in their lives, usually their parents. Mobiles may be fun to look at and blocks may be fun to bang, but as any babysitter will tell you, there is no substitute for a parent when a young infant is upset and needs to be comforted. Finally, during the first three years of life, infants gradually become aware that they too are members of a category they have been learning to deal with—the category of people. They come first to understand that they are like other people in size, gender, and familiarity, and in the process, they develop an initial understanding of themselves as distinct individuals.

This socialization process is cumulative, and it is interactive. Each participant influences and, at the same time, is influenced by every other participant. It is a lifelong process. The understanding of self and other never stops evolving.

Recall the discussion of Erikson's **psychosocial stages of trust and autonomy** at the beginning of this chapter. Erikson sees the primary task of early socialization as the development of a sense of mutuality between infant and parent. This sense of mutuality helps young infants detect the stable and predictable elements of their environment, and, as a result, develop a sense of trust. A sense of trust in self and others fosters the infant's further social encounters, which then allow the toddler to establish an initial sense of autonomy.

What do the parent and the newborn bring to their first encounter? Neither brings direct experience with the other, but they don't come empty-handed. The parent brings a set of expectations about infants and about the role of parent. These expectations reflect the culture's teaching about the roles of parents and children as well as the parent's direct experience with children and with others who have children. Certainly the first-time parent brings a set of expectations that are different from the fourth-time parent's. Finally, the parents bring the experience of their own childhood (Rutter, 1979).

The newborn comes with a relatively well-developed sensory system that shows definite sensory preferences. In particular, Schaffer (1977) notes that if we were to design an object containing all of the features that an infant's perceptual system finds maximally attention-worthy, we would end up with the human face. In addition, the physical appearance of the infant elicits very unique behavior from almost all adults but most especially from the parent.

> There is something about the appearance of the very young infant that is intrinsically appealing. The head is disproportionately large, by adult standards, and contains particular characteristic features: a protruding, large forehead; large eyes set below the midline of the head; and round protruding cheeks. In addition, the baby has thick, short extremities, a rounded body, and soft, elastic body surfaces. (Emde, 1980, p.89)

Most adults find that infants' behavior is particularly captivating and compelling. This is especially true of the infant's gaze and cry. Neither are intentional. They are simply responses to some form of internal or external source of stimulation. Parents, however, attribute meaning to these cues. They serve as one of the early means by which parents judge their influence on the young infant.

Development of Parent-Child Attachment

Most of the parent-infant interactions during the first few weeks are responses to the biological needs and the biological state of the newborn. Since newborns spend approximately two-thirds of their day sleeping, much of their waking time is spent on feeding, bathing, and diapering tasks. Within the context of these caretaking tasks, social interactions begin that eventually result in **attachment.** The parent and newborn begin to develop a kind of synchrony (Brazelton, 1980). Each seems to become increasingly finely tuned to the actions of the other. This synchrony forms the base for the infant's perceptions of the environment as stable and predictable. It is the earliest form of relationship.

During these first two months, a major determinant of the infant's perception of the environment as predictable is the responsiveness of the parent to the infant. Although infants' early behaviors are nonintentional, parents often interpret them as meaningful and respond accordingly. Through the use of gesture, voice inflection, physical contact, gaze, and facial expression, parents provide infants the means to understand the contingent relation between their behavior and that of another person. Gradually the infant comes to expect predictable responses from the parent and eventually assistance in carrying out actions that the infant can't do alone.

When infants are two to three months of age, a significant shift occurs in their interactions with their environments. Increasingly, their behavior seems to be regulated less by internal states such as hunger or fatigue and more by events in the environment. Behavior toward social objects becomes different from behavior toward nonsocial objects. Smiling and cooing are reserved for people and gradually over the next few months only for those special people who serve as the infants' primary caretakers. By six months of age, infants are not only quite able to distinguish their parents from other adults but are also increasingly selective in the manner in which they deal with other adults. Other adults become less able to comfort infants or to elicit a smile. Infants begin to show distress when separated from their parents and joy when reunited with them. This distress is termed **separation protest.** The parent has become the special person to go to when the infant is tired or upset. The parent has come to be a source of emotional support for the infant. In other words, the infant is demonstrating a definite attachment to the parent.

The attachment process continues during the second half of the first year. As infants learn to creep and crawl, they show increasing efforts to remain close to their parents. Infants' ability to move toward or away from a parent gives them a larger measure of control over the parent-infant relation. Previously, they could signal a parent with a cry but then had to wait for the parent to respond. With locomotion, the relation becomes somewhat more balanced. Now parent and infant can each signal the other and actively seek the other.

Another major addition to the attachment process is the infant's clear demonstration of **separation anxiety.** Separation anxiety occurs when a parent leaves an infant with an unfamiliar person or in an unfamiliar setting. The infant protests the separation by trying to cling to the parent, reaching out for the parent, and crying. In the absence of the parent, the infant cries and is unable to be consoled by other adults. Separation anxiety develops at about the same time as object permanence, and both reflect the

As every babysitter knows, separation protest is one of the more distressing manifestations of the parent-child bond.

infant's growing ability to retain an image of an object or person in the absence of that object or person. The distress is caused by the infant's inability to resolve the discrepancy between the image of the parent and the parent's absence.

Signs of separation anxiety increase until about fifteen to eighteen months of age and then begin to wane (Kagan, Kearsley, & Zelazo, 1978). The waning reflects an increase in both the instrumental and representational competence of the infant. Kagan and his colleagues suggest that the anxiety and fear produced by the inability to cope instrumentally with unfamiliar situations or to generate adequate representations to mentally resolve discrepancies are typical not only of the infant but of all individuals. That is, our ability to realize something is amiss appears to develop before our instrumental competence to deal with or resolve the discrepancy. Awareness of discrepancy thus becomes a significant factor in the development of instrumental competence.

The development of language and representational competence during the second year decreases the importance of physical contact and close proximity for emotional support. Infants' increasing fascination with the variability of their environments (coupled with the ability to feel secure at a distance) serves as the base for continued cognitive growth and emotional independence. Infants acquire a greater understanding of themselves as a source of independent action and of the parent as an independent adult. A further shift in the parent-infant relation results from these new awarenesses. The relation increasingly becomes a reciprocal meshing of two individuals, such that each partner takes account of the other's needs, actions, and plans (Maccoby, 1980).

Origins of the Self-Concept

The social interactions that provide infants the context in which to develop attachments to the primary care givers also provide the means for infants to develop an initial understanding of themselves as unique individuals, as unique sources of action, and as members of social groups that also include the other people in their lives.

Infants are able to recognize their own image as early as at nine months of age (Lewis & Brooks-Gunn, 1979). They are quite captivated by their image in a mirror. They will smile and vocalize and often point to themselves while turning to a parent. It seems as if they are telling the parent "look at me." This early recognition seems dependent on infants realizing that their image is doing exactly the same thing they are doing.

By eighteen months of age, self-recognition is no longer dependent on infants being able to make a one-to-one correspondence between their behavior and that of their image. When infants are now shown a videotape of themselves taken earlier in the day, they appear more interested in their own image, they smile more, and they respond more than when the image is of another infant of the same age and sex.

The increasing use of language provides the infant a second means of self-definition. Beginning around age two, toddlers first use the personal pronoun "I" correctly and then the pronoun "you." The increasing use of words to define possession (such as "my" and "mine") suggests that toddlers are becoming aware of themselves as "possessors of things."

Infants' sense of self is also reflected in their increasing awareness of themselves as a source of action. A first step occured when infants became aware of the contingency of events. This initial understanding evolved into the ability to distinguish means-ends relations and the awareness of intentionality. But it is only now toward the end of the second year, when infants first systematically modify their actions to see the corresponding consequences, that they really grasp the notion of themselves as independent sources of actions.

At about the age of one year, then, children begin to understand that they have an impact on their environment by responding and

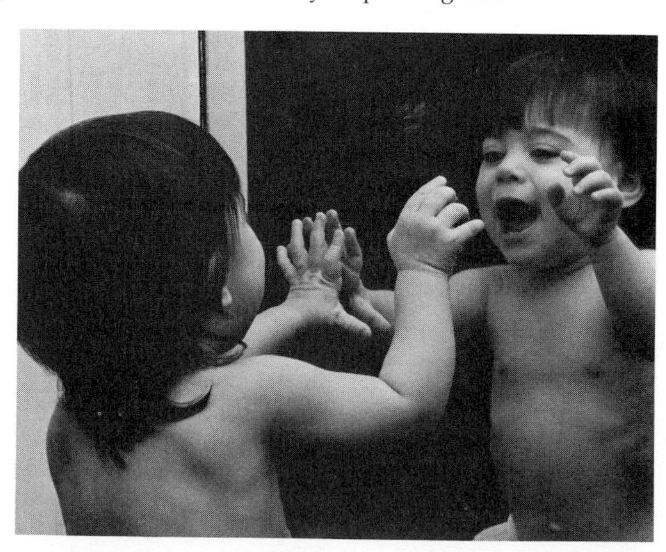

Infants are able to recognize themselves as early as nine months of age. This young lady seems very pleased with the image she sees in the mirror.

> not responding. They begin trying out their newfound powers on their caretakers. After a certain amount of time passes, children discover that withholding compliance may also provide them with a certain amount of power over their caretakers. Once this discovery has been made, the famous negative period is ushered in. (Maccoby, 1980, p. 284)

In developing social relationships with others and in developing an initial representation of themselves as unique social beings, infants begin to develop an understanding of themselves in relationship to others. This early understanding is evident in the increasingly instrumental, or deliberate, way infants deal with others, especially their care givers. These deliberate attempts by infants to influence the behavior of others are evident as early as three months of age. But they remain limited to specific interactions until infants' representational and symbolic skills are mature enough to make possible the beginnings of a generalized self/other conceptualization. This happens when infants are about two years old. For example, infants are much more likely to show distress when approached by an adult than when approached by another child. They seem to understand that they and the other child are in some way alike (most probably on the basis of size) and therefore that the child is more predictable in behavior. Since predictability is a major determinant of infants' reactions to new situations, it is not surprising that the approach of another child produces a positive affect while the approach of an adult causes them to look away (Lewis & Brooks-Gunn, 1979).

The impact of the identification process is also evident in the way toddlers deal with objects and situations. By age two, toddlers have begun to internalize adult norms and standards. That is, they are beginning to realize that their behavior should conform to the same criteria other people use. According to Kagan (1981), these internalized norms are evident when children "point to small holes in clothing, tiny spots on furniture, a doll with chipped paint, a torn cord, missing bristles on a broom, a missing button on a dress, or an almost invisible crack in a plastic toy, and utter a negative 'Oh, Oh'" (p. 48).

The normative sequence of cognitive and social developments that has been described doesn't simply happen, and it doesn't necessarily happen the same way in all infants. These developments reflect not only the continuing maturation of infants' biological systems, but the nature of their social and physical environments. They reflect the types of parenting styles encountered and the influence of other children and adults. These microsystem factors, in turn, reflect the influence of factors that exist at the meso-, exo-, and macrosystems level. The next section examines the impact of these contexts of development during the first three years of life.

THE CONTEXT OF DEVELOPMENT DURING THE FIRST THREE YEARS

From the moment that infants enter the world, they are immersed in a social and physical ecosystem that influences and regulates the course of their development. The system exists at many levels, with only the most immediate microsystem being apparent to the infant. Nevertheless, the fit between the different systems of the infant's world—especially the degree of continuity between the home, and when present, the child-care center—is a significant influence on the infant's behavior. Equally important are the settings that influence the parents' behavior (the infant's exosystem). Job satisfaction, for example, is a major determinant of marital happiness which in turn is a significant influence on parent effectiveness. And to one degree or another, we are all influenced by the broad social and cultural values that exist at the level of the macrosystem.

The various elements that exist in this social matrix are in continual interaction with each other. Therefore, the influence of each element must be considered in light of the other elements in the ecosystem. This perspective is reflected in Parke's observations on the role of the spouse in parent-child interactions (Parke, Power, Tinsley, & Hymel, 1980). Parke and his colleagues found that parents smiled more at their infant when the other spouse was present. They believe that this occurs because one spouse serves to reinforce and expand the infant contacts of the other spouse. By talking to each other about their infant, parents help each other become more aware of, and perhaps even more interested in, the infant's appearance and behavior.

Infant Influences

As already mentioned, evolution has made it likely that the parent and infant will become a pair. The infant seems particularly intrigued by the human face. The parent seems captivated by the infant's gaze. It is usually very difficult not to feel an emotional response to the young infant. How the interaction progresses depends on how well the parent can read the infant's physiological states. This reading, in turn, depends on the regularity of these states.

Newborn activity ranges from deep sleep to a vigorous cry. In between, it is possible to differentiate a restless sleep, a relatively passive awake state, and very active awake state (Schaffer, 1977). Levels of arousal determine how infants respond to stimulation. Sounds that produce a smile during irregular sleep or drowsiness evoke a startled response during regular sleep. A pat-a-cake game that elicits smiling when infants are content provokes crying when they are active or fussy.

This reading of states is essential because young infants are unable to modify their behavior to conform to the state of the care giver. Any factor that limits the parent's ability either to "read" the infant correctly or to respond accordingly adversely affects the interaction process and as a result the infant's development (Goldberg, 1977). Examples of such factors might include a handicapping condition that leaves the infant unresponsive, a difficult temperament profile that makes the infant's behavior unpredictable, a lack of knowledge about normal child development that makes the care giver unaware of the significance of the infant's behavior, and emotional stress in the parents' lives that leaves them less responsive to the infant. On the other hand, when the parent is able to successfully read and respond to the infant, the interactions become more frequent and more in step and as a result the infant's development proceeds optimally.

Readability is reflected in the relative distinctiveness of the young infant's states of arousal, the predictability or regularity of the young infant's behavior, and the responsiveness of the young infant to parental interventions. The more readable the infant, the better able the parents are to respond in a contingent and appropriate manner. The more able parents are to respond appropriately, the more likely their infants are to develop the expectation that their behavior will be effective, that is, that they are a source of action (Bell & Ainsworth, 1972).

Parent Influences

The temperament characteristics (Thomas & Chess, 1977) described earlier in this chapter are a prime example of the qualities that distinguish the readability of infant behavior. These temperament characteristics affect the parent's ability to anticipate the infant's behavior in order to provide appropriate responses and, in turn, to feel that their efforts are effective. This last item is particularly important because parents sometimes believe that their infant's behavior is more a reflection of their parenting skill than in fact it is—an assumption that can have potentially serious consequences.

Belief in the power of parenting can work in two ways. Parents who have a particularly manageable infant may give themselves more credit than is perhaps due. Just the opposite pattern is possible with the difficult infant. In this case, parents are likely to blame themselves exclusively for the intense and often unpredictable behavior of their infant. According to Thomas and Chess (1977), these parents may come to see themselves as inadequate or in some way rejecting of their infant. Although the potential consequences are worse for the second situation than for the first, they may be bad in both cases because the parents have a somewhat inaccurate perception of their role in the parent-child interaction process.

Fortunately most parents soon come to realize that the development of the parent-child relationship is a two-way proposition in which each learns to respond to the needs and characteristics of the other. The degree to which parents appreciate this fact and the degree to which they are able to translate their understanding into appropriate behavior are major determinants of the character and quality of the parent-child bond.

Brazelton's (1980) research provides an excellent demonstration of the way this interaction process evolves. The research involved filming mothers playing with and caring for their very young infants. The films were then analyzed on literally a frame-by-frame basis to determine the extent to which changes in the behavior of one influenced the behavior of the other. Figures 3-7 and 3-8 provide two distinct examples of the mother-child interaction patterns obtained from the frame-by-frame analysis. Figure 3-7 graphs the interaction of a pair in which the mother was sensitive to her infant's needs; Figure 3-8 graphs the interaction of a pair in which the mother was insensitive. Each graph shows the degree of similarity in the behavior levels of mother and child. In Figure 3-7 the mother's attempt (represented by the solid line) to draw the infant's attention (represented by the dashed line) is successful. As a result of the mother's increasing attempts to stimulate the infant, the infant's attention also rises. After a period, the infant's attention seems to wane and the mother, sensitive to this change in behavior, reduces her level of stimulation. After a time, the arousal cycle is begun again by the mother, the infant again responds accordingly, and again the synchrony between parent and infant is evident.

Figure 3-8 portrays a very different pattern. This mother seems relatively insensitive to her infant. Although this sequence starts like the other one, there is no change in the mother's behavior to accompany the change

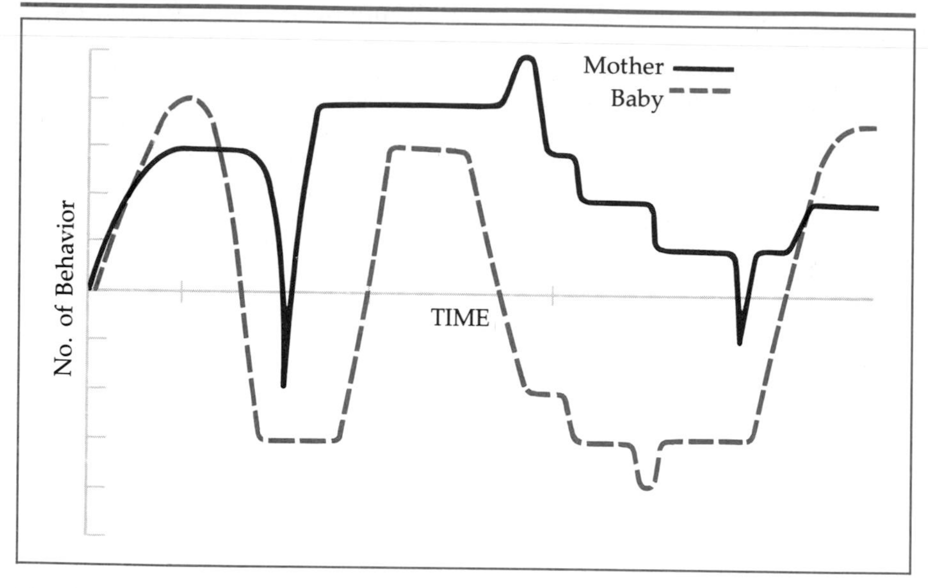

FIGURE 3-7
This mother's behavior corresponds to that of her infant. A high degree of synchrony is being maintained. Brazelton (1970) believes that such synchronous interaction facilitates good parent-child relationships (from Brazelton, 1970).

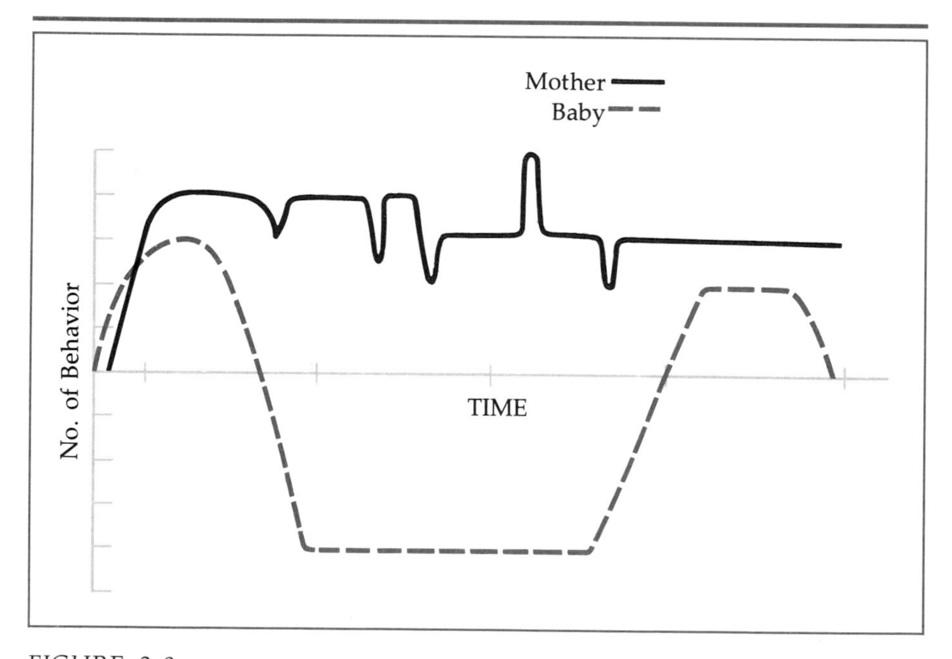

FIGURE 3-8
There is little synchrony evident in this interaction sequence. Apparently the mother does not appreciate the importance of maintaining behavioral levels consistent with her infant's for good parent-child relationships (from Brazelton, 1970).

in infant behavior. Mother and child are clearly out of step with each other. Brazelton believes that the degree of synchrony present in these early parent-infant interactions helps determine the nature and quality of the infants' affective, or emotional, development.

The importance of the parent's sensitivity to the infant is also evident in Ainsworth's research on the development of mother-infant attachment patterns (1969; 1973; Ainsworth & Bell, 1974; Ainsworth, Blehar, Waters, & Wall, 1978). Ainsworth found that the mother's perception of the parent experience was the single most significant predictor of the quality of the mother-infant relationship. The more realistic the perception was, the better the relationship. To approach parenting realistically, the mother had to have both appropriate and sufficient knowledge about how to care for the infant, a realistic understanding of the amount of time and energy involved in the care of an infant, and the awareness that the addition of a child to a family, especially a first child, results in a significant change in that family's life-style. These perceptions didn't prevent the mother from sometimes being out of step with the infant. They didn't prevent the fatigue and resentment that often accompany the arrival of a newborn. And they certainly didn't prepare the mother for all the surprises and changes that parenthood entails. Rather these perceptions fostered a willingness to accept the changes that accompany parenthood as legitimate consequences of parenthood, while at the same time they didn't stop the mother from thinking that her infant was the most wonderful child in the world—she even had the pictures to "prove" it.

Ainsworth was able to identify three relatively distinct **patterns of maternal behavior** and found that each corresponded to a distinct pattern of infant attachment behavior. The infant behaviors measured were responsiveness to the mother's attempts to comfort and willingness to explore the environment, wariness toward strangers and strange situations, and attachment to the mother.

Infants who showed the most preferred patterns in terms of these three behavioral dimensions seemed most positive in their relationships to their mothers. They were the most willing to comply with a request, demonstrated the least anxiety and crying when separated from their mother, and upon her return, actively sought her out, eager to be held and comforted. The mother's ministrations must have been effective because these infants were soon calmed and quickly returned to their explorations. Ainsworth noted that these infants were also less fearful of strangers, more outgoing and cooperative, and less easily frustrated when pursuing a difficult task. These infants seemed to know that they could count on their mothers, and that knowledge provided them with the confidence to approach new situations and persist in difficult tasks. Given their mothers' behavior, it becomes evident why these infants had such confidence.

The mothers of these infants were sensitive to their infants' needs and thus could be labeled **accepting parents.** If their infants were crying, these mothers were most likely to pick the children up and to do so in a tender, affectionate, loving manner. They may have viewed the task as a responsibility, but they also seemed to see it as an opportunity. The tempo of their behavior changed in response to their infants. Their movements were slower and more rhythmic. Their speech and action patterns were exaggerated—a quality infants find appealing. They fed their infants when they were hungry, at a rate that the infants found comfortable, and stopped when the infants appeared satisfied. These mothers took genuine delight in the development of their infants. They didn't resent the presence of the infant in their lives. Rather they saw it as a source of pleasure and satisfaction and further growth as an adult. They may not have had these positive feelings when trying to comfort the infant in the middle of the night or when changing a particularly dirty diaper, but the positive feelings were there more often than not, and they were of sufficient intensity to offset the naturally occurring negative feelings. Equally important, these mothers seemed able to maintain a healthy balance between meeting their infants' needs and meeting their own needs as adults.

White (1971) whose studies of early parent-child relations revealed a similar pattern of sensitive parenting found that accepting parents didn't devote the bulk of their day to their infants but rather served as designers and consultants. They provided an environment suited to meet the needs of an infant. They didn't spend long periods of time with their infants but rather had frequent short interactions that were focused around a shared interest or concern. They demonstrated a good balance between valuing possessions and safety, on the one hand, and granting the infant a reasonable amount of autonomy and exploration, on the other hand. Again, these were parents who enjoyed being with their children. They didn't take their children to the zoo because that was what parents were expected to do; they took them to the zoo because it was a wonderful place to have fun with them.

The most general characteristic about good parenting, according to both White and Ainsworth, was that it was appropriate to the developmental level of the child. Good parents used communication techniques that ensured that their infants understood the intent of their messages. When speaking to their infants, they repeated phrases, exaggerated key words, spoke slowly, and frequently used elaborate gestures (Schaffer, 1977). Their choice of words, sentence length, and sentence structure changed as the infants' ability to understand and respond appropriately increased. As the infants became better able to inhibit behavior or delay gratification, the parents became less likely to respond immediately to a request. As the infants became more able to do things for themselves, the parents encouraged them to do so.

A second group of infants identified by Ainsworth showed almost the opposite attachment pattern to that shown by the first group. These infants did not seek contact with their mothers after a short separation or try to establish eye contact with them when a stranger entered the room. These infants seemed detached from their mothers. Ainsworth and her colleagues found that these infants would first approach their mothers but then often back away. If contact between the infants and mothers was actually made, it was best described as passive; that is, the infants would do little to maintain the contact. They would not hold on to or assume a comfortable or "cuddly" posture against their mothers (Ainsworth, Blehar, Waters, & Wall, 1978).

Ainsworth found that this tendency to avoid the mother was not always present in these infants. Although it was clearly evident by one year of age, they were little different from other infants at three months of age. In other words, something in the nature of the parent-infant interactions resulted in this kind of behavior. The mothers of these infants could be labeled **rejecting parents.** They were especially likely to reject close physical contact. Compared with accepting mothers, they were less affectionate toward their infants, less likely to modify their behavior in response to their infants' behavior, and more often irritated about having to deal with their infants. They took longer to respond to their infants' cries and were more arbitrary in their dealings with their infants—their interventions were usually out of step with their infants' behaviors. For whatever the reasons, these mothers resented the intrusion of an infant in their lives. They acknowledged and accepted the responsibility of being a parent but perhaps did so in the same spirit that most of us feel when we file our income tax forms.

Ainsworth's third group of infants displayed a set of behaviors in between those displayed by the other two groups and, not unexpectedly, so did their mothers. These mothers could be labeled **ambivalent parents.** They were less responsive to their infant's cry than the first group but did not reject their infants as the second group did. They were not averse to holding their infants but only did so when routine caretaking required them to. The ambivalence or uncertainty of these mothers was reflected in the infants' behavior toward them. The infants showed more separation anxiety when their mothers left. They were more likely to seek close physical contact with their mothers when a stranger entered the room. However, unlike the infants in the first group who only needed to establish eye contact to regain a sense of security, the infants in the third group sought close physical contact when anxious but were less soothed by this contact.

The mothers in this group seemed insensitive to their infants' changing developmental needs. Many still held their infants when feeding them even though by one year many of the infants were holding their own bottles, were able to sit when being fed, and were able to feed themselves finger foods. The mothers seemed unable or unwilling to grant their infants the autonomy they were requesting through their behavior. When this happened, the infants actively resisted the nurturing attempts of the mothers. As a result, feeding times became increasingly unpleasant occasions for both the mothers and the children. These mothers were not rejecting or resentful of their infants as was the second group, but they did seem insecure or inept in their dealings with their infants. This insecurity interfered with their ability to be sensitive to their infants' changing needs.

The developmental consequences of these three interactions patterns are distinct but not irreversible. Although a secure attachment fosters exploration of novel situations and a willingness to persist in the face of frustration, it would be incorrect to conclude that infants with insecure attachments inevitably face a less optimistic future. Developmental experiences are cumulative but not irreversible. The influence of these early experiences is often dependent on the nature of subsequent developmental experiences (Kagan & Klein, 1973).

In most cultures, mothers are the primary caretakers of infants. But they are not the only caretakers. Fathers also establish relationships with their infants, but the research indicates that the nature of the interactions between father and child as well as the resulting bond is often quite different from that between the infant and the mother.

Fathers are just as intrigued with their newborns, especially firstborns, as mothers are. During the first few weeks, they are as likely to hold the infant, to smile, to vocalize, to touch the infant, to explore the infant's body, and to feed the infant (Parke, Power, Tinsley, & Hymel, 1980). Fathers are as likely as mothers to modify their speech when talking to their infants. Both fathers and mothers are likely to talk slowly and to use exaggerated speech.

Over the course of the first few months, however, a distinct difference evolves in the way most fathers and mothers deal with their infants. Fathers become more selective than mothers about the kinds of activities they engage in with their infants. They are much more likely than mothers are to restrict their interactions to playful, nonstress situations. Mothers usually perform more of the caretaking activities such as feeding, bathing, and diapering. Fathers also differ from mothers in the manner in which they interact with their infants. Fathers are more likely to be physical in dealing with their infants; while mothers are more likely to be verbal. Fathers are more likely to have direct contact with their infants; mothers are more likely to use a prop such as a toy or to play a game such as pat-a-cake or peek-a-boo. Finally fathers are more likely to differentiate between male and female infants in their choice of play patterns. Mothers are more likely to view their male and female infants simply as their infant children. The fact that the sex of the infant has more of an influence on the father's choice of play patterns suggests that fathers may be more instrumental in influencing early sex-role development than mothers are (Parke, Power, Tinsley, & Hymel, 1980).

The different interaction patterns displayed by mothers and fathers do not appear to influence the degree of attachment the infant demonstrates toward them (Lamb, 1978), but they do influence the value each parent comes to have for the infant. By two years of age, most infants tend to seek out the mother in stressful situations and the father when they want a playful companion. This pattern probably has more to do with the associations the infants form between the mother and stress reduction and between the father and play than with the actual sex of the parents (Clarke-Stewart, 1978). In other words, if the parent-infant interaction patterns were reversed, it would be reasonable to expect the infant to seek the father when distressed and the mother when playful.

The association of care giving with nurturance and comfort probably reflects the fact that caretaking activities such as feeding and diapering are performed when the infant is in need of them, and for this reason, the mother comes to be seen as the person to go to when in need. The father's interactions with the infant are less contingent on the state or need of the infant and in many cases may reflect the father's availability. Because of the relatively noncontingent nature of the father-infant interactions, the infant sees the father as the one to go to when he is available rather than when the infant is in need.

The special quality parents come to have for infants is also evident in comparisons between infants who are home reared and those who attend full-time day-care programs. A common concern expressed about the effects of such programs is that the reduced amount of contact between the infant and the parent will in some way adversely affect their attachment to each other. The evidence indicates otherwise. Kagan, Kearsley, and Zelazo (1978) compared longitudinally the developmental patterns of infants between three and thirty months of age. Half of the infants were in high quality group day-care settings; the other half remained with a parent at home. The researchers found virtually no difference between the two groups on measures of cognitive competence, language development, attachment, separation protest, and social interaction with peers and adults. Further, when compared on objective assessments of their infants, mothers of infants in the day-care setting were as accurate and complete in their descriptions as the mothers of home-reared infants. The nature of the relationship rather than the frequency of interaction is the primary determinant of the bond between the infant and the parent. Day-care providers, grandparents, extended family, and babysitters may all be valued and enjoyed by the infant, but they do not replace the parents as the primary people in the infant's world. The bond between the infant and the parent seems to have the highly canalized quality discussed in Chapter 2. It is intense, relatively permanent, and relatively independent of frequency of contact.

Peer and Sibling Influences

As important as adults are in the infant's world, they are not the only source of social contact. Infants also encounter peers and siblings. Peer and sibling interactions are rather limited during the first three years, but they are important in defining the pattern of early social interactions and in the infant's early development of a self-concept (Lewis & Brooks-Gunn, 1979). Peer contacts during this period are less preferred than familiar adult contacts but more preferred than contact with unfamiliar adults.

Before infants reach one year of age, there is little if any indication of peer interaction. Even in a situation where three or four infants are placed in a play area with a variety of toys (with parents close by), each infant will play as if the others were not there. By the first birthday, this pattern begins to change (Mueller & Lucas, 1975). The following account is typical of one-year-old peer interactions. It centers on a toy train that infants at a child-care center like to play with.

> One child discovers the engines, squats down, and makes the engine whistle sound repeatedly. All other children in the room immediately toddle over to the engine and either sound the whistle or manipulate the steering mechanism, an action introduced by a second child. During the 68 seconds that several children are at the train three children never look at their peers' faces and the other two children look only once or twice, and get no response to their looks. (Mueller & Lucas, 1975, p. 22)

Although there is still no direct contact, the similarity between the peers' actions in sounding the whistle and playing with the steering wheel indicates that the infants are aware of each other's actions and that they are inclined to mimic them. This is the beginning of a common play pattern of infants and preschoolers—parallel play, which is discussed more fully in the next chapter.

By the time infants are fifteen to eighteen months old, the wariness that is typical of infant-adult interactions is also evident in peer situations. A typical situation is one in which a group of eighteen-month-olds spend a great deal of time in physical contact with their parents while at the same time observing the other peers in the room. This wariness reflects the infant's gradual awareness of others as a unique source of action and at the same time a limited ability to predict what the others will do. The standoff is often ended when one brave infant makes a gesture or offer to another. Such an offer apparently resolves much of the uncertainty about the other infant and such episodes often lead to more active play (Kagan, Kearsley, & Zelazo, 1975). Wariness is not unique to infants. It is the almost universal first step taken by a child of any age when joining a new group. What seems equally universal is the ability of a welcoming gesture to substantially remove that uncertainty.

TABLE 3-5 Infant-Peer Interaction Patterns

TIME (seconds)	*BERNIE (13 months)*	*LARRY (15 months)*
	Stands about 6 feet from Larry; looks at ceiling light	Sits on Floor, mouthing toy and watching Bernie
0	TURNS HEAD TO LOOK AT LARRY AND VOCALIZES "DA"	
2½	VOCALIZES "DA," LOOKING AT LARRY	
		LAUGHS (VERY SLIGHTLY), LOOKING AT BERNIE
6	VOCALIZES "DA," LOOKING AT LARRY	
		LAUGHS, LOOKING AT BERNIE
12	VOCALIZES "DA," LOOKING AT LARRY	
		LAUGHS, LOOKING AT BERNIE
14	VOCALIZES "DA," LOOKING AT LARRY	
		LAUGHS, LOOKING AT BERNIE
17	VOCALIZES "DA," LOOKING AT LARRY	
		LAUGHS, LOOKING AT BERNIE
	Looks away at something off camera	
		LAUGHS, LOOKING AT BERNIE
23	VOCALIZES "DA" AND LOOKS AT LARRY	
		LAUGHS, LOOKING AT BERNIE
28	VOCALIZES "DA," LOOKING AT LARRY	LAUGHS, LOOKING AT BERNIE
32½	VOCALIZES "DA," LOOKING AT LARRY	Looks at adult and offers object to adult
	WAVES, LOOKING AT LARRY	
		LAUGHS AND LOOKS BACK AT BERNIE
34	VOCALIZES "DA," LOOKING AT LARRY	
		LAUGHS, LOOKING AT BERNIE
40	VOCALIZES "DA," LOOKING AT LARRY	

TIME (seconds)	*BERNIE (13 months)*	*LARRY (15 months)*
		LAUGHS, LOOKING AT BERNIE
43	VOCALIZES "DA," LOOKING AT LARRY	
		LAUGHS, LOOKING AT BERNIE
47½	VOCALIZES "DA," LOOKING AT LARRY	
		LAUGHS, LOOKING AT BERNIE
50	VOCALIZES "DA," LOOKING AT LARRY	
		LAUGHS, LOOKING AT BERNIE LAUGHS (IN FORCED WAY) 3 MORE TIMES, LOOKING AT BERNIE
54	VOCALIZES "DA," LOOKING AT LARRY	LAUGHS, LOOKING AT BERNIE
59	VOCALIZES "DA," LOOKING AT LARRY	LAUGHS, LOOKING AT BERNIE
62½	VOCALIZES "DA," LOOKING AT LARRY	
		LAUGHS, LOOKING AT BERNIE
66	VOCALIZES "DA," LOOKING AT LARRY	
		LAUGHS, LOOKING AT BERNIE
69	Turns and walks away	LAUGHS (IN FORCED WAY). LOOKING AT BERNIE, AND WATCHES HIM LEAVE ROOM

Note: Reprinted, by permission, from E. Mueller & T. Lucas, "A Developmental Analysis of Peer Interaction Among Toddlers," in M. Lewis & L. A. Rosenblum (Eds.), *Friendship and Peer Relations* (New York: Wiley, 1975), 242–243.

Peer encounters during the middle of the second year are contingent but not interactive; that is, each infant acts in response to others, but the actions themselves are quite independent. Bernie and Larry's interaction pattern (see Table 3-5) is a good example. Bernie and Larry each behave in response to the other, but their behaviors are separate and independent. Each infant simply serves as the prompt for the other. A few months later, however, Bernie's and Larry's behavior has become considerably more interactive (see Table 3-6). Now the behavior of one not only prompts the other but in addition influences the nature of the response. There is a cumulativeness to these interactions that was absent in the earlier set.

TABLE 3-6 Toddler-Peer Interaction Patterns

TIME (seconds)	*BERNIE (16 months)*	*LARRY (18 months)*
0	Puts paper to wall Releases paper Picks paper up from floor OFFERS PAPER TO LARRY	
11		RECEIVES PAPER FROM BERNIE Puts paper to wall Releases paper BACKS AWAY A FEW STEPS AND LOOKS AT BERNIE Puts paper to wall HOLDS PAPER AGAINST WALL AND LOOKS AT BERNIE Releases paper BACKS AWAY FURTHER AND LOOKS AT BERNIE
32	LOOKS AT LARRY AND PICKS UP PAPER Puts paper to wall Releases paper Picks up paper from floor OFFERS PAPER AND VOCALIZES TO LARRY	
51		RECEIVES PAPER FROM BERNIE Puts paper to wall Releases paper BACKS AWAY AND LOOKS AT BERNIE
65	POINTS AT PAPER AND LOOKS AT LARRY	
67		LOOKS AT BERNIE AND PICKS UP PAPER Puts paper to wall Releases paper BACKS AWAY AND LOOKS AT BERNIE
72	Departs	
81		LOOKS AT BERNIE AND PICKS UP PAPER Puts paper to wall Releases paper and departs, following BERNIE

Note: Reprinted, by permission, from E. Mueller & T. Lucas, "A Developmental Analysis of Peer Interaction Among Toddlers," in M. Lewis & L. A. Rosenblum (Eds.), *Friendship and Peer Relations* (New York: Wiley, 1975), 250–251.

Cumulative interaction episodes continue to grow for the remainder of the toddler period. As infants' and toddlers' cognitive skills of representation and symbolic functioning continue to develop, so does the length and complexity of their play. Peer interaction becomes more frequent, both positively and negatively.

The interaction pattern of siblings is similar to that of peers (Dunn & Kendrick, 1979). Frequency of contact is relatively low when the youngest child is under two years of age. When both siblings are under three years of age, the older is likely to imitate the actions of the younger. By two years of age, siblings show concern at the distress of their younger brother or sister. They will often try to distract their brother or sister with a toy. One should not, however, make too much of this sibling concern: as often as not the distress that the sibling is trying to alleviate was caused by him or her in the first place.

The role of early peer interactions in development is still unclear. Unlike the parent-infant bond, these early interactions are not necessary for survival, but their relevance to the development of a self-concept and their influence on subsequent peer interactions are only beginning to be studied. Since peer relationships increasingly characterize the nature of social contacts as we grow older, an understanding of their origins should help us understand their functioning during the adult years.

Cultural Influences

The development of infants is most directly determined by events that occur within their immediate environment, at the level of the microsystem. But the behavior of parents is often a reflection of events far removed from the infant. The parents' marital status is a significant factor. As discussed in Chapter 8, single parenting is a difficult endeavor. If the parents are married, the quality of their marriage is significant. So are the nature and satisfactions of their work.

Parenting behaviors are in part a reflection of broad social values and knowledge about development. Books offering advice to parents reflect these values, and these "how-to-do-it" books have become more popular as our society has become more fragmented. It is not surprising that the parents most likely to use these books are the ones least likely to have frequent contact with their parents or other close relatives (Clarke-Stewart, 1978). As our views on the importance of such topics as the relative influence of early experience, or demand feeding versus scheduled feeding, or the relative importance of nurturance and self-expression in comparison with discipline and conformity, or the role of the parent in the infant's development have changed, so too has the content of parent advice books and so too have parenting styles (Lomax, Kagan, & Rosenkrantz, 1978).

Differing views on preferred parenting strategies do not only result from different historical circumstances. Our culture is not so uniform that all individuals within it share the same value system. Rather individuals from different backgrounds reflect in their parenting styles the values of those backgrounds. Different value systems are often evident, for example, in the extent to which parents believe that they should modify their behavior to accommodate their infant and in the extent to which they believe that young infants benefit from being talked to. One good example of these parenting differences is offered by Nelson (1973). Using a longitudinal design, she observed the language usage and parent-child interaction patterns of eighteen infants between the ages of twelve and thirty months.

These observations were done both in the infants' homes and in a child-development laboratory setting. Videotapes of parent-child interactions were analyzed. Each infant was also given the Bayley Scales of Infant Development (a measure of cognitive competence) and the Peabody Picture Vocabulary Test (a measure of language competence). Nelson found that mothers differed in the extent to which they reinforced and encouraged their infant's use of language either in a social, directive fashion or in a descriptive, elaborative fashion. Examples of the first strategy are statements such as "Play with this doll," "Be nice to the little boy," "Don't throw the block," and "Stack the blocks in a tower." All of these statements are directive: they tell the infant what to do. On the other hand, consider the following examples of parent statements to an infant: "That is a very pretty picture." "What is the name of the dog?" "Which one do you want to play with?" These statements do not restrict the infant's behavior as much and are more likely to help infants become aware of their actions. Nelson's findings indicate that infants of mothers who are highly directive in their communication tend to have smaller vocabularies, a factor that Nelson believes hinders early concept formation. The directive mothers seem to be expressing the social value that the role of the parent is to teach the infant socially approved patterns; the nondirective mothers seem more intent on fostering self-expression.

Although all of Nelson's families came from middle-class backgrounds, these differences in parenting strategy (that is, the emphasis on self-direction or on social compliance) are generally seen as correlates of social class membership. In general, the more educated the parents, the higher their income is; and the more professional their occupations, the more likely they are to provide a greater variety of both verbal and nonverbal stimulation to their infant and the less likely they are to restrict their infant's behavior (Tulkin & Kagan, 1972). The issue is not one of love or concern or nurturance: there is no reason to believe these characteristics differ across social class. Rather the issue is how love and concern are reflected in the parents' behavior toward their children. In particular, people from more advantaged socioeconomic backgrounds are more likely to believe that they have a substantial degree of control over the direction of their lives than are people from less advantaged socioeconomic backgrounds (Kohn, 1977). Since life circumstances are different at different socioeconomic levels, both perceptions are usually accurate. The more advantaged the background, the more likely the individual is to have a job that encourages initiative and self-direction and that provides sufficient economic reward to finance the pursuit of individual interests and preferences. Given the difference in social orientation, it is not surprising that socioeconomically disadvantaged parents believe their role is to help their children fit into a world that seems beyond their control, while parents in the higher socioeconomic categories believe they are preparing their children to make a difference in the world.

The influence of social circumstance on parenting is also evident in our changing view of appropriate sex-roles for men and women. The increase in the demand for shared childbirth experiences, the now greater desire of fathers to assume a more direct role in their children's development, and the now more common desire of mothers to maintain an occupational as well as a parental role are all more evident in the economically advantaged segments of the population.

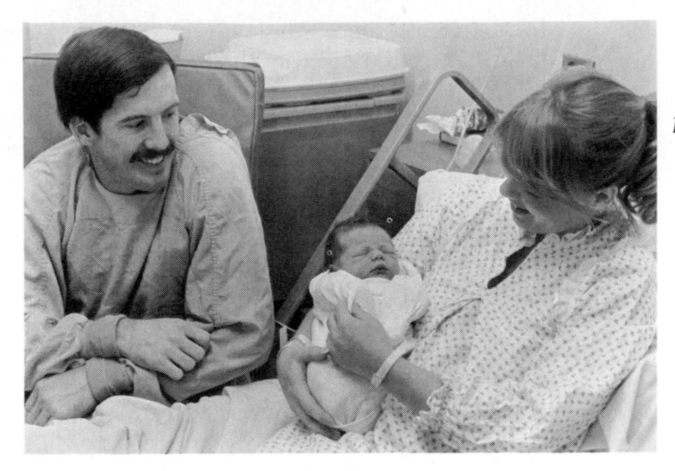

The trend toward fathers participating in the childbirth experience and both parents sharing child rearing responsibilities is much more evident in the economically advantaged segments of the population.

Not all changes in social circumstances are necessarily intentional, nor are they necessarily beneficial to infants and children. The significant increase in the divorce rate has swollen the numbers of single parent families, almost all of which are headed by the mother. Economic circumstances have caused many women who would prefer to remain home with their children to seek employment. Significantly, the effects of maternal employment on infants have much less to do with the number of hours the mother is away from home or with the nature of her work than they do with her perception of her work. The quality of the mother-infant bond is enhanced if the mother values her work and sees it as a further means of self-expression (Hoffman, 1977). It is equally important to note that this fact holds true for women who are full-time care givers. Those that value the role foster good relations with their infants; those that see it as an intrusion or a limit on their own development inadvertently hinder good development (Ainsworth, 1973). The evidence for fathers, although less substantial, shows a similar pattern (see Belsky, 1981).

Although infants are quite unaware of it, all these examples highlight the fact that the course of their development is determined by a wide variety of factors functioning throughout the various levels of the human ecosystem.

THE SIGNIFICANCE OF DEVELOPMENT DURING THE FIRST THREE YEARS

It is possible to think of the developmental events of the first three years of life in two ways. The first is simply to consider them on their own merits. From this perspective, you might learn what behaviors are typical of infants of different ages, what parenting strategies seem most effective, or what problems are typical and solutions appropriate during this time. This information is important, and for those who find themselves working with infants and their families, it is essential.

There is also, however, a second perspective from which to view these same events. From this perspective, these events are not in their own right of primary importance, but rather they acquire their importance because they are representative of or examples of certain patterns of development that are evident at various points across the life-span. It is to this second perspective that we now turn. The two perspectives are neither separate nor contradictory. Rather they are like looking at some microscopic object under two powers of magnification—the higher magnification provides a more detailed view of a limited area of the object, while the lower magnification brings into focus a wider area but provides less depth. In the case of human development, each power of magnification provides a different picture of the same event or element and in the process makes clear that a complete understanding of development across the life-span requires an examination both of individual stages and contexts and of the relationships among these various stages and contexts.

Patterns of Development

The most obvious characteristic of life-span development evident during the first three years is that infants are actively involved in influencing and directing the course of their development. They are active in that they significantly influence the behavior of others who directly or indirectly deal with them, and they are active in the sense that they construct or attribute meaning to the people, objects, and events they encounter.

This active involvement is certainly less intentional and far-sighted than that of the adult, but it is no less significant. The extent to which the infant fulfills the expectations of the parents—for example, by being a boy or a girl, an easy or difficult child—is a major determinant of the way the parents deal with the infant. The quality and consistency of the responsiveness of the infant tell the parents about their parenting, which in turn influences the nature and extent of their future contacts with the infant.

Infants influence their parents, regardless of the infants' degree of responsiveness. The unresponsive infant is as much a source of information as the very responsive one. In the special circumstances in which the infant's behavior is less a reflection of the parents and more a reflection of the state of the infant, parents need to be assured that they are competent care givers. This is especially true in the case of handicapped infants. Emde (1980), for example, has found that Down's Syndrome infants tend to be less responsive than normal infants. Their social smile is delayed; and when it does occur, it tends to be less intense and less a reflection of parental involvement. These handicapped infants are less likely to show excitement when the parents appear and are less able to maintain face-to-face contact. Parents who believe that their infant's behavior is a reflection of their parenting skills might well feel inadequate and be less likely to engage the infant. Such parents need sufficient support to maintain a high level of interaction and to correctly interpret the signals their infant is sending them.

Infants do more than merely present themselves as stimuli for others to respond to. They are actively involved in constructing an understanding of the people, objects, and events in their world. This construction process is evident in their behavior. The neonate prefers to look at some objects more than others. The three-month-old smiles at the approach of any person, but the eight-month-old smiles only at the parent. The appearance of a strange adult causes the eighteen-month-old to seek the comfort of the parent, but the approach of a same-age child only prompts a visual check of the location of the parent and then observation of the peer. The two-year-old correctly labels a number of common household objects; the three-year-old pretends that a block is a cup and proceeds to drink from it. In all of these instances, specific meaning has been attributed to an object or person, and that meaning is reflected in the infant's behavior. Through continuing interactions with the environment, infants construct an understanding of the nature of the objects, people, and events in their environment and, in the process, an understanding of themselves in relationship to that environment.

Although people remain actively involved in determining the course of their development, both in terms of acquiring information about the environment and in terms of exerting an influence on it, the manner in which they do so changes throughout the course of the life-span. The infant who finally manages to stand erect sees the world from a new perspective; the world looked very different when the infant lay prone. The infant whose walk is steady enough to free the hands and arms from efforts to achieve balance finds a new avenue for exploration. Infants who are able to find comfort in merely seeing the mother rather than in going to her are less restricted than other infants are in their continuing explorations. Infants who realize that actions and objects are separate, that an action can serve as the means of achieving some distant goal, and that objects exist independently of their direct awareness, are all discovering new means of acting upon the environment and in the process are gradually acquiring new levels of understanding about that environment. The "what" of development does not change across the life-span; it is the "how" that changes.

A second characteristic of development across the life-span that gradually becomes evident during the first three years of life is that there appears to be direction to the course of development. The changes that result from the continuing interactions of active individuals with each other as well as with the inanimate environment follow a general path. Werner (1957) characterized this development as a process in which people and events become distinct from each other and then gradually come to be understood in relationship to other people and events. He termed this process **orthogenetic development** (literally, correct development). In orthogenetic development, actions gradually become differentiated from objects. Only then does the infant begin to consider which actions create which results with which objects. Familiar people gradually are differentiated from the unfamiliar. Only then does the infant develop a unique relationship with each. The process continues to repeat itself as the infant and then the child and then the adult acquires new levels of understanding. Infants only view their parents in relation to their own wants and needs. The relationships that develop between infant and parent reflect this fact. As a child develops and comes to realize that the parent has other roles in addition to that of parent (such as, spouse, worker, citizen), this new knowledge requires the child to develop additional means of thinking about and relating to the parent. In Werner's words, orthogenetic development leads to a person's

> becoming less dominated by the immediate concrete situation; the person is less stimulus bound and less impelled by his own affective

states. A consequence of this freedom is the clearer understanding of goals, the possibility of employing substitute means and alternative ends. There is hence a greater capacity for delay and planned action. The person is better able to exercise choice and willfully rearrange a situation. In short, he can manipulate the environment rather than passively respond to the environment. (1957, p. 127)

Werner's description of development as a process that gradually frees the individual from the immediate environment is also evident in Piaget's (1970; 1975; 1980) equilibration principle. Development is a process of adaptation that involves balance, or equilibrium, within the environment. Balance is achieved through perspective: the broader the perspective the greater the balance. Broader perspective is achieved through the continuing maturation of cognitive structures in interaction with a stimulating and responsive environment. Infants' dawning interest in the external environment when they are two months old and their developing abilities to coordinate sensory and motor actions, to sequence means and ends, to act upon objects outside of their sensory awareness, to label objects, and to have one thing stand for or represent another all lead to broadened perspectives and therefore to greater equilibrium within the environment.

Erikson's (1980) emphasis on direction stresses the cumulative aspect of development. The degree and manner of resolution of each psychosocial task set the stage for encounters with subsequent psychosocial tasks. Trust in self and in the environment is a necessary prerequisite to autonomy; an initial sense of personal identity in adolescence is a necessary prerequisite to the establishment of intimate relations in young adulthood. Each step serves as the springboard for the next. And each step serves to broaden individuals' perspectives of themselves within their environment. The primary focus in the development of a sense of trust is the primary caretaker; in the development of a sense of autonomy, it is the parents; in the development of a sense of industry during middle childhood, it is the neighborhood; and in the development of a sense of ego integrity during old age, it is all humankind. Each step broadens the radius in which the individual is able to function effectively.

The Place of Early Experience

How important are the first three years of life? Do the events of the first three years have the same irreversible character as those occurring at the end of the first trimester of pregnancy? Or, could we argue that since the first three years comprise only $\frac{1}{24}$ of the average life-span, there is more than ample time to alter the course of development after age three? There are no simple answers to these questions. In fact, the more researchers learn about early development, the more divided they are as to the answer (Clarke & Clarke, 1976; Goldhaber, 1979; Kagan, 1984; Sroufe, 1977).

There are a number of reasons for this dilemma. One concerns the decision about when to test for the effect of some earlier event. For example, efforts have been made to accelerate the rate of development. The results show that children who experience such deliberate interventions show an initial advantage (that is, demonstrate the skill sooner than other children) but the relative advantage fades over time (Kamii & Derman, 1971). Studies concerned with adverse conditions during early develop-

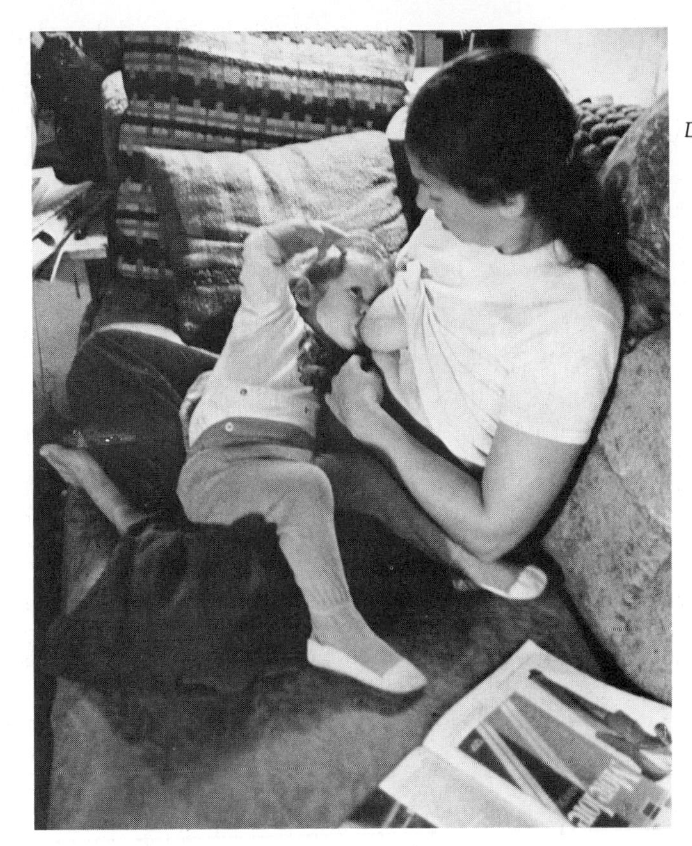

What effects do early experiences have on later development? Will this two-year-old be more secure because she has been breast fed for so long, or will she have a lessened sense of autonomy because of her prolonged direct dependence on her mother for nurturance? Arguments could be made for both of these viewpoints.

ment find a similar pattern. Infants experiencing moderate degrees of anoxia at birth score lower on tests of developmental level during the first three years, but by the time they reach age seven, the difference is virtually gone (Corah, Anthony, Painter, Stern, & Thurston, 1965). Children who are disadvantaged during the first year or two but who are subsequently placed in a good adoptive home show little evidence of their initial trauma (Kadushin, 1967). The issue is not one of simply determining if early experiences have a lasting impact. Rather the issue is to identify those factors that maintained the effect at one point and not at another.

A second reason reflects the multifaceted nature of human development. Developmental events have many consequences. These consequences are expressed at different times and in different modes. The effects of early parent-child relations influence ease of separation at age three, peer relations at age twelve, choice of a spouse in young adulthood, and parenting skills in adulthood. To make matters more complicated, these four kinds of behavior are also influenced by other experiences. The full ability to distinguish between multiple determinants and to link causal chains is still beyond the limits of our research capacities and theoretical formulations.

Third, the dilemma is complicated by the fact that development is more a relative than an absolute process. The significance of many events is not evident at the time they occur. Instead, their significance is defined in relation to subsequent events. Is it better to be a boy or a girl, to attend a day-care center or be cared for at home, to be the first born or the last born, to be an easygoing or an energetic infant? The answer given is usually "it depends." It depends on the resources and desires of the parents, the infant's likely future, broad economic and social factors, and a host of other things. It is in this sense that development is portrayed as a relative process.

It is probably best to think of development during these first three years as a necessary but not sufficient condition to guarantee full development. It is the starting point of postnatal life, and every effort should be made to be sure the start is the best one possible. If development starts on the wrong track, however, it is equally important to remember that the cognitive and social developments during the first three years of life are more responsive to change than one would think possible (Kagan & Klein, 1973).

HUMAN DEVELOPMENT AND HUMAN SERVICES

Many events happen during the first three years of life. Most are significant, but the pivotal event is the development of the parent-infant bond (Rutter, 1979). Such social changes as the increasing divorce rate, the number of mothers in the labor force, the growth of infant- and child-care centers, the decline in the birth rate and the isolation of the nuclear family (to name a few) are usually discussed in terms of their impact on the parent-infant relationship (Huntington, 1979; Kagan, 1984; Keniston, 1977; Lamb, Chase-Lansdale, & Owen, 1979; Scarr, 1984). Although you only need to visit a supermarket or a park or any public place to see that most infants have strong ties to their parents and that most parents are usually responsive to their infants, the concern of researchers and practitioners is that broad social circumstances often beyond the direct control or understanding of most parents are making the task increasingly difficult for many and even impossible for some.

Under most circumstances and for most individuals, the attachment between parent and infant develops. It is what has been termed previously a highly canalized event. There appear to be a number of elements (both in terms of the infant and in terms of the parent) that ensure its occurrence. However, when these typical avenues are unavailable, others must be sought. In some instances, these interventions are deliberate and systematic; in others, they may be little more than encouraging and supporting a parent's holding of a premature or physically handicapped infant (Emde, 1980).

The research of Fraiberg (1975) with blind infants provides a wonderful example of the way practitioners can help the development of attachments when normal channels are blocked. For most infants, vision is the primary avenue of attachment. Differential smiling, discrimination of mother and stranger, and separation and reunion behaviors unite the affective experience of the mother with sensory pictures, and the picture itself is the synthesizer of all sense experience (Fraiberg, 1975). But none of this is possible for the blind infant. The blind infants that Fraiberg studied demonstrated the same developmental sequence that sighted infants demonstrate but did so at a slightly slower rate and expressed the developments in different ways. For the sighted infant, the primary eliciter of the smile is the human face; for the blind infants, it was the sound of a familiar voice. Sighted infants discriminate between people on the basis of appearance; the blind infants discriminated on the basis of touch. Sighted infants protest separation when their caretaker is no longer in view; the blind infants protested when the caretaker could no longer be found through search of familiar places or was no longer responsive to a call. In circumstances such as these, when the typical infant's signals were not present, Fraiberg and her colleagues were able to help parents understand that it was still possible to establish the parent-infant bond, but that different avenues (such as making greater use of auditory and tactile cues) needed to be taken.

The use of infant development scales such as the Brazelton is another way practitioners help parents appreciate the development of their infants. Although these instruments are typically used to assess infant development, they are also useful in helping parents appreciate the diversity and competence of their infant's behavior. This is especially true when the parent is unable to understand the significance of the infant's behaviors and as a result is little motivated to provide the necessary and appropriate stimulation.

To the uninformed eye, the infant may behave in an apparently meaningless way and not seem to need any attention other than basic physical care. Since infants can't talk or even appear to understand words, parents may not only feel little need to talk to them but may even believe that if others saw them talking to an infant, their behavior would be considered silly or childish. Helping parents realize that infants do respond to a human voice, especially a familiar one, and that language development in infants is partially dependent on the vocalization of others can help parents better appreciate their infants. Parents who understand the importance of their efforts for the infant's development are more likely to take those efforts seriously.

The philosophy of helping parents appreciate the significance of their infant's behavior and at the same time providing parents more appropriate means of dealing with infants is evident in a variety of early intervention programs (Beller, 1979). These federally supported parent-child centers are located throughout the country (Andrews et al., 1982). They emphasize parenting skills such as showing affection, praising the child's accomplishments, avoiding overcritical judgments, using elaborate language, providing explanations and information, asking questions, encouraging the infant to talk, and being sensitive to and accepting the infant's needs. Since a major goal of these intervention programs is to maintain the improved parent-infant patterns even after the intervention program has ended, many of the programs help the parent understand how normal day-to-day contacts between parent and infant can be used to enhance development. Further, since a major determinant of parenting interest and ability is the parent's self-concept, these intervention programs are also designed to help parents come to terms with their own lives. In particular, they not only provide parents with training in good child development practices and principles but also in coping with such common adult crises as divorce, unemployment, remedial education, and personal finance (Huntington, 1979).

Direct intervention programs designed to enhance the quality of parent-infant relationships are based on the fact that there is some specific circumstance or set of circumstances that hinders the development of a normal parent-infant relationship. The problem may be some handicapping condition of the infant or limited parent education or inadequate adult coping skills. The programs are designed to provide new, more appropriate parenting skills and to help parents understand the behavior of infants and how to best foster their development. In all of these instances, the intervention is focused on the immediate environment, at the level of the microsystem.

There are, however, a number of circumstances beyond the immediate environment that interfere with what would be, under more desirable circumstances, typical, healthy infant development and parent-infant relations. Interventions at these broader levels are not necessarily provided only by practitioners. Indirect interventions focus on the factors influencing the family's ability to provide for the infant's basic psychological and physical needs (Garbarino, 1982). They might involve attempts to provide better communication between parents and child-care providers. They might involve increasing the availability and quality of child-care centers. They might involve forming parent support groups for single parents. They might involve urging directors of large companies to coordinate their needs to transfer workers with family needs. They might involve getting legislators to consider the potential effects of new laws and regulations on the family.

All of these indirect efforts reflect the fact that although parents bear the primary responsibility for their children, they do not bear sole responsibility. Our society has become more interconnected and interdependent and one of the consequences is a greater need to consider the impact of social change on the family, especially the family with young children. Although the focus of this text is development at the level of the microsystem, it is important to recognize that what happens within the immediate environment is in part a reflection of what happens beyond its border. A statement from the Carnegie Council on Children "report card" on the health of the family highlights this perspective and also serves as a transition from the world of the infant to the "magic kingdom" of the preschooler:

> There is nothing to be gained by blaming ourselves and other individuals for family changes. We need to look instead to the broader economic and social forces that shape the experiences of children and parents. Parents are not abdicating—they are being dethroned, by forces they cannot influence, much less control. Beyond today's uncertainty among parents lies a trend of several centuries toward the transformation and redefinition of family life. We see no possibility—or desirability—of reversing this trend and turning the clock back to the "good old days" for the price then was high in terms of poverty and drudgery, of no education in today's sense at all, and of community interference in what we today consider private life.
>
> At the same time, however, most American parents are competing on unequal terms with institutions on which they must depend or have taken over their traditional functions. To be effective coordinators of the people and forces that are shaping their children, parents must have a voice in how they proceed, and a wide choice so that they do not have to rely on people or programs they do not respect. Parents, who are secure, supported, valued and in control of their lives are more effective parents than those who feel unsure and

who are not in control. Parents still have the primary responsibility for raising children, but they must have the power to do so in ways consistent with their children's needs and their own values.

If parents are to function in this role with confidence we must address ourselves less to criticism and reform of parents themselves than to criticism and reform of the institutions that sap their self-esteem and power. Recognizing that family self-sufficiency is a false myth, we also need to recognize that all today's families need help in raising children. The problem is not so much to reeducate parents but to make available the help they need and to give them enough power so that they can be effective advocates with and coordinators of the other forces that are bringing up their children.

Note: Reprinted, by permission of The Carnegie Corporation, from Kenneth Keniston & The Carnegie Council on Children, *All Our Children: The American Family Under Pressure* (New York: Harcourt Brace Jovanovich, 1977), 22–23.

SUMMARY

THE COURSE OF DEVELOPMENT

1. The rapid gains in height and weight over the first three years change not only the size of infants but also their body proportions. These changes in body proportions, as well as the continuing development of the musculature, account for most of the improvement in motor skills during this stage.
2. Infants differ greatly in terms of temperament characteristics, that is, in their tempo, rhythmicity, adaptability, and mood.
3. Temperament characteristics are relatively enduring over the course of early development. They play a significant role in determining the nature and quality of the parent-infant relationship.
4. Newborns' sensory systems are considerably more mature than either their motor or cognitive systems are.
5. Over the course of the first year, infants demonstrate an increasing degree of intentionality in their behavior, an increasing ability to sequence very simple means-ends relationships, and a primitive understanding of object permanence.
6. Toward the end of the second year, infants begin to demonstrate the ability to form mental representations of objects, events, and people and to use mental symbols so that one object can stand for another.
7. Language development becomes increasingly evident toward the end of the second year. It reflects an early attempt by infants to understand or act upon (through labeling) the people, objects, and experiences they encounter.
8. The development of attachment to the primary care givers reflects an increasing ability to discriminate between individuals, an appreciation of the predictability of the environment, and the quality of the care givers' interactions with the infant. A variety of biological factors appear to facilitate the process.
9. The emergence of separation distress toward the end of the first year reflects the growth of attachment between the infant and the care giver. Its resolution, during the second year, reflects the quality of the attachment bond as well as the increasing cognitive competence of the infant.

THE CONTEXT OF DEVELOPMENT

1. The development of the care giver–infant bond is influenced by the degree of synchrony that is established between the two. Synchrony is dependent on qualities present both in the infant and in the care giver. It is also dependent on the context of the interactions.

2. Although fathers seem as engrossed in their newborn as mother's, they typically play different roles in the child-rearing process.
3. Social class differences in parenting strategies are largely a reflection of the degree to which certain circumstances foster self-direction or compliance in adult behavior.

THE SIGNIFICANCE OF DEVELOPMENT DURING THE FIRST THREE YEARS

1. From birth, individuals play an active role in influencing the course of their development.
2. The orthogenetic principle, indicates that individuals become increasingly able to distinguish among the various elements of their experience and, over time, become increasingly able to place these elements in various relationships to each other.
3. The events of early experience are best appreciated as the necessary but not sufficient conditions for full development. Depending on the circumstances and the nature of the behavior, subsequent life experiences can significantly offset the influence of early life events.

HUMAN DEVELOPMENT AND HUMAN SERVICES

1. From a long-term perspective, the best way to improve the lives of children is to improve the lives of the adults who care for them.

KEY TERMS AND CONCEPTS

PHYSICAL AND MOTOR DEVELOPMENT

Apgar Score
Brazelton Neonatal Behavioral Assessment Scale
Infant Temperament Characteristics
Gross Motor Skills
Fine Motor Skills
Cephalocaudal Progression
Proximal-Distal Progression
Mass-to-Specific Progression

COGNITIVE DEVELOPMENT

Piaget's Stage of Sensorimotor Development
Scheme
Primary Circular Reaction
Secondary Circular Reactions
Means-Ends Separation
Object Permanence
Tertiary Circular Reaction
Mental Representations
Deferred Imitation
Pretend Play
Telegraphic Speech

SOCIAL DEVELOPMENT

Erikson's Psychosocial Stages of Trust and Autonomy
Attachment
Separation Protest
Separation Anxiety
Ainsworth's Patterns of Maternal Behavior
- Accepting Parents
- Rejecting Parents
- Ambivalent Parents

THE SIGNIFICANCE OF DEVELOPMENT

Orthogenetic Development

SUGGESTED READINGS

Goldberg and DiVitto provide a very comprehensive discussion of biological and psychological aspects of prematurity.

Goldberg, S., & DiVitto, B. A. (1983). *Born too soon: Preterm birth and early development.* San Francisco: Freeman.

One of the more significant aspects of the resurgence of research on infant development over the past years has been the greater awareness of the behavioral competence of the neonate. A good review is provided by Brazelton.

Brazelton, T. B. (1980). Behavioral competence of the newborn infant. In P. M. Taylor (Ed.), *Parent-infant relationships.* New York: Grune & Stratton.

The degree to which biological factors limit, direct, or foster psychological development is a perennial question within the field of life-span human development. Weitz provides a good summary of the debate, especially as it relates to sex roles.

Weitz, S. (1977). *Sex roles.* New York: Oxford University Press.

There is a vast literature on the topic of early parent-child relationships. These five readings are representative. The ones by Goldberg and by Ainsworth and her colleagues examine patterns of early attachment. Fraiberg considers what happens to this attachment process when the infant is handicapped. Belsky considers what happens to the attachment process when family stress leads to abuse. Finally, Keniston considers the role society plays in fostering good parent-child relationships.

Ainsworth, M. D. S., Blehar, M. C., Waters, E., & Wall, S. (1978). *Patterns of attachment.* Hillsdale, NJ: Lawrence Erlbaum Associates.

Belsky, J. (1980). Child maltreatment: An ecological integration. *American Psychologist, 35,* 320–336.

Fraiberg, S. (1975). The development of human attachments in infants blind from birth. *Merrill-Palmer Quarterly, 21,* 315–335.

Goldberg, S. (1977). Social competence in infancy: A model of parent-infant interaction. *Merrill-Palmer Quarterly, 23,* 163–179.

Keniston, K. (1977). *All our children: The American family under pressure.* New York: Harcourt Brace Jovanovich.

The traditional wisdom in child development has always been that the early years play a disproportionately greater role in charting the course of development than subsequent life experiences do. The Clarkes suggest that this may not necessarily be the case.

Clarke, A. M., & Clarke, A. D. B. (1976). *Early experience: Myth and evidence.* New York: The Free Press.

4
EARLY CHILDHOOD

CHAPTER OUTLINE

This chapter surveys development between the ages of three and seven. Like infancy, early childhood is a period of major transition. Three-year-olds may look mature when compared to newborns, but compared to seven-year-olds, they are still very young. By the age of seven, children are able to converse as well as many adults, run, jump, skip, hop, ride a two-wheeler, and participate in sustained peer play activities. Their ability to organize information has become more systematic, their store of factual knowledge is extensive, and they are considerably less egocentric in their view of the world. They easily tolerate separations from their parents that last many hours, and some seven-year-olds are even brave enough to chance an overnight stay at a friend's house. Most seven-year-olds have a well defined sex role, a budding sense of morality, and sufficient stability of personality to make fairly accurate long-term prediction possible. Seven-year-olds have entered school and are on the threshold of entering the world independently of their parents.

Erikson (1980) sees early childhood as a time for developing a sense of initiative and avoiding a sense of guilt. For the toddler, autonomy seems to be a value in its own right. Toddlers don't value autonomy because it allows them to do anything in particular. They value it because it allows them simply to do. However, during Erikson's **psychosocial stage of initiative,** children add a goal or purpose to the desire to do. The goals are usually self-defined and often not completed. They nevertheless represent a major advance over the previous stage. Unlike autonomy, initiative involves purpose, decision making, and goal setting.

Preschoolers' growing sense of purpose coupled with their almost limitless energy tends to make them rather intrusive.

> The intrusive mode, dominating much of the behavior of this stage, characterizes a variety of configurationally "similar" activities and fantasies. These include the intrusion into other people's ears and minds by aggressive talking, the intrusion into space by vigorous locomotion, the intrusion into the unknown by consuming curiosity. (Erikson, 1980, p. 80)

The parent in particular and the society in general deal with this intrusiveness by channeling, limit setting, and censure. Erikson believes that the way these techniques are used determines the extent to which preschoolers feel guilt over their ideas and actions. Some sense of guilt is of course a necessary thing in a society. It is the basis for self-regulation and the foundation upon which conscience is built. But when parents tell their preschoolers that ideas and goals as well as actions are bad, then the result is a "sense of guilt." The children begin to doubt their competence to do anything and to doubt the value or goodness of their goals and expectations. A more appropriate balance is achieved when the parents are able to separate their evaluations of goals from their evaluations of efforts. When the task is unrealistic (when the child wants to make dinner alone, for example), the parents should assure the child that the goal is fine and that sometime in the future he or she will be able to do it. In the meantime, the parents' response might be "why don't we make dinner together." Erikson

While this child's goal of building a house for her new puppy will probably not be achieved, her attempt at setting goals and testing new skills is an indication of a growing sense of initiative.

believes that such procedures give the preschooler an optimistic view of the future. On the other hand, the parent who interprets the preschooler's sometimes unrealistic goals and desires as reflecting some enduring quality rather than the preschooler's present level of development fosters in the child a set of negative expectations about the future. When this happens, a self-fulfilling prophecy often results (Breger, 1974).

This chapter follows the same outline that was followed in the last. It reviews typical development in the physical, cognitive, and social domains and then discusses the various contexts in which this development occurs. Finally the chapter describes the place of early childhood in the life-span.

In reading this chapter, try to keep three questions in mind. First, how are preschoolers' developing competencies evident in the various activities they pursue? Second, how might development in the physical, cognitive, or social domain influence development in each of the other two domains? Third, how do parents and other adults and peers enhance the preschooler's growing sense of initiative within these three domains?

THE COURSE OF DEVELOPMENT DURING EARLY CHILDHOOD

The physical, social, and cognitive changes that occur during early childhood are truly remarkable. Between the ages of three and seven, children make great strides in their ability to use and comprehend language, to perform most simple motor tasks, to initiate and maintain peer activities, to respond to adult directives, and to use adults more effectively as sources of information and aid. Although the rate at which these various skills are acquired may vary considerably, the sequence of their development follows a fairly orderly course.

Physical and Motor Development

Children continue to gain in height and weight through early childhood but at a slower rate than the one they previously maintained. Boys and girls each gain about three to four pounds per year and grow two to three inches. A typical three-year-old weighs about thirty-two pounds and is just over three feet in height. By age five, the same child weighs about forty pounds and is forty-three inches tall (Stott, 1967).

Although preschoolers are certainly larger than infants, what really separates the two age groups in physical appearance is their body proportions. Toddlers are chubby, bowlegged, and almost cherublike in appearance. Their heads are disproportionately large; their abdomens are round and protruding. Watching toddlers walk leads to the impression that they could topple over at any moment. Preschoolers gradually present a different image. Growth of the arms and legs proceeds faster than growth of the rest of the body. As a result preschoolers' bodies appear more adultlike in proportion than infants' bodies do. Muscle growth accounts for a larger proportion of their total body weight, while fat tissue accounts for both a relative and absolute smaller proportion. Preschoolers look thinner than infants, and as their muscles become stronger, especially the abdominal muscles, they stand more erect, appear less top-heavy, and seem less likely to topple Humpty-Dumpty fashion at the slightest touch.

As their bodies become more proportional to their heads and their center of balance shifts upward, preschoolers become more proficient at tasks requiring good balance. As their legs become longer, they improve at tasks requiring good locomotor skills. As their shoulders widen and arms lengthen, preschoolers are able to throw more accurately and for greater distance. These changes in motor ability reflect not only changes in body proportion but also the continued integration of the perceptual and motor systems (Cratty, 1979). As a result of these changes, most six- and seven-year-olds can run, jump, skip, hop, climb, and balance almost as well as anyone. The further refinement of these skills in middle childhood, as well as developing proficiency in throwing, catching, and kicking, is largely a result of continued practice.

The development of these motor skills influences not only the performance of various physical-motor tasks. It also contributes to the preschooler's cognitive and social development. Movement serves as one of the foundations upon which the preschooler develops a basic understanding of the spatial and temporal characteristics of the physical world (Gerhardt, 1972). For example, preschoolers first understand **temporal relationships** in terms of sequence rather than duration, and their notion of sequence in turn is partly based on their awareness of the series of actions or steps involved in various motor activities. Out of this awareness of sequence evolves the awareness of activity patterns that punctuate their days, weeks, and months. Preschoolers know that Christmas comes after Thanksgiving long before they know by how many days the two holidays are separated. As soon as Thanksgiving is past, they start asking if it's Christmas yet.

Preschoolers' initial understanding of **spatial relationships** (up, down, left, right, in front of, behind, and so forth) is also partly a reflection of their continuing physical-motor development. As their movements become more sophisticated and varied, so too do their contact with and understanding of the physical world. Up looks much higher from the top of a ladder, down more "relative" when hanging by your knees from a tree limb. Far keeps on getting farther away as your throw keeps getting longer. In other words, as preschoolers' physical-motor skills mature, so

While not yet World Cup material, changes in body proportions and developing motor skills make it possible for these preschoolers to attempt the game of soccer.

too does their construction of the meaning of words and concepts that describe the physical world and their place within it.

During early childhood, physical-motor development also begins to exert an influence on social development. It does so in two ways. First, it serves as an initial criterion for self-evaluation. Preschoolers' early perceptions of themselves are largely based on their judgments about what kinds of things they can do and what kinds of things they can't do. These initial perceptions may not be very accurate, but they nevertheless influence preschoolers' decisions to attempt certain activities (for example, to climb a ladder or jungle gym) and their choice of forms of social interaction (such as quiet versus rough-and-tumble play). These initial perceptions are certainly not irreversible; they will change as preschoolers' physical-motor skills develop and as their perception of themselves becomes based on other factors. But they are nevertheless an initial perception, and to a large extent they reflect the influence of physical-motor development on social development.

The emergence of peer-group activity during early childhood is the second means through which physical-motor development influences social development. Much of the peer activity during early childhood is physical. It involves running, jumping, climbing, swinging, hopping, and skipping. Further, this physical activity serves as the initial basis for status in the peer group. Although certainly less of a factor at this time than during the school years, the ability to participate successfully in these active play patterns is an important determinant of preschool peer status.

Cognitive Development

During early childhood, the preschooler's cognitive understanding of the world continues to undergo differentiation and expansion. Increasing competence in the use of language as a means of symbolizing experiences and as a means of communicating with others helps the preschooler's constructions become less idiosyncratic, more stable and consistent, less egocentric, less dependent on immediate perceptions, and more reliant on symbolic representations of these perceptions.

The goal-directed, purposive behavior typical of a growing sense of initiative further propels the preschooler into the world of objects, people, and events. The resulting experiences compel preschoolers to continually reconstruct their cognitive understandings in more mature forms.

Piaget (Inhelder & Piaget, 1964) called this second stage of cognitive development the **stage of preoperational thought.** During this time, preschoolers' thoughts are to a large extent based on the appearance of things. Preschoolers are also likely to focus on one aspect of a situation to the exclusion of others. The adult who tries explaining to a four-year-old that a broken cookie still contains "just as much cookie as it did when it was whole" encounters a vivid example of the limitations of preoperational thought. The four-year-old is simply unable to go beyond the observation that the cookie is no longer the same as it was, even though the change in appearance has no influence on the amount.

Piaget's (Inhelder & Piaget, 1964) studies of preschoolers' ability to perform **seriation** and **classification** tasks provide good examples of preschoolers' developing ability to accurately and effectively process information during the preoperational stage. The ability to order, or seriate, a set of objects first requires the realization that the objects can be arranged along some common dimension such as length, weight, or hue, and then the ability to note the relative differences between each member of the set in terms of the relevant dimension. A typical task requires a preschooler to correctly order a series of sticks of differing lengths. Children below the age of four are rarely able to complete the task. Some place the sticks in random order. They understand that they are to do something with the sticks but little else. Some make small clumps of sticks. They seem to understand the notion of size but are limited in their ability to make use of it. Cowan (1978) notes that some three-year-olds are able to line the sticks in a staircase fashion but only because they fail to consider variations in the bottoms of the sticks. Their staircases look like this:

By five years of age, most children are able to solve the problem. The task is still not an easy one for them, however. Errors remain common, and solution is more a matter of trial and error than of an overall plan (Ginsburg & Opper, 1969).

Finally, by six or seven years of age, children solve the problem with little difficulty on the first attempt. They are even able to order two different sets of objects and then show a **one-to-one correspondence** between them. That is, they can put the biggest in each set together (for example, sticks and balls), the next biggest, and so on.

The preschooler's development of classification skills shows a similar sequence. A typical Piagetian task involves presenting a child a set of objects that differ in terms of a dimension such as shape or color or size (a more advanced task would have the objects differing on two dimensions simultaneously) and asking the child to put the ones that are alike together. Three- and four-year-olds are unable to complete the task. Typically, they either use the shapes to make some sort of picture (by putting a triangle on top of a square and calling it a house, for example) or they form groups that don't appear to have any common property. The criterion for placing a block in a group is idiosyncratic to the child and the moment.

By age five, most children can solve this simple classification task as well as a more difficult one requiring them to subdivide piles into more exclusive categories. For example, they can first divide a collection of pictures into two categories—animals and flowers. Then they are able to further subdivide the animals into dogs and cats and the flowers into red and yellow ones. This performance represents a clear improvement over the idiosyncratic sorting of the three-year-old, but to Piaget it still lacks an important quality of true classification. Five-year-olds are unable to understand the relationships between the different levels of the classification hierarchy. If for example the set of animals consisted of eight cats and two dogs, children, when asked if there were more cats or more animals, might say that there were more cats. Their perception of the pile of cats and the pile of dogs seems to prevent them from realizing that both are part of the original and necessarily larger grouping of animals. By age seven, most children can understand the inclusive relationships in a hierarchical classification if the groupings are as simple as the examples just described.

When the task becomes more complicated, the perceptual demands of the task again overwhelm their ability to comprehend it.

The relation between task demands and level of understanding is a common one in development. It is also evident, for example, when the infant who correctly finds the rattle under the first blanket initially fails to search for it under a second blanket. The pattern suggests that a child's understanding of the relationship between the elements in a task is influenced by the complexity of the task. It would be incorrect to assume that children can necessarily solve a complex task if they have already solved a conceptually similar but simpler version of it. What may first be required is the opportunity to use the new level of understanding with increasingly more complex problems.

General Characteristics of Preoperational Thought

The developmental sequences for seriation and classification reflect the three dominant characteristics of preoperational thought. They also reinforce the concept that children's thought is not simply an incomplete version of adult thought but rather a qualitatively distinct way of dealing with information.

Actions based on the appearance of things. The first general characteristic is that the preoperational child's view of the world is primarily based on the appearance of events. Preoperational children seem to believe that things are as they appear to be. They are not very likely to consider the possibility that there may be more than one explanation for an event or that looks can be very deceiving sometimes. Although controlled laboratory studies have shown that preoperational children, when provided props to help them remember specific pieces of information or the relationships among bits of information, can use inference as a basis for judgment (Bower, 1979; Hawkins, Pea, Glick, & Scribner, 1984), the fact remains that in most situations, the surface appearance of things most influences their judgments. For example, preschoolers are likely to report that there is more water in a tall, thin glass than there is in a short, wide glass that actually contains the same amount. If asked for an explanation, they reply that there is more water in the tall, thin glass because it "looks" like it contains more water. Although they are aware that it is the same water in both glasses, they fail to conserve the quantity of water (Elkind, 1976). That is, they fail to realize that the increase in the height of the column of water is offset by the decrease in its width. The ability to solve **conservation** problems of this type marks the preoperational child's transition to the third or concrete operational stage of cognitive development.

Actions based on limited amounts of information. The preschooler's explanation for the greater amount of water in the tall glass points out a second general characteristic of preoperational thought. It tends to be based on only a limited part of the available relevant information. Piaget refers to this tendency as **centration.** Because preschoolers center their attention only on the height of the water in the two glasses, they fail to consider the significance of the other available information, that is, the width of the water. If they considered both, their answer would be quite different. They might still not be able to solve the problem correctly, but their behavior would indicate that they knew something was amiss. This tendency to center on only one aspect of a task is also evident in the three-

In Piaget's conservation of liquids task, two short, wide glasses are filled with identical amounts of water. Once the child confirms that the amounts are equal, the water from one of the glasses is poured into a tall, thin glass, and the child is asked if one glass has more, less, or the same amount of water as the other. Because they are unable to conserve continuous quantity, preoperational children are likely to state that the tall, thin glass now holds more water.

year-old who forms a staircase seriation by disregarding the uneven bottoms of the sticks.

The tendency of preschoolers to focus or concentrate on only limited aspects of a task is also evident in their tendency to focus on the state of an object or event rather than on the process or events that led up to the present circumstance. For example, two four-year-olds are each playing with identical lumps of clay. The first child molds her clay into a ball. The second child sees the ball and tells the preschool teacher he wants the ball. The teacher tells him to make his clay into a ball, but he continues to be adamant that he wants the other child's ball of clay. The problem for the child (not to mention the teacher) is that he is focusing on the state of his clay as compared to the state of the other child's clay. He seems unable first to consider the process by which the little girl transformed her lump of clay into a ball and second to realize that he could apply the same transformation to his lump of clay. The ability to appreciate the relation between what is and how it came to be only gradually develops during the preschool years and is not clearly evident until middle childhood.

Lack of reversibility. A third general characteristic of preoperational thought is that, in Piaget's term, it lacks **reversibility.** Reversibility implies the ability to make reciprocal mental operations (Flavell, 1977). In the water example, preschoolers believe that the amount of water changes as it is poured from one glass to another. They seem unable to consider the relationship between the water in the first glass and the same water in the second glass; that is, they are unable to observe the water in one glass while mentally reversing the pouring operation. The same limit is evident in the five-year-old's failure to realize that the collection of pictures contains more animals than cats.

The development of mental reversibility is a necessary prerequisite to the feeling of necessity that is a component of mature logical reasoning. If you present the water task to a group of ten-year-olds and ask if the amount of water is greater in one glass than in another, you will probably get a look that suggests the ten-year-olds think "you have lost some of your marbles." They will tell you that "of course" the amount of water is the same. Both the certainty of their response and their assessment of your intellect reflect the presence of reversible mental operations. The absence of reversibility is one component of preschoolers' lack of coherence in their judgments, preferences, and behavior. The logical connectors that for the adult bind discrete events into a relatively coherent whole are not yet present in the thought of the preschooler.

These three general characteristics are not unique to the preschooler's understanding of seriation and classification. Rather, they are evident in all cognitive activity during the preoperational stage. Three of the more important areas in which they are evident are the ability of preschoolers to take another's perspective, their understanding of the reasons why things happen, and their use of language.

Egocentric Thought

Preoperational children are often described as egocentric. This means that they have a very limited ability to view situations from any perspective other than their own. As a result, they feel little motivation or need to explain themselves to others (thinking that if they understand something, so must others), assume that others are as aware of things as they are, and as any older sibling will tell you, are terrible secret keepers.

A four-year-old will tell you that she has a brother but, when asked if her brother has a sister, will answer "no." A six-year-old will be able to show you his right and left hand, but if you stand facing him and ask him to show you your right and left hands, he will label the hands incorrectly. The parents of a three-year-old get used to hearing their child's voice from another room ask "What's this called?" In all these instances, the preschooler simply fails to consider the other person's perspective. The four-year-old has never thought of herself as her brother's sister, the six-year-old fails to consider that one's right and left hands are relative to the person, and clearly the three-year-old never stops to consider that the object he sees on the television is not visible to the parent in the other room.

To quote Piaget (1974),

> If children fail to understand one another it is because they think they do understand one another. The explainer believes from the start that the . . . [other child] will grasp everything, will almost know beforehand all that should be known, and will interpret every subtlety. . . . It is obviously owing to this mentality that children do not take the trouble to express themselves clearly, do not even take the trouble to talk, convinced as they are that the other person knows as much or more than they do and will immediately understand what is the matter. (p. 116)

Being egocentric is not the same as being selfish or self-centered. **Egocentric thought** implies a limited ability to take another's perspective. Selfish people are able to understand, for example, that others have the same right to use the television as they do. They just don't care. The preoperational child isn't able to appreciate the rights of others. Parents who fail to appreciate this distinction may punish their preschoolers unjustly.

Egocentrism is not unique to preschoolers. Who is more egocentric than infants before they develop the object concept or young adolescents who don't believe that even Romeo or Juliet could understand the extent of their sadness over the loss of a first love? Each stage of the life-span involves some elements of egocentrism. What is unique about egocentrism in preschoolers is the extent to which it is evident in all aspects of their behavior.

Causal Reasoning

The preschooler's understanding of the reasons why things happen also reflects the three general characteristics of preoperational thought. Consider the following examples taken from interviews Piaget (1979) conducted with Swiss children:

> Cam (6) said of the sun: "It comes with us to look at us." (Why does it look at us?) "It looks to see if we are good." "The moon comes at night because there are people who want to work." (p. 216)
>
> Jac (6½) (What does the moon do when you are out for a walk?) "It goes with us." (Why?) "Because the wind makes it go." (Does the wind know where you are going?) "Yes." (And the moon too?) "Yes, it comes so as to give us light." (p. 216)
>
> Pug (7) (Does a bicycle know anything?) "No" (Why not?) "I mean it knows when it goes fast and when it goes slowly." (Why do you think it knows?) "I don't know but I think it knows." (p. 180)
>
> Van (6) (What is night?) "When we sleep." (Why is it dark at night?) "Because we sleep better, and so that it shall be dark in the rooms." (p. 293)

Responses such as these are certainly the types of responses that adults find humorous. They also tell us a great deal about young children's understanding of causal relations. First, they indicate that preschoolers have what Piaget calls an animistic view of the world. They believe that the sun, and fire, and stones, and wind, and virtually all other inanimate objects can have motives and intentions (Elkind, 1976). It is no wonder that so many preschoolers cling to blankets and teddy bears and require elaborate bedtime rituals before they can go to sleep. The world must appear sometimes a very scary place.

The responses also demonstrate that the logic of the preschooler tends to be transductive. Adults infer cause by either **inductive reasoning** (from the particular to the general) or **deductive reasoning** (from the general to the particular). **Transductive reasoning** is from the particular to the particular. It assumes that a causal relationship exists if two events are closely associated in time. The preschooler confuses correlation with causation (an error, as you remember from Chapter 1, also found among researchers). In some cases the reasoning is backward. Van and Duc provide

While the preschooler's animistic viewpoint sometimes makes the world a scary place, it also makes it possible for this boy's "best friend" to be a teddy bear.

good examples. Van seems to be saying that it is dark at night because darkness makes sleep easier; Duc believes that it becomes dark because it is time to go to bed. If questioned further, Duc might even agree to the idea that one way to make the hour at which darkness comes change would be to change one's bedtime. In fact, if there is any relationship between the two, it is the other way around. We go to bed because it becomes dark outside; it doesn't get dark so we can more easily fall asleep. In some cases the preschooler seems to believe that what was true on one occasion will always be true on others. Cowan (1978) relates an anecdote in which one of Piaget's own children drew the conclusion that since Piaget got a bowl of water whenever he shaved, that every time she saw him with water, he must be getting ready to shave again.

Preschoolers' unique view of the world is neither taught to them by adults nor unique to a particular culture. Rather it is typical of all preoperational children. Specifically, it reflects the developmental level of the preschooler's cognitive thought. As Cowan (1978) concludes, "Children's cognitive egocentrism at this stage reinforces their tendency to assume that the rest of the world is like themselves—filled with living things which move and have thoughts and feelings of their own" (p. 168).

Language

By the age of seven, most children have a production vocabulary of ten to twelve thousand words, and the grammatical structure of their speech is virtually identical to that of average everyday adult usage. Even by age three or four, preschoolers use plurals and past and future tenses, ask questions, give commands, make requests, and verbally respond to commands. By age five or six, children frequently use prepositions, conjunctions, and articles. In fact the major developments in the preschooler's use of language are not so much in terms of structural improvement as in terms of functional improvement. Specifically, preschoolers become increasingly proficient in using language as an effective means of communication and as a means of symbolically representing experiences (Bloom & Lahey, 1978).

The development of symbolic thought at the end of the sensorimotor stage provides preschoolers a very powerful tool for dealing with new experiences. It provides a means of relating past experiences to present ones as well as of using present ones to anticipate future events. It allows preschoolers to deal with events removed from immediate experience. It allows them to construct a broader perspective on past and future situations and settings far removed from everyday, direct experience. Language plays a pivotal role in this expansion because it is the most effective and efficient means of coding and expressing this expanding perspective. Two- and three-year-olds soon learn that although a gesture is sufficient to differentiate between two objects in sight, only a word will suffice for objects out of sight. The seemingly incessant questions of the preschooler are in part a growing awareness that, unlike an understanding of what, an understanding of how and why and when is best conveyed through language (Blank, 1974).

Because language is the most arbitrary of the various means that can be used to symbolically represent events, it is potentially also the most powerful. But not yet for the young preschooler. For three- and four-year-olds, names are not arbitrary designations, but are viewed as properties of the thing they represent. For example, Piaget questions a four-year-old about the properties of words. There is a cup on the table in front of the child, and he asks if it is possible to call the cup a table and the table a cup. The child replies "no," and Piaget asks why. The child confidently replies, "Because you can't drink out of a table." One of the major accomplishments of the early childhood years is the dissociation of words from things. When the eight-year-old announces that sticks and stones can break her bones but names can never hurt her, she is unknowingly proclaiming her graduation from early childhood.

A related language task for the young preschooler is learning that although language is arbitrary (the table could be called a cup), once a label is attached to a referent, it stays with the referent. Three- and four-year-olds sometimes have difficulty in learning the specificity of language. They don't necessarily know for instance that not all men can be called daddy.

Preschoolers' learning of the arbitrary but rule-governed use of language is evident in their learning of relational terms and conjunctions. The toddler's initial understanding of words is that they stand for or refer to something; that is, they name something. Since relational words are different, the toddler's understanding of them is less well developed. The same object can be big or little depending on what it's compared to. Two things that are both big (a big dog and a big building) can still be very different in size. One four-year-old when confronted with the problem of describing the same size glass in relation to a group of dolls and herself confidently and quite typically replied: "It's a little glass, big glass. . . .a little glass for me and a big glass for them. . . .and a 'larger size' glass for her" (pointing to the smallest doll at the table) (de Villiers & de Villiers, 1979, p. 135). Between the ages of three and six, preschoolers learn to use most of the spatial adjectives correctly. Big and little are usually mastered first, followed by long and short. Less frequently used modifiers such as deep and shallow are learned toward the end of the interval.

Other difficult words shift meaning not with the physical context but with the conversational circumstances: the word "I" refers to the speaker, "you" to the hearer, so that paradoxically preschoolers are always called "you" but must never call themselves that (de Villiers & de Villiers, 1979). By the end of the preschool years, children have mastered these pronouns.

The correct use of conjunctions (but, because, and, if, however, and so forth) requires an understanding of the relationship between groups of words. The sentence "The boy hit the girl because the girl ran away" takes on a very different meaning if the word "and" is substituted for the word "because." The elements of the sentence remain the same (the boy hitting the girl, the girl running away), but their relationship is altered significantly. Beginning speakers do not use conjunctions. Each element seems to stand on its own. By age three, "and" comes to be used as a universal conjunction. The preschooler seems to understand that elements exist in relationships but doesn't yet understand the specific forms relationships take. The conjunctions "but" and "because" are used increasingly during the preschool years, but the full use of conjunctions, as well as the use of relative clauses embedded within the sentence (an example of which you have just read), must wait for the school years.

The egocentric character of preoperational thought is very evident in the preschooler's use of language as a means of communication. The following excerpt comes from the conversation of two six-year-olds. The children were separated by a screen. They each had identical sets of pictures in front of them. The task of one child was to tell the other which picture he had selected.

Child 1: (pointing to a picture in front of him) "It's this one"

Child 2: (pointing to a picture from his set) "You mean this one?"

Child 1: (unable to see, answers confidently) "Yep" (de Villiers & de Villiers, 1979, p. 88)

It is clear from this small excerpt that each child had a specific message in mind, that each felt his words effectively communicated the message, and that each thought he completely and accurately understood what the other was saying. In fact, although each child had a message in mind, they misjudged the effectiveness of their communication. The two children were unable to appreciate the fact that each was viewing the situation from a different perspective and that as a result their communication had to be tailored to meet the needs of the listener. Although preschoolers become more proficient at communication, especially when they want something, it is the school years that witness the emergence of true social communication.

A more typical pattern, especially among young preschoolers, is parallel conversations, or **collective monologues.** Each child talks about something, but the two things are different. One may talk about a favorite toy, and the other may talk about a recent visit. There is no indication that the speaker tailors her message to meet the needs of the listener. In fact, it is even difficult to determine if one child is actually trying to tell the other something or if the other child's presence is merely a necessary condition for the first to begin her monologue.

By the time children are four or five years old, their conversations are increasingly likely to have a common theme (a favorite television character, a block structure they are building, and so forth) but may remain parallel. There is little or no cumulativeness to a conversation. Little explanation is given. Few questions are asked (Ginsburg & Opper, 1969). When children reach age six or seven, true conversations become more frequent. They are limited to matter-of-fact, immediate topics, but they nevertheless represent a true attempt to construct a message that is meaningful to the listener. As you can imagine given the relatively limited temporal and spatial organization skills of the six-year-old, these conversations are often less than perfect and are usually of short duration.

Social Development

The same cognitive structures that influence preschoolers' understanding of events in their physical world act to influence their growing understanding of events in their social world. In the social domain, preschoolers are actively constructing meanings to culturally defined events such as conceptions of right and wrong, appropriate behavior for males and females, acceptable ways of relating to peers and to adults, and an understanding of self in relation to others. As is true about the preschooler's understanding of the physical world, the process is cumulative and interactive.

The three characteristics of preoperational thought that influence the understanding of physical phenomena are equally influential in the social domain. The tendency of preschoolers to respond to the appearance of things, to focus on only one aspect of a situation, and to view things from an egocentric perspective helps explain why, for example, their emotions can fluctuate greatly, from utter misery and despair to absolute ecstasy in the shortest of time intervals (Cowan, 1978).

The parallels between cognitive and social development are evident in preschoolers' understanding of classification and seriation skills and in their understanding of fairness. In both cases, preschoolers' early groupings are idiosyncratic and fluctuating. If there is any order or logic to them, it is not evident to others. Judgments are based solely on personal desire. No consideration is given to merit, need, equality, or past or present events (Damon, 1977).

By the time preschoolers are able to construct stable collections based on one dimension, their understanding of social justice has made a parallel advance. Five- and six-year-old children are able to consider such factors as effort, quality, intellect, and good behavior in their judgments of merit and fairness. But as is true of their collections, their judgments are limited in scope and are not easily considered from other perspectives. Once they have come to a decision, there is little likelihood that they will change their minds (Damon, 1977). This is a system of justice that offers no room for appeals based on new evidence. It is not until middle childhood that the classification strategies and the social justice concepts become structured into a hierarchy of subordinate and superordinate groupings. Notions of fairness are now made irrespective of the person. For the first time, at least at the intuitive level, children become able to understand the notion that no one is above the law. Further, it is not until school age that children seem able to consider the degree to which the specifics of a situation might influence judgments about punishment and fairness. This same logic allows our legal system to distinguish between first-degree murder and manslaughter. In both cases, one person kills another. But since the circumstances differ, so too must the consequences.

The remainder of this section surveys the course of development in four social domains and considers how specific developments are integrated into the preschooler's concept of self. The four domains are sex-role development, friendship and peer relations, moral development, and the understanding of legitimate authority.

The Construction of Sex-Role Identity

Sex-role identity is one of the major components of our self-concepts. It defines us as belonging to one of two distinct groups, and it defines a number of behaviors and attitudes that society considers appropriate for members of each of the two groups. There are few topics that have generated as much debate as sex-role development. Most of the debates have concerned discrimination on the basis of sex, the origin of sex differences, and sex-role models appropriate for children. For the purposes of this text, the perspective defined by Maccoby (1980) will be used:

> When sex differences are found, they are only averages. A great deal of variation exists within each sex, and the characteristics of the two sexes overlap greatly. Certain bodily features, of course, are distinctly male or female, and except for occasional ambiguous cases, qualitative sex differences in physical characteristics do exist. In terms of psychological characteristics, however, the sexes are more alike than different and any average sex difference must be seen from the perspective of overall similarity. (p. 204)

Three-year-olds, however, don't seem too concerned about the need for such perspectives. If they aren't happy about being a boy or a girl, they are quite certain they can change to the other sex. Like most of their constructions, preschoolers' understanding of sex-role identity is initially fluctuating and idiosyncratic.

Sometime between the ages of two and three, most children become able to identify themselves correctly as a boy or a girl. This is the first step in acquiring a sex-role identity. Like most words, however, the words boy and girl mean different things to three-year-olds from what they mean to adults. To three-year-olds, the label boy or girl does not indicate class membership. Rather it is like their names. It is simply something that belongs to them. They know that they are a boy or a girl, but they may not be able to decide if their picture belongs in the "girl picture box" or the "boy picture box" (Thomason, 1975). They may say that some toys like trucks are for boys and some toys like dolls are for girls, but because they aren't yet able to view themselves within the context of a group, they don't see the information as relating to them. A boy would see no conflict in saying "boys don't play with dolls but I do"; a girl would see no conflict in saying "girls don't play with trucks but I like to."

Although sex-role socialization is becoming more androgynous, the socialization patterns depicted by the boy "mowing" the grass with his father or the girl putting on "makeup" with her mother are still more common than the one depicted by the boy and his doll.

By the age of four or five, most children correctly generalize sex-role labels. They associate themselves with other boys or girls, begin to show typical preferences in play partners and play activities, and are aware of some of the characteristics that identify people as male or female. Typically, hair length and dress are their most common criteria for assigning sex. However, they still have trouble understanding the permanence of a sex-role label. Some children even as old as six still believe that a boy could become a girl if he really wanted to (or vice versa) and that changing one's clothes and hair length can change a boy into a girl or a girl into a boy (Emmerich, 1977).

The acquisition and generalization of an appropriate sex-role label seem to enhance preschoolers' participation in same-sex play groups. Same-sex play groups precede gender labeling. They begin to become evident by age two. This differentiation reflects both social and biological factors. Even at this age, there is more of a rough-and-tumble quality to the play of boys (Maccoby, 1980; Ullian, 1981).

Once preschoolers identify with their same-sex group, there is a strong increase in their desire to do sex-role appropriate activities and to assign sex-role appropriate labels to others. What follow during the late preschool period are a relatively self-imposed segregation by sex and the development of a very stereotypic view of appropriate sex-role behavior. "Focusing on concrete perceptible cues such as size, strength, child bearing capacity, voice and body build, young children conclude that men are powerful, dominating, scary and tough, whereas women are kinder, gentle and more fragile" (Ullian, 1981, p. 179). These two patterns last through much of middle childhood and early adolescence and are only appreciated as stereotypic when the adolescent is able to appreciate the distinction between what is possible and what is probable.

The emergence of these stereotypic views of sex-role behavior is evident in these excerpts of interviews taken from Damon (1977). The children were read the following story and then asked a series of questions concerning the story:

> This little boy, George, likes to play with dolls. But his friends think that he's silly to play with dolls, and his parents always tell him that little boys shouldn't play with dolls, only little girls should. His mother bought him all kinds of other toys, like model airplanes, trucks, baseballs, and so on. But George still liked to play with dolls the best. (p. 242)

Two of the younger children, Jack and Alvin, didn't see any problem in George's playing with dolls. Jack said that it's okay for boys to play with dolls, that there is no rule against it, and that the decision is up to George. It wouldn't be fair for his parents to punish him because then George wouldn't be able to have any fun. Besides, Jack said, he likes to play with his teddy bear. Alvin believed that George should be able to play with anything he wanted to. Alvin didn't believe that George's parents would be angry with him for playing with dolls because George was doing something he wanted to do. When asked if it would be okay for a boy to play house, Alvin answered, "That would be okay, because I play house, and no one laughs at me" (Damon, 1977, p. 250). His answer makes clear the kind of reasoning he and Jack are using.

Michael and Eugene, a few years older than Jack and Alvin, had come to see the world in a very different light. Michael believed that boys should only play with "boy stuff." Dolls are "girl stuff." It would therefore be all right for George's parents to tell him not to play with dolls. Michael, who wasn't quite six at the time of the interview, did leave George one out: "If he starts playing with G.I. Joe and that stuff then he wouldn't be that bad" (Damon, 1977, p. 255). Eugene, who was seven, seemed much clearer in his thinking. "Because it's not right for boys to play with dolls because boys play with other toys like cars and girls play with dolls. Cause girls like dolls and boys like cars" (Damon, 1977, p. 256).

The highly stereotypic sex-role values typical of children from the ages of six to eight fortunately do not represent their final view of the subject. But these values do seem to be a necessary step in developing a mature sex-role identification. As Maccoby (1980) suggests, children may need to exaggerate sex roles in order to make them cognitively clear. Clarity may be the necessary first step in developing a more flexible and less sex-typed conception of social roles and relationships. From this perspective, parents would do better to judge the effectiveness of their sex-role and socialization efforts when their children were approaching adulthood than when they were approaching first grade.

The Development of Peer Relations

Learning how to be a friend is a difficult task for preschoolers. It is nevertheless an important skill. The egalitarian nature of friendship probably serves as the foundation for our moral development and certainly as the foundation for the development of intimacy in adulthood.

> Through their interactions with peers, children discover that other children are similar to them in some respects and different in others. And as children attempt to cooperate with one another, they discover that the coordination of behavior requires an appreciation of the other's capabilities, desires and values. At first these discoveries remain implicit and unexamined. Gradually, however, children integrate and organize what they have learned, leading to an increasingly sophisticated understanding of social relationships.
> (Rubin, 1980, p. 41)

The egocentric nature of young preschoolers limits them to the most temporary and activity-centered peer interactions. Between the ages of two and three, children have a difficult time realizing that peer relations are different from parent-child relations. "Accustomed to being on the receiving end of a nurturant relationship, to being favored and given special consideration, they are naturally reluctant to surrender their privileged positions or consider the needs of others" (Ausubel, Sullivan, & Ives, 1980). Accordingly, much of the play of the three-year-old is solitary and object centered. The more "brazen" may engage in parallel play (two or more children simultaneously but independently involved in the same or a similar activity, such as sand or water play) or may observe from a safe haven the group play of older preschoolers.

The friends of four- and five-year-olds are best described as playmates. They are friends because they play together, and they play together because they have similar interests or similar temperaments or simply because they live next door. Like preschoolers' other judgments, their criteria for defining friends are based on the physical rather than the psychological qualities of the person, and they tend to be specific to a time and place. The

Unlike older children, the play of three-year-olds tends to be parallel, even when several children are engaged in the same activity.

concept of the enduring friendship that lasts for years through thick and thin is still many years away. Nevertheless, the play groups of four- and five-year-olds do increase in size (up to five or six members) as communication and conflict-resolution skills improve. As children at this age become better able to define a common interest and engage in a division of labor, their activities become less parallel and more cooperative and cumulative. Projects started on one day are continued on following days. The decline in egocentric thought allows children to be more conscious of the qualities of other children. As a result they become better able to find other children who share their interests and their style of play.

Out of these early playmate friendships come the first real friendships between the ages of five and seven. Whereas frequency of contact was a primary determinant of playmate associations, now children want friends who are willing to share, helpful, nice, and easy to get along with. Friendships are still easily terminated by some specific incident (not sharing a treat, being mean, and so forth), but for the first time, young children are beginning to understand the reciprocal nature of peer relations. That is, they are beginning to appreciate the fact that if they want other children to be nice to them, they have to be nice to other children. Out of this early sense of reciprocity, most commonly expressed as sharing and turn taking, develop the trust and giving of self that will characterize friendships during middle childhood (Rubin, 1980).

The friendship patterns of early childhood serve not only as the foundation for more mature peer relations but also as the basis for developing a sense of social justice. Specifically, it is from preschoolers' early peer interactions that they first develop the concepts of equality and reciprocity—the two key ingredients of early social justice.

The Development of Social Justice

Social justice skills concern the understanding of right and wrong. For the preschooler, right and wrong usually concern fairness. Being fair requires the preschooler to consider and appreciate another person's point of view—

a skill not often found before the age of three or even four. Although it is true that children below the age of two react to the emotions of others and show distress when another is experiencing difficulty (Hoffman, 1975), it is probably also true that these very young preschoolers are not able to appreciate what the other person is experiencing (Chandler, 1977). Since they are unable to take the other's perspective, it does not enter into their social justice decisions. There is no distinction made between personal desire and fairness. What is fair is what they want (Chandler, 1977). This view is evident in three-year-old Sean's very clear determination of the way to divide ten candy bars among three children. He said, "I want seven. Then one for you and one for you and one for Tina" (Damon, 1977). The four-year-old's confusion of justice and personal desire is evident in James's explanation as to why he should get more toys than Sammy ("These are the ones I like") and in Mary's reason for getting more ice cream than the other children ("Because I like ice cream"). Unlike Sean who seems unable to consider another perspective, James and Mary acknowledge that other children may also have a right to the toys or ice cream. But like other four-year-olds, they resolve the problem. They simply assume that the others won't mind if they get more (Damon, 1977).

Probably as a result of preschoolers' continuing peer interactions, they become increasingly aware of the conflict between their own needs and desires and those of others. Preschoolers become very direct and explicit in telling other children that they disapprove of their behavior. As a result they begin to feel the need to justify their decisions using a more commonly accepted rationale than personal desire (Damon, 1977). Jack, in explaining why he would keep the most toys, justifies his decision by saying that he is bigger than the other boy. Miriam, in explaining why she would keep four blue chips and give Jenny only two, offers a whole list of justifications. She says that she likes blue more, would play with the blue chips, has a blue dress at home, and besides that "she should get four because she is four." These justifications are still very self-serving. Nevertheless, they indicate that by the time preschoolers reach age five, they have enough understanding of others' legitimate rights to feel the need to offer an objective, though often irrelevant or self-serving justification for their behavior.

By the time children are six, their understanding of social justice acquires the same absolute quality that is evident in their sex-role perceptions. What is fair is simply what is equal. Everyone should get the same treatment. It is as if the child, in finally recognizing that personal desire is not an acceptable criterion for justice, rejects any and all possible qualifiers.

By age seven, the first indications of reciprocity in social justice are evident. When Alison is asked what she would say to her mother if her mother told her she couldn't play with her friend anymore, she replies, "But he's my friend, because he shares his things with me, so I've got to share my toys with him" (Damon, 1977, p. 83). Alison's sense of obligation in the exercise of social justice is very clear: since he shared with her, she is obliged to share with him. Although Alison is light-years ahead of Sean, who simply took seven candy bars, and Jack, who seemed to think that might makes right, she still doesn't view fairness as a general principle. Fairness is still tied to reciprocity; it is still a conditional justice. The notion that one should be fair because it's the right thing to do must wait at least for middle childhood.

The Construction of the Concept of Authority

Preschoolers' understanding of social justice—of what is fair—evolves from their relationships with peers. However, many of the social encounters of childhood are not between equals but between unequals; that is, they are encounters between children and adults. Because children in our culture rarely have the same power that adults have, these encounters are less effective than encounters with peers are in facilitating the preschooler's understanding of equality and reciprocity. They are very effective, however, in facilitating the preschoolers' understanding of authority and obedience. From encounters with adults and to a lesser extent with older children, preschoolers begin to understand who has a legitimate right to tell them what to do and why they should do what they are told to do.

> The question of legitimacy—one of the two crucial issues in authority knowledge—has profound developmental implications. Not only is there development in the kinds of social power traits which children recognize and respect, but there is also development in the child's ability to apply different legitimizing social power traits to different types of authority relations. The second crucial issue in authority knowledge is the rationale for obedience. . . .For a child, the task is to construct a basis for obedience as well as for disobedience, beginning with an understanding of the most basic issue: why should one obey at all. (Damon, 1977, pp. 172-173)

Just as egocentrism makes it difficult for three- and four-year-olds to understand social justice, it also makes it difficult for them to understand authority and obedience. Although they certainly understand that they are different from adults, they are still very limited in their ability to appreciate the fact that their wishes, desires, needs, and demands are not only sometimes different from those of their parents, but in fact may actually be in conflict with each other. The young preschooler's confusion between independent authority and personal desire is reflected in Marisa's response to a story about a boy named Peter who wants to go to a picnic with his friends but whose mother wants him to clean his room first.

Q: What do you think Peter should do?

A: Go to the picnic.

Q: Why should he do that?

A: Because he wants to and all his friends are going.

Q: But what if his mother says, "No Peter, you can't go until you clean up your room first."

A: He would do what his mama says.

Q: Why would he do that?

A: Because he likes to.

Q: What if Peter really wants to go on the picnic and he doesn't want to clean up his room at all because if he does he'll miss the picnic?

A: His mama will let him go out with his friends.

Q: But what if she won't let him?

A: He will stay home and play with his sister and clean up all his toys in his toy box.

Q: Why will he do that?

A: He wants to. (Damon, 1977, p. 182)

Marisa's responses are typical of three- and four-year-olds. She seems to believe that authority should simply conform to personal desire. Peter should conform to his mother's desires, and at the same time his mother should conform to Peter's desires. The fact that these two acts are contradictory doesn't seem apparent to Marisa. She still seems to be dealing with events one at a time.

When preschoolers become able to see another's perspective as distinct from their own, they often first see it in opposition to their own wishes and desires. This oppositional perspective is still somewhat egocentric, in that another's view is defined only in relation to that of the preschooler. However, now at least the preschooler understands that there can be more than one perspective on a topic. Because the parent's view is seen as merely existing in opposition to personal desire rather than as an independent legitimate position, four- and five-year-olds' logic is often directed at ways to get around the obstacle.

Billy's solution to Peter's predicament is different from Marisa's. Billy thinks that Peter shouldn't sneak out to the picnic because his mother might catch him and punish him. But, "Suppose that the door was open and his mother was taking a nap and she couldn't find him. Would it be ok then?" Peter answers: "If he was going to the picnic and his mother was cleaning then he should run down the stairs and out the door so that his mother can't see him" (Damon, 1977, p. 185). Unlike Marisa, Billy understands that Peter's mother isn't going to change her mind just because Peter wants her to. However, Billy's sense of legitimacy seems to end at the bottom of the stairs.

By the time children reach the age of six or seven, they begin to understand that there are people who have the right to be obeyed. They have this right largely because they have the power to enforce their demands for obedience. Most typically, this power reflects size, strength, or social position. And it would seem from Heather's observations of Peter's predicament that these powers are very powerful indeed.

Q: If he sneaks out and goes on the picnic would that be ok?

A: No, because his mother said to pick up his room or you can't go out. So he'll just get in more trouble.

Q: But what if the mother never finds out, is that ok?

A: She would probably hear the door opening when he left.

Q: What if Peter were real quiet and she didn't hear anything.

A: No, because she would see dirt on his clothes when he got home. (p. 189)

It will not be until middle childhood that Marisa, Billy, and Heather will consider the possibility that the nature of the request or the reciprocity between the parent and the child is as much reason to obey a request as the consequence for not doing so is.

The Self-Concept During Early Childhood

The preschooler's concepts of sex-role, friendship, social justice, and authority develop along parallel courses. The parallels are more than simply coincidence. They reflect the preschooler's continuing development of a concept of self—in particular, an understanding of self in relation to others. This understanding began in infancy. The concept of object permanence, the knowledge that the self is distinct from other people and other things, was its first prominent milestone. The process continues during early childhood as it will for the rest of the life-span.

Preschoolers before age five seem very limited in their ability to appreciate the perspective of another (Selman, 1976a, 1976b). Although infants and certainly toddlers show distress at the discomfort of another and indicate, through their use of language, that they don't always think that others are aware of their wants and needs, the realization that others may perceive and react to a situation differently is fairly late in developing. Four-year-olds simply assume that others will respond to a situation as they would; that others feel what they feel; and that others share their personal point of view. These assumptions are not explicit in preschoolers, but their surprise when one is contradicted clearly indicates their implicit presence.

This inability to appreciate another's perspective helps to explain why early childhood friendships are typically situational. Preschoolers are simply unable to appreciate the unique character of another person. Similarly, there is no compulsion to justify their frequent unfair behaviors. There is no need for justification because there is no conflict, and there is no conflict because no other perspective is yet recognized. This failure to recognize potential conflict is equally evident in preschoolers' dealings with authority figures. From their perspective, what they want is obviously what others want them to have or to do, so they don't see there should be any problem.

It is useful to notice the parallel between the young preschooler's level of social perspective taking and the young infant's object concept. Neither involves any degree of differentiation. Preschoolers can't conceive of perspectives other than their own; young infants can't conceive of objects existing outside of their immediate sensory awareness.

This preschooler's attempt at a self-portrait is indicative of his growing sense of self.

Preschoolers and infants first encounter difficulty in their dealings with others when they finally become aware that others' thoughts and others' bodies, respectively, are distinct from and potentially independent of their own. Once these discoveries are made, the process of integration can begin. Such parallels as these between preschoolers and infants are frequently encountered at various points across the life-span. They show that development is a process of constructing and reconstructing an understanding of the meaning of events in the physical and social worlds.

By five or six years of age, preschoolers are often able to appreciate the fact that others do hold a different perspective from theirs. However, they are still limited in their ability to relate these perspectives. In particular, they are still unable to see themselves as others see them, and they do not yet appreciate the fact that their view of another is partly determined by their perception of the other's view of them. But because they now appreciate the fact that others can see things differently, they begin to sense the need to justify their behavior to others; or in the case of authority figures such as parents, the need to follow their directions. The shift in criteria for friendship from circumstance to similarity of interests also reflects this increasing awareness of the uniqueness of others. The further evolution of social perspective taking, specifically appreciation of the reciprocal nature of interpersonal relations and the subjective quality of personal evaluation, must, however, wait for middle childhood.

Play: The Integrating Experience of Early Childhood

If you have had the opportunity to spend much time with preschoolers, you no doubt quickly made the discovery that preschoolers spend most of their time playing. From the preschoolers' perspective, they're just having fun. From the adult's perspective, play should be seen as a source of delight to preschoolers and as the primary means by which they gain a sense of mastery over their environment, exercise newly acquired cognitive skills, rehearse their new understanding of social roles, foster divergent thinking, facilitate decentration, and acquire social interaction skills (Fein, 1978; Garvey, 1974; Vandenberg, 1980).

In one sense the play of the preschooler is a continuation of the tertiary circular reactions that characterize the last phases of the sensorimotor stage. There is still the quality of curiosity, the "I wonder what would happen if" kind of exploration, but now, in addition, preschoolers make much greater use of their emerging symbolic and social interaction skills. The play of the young preschooler is primarily solitary. The play themes reflect the materials available. A car prompts a driving game; an airplane, a flying game; a doll, a nurturing game, and so forth. The preschooler makes up a story and acts it out. It probably has little in the way of plot and sequence and almost certainly no end point, but nevertheless it reflects an early attempt to go beyond the limits of the immediate environment.

When children reach the age of four or five, the quality of play begins to change. Activities become more social in nature; pretend play becomes a group activity. Make-believe becomes less restricted by the availability of props. Almost any object comes to serve as the car, boat, doll, or hat. Sometimes just the action of pretending to put on the hat is sufficient. Play takes on an increasingly social quality as the activity of one child reflects

BOX 4-1
THE SCHOOL FOR CONSTRUCTIVE PLAY

How would you go about freezing motion? How could you show children what the back of a picture looks like (not the back of the piece of paper the picture is drawn on but the back of the picture itself)? How can preschool children learn to better appreciate the fact that the same object or experience can be seen from more than one perspective?

These questions and many more like them form the core of a unique preschool curriculum developed by George Forman and his colleagues at the University of Massachusetts at Amherst (Forman & Hill, 1984; Forman & Kuschner, 1977). The curriculum is in use at UMass as well as at the Early Childhood Development Center at the University of Vermont and other early childhood centers. It is a program that has been shown useful for both normal and exceptional children.

The constructive play curriculum draws on Piaget's notions about the way preoperational children view their world. In particular, its focus is to help children make connections between what often appear to them as discrete and unrelated events. In the case of the first question posed, the question about freezing motion, preschooler's often have difficulty in seeing how some-

The constructive play curriculum uses learning encounters, such as the sand pendulum, to enable preoperational children to see how things change state.

Sources: Forman & Hill, 1984; Forman & Kuschner, 1977; Goldhaber, Goldhaber, Ishee, & Thousand, 1980.

that of another. However, with the exception of games such as "house" or "doctor" in which the children are playing stereotyped roles, many of these social games are still very loose in organization.

A game for two players may be played alone; when played with someone else, each child operates according to a different set of rules and little or no attempt is made to recognize and reconcile the differences. Four year olds can take turns and respond to each other's actions, but as often as not, they fail to watch each other, and new turns have little relation to whatever came before. Games have no beginning and no end. Everyone can win; winning does not

thing gets from one place or state to another. One solution is the "sand pendulum." A hollow cone is filled with sand and suspended over a sheet of paper. The tip of the cone has a small hole in it. As the child makes the pendulum swing, the sand flows out of the cone onto the paper. The resulting sand picture is a frozen representation of the route of the pendulum. Seeing the picture is one way preschoolers are helped to understand the relationship between product and process. In fact, the importance of this notion is reflected in the curriculum's motto: "change, don't exchange."

Activities such as the sand pendulum are called learning encounters. On a typical day a number of different encounters might be available to children. The children are not "taught" the message in each encounter but are free to approach the setup, play with it, talk about it, and use it in different ways. The teachers take their cues from the children. Through questions, comments, modifications of the setups, and even through modeling, teachers help children become more aware of the relationships between process and product, between their actions and the words used to describe them. In a sense, teachers serve as "supportive troublemakers," always asking the question or presenting the situation that leads children to think a little more about the relationships between the different elements of the encounter they are pursuing.

How do you see the back of a picture? You could try painting a picture on a piece of plexiglass. How do you teach perspectives? You could try the silhouette sorter. The silhouette sorter is a variation of a common infant toy. The toy is a puzzle box with a variety of differently shaped openings on its sides. Infants enjoy learning which of the different objects that come with the box goes with or fits into each opening. Each object has its own opening. The silhouette sorter uses only one object, a very complex one that has to be turned in a variety of different ways to fit into its various silhouettes cut on the sides of the box.

mean being better than others according to agreed upon criteria, but rather, playing on one's own and achieving one's own goals.
(Cowan, 1978, p. 169)

How does all this playful activity foster the long list of accomplishments mentioned at the beginning of this section? First, because play places greater emphasis on process than on product (the pretend games don't so much end as "dissolve"), it provides preschoolers an opportunity to practice a variety of problem-solving strategies that prove very useful in middle childhood (Sylva, Gruner, & Genova, 1976; Weisler & McCall, 1976). Without having to worry about the pressure of success or failure,

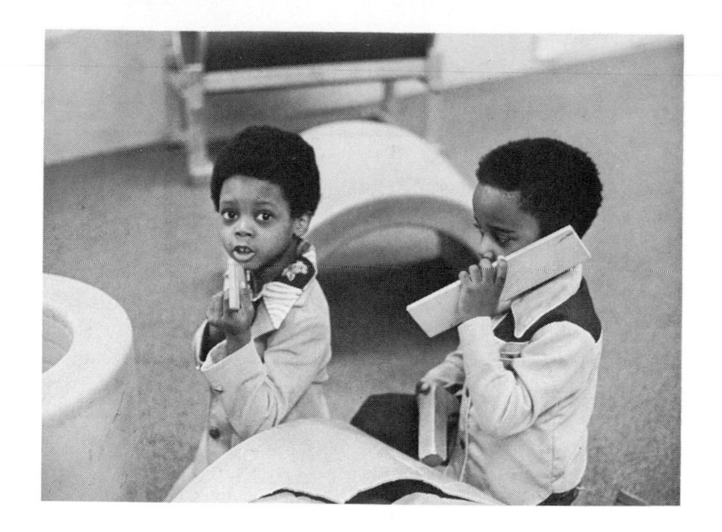

These four-year-olds have no trouble turning their blocks of wood into walkie-talkies and guns for a serious game of cops and robbers.

preschoolers are free to try a variety of ways of doing things, and in the process, flexibility in thought is enhanced. Starting an activity, organizing its content, and regulating its direction are all skills that foster the more goal-directed activities of middle childhood as well as the pretend play of the preschooler.

More generally, play provides preschoolers an opportunity to nurture a variety of cognitive skills. Garvey (1974) notes that play helps preschoolers separate reality from fantasy. Preschoolers seem very proud of the fact that they can assure parents or teachers that they are "just pretending." This accomplishment may not immediately ease the preschooler's fear of the dark or remove the necessity of elaborate bedtime rituals, but it is nevertheless a definite step in that direction. Further, social play helps preschoolers decenter, that is, to become less egocentric (Fein, 1978). Playing different roles in different plays or even in the same play and having to coordinate your actions with those of another child all help you begin to realize that there is more than one way to view a situation and that sometimes those views may be in conflict.

At the same time that play helps preschoolers become less egocentric, it also helps them gain a better sense of themselves as unique individuals (Rubin & Pepler, 1980). Preschoolers are not only proud of being able to assure the adult that "we're just pretending," they may also feel the need to assure the adult that "it's still really us even though we're pretending to be different people" or wild animals or whatever. Preschoolers are demonstrating an early form of conservation—in this case, **conservation of social identity.** They are demonstrating a first awareness of the fact that even though something can change at one level (dressing up like another person), at another level it remains the same. Again this awareness may be very situational and not easily transferred (preschoolers, for example, often have trouble recognizing their preschool teachers outside of the classroom), but like the reality-fantasy differentiation, it is a step in the right direction.

Although play is an important element in the young child's development, its occurrence can be very situational. Children, both preschool and older, play when they feel comfortable in a situation. A toy becomes a prop in pretend play only after it has been fully explored, only after it offers no surprises (Hutt, 1976). The same is true of settings. The first day at a new preschool is a time for observation from a safe vantage point (from the child's point of view, a teacher's lap will do just fine), not a time for games. Since objects are more predictable than other people are, it is reasonable that from both a developmental and a situational perspective, children play with objects before they play with other children. Children who are hungry or tired or sick don't play. In fact, most parents quickly learn that their children's play patterns are a very good indicator of their state and mood. Children from privileged backgrounds, both in the emotional and physical sense, are more likely to play than those less fortunate. In this sense, preschoolers' play, like other aspects of their development, is highly reflective of the context in which it occurs. And it is to the contexts of development during early childhood that we now turn.

THE CONTEXT OF DEVELOPMENT DURING EARLY CHILDHOOD

Parents continue to be the dominant influence on children's development during the early childhood years. But they are not the only influence. Peers begin to play a significant role in influencing the course of development. For many children, early childhood marks the first introduction of the child into a group educational setting—either full-time day-care programs, part-time preschool programs, or home-based child-care centers. And, there's television. Although the influence of television is still a topic of much controversy, we do know that the average preschooler spends two to three hours each day in front of a television and that by the time children reach the age of eighteen, they have spent more hours watching television than attending school (Liebert, Sprafkin, & Davidson, 1982).

Parent-Child Relations

The tasks of the parents of a preschooler can often seem contradictory. They need to continue to be responsive to the biological needs of the child while at the same time encouraging the child to assume some of the responsibilities of dressing, bathing, brushing teeth, eating, and so forth. They must be certain that their efforts are appropriate to the present needs of the child while at the same time considering the potential long-term consequences of such activities as piano lessons and the like. They must regulate and direct the child's behavior while at the same time providing the child increasingly greater opportunities for independence and self-direction. Maintaining the appropriate balance between each of these attitudes can, at times, be a difficult task, especially with a child who can like string beans on Monday but hate them on Tuesday, or one who can ask for help but then get mad when you provide it, or one who is not able to appreciate how hard you really tried to pour the same amount of milk into the two differently shaped glasses.

As was true during infancy (Ainsworth & Bell, 1974), the perceptions parents hold about preschoolers are an important influence in determining how they deal with their children. Parents who have a realistic perception of the expectations one can have of preschoolers, of the types of control techniques that are most effective in regulating their behavior, and of the behaviors that a preschooler has a reasonable right to expect of a parent are most likely to have preschoolers who are generally happy, self-reliant, and controlled. Not surprisingly, they are also the parents whose preschoolers are beginning to realize that parents' actions are not arbitrary but deliberate and well intentioned. Children who view their parents' actions as justified in terms of their own welfare are much more likely to accept parental directives and to behave in a mature and independent manner than are children who see their parents' actions as arbitrary and overprotective (Baumrind, 1973).

Dimensions of Parent Behavior

Research on parenting techniques (Baumrind, 1971; 1973; 1978) has identified four dimensions along which parenting behavior can be described and compared. The first is *parental control.* It defines the specific techniques that parents use to regulate the preschooler's behavior, to shape goal-directed activities, and to foster internalization of parental standards. Control is not the same as punitiveness. Rather it is a measure of "the parent's ability to enforce directives when the child initially does not obey" (Baumrind, 1973, p. 7).

The second dimension of parent behavior is *maturity demands.* It measures the degree to which the parent holds age-appropriate and ability-appropriate expectations concerning the preschooler's behavior. In a sense, maturity demands measure how soon and in what domains parents expect their preschoolers to assume responsibility for their own behavior and decision making. A parent making high maturity demands would be unlikely to accept a temper tantrum from a six-year-old. If told by the child to zip a zipper or buckle boots, the parent would either encourage the child to try again, suggest another way to solve the task, or perhaps simply get the zipper started. The parent making low maturity demands would be more likely to accept children's statements that they couldn't dress themselves and do the zipping and buckling for them.

The third dimension concerns the *nature of the parent's communication* with the preschooler. It measures the extent to which parents use reason and explanation in their efforts to regulate their child's behavior. It also measures the extent to which parents elicit preschooler's opinions in decisions concerning them as well as their reactions to parent requests and demands.

The fourth dimension of parent behavior is *nurturance.* It is expressed in a variety of ways. It is evident in the caretaking aspects of the parenting role. Are bathtime and mealtime more often seen as opportunities to spend time together or merely as tasks that must be done? Nurturance is reflected in the efforts parents make to protect the physical and emotional well-being of the child. Is the physical environment safe for young children? How available are parents for support and comfort? Nurturance is demonstrated through the parents' attempts to spend time with their child. It is reflected in the pride parents take in the accomplishments of their children, but it is also reflected in the various ways that parents conscientiously protect their child's welfare. That is, nurturance is not always shown by allowing children to do what they wish. Sometimes nurturance is shown by *not* allowing children to do what they wish.

Parental Authority Patterns

Ratings on the four dimensions of parental behavior cluster into three relatively unique patterns. The patterns, collectively referred to as **Baumrind's parenting patterns,** include **permissive parenting, authoritarian parenting,** and **authoritative parenting.** Each unique pattern is in turn associated with a particular pattern of preschooler behavior. Table 4-1 presents the three patterns of parental authority as defined by Baumrind (1968).

Perhaps Baumrind's most interesting finding is that there are not three unique patterns of child behavior associated with the three unique patterns of parent behavior. There are only two unique patterns of child behavior. The preschool children of authoritarian and permissive parents are actually quite similar to each other, and both are quite distinct from the preschool children of authoritative parents.

When compared to the children of permissive or authoritarian parents, children of authoritative parents are more likely to be socially responsible, independent, purposive, and achievement oriented. Social responsibility concerns how well children get along with peers, cooperate in adult-directed activities, and help rather than hinder others' work. Socially responsible preschoolers would, for example, usually agree to a parent's request to put away their toys, would be less likely than other children to destroy another child's block structure in a preschool, and would usually be willing to take turns or share crayons or paints with other preschoolers.

Independent preschoolers are usually able to find something to keep them busy. They are least likely to constantly ask adults for suggestions for activities. Their ability to keep themselves occupied isn't simply a willingness to follow the suggestions of others. Rather, they are usually able to decide on a task for themselves. They might ask an adult for assistance in defining the task or in obtaining the necessary materials, but the idea itself is typically theirs. This sense of self-direction is apparently noticed by their peers. These preschoolers are viewed as leaders and as favored playmates.

The preschool children of authoritative parents like to set tasks for themselves that are challenging. They are very achievement oriented. They don't shy away from new tasks but pursue them actively and efficiently. The preschool children of both permissive and authoritarian parents tend to be submissive to others, demonstrate little social responsibility, are more likely to demonstrate immature forms of attention getting, dependency, and disapproval of adult sanctions, and are less interested in intellectually challenging tasks (Baumrind, 1973). Why are these parent-child patterns as they are? How is it possible that two distinctly different forms of parenting (authoritarian and permissive) should produce such similar children?

Baumrind (1973) believes that both the authoritarian and the permissive parent lack confidence in their parenting skills. This lack of confidence limits the parents' ability to formulate a coherent child-rearing policy. As a result, their parenting behaviors lack a balance between what is offered the child in the way of support and what is demanded in terms of compliance. The balance is in the opposite direction for the two groups of parents (high support and low compliance for permissive parents; low support and high compliance for authoritarian parents), but the direction of the imbalance seems to be of less significance than its presence. The authoritative parent offers, in contrast, a more balanced and coherent child-rearing strategy. The high demands for maturity are balanced with a high level of support to help achieve the goals.

TABLE 4-1 Current Patterns of Parental Authority

AUTHORITARIAN
The authoritarian parent attempts to shape, control, and evaluate the behavior and attitudes of the child in accordance with a set standard of conduct, usually an absolute standard that is theologically motivated and formulated by a higher authority. The parent values obedience as a virtue and favors punitive, forceful measures to curb self-will at points where the child's actions or beliefs conflict with what the parent thinks is appropriate conduct. The parent believes in inculcating such instrumental values as respect for authority, respect for work, and respect for the preservation of order and traditional structure. The parent does not encourage verbal give-and-take, believing that the child should accept the parent's word for what is right.
AUTHORITATIVE
The authoritative parent, by contrast with the authoritarian parent, attempts to direct the child's activities but in a rational, issue-oriented manner. The parent encourages verbal give-and-take and shares with the child the reasoning behind the policy. The parent values both expressive and instrumental attributes, both autonomous self-will and disciplined authority. Therefore, the parent exerts firm control at points of parent-child divergence but does not hem in the child with restrictions. The parent recognizes his or her own special rights as an adult but also recognizes the child's individual interests and special ways. The authoritative parent affirms the child's present qualities but also sets standards for future conduct. The parent uses reason as well as power to achieve objectives. The parent does not base his or her decisions on consensus or the individual child's desires but also does not consider him or herself infallible or divinely inspired.
PERMISSIVE
The permissive parent attempts to behave in a nonpunitive, acceptant, and affirmative manner toward the child's impulses, desires, and actions. The parent consults with the child about policy decisions and gives explanations for family rules. The parent makes few demands for household responsibilities and orderly behavior. The parent presents him or herself to the child as a resource to be used as the child wishes rather than as an active agent responsible for shaping or altering the child's ongoing or future behavior. The parent allows the child to regulate his or her own activities as much as possible, avoids the exercise of control, and does not encourage the child to obey externally defined standards. The parent attempts to use reason but not overt power to accomplish his or her ends.

Source: Adapted from Baumrind, 1968.

The balance between high expectations and high involvement convinces preschoolers that their parents are truly concerned about their welfare. As a result, the children cooperate. The high parental expectations set clear goals for the preschoolers while at the same time the high parental involvement provides a visible and supportive model for the children to imitate.

The value of authoritative parenting isn't simply that it produces a well-behaved child. This parenting approach also has a more long-range benefit: it fosters cognitive growth and a sense of self-worth in the child. By behaving in a rational, issue-oriented manner, by allowing children to express opinions, and by sharing with children reasons for parental actions, parents help their children develop a more realistic perception of themselves in relation to events and other people. Unlike the authoritarian parent whose techniques get the preschooler to focus on compliance or the permissive parent who provides little for the preschooler to focus on, the authoritative parents' techniques help preschoolers focus on the consequences of their behavior in relationship to themselves as well as others. Further, the authoritative parents' granting of a reasonable degree of autonomy gives preschoolers sufficient independence of action and choice to foster the development of self-control (Hoffman, 1975). The trust that the parent conveys to the preschooler through the granting of independence fosters a sense of responsibility in the preschooler. The verbal give-and-take between parents and preschooler as well as the parents' willingness to justify their actions to the preschooler fosters a sense of self-worth and self-esteem in the preschooler.

Sex-Role and Social Class Differences in Parenting Strategies

Baumrind's research provides the most complete overview of parenting styles and their consequences during early childhood. However, it does not focus on the interactive and individual quality of the parent-child relationship, nor does it consider factors that prompt the choice of parenting style. For example, there continue to be differences in the parenting styles of mothers and fathers during early childhood. In particular, fathers appear to trade their playful role for a very instrumental one (Clarke-Stewart, 1977). Fathers increasingly become more task oriented than mothers, while mothers continue their nurturant, care-taking, expressive roles.

This instrumental role assumed by fathers is most evident in their increasing expectations that their sons and daughters each act in sex-appropriate ways. During the preschool years, it is fathers rather than mothers that through both modeling and reinforcement foster the differentiation of male and female behaviors. Given the pivotal role that fathers play, any efforts to make male and female behavior less sex typed will have to take into consideration ways to change the attitudes of fathers concerning appropriate sex-role behavior in children.

Clarke-Stewart (1977) also notes that socioeconomic factors influence parenting styles. She notes that the higher the educational, occupational, and income level of the parents the more likely they are to use parenting techniques emphasizing the preschooler's need for companionship, affection, and intellectual stimulation. Further, these parents are more likely to influence their children's behavior through the use of requests, consultations, and explanations. On the other hand, parents from less economically advantaged backgrounds are more likely to influence behavior through the use of coaxing, command, threats, and punishments. If you think that these different patterns sound very much like authoritative and authoritarian parenting strategies, you are right. Considering the social class perspectives noted by Kohn (1977), it is not surprising that economically disadvantaged families, perceiving themselves as having little power to influence the social structure, employ parenting techniques that emphasize conformity; whereas those from more economically advantaged backgrounds, who believe that they do have the opportunity to make a difference, employ techniques stressing independence, self-control, and self-esteem.

Child Influences on Parenting Strategies

What influence does the child have on parenting techniques? As was true during infancy, children's behavior serves as a barometer of parent effectiveness. Techniques that work are continued; those that don't are eliminated or modified, or in some instances merely intensified. The same temperament issues that were shown to influence parent-infant interactions also influence parent-preschooler interactions (Thomas & Chess, 1977). Parents who report the greatest difficulty in getting their children to conform to the various socialization demands that become increasingly common during early childhood have children who tend to maintain a high level of activity, are intense in their reaction to situations, are very persistent in self-directed activities but easily distracted from activities initiated by others, and have difficulty adjusting to new situations. These hard-to-manage children are also more likely to be referred by parents to child guidance clinics.

The hard-to-manage child pattern was also noted by Patterson (1980) in his studies of the negative interaction patterns into which parents and children can fall. Patterson and his colleagues developed a comprehensive observation system that was used in the homes of families with hard-to-manage children. The observer recorded not only the actual behaviors but also the sequences in which they occurred. Through this procedure, it was possible to analyze both the actual behaviors involved in conflict and the degree to which any one particular behavior was likely to be followed by any other behavior. This same coding system was used following participation in the parent training program developed by Patterson in order to determine the effectiveness of the intervention procedures.

Hard-to-manage preschoolers quickly learn that negative behaviors can be more effective than positive behaviors in gaining the attention of a parent or teacher. Parents find such behavior unacceptable and respond with some form of punishment. Although this pattern is common in all families, with hard-to-manage children, the effect sometimes backfires. Rather than reducing the undesirable behavior, it may actually increase it. Left unchecked, the cycle perpetuates itself. Parents become less likely to initiate positive activities while at the same time actually becoming increasingly critical of many of the child's initiatives, even those that were once considered acceptable. Parents are often aware of the ineffectiveness of their techniques but knowing no alternatives, simply intensify their efforts. When this intensification is met with a corresponding escalation by the child, parents feel trapped, bewildered, and depressed. The programs that Patterson has developed to help parents out of this trap teach such skills as the effective use of both positive and negative reinforcement, role play and modeling techniques to help parents become more aware of how situations can get out of hand, and child observation techniques to help parents note and appreciate the good qualities of their children.

Cultural Influences on Parenting Strategies

Factors that influence parent-preschooler interactions are not restricted to the immediate environment. The considerable body of research (e.g. Belsky, 1980; Garbarino, 1976; Gellas, 1978; Kempe & Kempe, 1978; Parke & Lewis, 1981); on child abuse makes very evident that factors existing at all levels of the ecosystem affect the likelihood that parents will abuse or neglect their children (see Box 1-2 in Chapter 1). This multilevel perspective is well illustrated by Parke and Lewis (1981) in Figure 4-1, and as Garbarino (1976) notes in the following quote, the precipitating factors may often go beyond the control of the individual family members:

> The results suggest that the ecological context generated by economic and educational resources is an important factor in the etiology of child abuse/maltreatment. The analysis suggests a hypothesis for further investigation: economic stress and inadequate educational resources undermine the functioning of parents, particularly mothers, and out of such stress comes child abuse/maltreatment. Overall economic distress appears to be important through its impact on mothers as well as directly on the community and neighborhood. In poor areas more mothers with children work, mothers heading households have lower incomes and less education, and fewer children are enrolled in educational programs. Such appears to be a major feature of the human ecology of child abuse/maltreatment: economically depressed mothers, often alone in the role of parent, attempting to cope in isolation without adequate facilities and resources for their children. These findings are hardly surprising, but they are important because they point to the pervasive and insidious effects of the combination of being poor and attempting to manage a household with only one parent. (p. 183)

The degree to which factors influencing child abuse and neglect are beyond the immediate control of the parent is evident in Parke's multilevel social interaction model shown in Figure 4-1. Parents have the most control over events occurring at the family level, less over events at the community level, and still less over events at the level of the culture. Parents can be much more effective in changing their child-rearing tactics than they can be in increasing the availability of employment, health care, or child-care services. Changing attitudes and values is even more difficult. Representatives of local government might at least agree that better health care facilities are needed in a community (even though they may be unwilling to fund such projects); they may be less willing or able to see a relationship between the incidence of abuse in an economically depressed neighborhood and the set of toy guns one of the representatives just purchased for his son as a birthday present. Parke is not suggesting that parents do not bear the ultimate responsibility for the welfare of their children; rather, he is arguing that society must assume more of the responsibility of aiding parents in carrying out the tasks that society has assigned to them.

BOX 4-2
MOTHERS: THE UNACKNOWLEDGED VICTIMS

The title of this box is the title of a monograph prepared by Gerald Patterson (1980). It is a report of the work that he and his colleagues have been conducting over a period of many years concerning the origins and evolution of self-defeating parent-child relationships. Consider this simple episode. A mother wants her child to clean his room. The mother makes the request; the child does not comply but instead begins to whine. The mother repeats the request, the whining continues, gradually increasing in intensity. Eventually, the mother withdraws the request, and the child stops whining. Patterson sees this all too common pattern as a reinforcement trap. Not only does the room not get cleaned, but of greater importance, the very behavior (withdrawing the request) that in the short run reduces the child's irritating behavior (whining) in the long run actually increases it. The child will undoubtly be more likely to whine on the next occasion. The parent is not unaware of this dilemma. She might well respond, "I know, I know, but I just can't stand it when he whines like that." If this pattern continues over an extended period, Patterson finds that it takes on additional qualities. The child's aversive behaviors become more frequent and more intense, gradually taking place not only in response to a parent request but also as "leverage" to get what he wants. Parent behavior also escalates, becoming more punitive but at the same time less effective in regulating the child's behavior. To make matters worse, the frequency of nonaversive interactions—the good times together—gradually decreases. All that remain are the increasingly aversive, increasingly punitive, decreasingly effective episodes.

To help parents remedy these destructive patterns, Patterson has developed a variety of family management skills training programs. These programs first help parents become more aware of the sequence of events that often ends in a punitive encounter. In many cases, parent-child interactions are

Source: Patterson, 1980.

Certainly not all cultural influences are negative. Those communities that provide adequate social, educational, and recreational services enhance the parenting process. Communities that instill a sense of civic or ethnic pride in individuals give both children and adults a set of guidelines and role models that foster the process of development. If I and others in the social sciences dwell on the negative, it is not because we do not see any examples that are good. Rather, it is because we see too many examples that are bad.

In most families most of the time, preschoolers do not spend their time designing more ingenious ways to hassle their parents, and parents do not resort to abuse and neglect as management (mismanagement actually) techniques. The openness and inquisitiveness of preschoolers, their genuine love for their parents, and the rather unique way they have of seeing their world usually make them a delight to be with.

In many families parent-child interactions fall victim to the reinforcement trap: parental requests are met with negative behavior on the part of the child. To eliminate the behavior parents withdraw their requests thereby reinforcing the negative behavior. Thus the room never gets cleaned, the garbage is never taken out, and the child learns that refusal is an effective tactic.

videotaped and then discussed with the parents. Parents are also given alternative management techniques. These techniques are especially helpful for parents who don't like the way they deal with their children but have never had the experience to learn less punitive strategies. Role playing is often used as a means of practicing these new skills. Parents are also given the opportunity to learn ways to more clearly define to themselves as well as their children exactly what it is they want the children to do. Telling a child to clean his room is a much less efficient direction than telling him to put all his toys in his toy box. Patterson reports that these techniques work, that the incidence of aversive and punitive interactions decreases. Perhaps of greater significance is the finding that as the interaction patterns become more positive, the parents' sense of competence in their parenting increases, ensuring that the benefits are likely to continue even after the training program has ended.

Peer Relations in Early Childhood

Although the egocentric nature of preschoolers often prevents them from fully appreciating the fact that others may see the world from a perspective that is different from theirs, it certainly doesn't prevent them from wanting to be with others, especially children of their own age. Between the ages of two and six, children become increasingly interested in playing with other children, in being able to sustain group play activities, and eventually, in participating in complex play situations that involve cooperation and divisions of labor.

Peers provide preschoolers learning opportunities that the adult can't provide. In general, preschoolers learn from other preschoolers patterns of social interaction that are reciprocal rather than complementary in nature

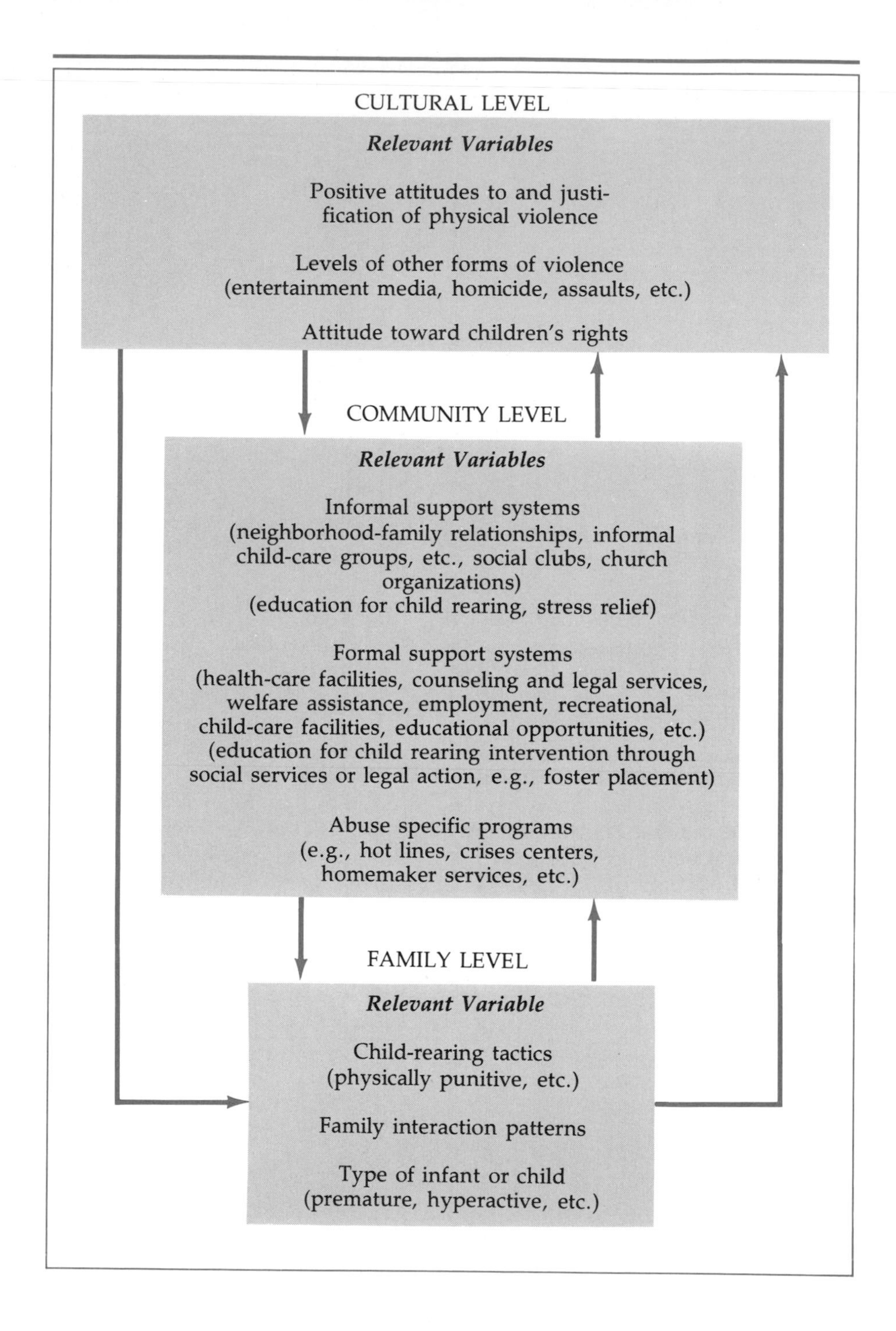

FIGURE 4-1
The factors that influence child abuse and neglect are not restricted to the immediate environment. Cultural, community, and familial factors work together to affect parent-child interaction patterns (from Parke & Lewis, 1981).

(Youniss, 1980). That is, they learn that unlike adults, children often hit back. They also learn the social skills necessary to be accepted into a peer activity and the fact that preschoolers can at times be very exclusive and restrictive in their dealings with peers. Consider the following example cited by Corsaro (1981):

In the following sequence drawn from field notes, Linda (3.8) attempts to gain access to an ongoing episode involving Barbara (3.0), Nancy (3.1), and the researcher (R). Barbara (B) and Nancy (N) had been playing with toy animals and building blocks for around ten minutes when Linda (L) attempted to gain entry. Linda watched at a distance for one minute, then entered the block area, sat next to Barbara, and began playing with the animals.

B-L: You can't play!
L-B: Yes, I can. I can have some animals too.
B-L: No, you can't. We don't like you today.
N-L: You're not our friend.
L-BN: I can play here, too.
B-N: No, her can't-her can't play, right N?
N-B: Right.
L-R: Can I have some animals, Bill?
R-L: You can have some of these. (R offers L some of his animals.)
B-R: She can't play Bill, cause she's not our friend.
R-B: Why not? You guys played with her yesterday.
N-R: Well, we hate her today.
L-All: Well, I'll tell teacher. (L now leaves but then returns with a teaching assistant (TA).)
TA-BN: Girls, can L play with you?
B-TA: No! She's not our friend.
TA-B: Why can't you *all* be friends?
B-TA: No!
N-B: Let's go outside, B.
B-N: Okay. (B and N leave. L remains shortly and plays with animals then moves into juice room).[1]

At other times, peers provide preschoolers social support and encouragement as well as information about how other children behave (Oden, 1982). Over the course of the preschool years, children become increasingly effective in finding friends who share their interests and styles and avoiding those whose behavior is quite different from theirs. As preschoolers become better able to make and sustain friendships, they place more emphasis on their play-partner preferences, a fact that poor Linda learned the hard way.

The Influence of Television on Young Children

Television serves many roles in the lives of young children. It is an educator, a source of amusement, a seller of goods and services, an informant about the lives of other people, and as most parents will reluctantly admit, an often effective baby-sitter. Because television has the potential to serve so many roles so very effectively, the extent of its influence has become a topic of much research and debate (see Liebert, Sprafkin, & Davidson,

[1]Reprinted, by permission of the publisher, from W. A. Corsaro, "Friendship in the Nursery School: Social Organization in a Peer Environment," in S. R. Asher & J. M. Gottman (Eds.), *The Development of Children's Friendships* (Cambridge, England: Cambridge University Press, 1981), 214–215.

1982, for a comprehensive review). A great deal of this inquiry has focused on preschool children because of their limited ability to effectively sort information and their tendency to imitate the behavior of others. The debates separate into four categories. The first concerns the impact of the role models that preschoolers see on television; the second, the impact of commercials; the third, the impact of the violent content of many television programs; and the fourth, the potential of television as an educational tool.

Television programs, both those directly aimed at children and those intended primarily for adults, present a distorted view of the roles males and females portray in real life and of the proportion of individuals from racial and ethnic minorities. If people were to draw an image of the world from what they saw on television, they would conclude that there are twice as many men in the world as women (Sternglanz & Serbin, 1974); that males assume all the dominant and action-oriented roles in society while women tend to provide support services and to be observers rather than participants; that members of racial and ethnic minorities are relatively rare in this country and those that are present often tend to have menial or secondary roles in society; and that almost everybody is young (Liebert, Sprafkin, & Davidson, 1982). None of these statements is true, but television continues to maintain these stereotypes in spite of frequent criticism from parent and professional groups. Even Big Bird is a male! Such stereotypes may make for good business, and it is important to acknowledge that commercial television is a business, but their potential impact on young children is generally considered negative (Mussen & Eisenberg-Berg, 1977).

Preschool children seem particularly vulnerable to television advertising. There is now clear evidence (see Liebert, Sprafkin, & Davidson, 1982) that young children's toy and food preferences (toys and food being the two most common kinds of merchandise in advertising aimed at children) are directly influenced by the products they see displayed on television. Further, preschoolers are often unaware of the various qualifiers that are included in advertising (since most are unable to read) concerning the need for assembly, the additional purchase of batteries and other accessories, and the fact that the pieces displayed in a panorama are usually sold separately. None of this should be surprising since the evidence indicates that preschoolers are often even unaware that a program has stopped and a commercial begun. In fact many preschoolers are so trusting—or naive—that they think advertising merely provides information about product quality and availability in the same spirit as publications such as *Consumer Reports*. When interviewed about the motives of advertisers, they are considerably more likely to say that the people who make the product want you to know what it is rather than to say that their primary motive is to get you to buy the product. Fortunately this naiveté is short-lived. By the age of ten or eleven, most children have become very skeptical about the claims made in advertising. However, since preschoolers are impressionable, rarely have their own money, and are a means advertisers can use to reach parents, recommendations to remove advertising from programs primarily aimed at young children are probably quite correct. In effect, preschoolers are being exploited to get parents to buy the advertiser's product.

Advertising pressures are not the only problems presented by television. By the time the average television viewer reaches age eighteen, he or she will have witnessed approximately eighteen thousand murders portrayed on all forms of television programming including children's cartoons (Brody, 1975). That's a lot of carnage, and as is true in the case of advertising, preschoolers seem particularly influenced by the depiction of violence on television.

The most convincing documentation of this pattern is provided by Friedrich and Stein (1973). It remains an important study because it demonstrates a relationship between television viewing and typical, everyday behavior as opposed to laboratory behavior. Friedrich and Stein observed that preschoolers who tended to be more aggressive in the preschool classroom actually increased their frequency of aggressive behavior when over a period of weeks they were shown regularly scheduled television shows that had a high aggressive content. On the other hand, children who were relatively nonaggressive were not influenced by the same programs. Their observed level of aggressive behavior did not increase. Equally important was the fact that children who were initially highly aggressive and who viewed, over the same time period, programs in which aggression was absent actually decreased their frequency of aggressive behavior toward other children in the classroom. Friedrich and Stein also observed that preschoolers who viewed the programming that was low in aggression were considerably more likely to respond to teacher requests, had a greater tolerance for delay, and were more able to persist at a task than were the children exposed to programming that was high in aggression. Although the findings provide a clear statement as to the potential impact of program content on preschoolers, it is important to remember that television content does not act independently of other variables in influencing behavior. It is no surprise that children who prefer watching violent television programming have parents with the same viewing preferences.

Isn't there anything good to be said for television? The answer, of course, is yes. Many television programs provide children information about the biological and physical worlds that is simply unavailable elsewhere. Further, by showing children how people in different parts of the world live, television serves to broaden horizons. Specific television shows, such as "Sesame Street," "3-2-1 Contact," "Mister Roger's Neighborhood," and the now retired (but still fondly remembered) "Captain Kangaroo," that are designed to provide deliberate educational content have been shown to provide children with information they might other-

Researchers, such as Friedrich and Stein (1973), have found a positive relationship between the level of aggressive content in television programs and aggressive behavior on the part of young viewers.

wise not acquire. The most extensive evaluations concerning educational television have been studies of "Sesame Street," and they have found that preschoolers who regularly watch the show demonstrate better letter and number recognition, alphabet recitation, and basic reading readiness skills than other preschoolers do (Ball & Bogatz, 1972; Lesser, 1974).

Educational Experiences during Early Childhood

There have always been preschool programs for young children, but during the past twenty years the justifications for these programs and the number of children attending them have changed drastically. Four factors are reflected in the increasingly larger percentage of preschool-age children attending early childhood programs (Evans, 1975; Morrison, 1980). Preschool children from economically disadvantaged backgrounds are enrolled in various types of early intervention programs such as Project Head Start in the belief that they will acquire the educational skills and experiences that are common among children from more economically advantaged backgrounds and have been shown necessary for early elementary school achievement. A second group of preschool children, usually from economically advantaged backgrounds, attend early childhood programs because their parents believe that if children are provided the opportunity to acquire academic skills sooner than might otherwise be the case, they will have an advantage over other children in a world that some of these parents probably see as very competitive and perhaps even hostile. A third group of children attend early childhood programs because their parents or parent (many of the children in this group come from single parent families) is either forced to seek employment for economic reasons or chooses to pursue a career for more personal reasons. A fourth group of preschoolers attend early childhood programs because their parents want them to have the opportunity to be with other children, to learn to function effectively in group settings away from the parent, and to have the opportunity to do things that might not be possible in a home setting.

There is considerable variety in the types of preschool programs available for young children. This diversity reflects the fact that preschool programs are often independent of the local school systems and as such are less confined as to the nature of their programs. Preschool programs range in length from half-time to full-time care (day-care centers often have children arrive as early as seven in the morning and stay as late as five or six in the afternoon) and from two or three times a week to every day. Some programs are child centered and primarily provide children an opportunity to pursue their individual interests in a safe, stimulating environment. Others are very deliberate in nature and have an academic format similar to that in an elementary school classroom. Some programs emphasize the learning of specific content, others the nurturing of cognitive skills, and others the fostering of a healthy self-concept, effective control of the emotions, and more effective ways of dealing with peers and adults.

What effect do these various preschool programs have on children? For disadvantaged children, the evidence seems positive (Brown, 1978; Lazar & Darlington, 1982). Children involved in programs such as Project Head Start do show initial gains in achievement when compared to similar children not participating in the programs. Although the initial advantage tends to decline (a finding whose interpretation is hotly debated; see Bronfenbrenner, 1975; Goldhaber, 1979; and the last section of this chapter), even in the middle grades children who attended educational intervention programs as preschoolers are more likely to be in the correct grade for their age and less likely to be in special learning classes than comparable children who were not involved in these programs. Further, since many early intervention programs provide nutritional, medical, and dental services as well as parent education components, they have other potential benefits besides the child's academic standing.

It may seem reasonable to suppose that if early intervention benefits preschoolers from economically disadvantaged homes, then a similar strategy will be equally beneficial for children from economically advantaged backgrounds. The logic may be good, but the fact of the matter is that there is little evidence to support such a view (Rohwer, 1971) and considerable reason to think that such early acceleration may actually have harmful effects on children (Elkind, 1982). It is certainly true that preschoolers can demonstrate behaviors more typical of older children. It is equally true that many could probably learn to read and do other academic activities (consider the mushrooming of computer camps for young children for example), learn to play musical instruments, and develop early athletic competence. But these early gains are short-lived, do not readily transfer to more mature forms of learning, often put pressures for achievement on young children that they cannot easily cope with, and deprive them of activities that most people would consider more age appropriate. If television advertising aimed at young children exploits their inability to effectively make judgments, then preschool programs that extol the virtues of accelerated early development exploit the insecurities of parents who want the best for their children but don't know how to obtain it.

Most parents who work require full-time day-care programs. What effect do these programs have on children? In particular, since children are away from their parents for extended periods, is the parent-child relationship harmed? The evidence generally supports the view that preschoolers in good quality full-time day-care programs are little different from children of the same age who spend the day at home with a parent (Caldwell, 1967; Clarke-Stewart, 1977; Fein & Clarke-Stewart, 1973). There is no evidence to suggest a decline in the parent-child bond or in the general quality of the relationship. For many parents, however, problems involving the availability and affordability of a good day-care program (full-time care can easily cost as much as college tuition) restrict their options and sometimes force them to place their children in less than ideal settings. Although these settings are rarely harmful, they nevertheless often fail to provide the preschoolers the adult contact and the variety of materials and experiences that all preschoolers are entitled to (Roby, 1973). Many companies have learned that by providing a day-care program for their employees' children, they reduce the anxieties of parents who are unable to find quality day-care, and can increase productivity.

Traditional half-day preschool or nursery school programs have always been considered a good experience for most preschool children (Day, 1983; Evans, 1975). These programs are able to provide diverse materials and experiences that are often unavailable in the home as well as group experiences that help preschoolers develop early social skills and facilitate the initial transition to more formal educational settings. The only "complaint" voiced by parents about such programs is that they find their children are often having such a good time that they are reluctant to leave at the end of the session.

THE SIGNIFICANCE OF DEVELOPMENT DURING EARLY CHILDHOOD

There is a tendency to view the skills of preschoolers in somewhat limited terms. Their abilities are typically compared to those of the school-age child and, of course, the preschoolers come out on the short end. This is an unfortunate tendency (Flavell, 1977) because the accomplishments of preschoolers have taken them a great distance from the limited competencies of the infant and toddler. The very fact that preschoolers are compared in any fashion to school-age children is an indication of the distance they have come.

Patterns of Development

Between the ages of three and seven, preschoolers make great advances in their ability to use language as a means of communication and as a means of processing information. The competence they develop in symbolically representing experiences allows them to extend their cognitive and social horizons far beyond the here and now. Further, they are beginning to understand that others see the world from different perspectives. Finally, they are becoming considerably more skilled in directing and controlling their own behavior. Each of these accomplishments can be viewed in relation to the developmental dimensions defined in the previous chapter.

Preschoolers continue to be active participants in directing the course of their own development. The nature of the participation, however, is considerably more deliberate than it was during the first three years. As their communication skills improve, preschoolers become more effective in expressing their wants and needs, justifying their actions, seeking information from others, and asking others to explain their behavior toward them.

As their cognitive competence increases, preschoolers become better able to plan and effectively carry out an activity; group objects in conventional categories such as size, number, weight, shape, and color; and view objects in relative as well as absolute terms. As their physical competence increases, they become better able to carry out their intentions. As their social competence increases, preschoolers become better able to resolve conflicts, find friends with similar interests and styles, understand and comply with social conventions such as sharing, and cooperate with others. In general, as preschoolers become more competent at directing the course of their actions, so too do the scope and depth of their efforts increase. And as the scope and depth of their efforts increase, so do their sense of competence and their willingness to confront new situations (White, 1972).

The quality of preschoolers' active involvement is not solely a reflection of their new competencies. It is also a reflection of new levels of interaction possible with both adults and peers. A responsive environment is still the crucial element, but the interaction becomes more balanced during early childhood. It is now easier for parents to interpret their child's behavior, and the child now begins to interpret the parent's behavior. Each is gaining better insights into the motives and reasonings of the other. Whether or not these insights lead to more effective parent-child relations depends on the compatibility of the behaviors and the techniques each uses to influence the other. Authoritative parenting, for example, is certainly going to be more effective with a temperamentally easy child than with a hard-to-manage child.

Even though parent-child interactions are more "balanced" than they were during infancy, parents and children are still far from being on an equal footing. Parents are clearly in control of more resources than preschoolers are and have more effective means of processing information and reflecting on experiences. This relative imbalance is evident when you compare peer interactions and parent-child interactions. The former are more limited in scope, of shorter duration, and more conflict prone. The parent's ability to anticipate and therefore avoid problems, to effectively interpret and comprehend the preschooler's message, and to use past experience to make current encounters more productive in large part accounts for the success of parent-child interactions during early childhood.

Infants deal with the world in a relatively global and undifferentiated fashion. Only gradually do infants come to understand the distinction between self and other. Toddlers are aware that others are physically distinct from them but still seem unaware of the psychological distinctions that exist between people. They are only beginning to develop a sense of social perspective.

Differentiation becomes a major cognitive task during early childhood. This process of separating and sorting is evident in terms of the preschooler's understanding both of the physical world and of the social world. It is evident in the preschooler's growing understanding of classification and class inclusion as well as sex-role development, peer relations, social justice, and obedience to legitimate authority.

A first step in the differentiation process is preschoolers' implicit recognition (as reflected in their behavior) of such social concepts as peers or sex role. Such distinctions do not exist for the infant and are, at best, words with little meaning for the toddler. The words acquire meaning gradually for the preschooler. Specific behaviors become associated with the words. A boy or a girl is one who does this or that, who plays with these boys and not those toys. A friend is someone "I" play with. The behaviors are localized to time and place. They are not viewed as enduring. Their meaning does not always carry over from situation to situation. They are not necessarily seen as extending to other people. But they are initial attempts to establish some degree of invariance or permanence in the construction of meaning.

Gradually this idiosyncratic pattern gives way to one more general. Common categories are used for grouping objects. The concept boy or girl is understood not only with reference to self but also with reference to others who have the same label. This commonality is still, however, limited in scope. Only one way of grouping objects is possible at any one time. The inclusive relationship between levels of grouping is poorly understood. The generalized category is viewed somewhat rigidly and one-sidedly. Boys should "never" play with dolls. Girls "always" have long hair. You "must" do what your mother says. You "have" to give me some candy because you are my friend. The qualifiers that would make each of these statements more realistic are yet to come. It is as if the preschooler, in order to maintain some sort of internal equilibrium, feels the need to shift from one extreme (the idiosyncratic view) to the other extreme (the stereotypic view). Only later does the stereotypic view give way to a more balanced view.

The sequence of idiosyncratic, stereotypic, and then integrated thought (actually another way of describing the undifferentiated, differentiated, integrated sequence) is not unique to early childhood. It is continually repeated throughout the life-span. It describes how people incorporate or assimilate new information or perspectives into their understanding of the physical and social world.

At each stage of the life-span, people tend to follow this sequence as they come to understand the issues, problems, and opportunities typical of each developmental level. At a subsequent level the issue may reappear. But when it does, it will be in a more sophisticated form. Preschoolers are very efficient at sorting blocks by color or shape. Limited to this context, their sorting skills are as good as any adult's. But this doesn't mean that they are as adept at sorting out the complexities of adult life. The process is the same, but its level of application is different. Ten-year-olds usually have a very clear sense of identity. They are well integrated into a peer group. They can function comfortably in all the domains that are typical to the ten-year-old. But well-integrated ten-year-olds usually become very "unintegrated" twelve-year-olds, as they begin to reconstruct a sense of personal identity at the adolescent level.

The Role of Early Experience

As is true of infancy, the early childhood years are generally considered (e.g., Brown, 1978; Evans, 1975) disproportionately more important in defining the course of development than subsequent life experiences. "As the twig is bent, so goes the tree" is still a common perspective both among professional and general audiences. This view is not without its critics (see Brim & Kagan, 1980), but if you will tolerate just one more aphorism, "an ounce of prevention is worth more than a pound of cure."

Actually, in some ways, life experiences during early childhood are considered more significant than those of the infant years. This is true for two reasons. The first concerns the nature of early development. The chapter on prenatal development introduced the concept of canalization. In essence, the concept suggests that there are certain aspects of development that are relatively resistant to interfering influences. The concept, as originally defined by Waddington (1957), concerned the actions of genes in regulating biological functions. But a number of developmental theorists (e.g., McCall, 1981; Scarr-Salapatek, 1976) have extended the notion to include various aspects of psychological development.

McCall (1981) believes that since the developmental events of the first year or two of life are more canalized than those that follow, the preschool years represent the first stage at which children are increasingly vulnerable to adverse environmental circumstances. Because the self-righting tendencies typical of canalized behaviors are less present during the preschool years than during infancy, adverse circumstances are more likely to leave a lasting impact on the course of development. If McCall is correct, the early childhood years rather than infancy are the first point in the life-span at which differential life experiences correlate with specific developmental

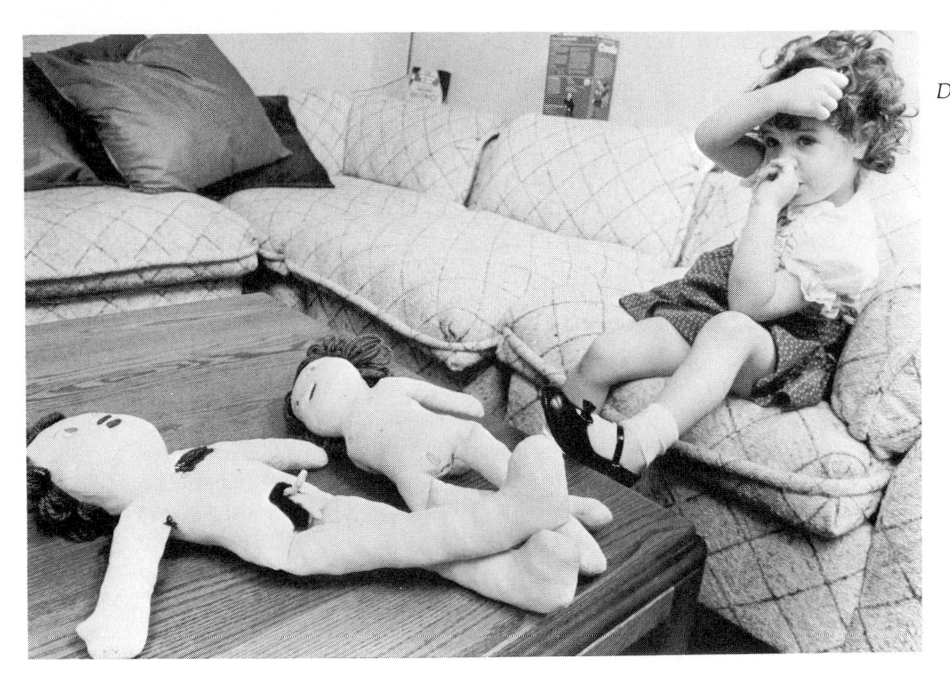

Research indicates that the preschool years are a time when children are particularly vulnerable to adverse environmental conditions. This heightened vulnerability makes funding for treatment and intervention programs for preschool children extremely important. One area that represents special concern is that of treatment programs for the young victims of sexual abuse.

patterns. One note of caution is necessary, however. The fact that such correlates are not evident during infancy does not necessarily mean that different experiences leave no lasting imprint.

Early childhood is important for a second reason. This is the *pre*school period. We call these children preschoolers because we often tend to think of them in terms of their future role as students. And since we live in a society where, to a large extent, merit reflects achievement, any opportunities to foster achievement or to correct situations that hinder achievement receive special attention. The preschool years are generally considered such a time (Hunt, 1969).

HUMAN DEVELOPMENT AND HUMAN SERVICES

The family is the dominant context regulating the course of development during early childhood. Therefore, any attempts to influence the course of the young child's development should take place either within the context of the family or in a manner that is compatible with family functioning. In particular, interventions need to be structured in such a way that the family can maintain the benefits derived from a program after the program itself has ended. The achievement of such broad basic goals requires therefore an understanding both of individual development and the ecological factors that influence family functioning. Two practical examples of the importance of this chain of influence are the efforts to eradicate child abuse in this country and the various educational intervention programs such as Project Head Start that are designed to offset the educational disadvantages some children demonstrate during the early school years.

Ecological Perspectives on Child Abuse

Although child abuse occurs within the family setting, factors that contribute to child abuse and neglect occur at all levels of the social ecosystem, as Figure 4-1 clearly demonstrated. They reflect our society's general tolerance of violence in the media and the ease with which we can obtain weapons. They reflect the widespread belief that physical punishment is not necessarily an inappropriate means of regulating children's behavior. They reflect the legal system's strong reluctance to sever a parent's custody rights. They reflect our culture's strong emphasis on independence and self-reliance, both for the individual and the family. They reflect the social and technological forces that have made us the most mobile society in the history of the world (more than half of us move every ten years). None of these broad macrosystem factors "causes" child abuse and neglect. In fact some, such as the legal system's strong reluctance to sever a parent's custody rights, in most cases function in a beneficial manner in regard to the family. Instead of "causing" child abuse, these factors define the context in which the family functions and therefore influence the way the young child develops.

This perspective makes quite clear the necessity for considering both the direct and indirect influences that exist at the different levels of the social ecosystem and the ways in which each serves to influence the course of development. When such an approach is taken (e.g., Garbarino & Crouter, 1978; Parke & Lewis, 1981), the interventions that result are also multilevel.

These interventions are designed to relieve the pressure that most poor or single mothers face. They include not only services for the children, such as child care, but also services to help the parent become more self-sufficient. These services might include job training programs, high school degree programs, and tax incentives for employers to locate their businesses in disadvantaged neighborhoods.

Multilevel interventions such as these have three positive outcomes. First, by relieving the social and economic pressures felt by many disadvantaged families, they cause parents to feel less of the helplessness, anger, and resentment that can fuel child abuse and neglect. Second, by providing opportunities for self-sufficiency, they foster the types of parenting strategies that reduce the incidence of undesirable child behaviors that can prompt abusive behavior from a parent. Third, by providing quality child care services, they free parents from worry about the safety of their children during the work day.

Ecological Perspectives on Early Educational Intervention

In 1965, the federal government began to fund preschool programs for children from economically disadvantaged backgrounds. Their backgrounds were disadvantaged in the sense that these children did not do as well on the various achievement and aptitude tests that are designed to predict success in school and in fact didn't do as well in school as children from more advantaged backgrounds. What began in 1965 as a half-day six-week summer program has today evolved into full-year programs that enroll young children for as many as two to three years before they enter public school.

These early intervention programs teach a variety of preacademic skills such as color and shape recognition, counting and letter naming, and vocabulary. They also model and reinforce the various social skills that are necessary for children to function with adults and peers within a classroom setting. The programs often involve nutritional, medical, and dental components as well. In addition to providing benefits for young children, these programs provide job training and employment for the adults in the community. Since "disadvantaged" background is often a euphemism for simply being poor and having a limited education and limited job skills, this second component is often of equal benefit to a community.

Being poor does not in and of itself make a person intellectually less competent. It simply means that the person has less money than more privileged people have. But money buys access, and this lack of access has been assumed to explain why disadvantaged children perform poorly in school (Hunt, 1964; 1969).

Attempts to evaluate early intervention programs have been continual and controversial. Virtually all programs can demonstrate that children who complete them score higher on various achievement and aptitude tests such as IQ tests than children from comparable backgrounds who were not involved in the programs. Follow-up studies (e.g., Cicirelli, 1969) have found that the initial advantage enjoyed by children who complete a program fades by the time they are in third grade. This does not mean that the children become less competent; rather it means that the degree of difference between children in the program and comparable children decreases during the first three years of school. Explanations for this apparent fade-out range from the argument that intervention programs create a "hothouse" effect (that is, they temporarily accelerate the rate of learning but cause no lasting improvement) to arguments blaming school systems for not building on intervention programs' gains to arguments questioning the validity of the program evaluations (Smith & Bissell, 1970).

As mentioned earlier in this chapter, Lazar and Darlington's (1982) follow-up research on children having early intervention experiences found that these children, who were at least ten years old at the time of the study, were more likely to be in the correct grade level for their age and less likely to have been placed in a special education classroom than comparable children were. So it would appear that at least in terms of grade level and special education placement, early intervention programs had a lasting effect on the educational achievement of these children. But why should this be so? What process or event was operating to maintain the initial

Research findings on the long-term benefits of early educational intervention programs such as Head Start are controversial. Follow-up studies have found that achievement differences between children completing such programs and comparable children who have had no preschool experience equalize by third grade. However, research has also shown that program graduates are more likely to be in their correct grade and are less likely to be in special education classes.

advantage after so many years (and why did it seem to disappear on earlier evaluations)?

There are no direct answers to the question of why these initial advantages are maintained; but there is evidence to suggest that where long-term gains such as these are evident, they are probably reflecting a social context, most typically the family and later the school, that maintains them. Bronfenbrenner (1975), who was involved in the early evaluation of these intervention programs, observed that "the family seems to be the most effective and economical system for fostering and sustaining the child's development. Without family involvement, intervention is likely to be unsuccessful and what few effects are achieved are likely to disappear once the intervention is discontinued"(p. 470). Also indicating the importance of social context, studies such as those by Wolff and Stein (1966) and Shipman (1976) have evaluated the impact of preschool-to-school continuity on the maintenance of early intervention gains. Wolff and Stein found that the stability of Head Start gains, as measured by teacher ratings and standardized testing, is partially a function of the percentage of children in the kindergarten class who were Head Start participants. Gains were maintained through the kindergarten year if at least 25 percent of the group had shared the Head Start experience.

Shipman was interested in determining what, if anything, it was about the school and family experiences of those Head Start children who maintained their initial advantage that might account for this stability. Shipman and her colleagues developed a very elaborate procedure to answer her question. Children were followed longitudinally for a period of six years through their experience with Project Head Start and their early elementary school years. Over this six-year period, children took a variety of standardized tests and were observed in classroom and free-play situations; mothers were interviewed about their parenting and were observed with their children; and teachers at all levels were interviewed about their teaching strategies and were observed in their classrooms. Preschool center directors and elementary school principals were interviewed about the overall educational and social climate of their schools, and finally, community agency representatives were interviewed about the social and economic character of the children's neighborhoods. Shipman's conclusion generated from all of these data speaks directly to the importance of the context and continuity of early developmental experiences:

> For those children who showed the greatest gain in academic skills between ages four and nine, following Head Start attendance, there was a continuity of facilitating school experiences; at each primary grade level these children had enthusiastic, warm, positively motivating, cognitively stimulating teachers who taught in one-to-one or small group settings. Such continuity in facilitating school experiences is particularly non-existent for a sizable minority of low-income children who move frequently between schools, a situation common to many urban areas. Moreover, this continuing warm and stimulating school environment was combined with a home environment that provided the child emotional support in general and support for school activities in particular. (p. 50)

For all of us, the "moral" is clear. The continuation of the developmental advances made during early childhood requires a supportive social context, most commonly the family. Therefore, any of us whose decisions directly or indirectly influence families also influence the development of young children.

Fraiberg (1959) called the early childhood period the "magic years." In many ways they are just that. They are a time of supernatural powers and fears. They are a time of idiosyncratic and seemingly illogical perspectives and reasoning. They are a time of emotional excess and limited self-control. But they are also a time of wonderment and discovery, of unconditional love and affiliation, and of growing initiative and self-esteem. What ever else early childhood may be, it is certainly a dramatic contrast to the conventional and deliberate behavior of the stage to follow—the stage of middle childhood.

SUMMARY

THE COURSE OF DEVELOPMENT

1. Significant changes in height and weight as well as in body proportions allow preschoolers to become proficient at a variety of motor skills. These new motor competencies have a positive effect on the cognitive awareness of spatial and temporal concepts.
2. Preoperational children seem particularly impressed by the events of the moment. Their ability to integrate these events with previous ones or to mentally compare and contrast events is therefore limited.
3. Preoperational children tend to be egocentric in their thought; that is, they find it difficult to view events from a perspective other than their own immediate one.

4. By age seven, most children have a production vocabulary of ten to twelve thousand words, and the grammatical structure of their speech is virtually identical to everyday adult usage.
5. The preschool period marks the initial attempt at a sex-role identity, at developing peer social skills, at comprehending the notion of authority, and at developing a sense of social justice, or fairness. All of these developments reflect the egocentric nature of the preschooler.
6. Play serves as the primary means by which preschoolers gain a sense of mastery, exercise cognitive competencies, rehearse social roles, and practice social interaction skills.

THE CONTEXT OF DEVELOPMENT

1. Parenting strategies differ in terms of control, nurturance, maturity demands, and communication.
2. Parents who set age-appropriate maturity demands, who use clear and consistent control techniques, who provide ample opportunity for communication with their children, and who show a genuine concern for their children's happiness and welfare are most likely to have children who act in a socially responsible manner.
3. Peer contacts during the preschool stage evolve from ones that are situation specific and parallel to ones that reflect the beginnings of stable friendships.
4. Preschoolers are strongly influenced by the content of television programming. Their food and toy preferences are closely linked to their viewing preferences.

THE SIGNIFICANCE OF DEVELOPMENT DURING THE EARLY CHILDHOOD YEARS

1. The preschool period marks the first time that children effectively use elementary concepts and dimensions.
2. Intervention programs with preschoolers are most likely to demonstrate long-lasting benefits if the programs involve parents and if the intervention practices are compatible with family life experiences.

HUMAN DEVELOPMENT AND HUMAN SERVICES

1. From a long-term perspective, the best way to improve the lives of children is to improve the lives of the adults who care for them.

KEY TERMS AND CONCEPTS

Erikson's Psychosocial Stage of Initiative

PHYSICAL AND MOTOR DEVELOPMENT

Temporal Relationships
Spatial Relationships

COGNITIVE DEVELOPMENT

Piaget's Stage of Preoperational Thought
Seriation
Classification
One-to-One Correspondence
Conservation
Centration
Reversibility
Egocentric Thought
Inductive Reasoning
Deductive Reasoning
Transductive Reasoning
Collective Monologues

SOCIAL DEVELOPMENT

Conservation of Social Identity

PARENT-CHILD RELATIONS

Baumrind's Parenting Patterns
- Permissive Parenting
- Authoritative Parenting
- Authoritarian Parenting

One of the few things more difficult than understanding the logic of the preoperational child is understanding Piaget. His own writing is generally incomprehensible to all but those well versed in the theory. One book that manages to present Piaget's ideas in a readable yet sophisticated form is by Cowan.

Cowan, P. A. (1978). *Piaget with feeling.* New York: Holt, Rinehart & Winston.

Over the course of the preschool years, children spend hundreds of hours in front of the television. Liebert and his colleagues provide a very detailed account of the effects of that exposure.

Liebert, R. M., Sprafkin, J. N., & Davidson, E. S. (1982). *The early window: The effects of television on children and youth (2nd ed.).* New York: Pergamon Press.

All parents want what is best for their children but aren't always sure what that is. The article by Caldwell discusses the advantages of preschool programs for young children and the one by Elkind the advantages of not letting a child grow up too fast.

Caldwell, B. M. (1967). What is the optimal learning environment for the young child? *American Journal of Orthopsychiatry, 37,* 8–20.

Elkind, D. (1982). *The hurried child.* Reading, MA: Addison-Wesley.

The parent-child relationship is unique. It can provide the most rewarding moments but also the worst. These readings provide insight into why and how parent-child relationships fail and suggest what can be done to reverse the process.

Baumrind, D. (1978). Reciprocal rights and responsibilities in parent-child relations. *Journal of Social Issues, 34,* 179–196.

Gil, D. B. (1975). Unraveling child abuse. *American Journal of Orthopsychiatry, 45,* 346–356.

Kempe, R. C., & Kempe C. H. (1978). *Child abuse.* Cambridge: Harvard University Press.

Patterson, G. R. (1980). Mothers: The unacknowledged victims. *Monographs of the Society for Research in Child Development, 45* (No. 5).

5
MIDDLE CHILDHOOD

CHAPTER OUTLINE

This chapter surveys development between the ages of approximately six and twelve. Middle childhood is both a time of consolidation and integration of the competencies first demonstrated during early childhood and a time of exposure to a world that is separate from and independent of the family. It is a time that sees the emergence of the first true peer culture, the first direct exposure to the school system and the community, and the first objective evaluations of the child's status vis-a-vis other children. For most children, middle childhood is a happy time (Elkind, 1978). For most parents, middle childhood is a time of relatively easy parenting. The sudden emotional shifts of early childhood are past; the turmoil of early adolescence, yet to come.

Middle childhood is a period of great academic accomplishment. By age twelve, most children have acquired sufficient competence in reading and mathematics to successfully accomplish virtually all of the day-to-day tasks encountered in adulthood. It is also a period of great interpersonal accomplishment. By age twelve, most children have a well-defined concept of right and wrong, of typical sex-role behavior, and of social conventions. In fact, it is probably not too much of an exaggeration to say that twelve-year-olds, within their world of middle childhood, have their "act more together" than most of us ever will within our adult worlds.

The conception of middle childhood as a period of accomplishment is evident in Erikson's (1980) description of his fourth stage of development—the **psychosocial stage of industry.** For Erikson, middle childhood is a time of developing a sense of industry and avoiding a sense of inferiority:

> One might say that personality at the first stage crystallizes around the conviction "I am what I am given," and that of the second, "I am what I will." The third can be characterized by "I am what I can imagine I will be." We must now approach the fourth: "I am what I learn." The child now wants to be shown how to get busy with someone and how to be busy with others. (Erikson, 1980, p. 87)

Children come to see themselves as producers, doers, and makers of things. They are busy, involved, active in projects or groups. Erikson sees their industriousness as an indication of their awareness that they are now in a larger world and of their desire to be accepted by it. This desire to be accepted often translates into a strict adherence to group rules—pity the poor parent whose eight-year-old child's teacher said he had to bring "exactly" one quarter to school and who has only two dimes and a nickel.

Children are proud of their accomplishments. They are eager to show others what they have learned. Implicit in this eagerness (or lack of it) is the awareness that their value to others is often in terms of what they can produce. The degree to which children define themselves in terms of their ability to produce determines the ease of resolution of this stage. Children who acquire what Erikson considers a healthy sense of industry demonstrate a balance between meeting the demands placed on them by others and meeting those they place on themselves, a balance between enjoyment in the process of being industrious and enjoyment of the products of that effort, and a balance between effort and accomplishment.

Children who fail to develop a sense of industry develop, instead, a sense of inferiority and inadequacy. They believe that their products have

Children who develop a healthy sense of industry during middle childhood are able to strike a balance between meeting the demands that other people place on them and meeting those they place on themselves. Equally important to future development, they are aware that productivity is only one source of gratification.

no value, and as a result they believe they have no value. Erikson thinks that such feelings are most likely to develop when there is little continuity between the values of the home and the values of the school. It is not a question of the child's having no skills. Rather, it is a question of the degree to which the school or the peer group values the skills nurtured in the home. The problem of feelings of inferiority is compounded when children, seeing little value in their products, stop trying to produce. Less effort results in less output. Less output means lower achievement and less reward. In other words, a self-fulfilling prophecy is created. And once created, it is hard to erase.

For Erikson, too much industry is as undesirable as too little. Viewing productivity as one's only means of gratification is as harmful to development as not being able to derive any gratification from the fruits of one's labor. Overindustrious children see themselves as having value only to the extent that others value their products. The basis of worth is always in terms of someone else's opinions, never in terms of their own. They cannot feel good about their effort unless someone tells them that they have done well. Erikson sees in these individuals an easily exploitable population.

As is true of all of Erikson's tasks, developing a sense of industry is primarily a matter of balance. In early childhood, balance is primarily a function of parent and child. In middle childhood, the equation also includes the peer group and the school. As a result of these additional components, the relative influence of the parent declines, and as children learn to juggle the often conflicting demands of these three groups, they develop a broader perspective from which to view themselves.

Chapter 5 continues to follow the outline used in Chapters 3 and 4. As you read the chapter, keep in mind four questions. First, in what ways are the cognitive competencies of middle childhood an advance over those of early childhood? Second, how do children's relations to parents change as they become immersed in the school and the peer group? Third, how do children's new cognitive competencies influence their ability to function in the social world? Fourth, what is it about the events of middle childhood that prompts researchers to view it as a period of consolidation?

THE COURSE OF DEVELOPMENT DURING MIDDLE CHILDHOOD

In many ways, the developmental accomplishments of middle childhood serve to refine and integrate skills first evident during early childhood. Gains in physical and motor development are primarily improvements on already existing large and small motor patterns. Advancements in cognition provide the school-age child a common and consistent frame of reference in which to integrate everyday experiences. Developments in the social domain make possible the first true peer-group culture.

Physical and Motor Development

For the most part, the changes in physical status and motor ability during middle childhood are extensions of the patterns established in early childhood. Gains in height and weight continue to be gradual and constant for both boys and girls. The continued growth of the leg bones further accentuates the lean look of seven- and eight-year-olds (Cratty, 1979). By age nine or ten, children begin to "fill out" as muscle development becomes more pronounced. The shape of the face becomes longer and more angular as the lower jaw expands to accommodate the eruption of the permanent teeth. Most of the child's biological systems become stable during middle childhood. As a result, children are less prone to infectious illnesses, especially colds and earaches. Accidents now become the primary threat to life. They account for approximately one-half of all deaths during middle childhood (Williams & Smith, 1980).

Motor development is evident in the improvement of already existing skills. Children run faster, farther, and longer. They jump higher. They throw farther and with greater accuracy. They show better coordination and balance, and they are stronger.

There is relatively little difference in either the physical status or motor abilities of boys and girls during much of middle childhood. Where differences are noted, boys tend to be more competent in tasks requiring strength and power, and girls tend to excel in tasks requiring good coordination. In all instances, however, the improvement with age for both boys and girls is considerably greater than are the differences between them at any particular age. Since there is relatively little difference in physical status, it seems most likely that where differences exist, they reflect opportunity for practice, which in turn reflects culturally valued patterns of male and female activity.

The relation of physical-motor development to social and cognitive development is now considerably different from what it was during early childhood. There is a much greater integration among the three domains. Physical status and motor ability are major determinants of a child's peer-group status, and the rule-governed nature of the games of middle childhood require both cognitive and physical competence for successful participation.

Watch a group of school children choose teams. The two team captains are probably fairly good athletes. The first few players they pick are probably the most skilled at the game being played. They are probably also friends of the captains. Watch as the other players are chosen. As the unselected become few, the criterion for selection seems to change from "he is really good" to "he's not as bad as . . . ." The last few to be chosen are probably the smallest children in the class. They are also more likely than they would be by chance to be overweight, have some sort of handicapping condition, or in some other way be different.

Captains don't pick their friends first because they are their friends. They pick them because they are the best players. The fact that they are also the captains' friends is not mere coincidence, however. During middle childhood, friendship patterns and peer status are heavily influenced by physical ability, size, body build, and physical attractiveness. Being nice, being kind, and being smart are considerably less important.

Because the emphasis is now on physical ability, the nature of play changes in middle childhood. For preschoolers, play is idiosyncratic. Each child plays. Even when projects are cooperative and require a division of labor, the pieces are worked on independently. Truly cooperative efforts during early childhood often require adult support. However, the pattern in middle childhood is very different. Children don't simply play; they play games. Games have goals, and they have rules. The ability to play the

Differences in physical and motor abilities between school-age boys and girls are largely a reflection of opportunities for practice and encouragement for participation.

games of middle childhood requires not only physical competence but also cognitive competence. Children must be able to keep track of the flow of the game, remember the rules, and play as team members. It's one thing to be able to catch a baseball hit to center field; it's another thing to know what to do with it once you catch it—especially with all your teammates yelling "throw it, throw it."

The ability to function as a member of a team develops throughout middle childhood. Watch seven- or eight-year-olds play games like football or volleyball or soccer. All have a position to play, but they probably aren't playing it. They are where the ball is. The ball seems to act like a magnet. It draws all the players to it. On the other hand, watch the same children at a baseball game and you may well see the center fielder contemplating a dandelion, the child at second base waving to a parent or friend, the child at third base watching the game in the next field, and so on. Unable to be near the ball, they act as if they simply don't conceive of themselves as part of the game. To them, the game is one of catch between the pitcher and the catcher with the batter in between. The coaches' sometimes frantic exhortations to "be ready" and to keep "heads up" are not sufficient to help seven- or eight-year-olds appreciate the fact that even though they aren't near the ball, they are still playing the game. (This awareness usually does not occur until the ball is about five feet past them.) The reason why every soccer player is chasing the ball and the outfielders bear a closer resemblance to Ferdinand the Bull than to a major league player is that they don't seem able to picture themselves as members of a group effort. In terms of a soccer or basketball team, for example, they don't yet understand what it means to position themselves to receive a pass.

Gradually during middle childhood, the cognitive competence necessary to play as a team emerges. Children come to see their role in terms of some larger strategy. This does not mean that they simply become better ball handlers. There are many nine-year-olds who are as good, individually, as many ten- or eleven- or even twelve-year-olds. The change in these children isn't in physical ability alone; it's in their ability to see themselves in relation to the other players on the field. This change in perspective is one of the major cognitive achievements during middle childhood.

It is interesting to watch children learn how to count. Preschoolers are able to do some counting activities. They are usually able to recognize and correctly name the numerals, they have often memorized the sequence of numbers from one to ten or one to twenty (or sometimes even higher), and they can tell you, if you place a small number of beads or blocks in front of them, how many there are. This description seems to suggest that preschoolers already know how to count. In a sense they do, but there is an important piece of understanding missing—the complete understanding of what they are doing. The acquisition of this last piece of understanding marks the beginning of the transition from preoperational thought to what Piaget (1967) termed the **stage of concrete operational thought.** Concrete operational thought characterizes cognition during the middle childhood years.

The last piece of understanding is, at first, acquired slowly. Four- and five-year-olds do a lot of simple counting. If you ask them if they can count, they usually say "yes." If you then ask a somewhat strange question—"what can you count?"—they are likely to answer straightforwardly and list the various things that they have counted. If you then ask if they can count something not named, they may say, "I don't know, I've never tried."

When children are about five or six, there is a change in their counting behavior. Sets of objects are counted, rearranged, and counted again. The members of the sets become more varied. Children may start counting "things" or "stuff" rather than all beads or all blocks or all glasses. This process, paralleling the infant's behavior during the sensorimotor stage, ends with a startling revelation: "I can count." If you now ask children what they can count, they may say "I can count anything!" What has happened is that the children have separated a mental action (what Piaget called an *operation*) from the content of that action. And as a result of this discovery, they have begun to see the world in a very different light.

The Conservation Experiments

The classic demonstration of the transition from preoperational to concrete operational thought is the **conservation experiment** (Piaget & Inhelder, 1969; Piaget & Szeminska, 1952). The basic question asked in the conservation experiment is whether or not children's reasoning is based primarily on the appearance of things or on their ability to use mental operations (see Table 5-1). In the conservation of continuous quantity task, a child is shown two glasses containing equal amounts of liquid and asked if the two amounts of liquid are the same. When the child is satisfied that they are the same (sometimes after requiring a slight addition to one or the other glass), the liquid from one glass is poured into a glass of a different shape. The child watches the liquid being poured or may actually do the pouring. The child is now asked if the new glass contains more or less liquid than the other glass (which is still in view) contains or if the amounts are the same. The child answers the question; an explanation is requested by the experimenter, and the procedure is then repeated using another differently shaped glass. If, for example, the liquid was first poured from the original

TABLE 5-1 *Judging the Levels of Children's Conservation Responses*

PRECONSERVER	TRANSITIONAL	CONSERVER
The children center on only one of the dimensions and state that the tall glass has more or less water than the short glass.	Transitional children are inconsistent in their replies to two related tasks. They may conserve liquid amount in one situation but not in the other. This inconsistency may be noted even after the children logically justify their first conservation response.	Children judge that the amount of water is conserved regardless of the container(s) involved. One logical justification is sufficient for each variation of the task. Piaget considers a logical justification to be critical to judging a conservation response.

CONSERVATION TASKS:	ESTABLISH EQUIVALENCE	TRANSFORM OR REARRANGE	CONSERVATION QUESTION AND JUSTIFICATION
Conservation of Number Number is not changed despite rearrangement of objects.	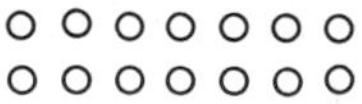	Rearrange one set	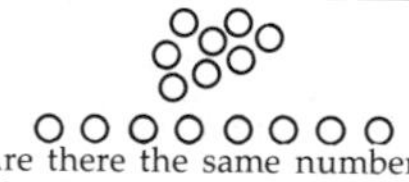 Are there the same number of red & green chips or ...?
Conservation of Length The length of a string is unaffected by its shape or its displacement.		Change shape of one string	 Will an ant have just as far to walk, or ...?
Conservation of Liquid Amount The amount of liquid isn't changed by the shape of the container.		Transfer liquid	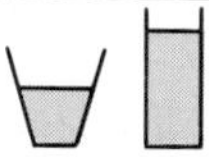 Do the glasses have the same amount of water, or ...?
Conservation of Substance (Solid Amount) The amount of substance does not change by changing its shape or by subdividing it.		Roll out one clay ball	 Do you still have the same amount of clay?
Conservation of Area The area covered by a given number of two-dimensional objects is unaffected by their arrangement.	Grass Garden	Rearrange one set of triangles	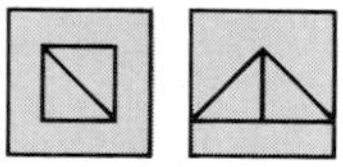 Is there still the same amount of "room" for planting, or 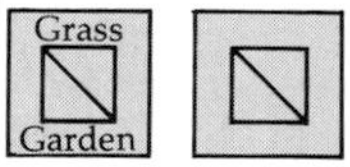 ...? Is there still the same amount of grass to eat, or ..?
Conservation of Weight A clay ball weighs the same even when its shape is elongated or flattened.	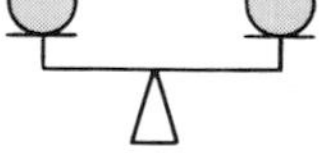	Change shape of one ball	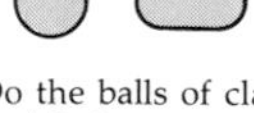 Do the balls of clay still weigh the same, or ...?
Conservation of Displacement Volume The volume of water that is displaced by an object is dependent on the volume of the object and independent of weight, shape or position of the immersed object.	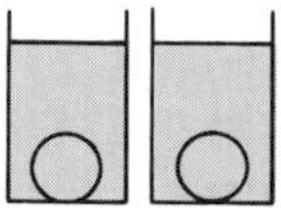	Change shape of one ball	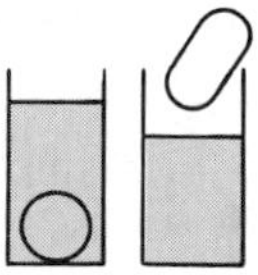 Will the water go up as high, or ...?

Note: Adapted, by permission of the publisher, from E. Labinowicz, *The Piaget Primer* (Menlo Park, CA: Addison-Wesley Publishing Company, Inc., 1980), p. 94.

glass into a taller and thinner glass, this time it might be poured into a shorter and wider one. Another variation might be to pour the water from the original glass into three or four smaller glasses. After each variation, the child is asked to make a judgment concerning the quantity of liquid and to provide a reason for the judgment.

The conservation of substance task follows a similar procedure but uses different materials. Instead of liquid, clay or some other pliable material is used. Instead of a liquid's being poured from one glass to another, the shape of the clay is changed. Once the child agrees that the two balls of clay contain the same amount of clay, the shape of one might be changed to that of a pancake or a hotdog. The questions would be the same: "Does the hotdog contain more or less clay than the original ball contained or are the amounts the same?" and "Why do you think so?"

The children's responses to the conservation problems take one of three forms. Most children below the age of five believe that the change in the physical appearance of the liquid or the clay has resulted in a change in the amount of liquid or clay. They might say there is more or they might say there is less (usually depending on what part of the new shape they center on), but in either case they have failed to conserve amount. That is, they have failed to realize that the amount of a substance is not changed by transformations in the physical appearance of that substance. When asked why they think the amount is now greater or smaller than it was, they might say "because it's taller" or "because now there are more glasses" or "because a pancake is flat." These answers share two characteristics. They are all based on the child's perception, and they only take into consideration a limited amount of the available, relevant information.

Five- and six-year-old children also usually fail to conserve, but their justifications suggest that they view the problem from a perspective that is different from the four-year-old's. Their answers are more tentative. Whereas four-year-olds always focus on only one aspect of the problem, five- and six-year-olds sometimes focus on one dimension (it's more because the liquid is higher) and at other times focus on the other dimension (it's less because the glass is thinner). They seem to be aware that all of this information is somehow relevant to the problem, but they aren't yet able to put it together to come up with the correct answer. They are clearly in a transition between two levels of understanding, and as is true of behavior during all such transitions throughout the life-span, their behavior is less coherent than it was earlier.

By seven or eight years of age, children have no trouble with the problem. They know the amount is the same no matter what you do to its shape. They will justify their answer by saying "you didn't add any and you didn't take any away" or "even though it's taller, it's also thinner" or "since it was the same when it was in the original glass, it must be the same now." In fact, so certain are children of the correctness of their answers that even attempts to dissuade them are rarely successful. Cowan (1978) offers a good example:

> I told one seven year old girl, who insisted that there was still the same amount of water in the taller glass: "But a girl your age was here yesterday and she said that there was more water because it's higher." (The girl's eyes widened). "Well, she was just silly, that's all." (p. 188)

Other Evidence of Cognitive Change

Changing responses to the conservation tasks are not the only changes that are evident during the five- to seven-year-old period. A variety of other changes are also evident during this transition (White, 1965; 1970). First a few examples, then the explanation. Children's word-association patterns begin to change during this period. When preschoolers are asked to indicate the first word they think of when they hear the words hot, fast, loud, and wild, they usually answer with such words as dog, car, noise, and lion. When the same question is asked of a seven- or eight-year-old, the responses are most likely to be cold, slow, quiet, and tame (Entwisle, 1966).

Simple discrimination learning tasks show a parallel pattern. Kendler and Kendler's (1962) **reversal-nonreversal shift task** provides an example of this pattern. In this task, children are required to learn which member of each of two pairs is the correct choice (see Figure 5-1). Pair 1 might consist of a large white block and a small black block; pair 2, of a large black block and a small white one. In pair 1, the large white block is designated as correct, and in pair 2, the large black block is so designated. The presentation of the two pairs is randomly altered until the correct block in each pair is regularly chosen. The combinations are then switched, and the children are again asked to learn the correct member of each pair. An interesting difference emerges in the behavior of the two age groups. If the two pairs now consist of the large black block and the large white block in the first pair and the small black block and the small white block in the second pair (called a nonreversal shift) and if the correct member of each pair is the black block, the preschoolers have an easier time learning the new correct answers than the school-age children do. However, if the new pairs consist of the large black block and the small black block in pair one and the large white block and the small white block in pair 2 (called a reversal shift) and if the small block in each pair is designated as correct, just the opposite is true (Kendler & Kendler, 1962; Kendler, 1979).

Two more examples should suffice. School-age children are more affected by delayed auditory feedback than preschool children are. That is, the older children encounter greater difficulty than preschoolers in talking if, at the same time, they are also listening to a tape recording of the words they said just a few moments earlier. Finally, school-age children are considerably more likely than preschool children are to use speech as a means of planning, rehearsal, and self-control (Vygotsky, 1962; 1978). These behavioral transitions occurring between the ages of approximately five and seven (as well as a number of others that also occur at this time) have been summarized by White (1970) as reflecting four basic shifts in the way children cognitively conceptualize and respond to the events in their environments. More specifically, children exhibit

1. a sharply enhanced ability to behave in a manner consistent with a set of expectations established for them;
2. an increase in the ability to use self-regulatory mechanisms;
3. an increased awareness of their own learning and memory processes; and
4. the ability to superimpose dimensionalization on the concrete situation, so that they do not so much deal with events as with events in context.

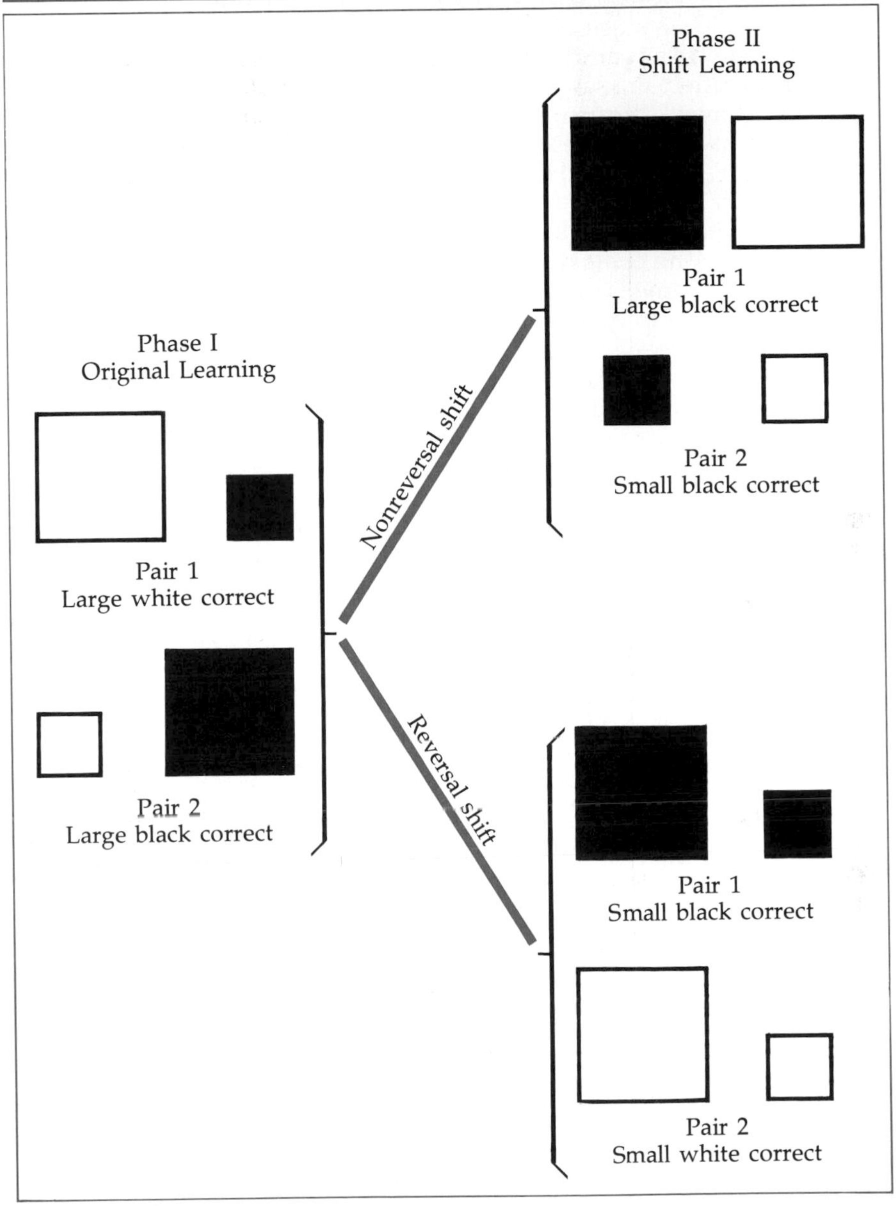

Phase II
Shift Learning

Pair 1
Large black correct

Pair 2
Small black correct

Phase I
Original Learning

Nonreversal shift

Pair 1
Large white correct

Pair 2
Large black correct

Reversal shift

Pair 1
Small black correct

Pair 2
Small white correct

FIGURE 5-1
In the reversal-nonreversal shift task children are required to learn which block in each of two pairs of blocks has been designated as correct. Once the children are able to choose the correct block in each of the randomly presented pairs, the combinations are switched and new blocks are designated as correct. Kendler and Kendler (1962) found that preschoolers had an easier time learning the new correct answers when the changes were nonreversal shifts, while school-age children had an easier time with reversal shifts.

Therefore, the changes in word-association patterns reflect school-age children's use of conventional categories or dimensions of experience to code, store, and retrieve information. The discrimination learning patterns reflect the same process. The younger children do better on the first transfer task because they seem to view the blocks in each pair as independent, isolated pieces of information. For this reason, they only have to relearn one piece. However, if the pieces of information are seen in a broader context, that is, as specific instances of elementary dimensions (size and brightness), then the second transfer is easier. In the second transfer, the context remains the same (that is, the solution is defined in terms of dimensions) and only the specific form has to be relearned. The delayed auditory feedback task presents more problems for older children because they have a greater sense of self-awareness than preschoolers do. Although self-awareness is usually seen as an asset, in this particular task, it becomes a liability. Finally, their greater tendency to use private speech as a means of self-control and rehearsal allows school-age children to remember more information and to stay on-task longer than preschool children can.

The most detailed analysis of the set of cognitive changes that marks the transition from preoperational to concrete operational thought is that of Piaget (1967; 1975). The key to this transition is school-age children's development of **concrete operational thought** and their use of this logic to integrate the various experiences of their lives.

Characteristics of Concrete Operational Thought

School-age children participating in conservation experiments are not only aware of the appearance of the liquid or the clay in its initial state and in its new state; they are also aware of what took place in transforming the ball into a pancake or hot dog. This **awareness of transformations** tells them that state 2 is simply a variation of state 1. Preschoolers, insensitive to the pouring or the molding, see no conflict in first judging two amounts to be the same and then when one has been given another shape to judge the same two amounts to be different. They do not consider the relationship between the two judgments because they do not consider the relationship between the two states. And they do not consider the relationship between the two states because they seem to lack the "mental operations" necessary to appreciate the transforming acts.

The ability of school-age children to consider all of the information in the conservation task as well as their understanding that states are related through transformation makes possible a different basis for judging the amounts of liquid or clay. Whereas preschoolers are limited to judgments based on the appearance of things, school-age children increasingly base their judgments on **inference.** They are able to consider the relation between the different elements of the task as well as the actions that have affected them. On the basis of all this information, they draw a conclusion. The children are able to feel certain about their answer because they know it is the only answer possible. They may agree that one amount looks like it's more than another because it's taller but, they will quickly add, it really isn't. This awareness of what "really" is reflects their ability to base their judgments on the events in context. Concrete operational children's awareness of transformations enables them to see a connection between two states.

An additional mental operation, which Piaget calls **reversibility,** allows concrete operational children to understand the nature of the relationship between the two states. In essence, reversibility is the ability to mentally represent some action on an object and then to reverse the action. Consider the three explanations typically given by children who successfully solve the conservation task. Each represents a form of reversibility. Children who say "you didn't add any and you didn't take any away" are demonstrating a form of reversibility called an **identity.** They consider the two actions (adding and taking away), realize that one cancels, or reverses, the other, and conclude that there must still be the same amount. Children who say "even though it's taller, it's also thinner" are demonstrating a type of reversibility called **compensation.** Again, two actions are considered (becoming taller, becoming thinner), one is seen as canceling, or reversing, the other, and the same conclusion is reached. The third explanation "since it was the same in the original glass, it must be the same now" is a type of reversibility called **negation.** The act of pouring a liquid into a new glass can be negated or reversed by pouring it back into the original glass. Since the two pouring actions cancel each other, the products of the two actions must be equal.

Children's ability to mentally reverse their actions is also a reflection of another aspect of concrete operational thought—**decentration.** Preschool-age children focus, or center, on specific states. It was centration that prevented the preschooler in the example in the previous chapter from realizing how his ball of playdough could be transformed into the same shape as that of the playdough belonging to the girl next to him. Decentration marks a shift in the relationship between perception and cognition. The abilities to seriate and classify as well as the ability to make a reversal

The mental operation of reversibility enables these children to understand the relationship between these seeds and the mature plant.

shift faster than a nonreversal shift are all examples of decentration. In all three instances, specific perceptions are seen within the context of a mental operation.

The various aspects of concrete operational thought (awareness of transformations, inference, reversibility, and decentration) do not exist separately from each other. Rather they function in an integrated, systematic fashion. They allow children to see elements in their environment not as isolated units but in relationship to other elements. This ability to recognize relationships is not restricted to the ones between physical objects. Family roles such as brother, father, and uncle are no longer seen in isolation but as parts of a larger system. Children become able to define themselves in relationship to all other members of the family and to see the relationships between other family members in similar ways. They now begin to understand how it is possible for one person to be, at the same time, a brother, a father, and an uncle. They also now realize that they are their sibling's sibling. They begin to demonstrate the same sort of understanding of geopolitical units such as towns, counties, states, countries, continents, and so forth. It is this ability to see themselves from many different perspectives, to realize which are jointly possible (being American and being Catholic), which are mutually exclusive (being a boy and being a girl), and which are independent (liking baseball and checkers), that gives school-age children the consistency and sense of rootedness that is lacking in the preschool child. School-age children are considerably less overwhelmed by the appearance of things than preschool children are. School-age children are, in a sense, in better balance, or equilibrium, with their environment, because they have a system of mental operations that allows them to act upon their environment rather than simply reacting to it.

Concrete operational thought allows children a much more sophisticated understanding of their world than was previously possible. In part this is so because by middle childhood children come to respond less to their direct impressions of the world than to their organization and interpretation of those impressions. For example, there are significant differences in the way preschool and school-age children remember information. Paris and Lindauer (1977) found that school-age children are more likely to identify a picture as previously seen (although in fact it was not seen) if it is an example of the category of pictures (such as fruits or cars or animals) that was actually shown to them. In a similar vein, Liben (1977) showed a group of preschool children the unlikely picture of a tilted water bottle in which the water was parallel to the bottom of the bottle (see Figure 5-2). When she asked the children to reproduce the picture from memory several months later, they drew the bottle as it was originally shown to them. School-age children presented with the same task reproduced the bottle as in reality it would have to be, that is, with the surface of the water parallel to the ground, not to the bottom of the bottle.

The significance of the Paris and Lindauer research and the Liben research is not that the older children behaved differently from the younger ones but that their behavior indicated that they had done something to the information rather than simply storing it: they had systematically acted upon it. In the first case, they categorized it; in the second, they retained the image of the bottle as their experiences with physical objects dictated. Although in these laboratory demonstrations their active memory processes were more of a hindrance than a help, in everyday activities just the opposite result is typical.

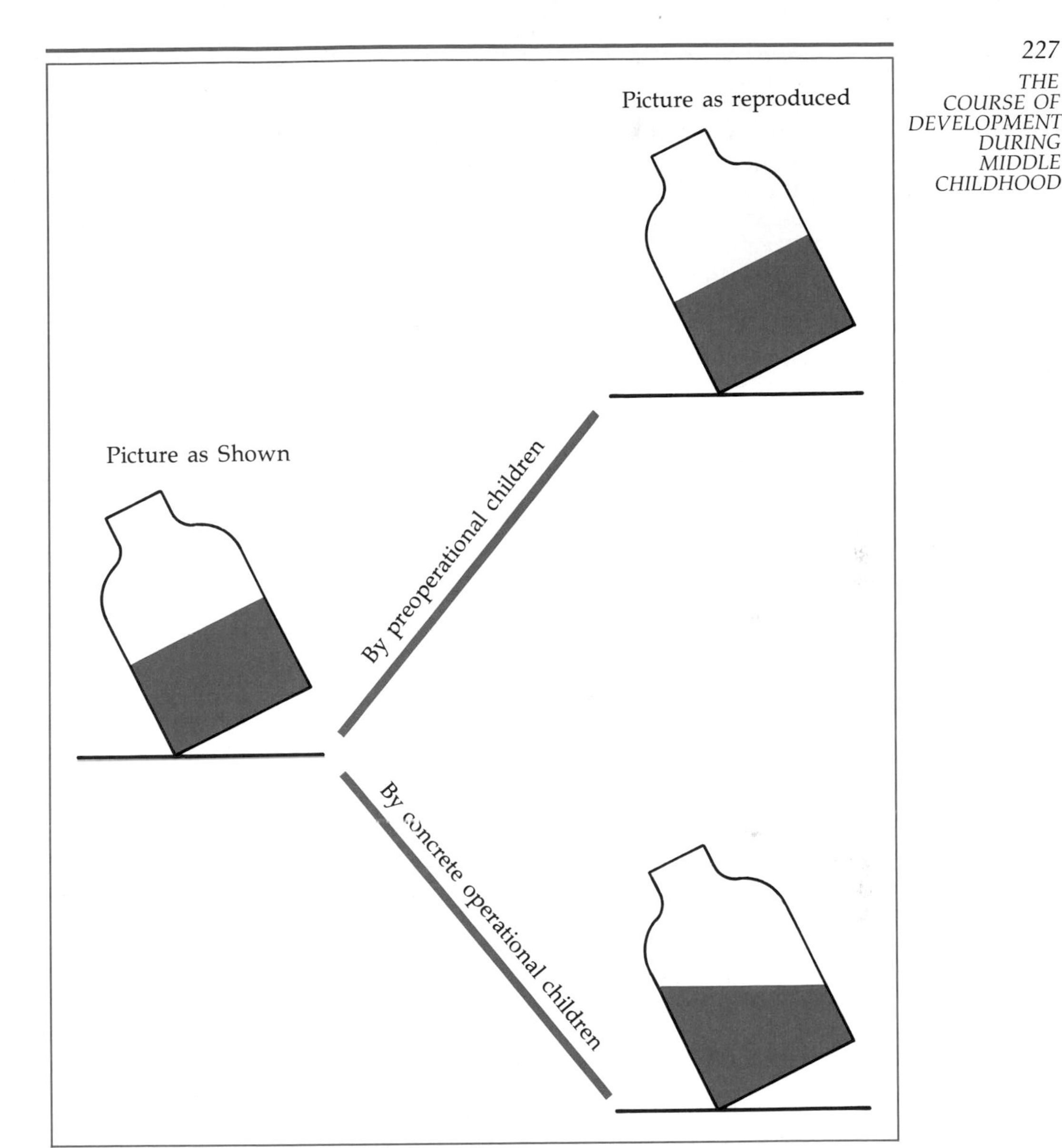

FIGURE 5-2
Liben (1977) found that when children were shown a picture of a tilted water bottle with the water surface parallel to the bottom of the bottle and were asked to reproduce it months later from memory, preoperational children drew it as it had existed in the original picture while concrete operational children drew it as it would look in reality—with the water surface parallel with the table.

Limitations of Concrete Operational Logic

Lest you become too impressed with the competence of the school-age child, it is important to realize that there are a number of very significant limitations to concrete operational reasoning (Cowan, 1978). Probably the most important is that these cognitive operations are applicable only to concrete, real events. They are limited to the world as perceived rather than the world as imagined. They are limited to a world that can be experienced and literally acted upon. Ask nine-year-olds to multiply 6 times 5, and they will tell you the answer is 30. Ask them how many heads there would be altogether if there were six dogs and each dog had five heads, and they will tell you that dogs don't have five heads. Learning to deal effectively with the abstract and the hypothetical and with situations contrary to fact must wait for adolescence.

Second, the ability to use concrete operations does not emerge all at once. It evolves gradually over the age range of seven to twelve years. When asked to spell "grandmother," seven-year-olds who can count anything may reply that they have not yet learned that word. They have learned to conserve quantity but have not yet learned to "conserve" spelling. That is, they have yet to generalize their discovery across domains that actions on objects (such as counting or spelling) are independent of the objects acted upon. It is not until children reach age nine or ten, for example, that they can successfully conserve weight.

The conservation of weight task is similar to the conservation of substance. The major difference is that now when the shape of one object is changed from, for example, a ball to a pancake, the child is asked if the pancake still weighs the same as the ball rather than if there is still the same amount of playdough. Although the procedures do not differ significantly, seven- and eight-year-olds often fail to realize that if the amount of substance does not change, then there also cannot be a change in weight. In other words, they fail to generalize across specific situations.

A third limit concerns applicability. It is one thing to perform an operation; it's another to know when to use it. The question of applicability concerns experience. It concerns having enough contact with various materials to realize that changing the shape of an object does not change its amount or weight.

Murray (1979) emphasizes the importance of both experience and logic in determining applicability by describing a hypothetical conservation of weight task in which playdough is transformed through exposure to X rays rather than by changing its shape. Unless children understand something about the effects of X rays on playdough, all the concrete operational logic in the world isn't going to help them reach the correct conclusion; both experience and logic are necessary to understand the transformation. The necessity of having both will again become apparent when adult cognition is discussed in Chapters 8 and 9.

Metacognitive Skills

The cognitive skills of school-age children are not only evident in the integrated, active manner in which they function; they are also evident in the fact that school-age children are considerably more aware of their cognitive and memory activities than preschool children are; that is, school-age children demonstrate what are known as **metamemory** and **metacognitive skills** (Flavell, 1977.)

Metamemory refers to the various techniques used to retain information. These techniques include such mnemonic devices as rehearsal, chunking, coining phrases, creating interacting images, inventing stories, and devising rhymes. Chunking, for example, refers to grouping items into categories, while creating interacting images involves producing mental images that link together the items that need to be remembered (for instance, creating an image of a monkey driving a car in order to remember the paired word combination monkey–car).

The basic activities of metacognition include predicting the consequences of an action or event, checking the results of one's own action (Did it work?), monitoring one's ongoing activity (How am I doing?), testing reality (Does it make sense?), and coordinating and controlling by a variety of methods deliberate attempts to learn and solve problems. The acquisition of these skills is one of the basic characteristics of efficient thought, largely because they reduce the possibility of error and because they are appropriate in any situation. Children who make this discovery, who at least from their behavior seem to implicitly understand the value of metacognitive skills, usually have a clear advantage over those children who have yet to make the discovery (Brown & DeLoach, 1976).

The advantage provided by the acquisition of metacognitive skills is well demonstrated in an experiment by Markham (1977). Children in grades one through three were told that they were going to be taught a game. First an equal number of cards with alphabet letters glued to them was dealt to each child. Then the following instructions were read: "We each put our cards in a pile. We both turn over the top card in our pile. We look at the card to see who has the special card. Then we turn over the next card to see who has the special card this time. In the end, the person with

This child's developing metacognitive skills enable her to fill out and understand this organizational chart.

the most cards wins the game" (p. 988). The children were then asked a series of questions to see if they understood how to play the game and if they believed they had all the information they would need. You have no doubt realized by now that the study didn't concern the playing of games but the degree to which children understood that the instructions given were inadequate—the specialness of the special card was never defined. The third graders were significantly more likely to ask the necessary questions than the younger children were. In fact, most of the younger children actually had to start playing the game before they realized that something was amiss.

This level of sophistication in problem-solving strategies is also evident in the way school-age children play the familiar game of twenty questions. Mosher and Hornsby (1966) found that school-age children progressively required fewer questions to arrive at the correct solution because of the way they went about constructing their questions. Younger children were more likely to ask if a specific choice was correct. Since each subsequent question was simply another guess, their probability of success actually improved very little after each question. In fact, for preschoolers it often did not change at all since they sometimes forgot that they had already made a particular guess. By the time children reached fifth or sixth grade, their early questions were rarely specific guesses but were designed to eliminate a group of possible choices; in a guess-the-number game, rather than asking if the correct number was a "three," they might ask if the correct number was an even number. The efficiency of such a strategy was obvious to these older children but not to the younger ones. Children in this age range seem to be realizing that they can take active steps to learn something better (rather than just guessing) and, perhaps more important, that an indirect strategy is often more effective than a direct one.

Metacognitive skills are particularly important in problems that require inferences for effective learning, that is, in problems that require children to go beyond the specifics of the task and to develop some general rule that is applicable to all examples of the problem (Karmiloff-Smith & Inhelder, 1974–75; Siegler, 1976). Consider two types of balance tasks. One requires children to balance a beam by learning how different amounts of weight placed at different distances from the center can serve to balance the beam. The other task requires children to move a block along a narrow fulcrum until the block balances. Children with good metacognitive skills, who are good at inferring general problem-solving strategies should not only make fewer errors with practice, but should also make successive approximations to the correct solution that are consistent with the logic of the problem.

For example, in the first task, if a child tried to get his side to balance the other side by putting three weights on the peg and if, when he released the arm, his side went up, he would be demonstrating better inference skills if he then tried four weights on the peg rather than two. The balancing task that requires children to move a block along a fulcrum presents similar opportunities for evaluation. If a child positions the block and releases it and it falls to the right, and if she has the ability to make some general inferences, she should reposition the block farther to the left on the next attempt.

The increasing ability to solve such problems efficiently during middle childhood reflects an increase in the ability to draw inferences from old experiences and apply these inferences to new ones as well as a shift in children's perception of themselves as problem solvers. Problems gradually come to be viewed less as isolated instances of success or failure and more in terms of opportunities to test or confirm either implicit or explicit theories or inferences. Only when children are able to view situations as opportunities to test ideas are they able to develop generalized expectations or inferences about the way their physical and social environments function.

Influence of Concrete Operational Logic on Everyday Events

The primary reason that children are able to develop these inferences is that the system of concrete operations allows for the quantification of experience. That is, it allows school-age children to have at least an implicit understanding that a whole can be broken into a number of segments. This quantifying, or unitizing, quality of concrete operational thought gives middle childhood its unique flavor.

The idea that a whole can be divided into a number of segments, or units, is a fundamental aspect of mathematics. It is also a fundamental aspect of children's ability to understand fractions, to use a ruler for measurement, and to understand the concept of time. Children's increasing understanding of time and their increasing understanding of such physical phenomena as shadows provide two good examples of the general course of this quantification of everyday experiences.

Many preschoolers can tell time (especially if they use a digital clock); very few understand time. Because they can read the clock, they can tell you what time it is. They know that certain events such as dinner or a visit from grandma and grandpa or a television cartoon happen at a particular time. But their ability to understand time as sequence or as duration or as a constant is, at best, very limited. These understandings occur during middle childhood.

The first component of the child's understanding of time is the realization that duration is independent of activity. Piaget (1927/1969) reported that five-year-old children believed that the clock they were observing moved faster when they were asked to work harder and faster. It was not until the age of eight or nine that they understood that the two kinds of action were actually independent.

A second component of the child's understanding of time is the realization that two intervals are of equal duration, irrespective of the nature of the events within them, if they both start and stop at the same time. Piaget found that before the age of eight most children believed that if two equal streams of water flowed for the same amount of time into two bottles of different sizes, the smaller bottle filled at a faster rate than the larger bottle. Although it is true that the larger bottle would take longer to fill, it is not true that the rate of filling differs. This distinction is very hard for the young school-age child to appreciate.

The confusion of rate and amount is also evident in children's understanding of age—the cardinal or cumulative aspect of time. Young school-age children often confuse age with size; that is, they think that a person keeps getting older as long as he or she keeps getting bigger and that after that, everyone somehow becomes as old as everyone else. Consider the following interview:

Vet (age seven) I have a little sister, Liliane, and a nine month old brother, Florian. (Are they the same age?) No, first of all there is my brother, then my sister, then me, then Mama and then Papa. (Who was born first?) Me, then my sister and then my brother. (When you are old, will Florian still be younger than you?) No, not always. (Does your father grow older every year?) No, he remains the same. (And you?) Me, I keep growing bigger. (When people are grown up, do they keep getting older?) People grow bigger and then for a long time they remain the same, and then quite suddenly they become old. (Piaget, 1927/1969, p. 208)

By age eight, most children are able to understand that physical growth and chronological age are independent and that people still get older even if they don't get bigger. Further, they realize that the number of years separating two people will always remain the same, no matter how old the people are. These understandings again indicate children's growing ability to quantify physical dimensions into equal units and to objectively consider the relationship between the two dimensions.

Long before children are able to enjoy the adventures of Peter Pan and the Lost Boys, they are fascinated by their own shadows. They want to know where shadows go when they disappear and how they know which way a person is going to turn. The answers to these questions become evident during the school years. They reflect children's growing ability to see themselves in relationship to other things and to draw conclusions from their specific experiences. The "correct" answers are certainly more mature than children's early speculations, but they can't be as much fun as thinking that your shadow is your very special friend.

Children who are four to six years old know what a shadow is but have very little understanding about how it works. They think that it comes from a specific place in the same sense that rain comes from a cloud. They think that shadows still exist at night even though they can't be seen. They are rarely able to predict where a shadow will appear. For these children, a shadow is a real thing with, apparently, a mind of its own. This animistic view is evident in Stei (age five):

(You know what a shadow is?) Yes, it's the tree that makes them, under the tree. (Why is there a shadow under the tree?) Because there are lots of leaves, the leaves make it. (How do they do it?) Because they are red. (What does that do?) It makes a shadow. (Why?) Because inside the shadow it is night. (Why?) Because it's day on top. The leaves are big because it is night inside them. (Piaget, 1960, p. 184)

By the time children reach seven or eight years of age, they have acquired a much more realistic view of shadows. They know that shadows have something to do with light, and they can usually predict where a shadow will fall. They understand that a shadow will always fall opposite the source of light. But they may still believe that shadows exist at night. The explanation for this belief is appropriate enough: "You can't see them because it's too dark." In other words, for the eight-year-old, shadows still seem to have at least a little life of their own.

Only toward the end of middle childhood do children reach a realistic understanding of the source and nature of shadows. They understand that shadows have no life of their own, that they don't exist at night, and that they don't exist at night because the condition necessary for their existence is absent; in other words, there is no source of light. Children are finally able to arrive at the correct conclusion when they change their focus from the shadow to the light source and the object it falls upon. They finally come to understand the relationship between objects and light and as a result must, perhaps sadly, conclude that shadows aren't real after all; they are merely the passive consequence of the action of light falling on an object.

The skill resulting from the decline of egocentrism, the awareness of transformations, the use of reversible mental operations, and the quantification of experience are skills whose uses are not restricted to the physical world. They are also used in the child's understanding of interpersonal relationships. It is to this topic that we now turn.

Social Development

This section continues the survey of children's developing understanding of social relations. Specifically, the section surveys children's understanding of peer relations, social justice, sex-role identity, and relationships to authority figures, and their perceptions of themselves as unique individuals. In addition, since the first direct independent learning about the working of society takes place during middle childhood, this section also surveys the child's initial understanding of such social concepts as money, government, and adult social roles.

During early childhood, preschoolers' understanding of social roles and behavior demonstrates two relatively distinct periods. The first reflects the strong influence of egocentric thought. Concepts are fluctuating and idiosyncratic. Understanding is tied to specific situations and not easily generalized. Toward the end of early childhood, as preschoolers become less egocentric, they begin to consider themselves in relationship to others. This new perspective brings a degree of continuity to their understanding of social relations. In fact, it brings more than a degree of continuity: it brings a stereotypic, rather rigid adherence to this new view of social relationships. The older preschooler might think, for example, "If I play with you, you have to play with me." This stereotypic quality continues to dominate the early part of middle childhood. It gradually gives way to a more relative perspective as school-age children further immerse themselves in the peer culture and as they learn about the world beyond their own "backyards."

Conceptions of Friendship

Friendship acquires great importance during middle childhood. It serves as the keystone of children's social world, changing the nature of their relationship to their parents and establishing patterns of interpersonal relations that are continuous well into adulthood.

During the early school years, peer relations take on a reciprocal quality (Damon, 1977; Selman, 1980; Youniss, 1980). Friends are seen as people who help each other. They do things for each other, give things to each other, and share their possessions willingly. Friends are no longer viewed as simply people whom you play with or who happen to live next door. Friends are now people whom you like. In other words, the friendships of eight- to ten-year-olds begin to reflect an awareness of the psychological qualities of the other person.

Just as good deeds establish these early friendships, unkind deeds terminate them. In a sense, these are truly fair-weather friendships (Selman, 1980). This fact is well reflected in the comments of Betty, a ten-year-old:

> (Who are your friends?) Karen, Dahlia, Martha, Barbara. (Who is your best friend?) I don't know, Karen at school and Martha in the neighborhood. (Why do you like Karen?) Because she's nice; she gives me jewelry and candy and I give her things too. (Suppose Karen didn't give you anything?) I'd still like her because she is nice. (What makes her nice?) I don't know. (Do you ever fight with Karen or Martha?) When Martha screams we fight, and she fights dirty. She pulls my hair and pinches me. (How do you get to be friends again after fighting?) I call her up and ask her if she wants to play but if I'm still mad I don't call her. Then she calls me and says she's sorry and I say I'm sorry to her. Then we make up and play again. (Is there anyone who is not your friend?) Yes, Bernadette. (Why don't you like her?) Because she picks fights and then acts cool, like saying "dynamite." And she's stupid. (How do you get to like someone?) I like them if they like me but if they don't like me I don't like them and I don't like her because she doesn't like me. (Damon, 1977, p. 158)

Betty's comments show that her conception of friendship is still tied to specific situations and behaviors. For her, a fight or argument ends the friendship. It then must be restarted through mutual apologies. When she and a playmate are fighting, they are no longer friends. When the fighting is over, they again become friends. It is the realization that people can still be friends even though they sometimes happen to be mad at each other that marks the transition to late middle childhood friendship patterns.

Preadolescent friendships (at approximately ten to thirteen years of age) have a special quality. Friends at this age are very close and very personal. They are the friends to share something special with or to talk to when you just can't tell anyone else because no one else would understand. They are clearly the ones you talk to when your parents don't, won't, or can't understand. Whereas the friendships of children between the ages of seven and ten are based mostly on common interests or shared activities, these friendships are increasingly based on shared feelings and the shared ability to comfort. Because there is such an intimate quality to these friendships, they are neither easily started nor easily terminated. Nine-year-olds may come home and announce that they have a new best friend, but twelve-year-olds are more cautious. It is important to get to know other people, to find out if they are the kind of people you can trust and the kind of people you want to trust.

Because there is so much investment in these friendships, frequent breaches of trust or disagreements are necessary before the two people conclude that they can no longer be friends. The special flavor of these friendships is evident in the following definitions of friendship obtained through interviews by Youniss (1980):

> (Male 11) Someone who will do things for you; help you when you can't understand something. (p. 177)
>
> (Female 11) Someone who really cares about you and doesn't want to betray you . . . a person who would want to help me out if I was in trouble; just to help me out when you need her help. (p. 178)

During middle childhood, friendships take on the tone of adult friendships. Unlike those of early childhood, these friendships are entered into carefully and are valued as sources of mutual support and trust.

(Male 13) Someone who can help you and stay with you when you're in trouble. Plays with you when there is no one else around. (p. 178)

(Male 12) A person who will help you, like if you have problems. Depend on you and won't leave you if you get in trouble . . . tries to console you if you have problems. Tries to relieve you of some of your problems. (p. 179)

(Female 12) Someone you like. Someone you can depend on. They tell you their secrets and you tell them yours and you talk about things you wouldn't tell other people. (p. 180)

(Female 11) A person you can really tell your feelings to. When you're lonely, you always have a friend to tell your problems. I know her so much I can do anything with her. (p. 182)

(Male 12) A person you can trust and confide in. Tell them what you feel and you can be yourself with them. (p. 182)

Perhaps the ultimate definition comes from an apparently early maturer (Selman as quoted in Damon, 1977): "A friend is someone that you can share secrets with at three in the morning with Clearasil on your face" (pp. 164–165).

During late middle childhood, children first begin to conceptualize friendship much in the way that adults do. Clearly the secrets of twelve-year-olds and the problems they face are less severe and less consequential than those of adulthood (at least in an absolute if not a relative sense), but the pattern is similar. The friendships are seen as intimate, they grow gradually, they are terminated reluctantly only after a number of conflicts, and their termination is accompanied by a sense of personal loss.

Development of Social Justice

The close association between peer relations and social justice continues during middle childhood. Through children's involvement with peers, they develop a notion of fairness; and from this understanding of fairness, the more abstract concept of equality evolves.

By the end of early childhood, children have developed what Damon (1977) calls a sense of **reciprocal obligation.** It is the belief that if I do something nice to you, you have no choice but to do something nice to me. The original act of kindness may not have been requested or even needed but since it occurred, the person who benefited from it must repay the kindness. When Damon asked seven-year-olds about fair ways to distribute rewards and to share, this sense of obligation was reflected in their strong belief in merit. They thought that distribution should be based on merit, that is, that those who were most deserving should receive the most rewards. There is a very instrumental quality to this level of understanding. If a person does something good, then that person merits, or deserves, to receive some appropriate reward or compensation.

There is also a "law of the jungle" quality to this sense of reciprocal obligation. Merit is viewed solely in terms of outcome. Those who do the most should get the most. But what happens when need is added to the distribution equation? Need is often in conflict with merit. This conflict is first confronted during middle childhood. And as children grapple with the task of integrating the two sometimes conflicting perspectives, their conceptions of fairness and justice and equality gradually broaden and mature. Eight-year-old Dan is asked to decide how to distribute rewards to children who worked on a project together. The merit and need of each child is evident in the dialogue:

> (So, would you give Rebecca, who made more things, more of the money?) Well, she would get a little more. She would be getting about three dollars more. (Billy says that he should be getting more too because he doesn't get any money at home.) I'd give him a little more. Because everybody deserves an allowance. (What about Melissa, the best-behaved kid? Would you give her more?) Probably. (You would?) Yeah, but not too much more, because being a real nice kid isn't much, you know. I'd probably give her the same as Billy, but not as much as Rebecca. (Damon, 1977,P. 85)

Dan's resolution is typical of eight- to ten-year-olds. He realizes that need and merit both are relevant considerations. Therefore, he is acknowledging that there is more than one basis for judging fairness. Compromise is the way to resolve the conflict. All the children get something. How much they get depends upon the value of their claims.

There are some shortcomings to Dan's solutions, however. First, the criteria for reward are defined solely in terms of the particular situation. No broad principle of social justice is used as a reference point. Since each situation is judged solely in terms of its individual elements, there is little likelihood that Dan has a general principle for resolving the conflict of need and merit. Faced with a similar situation in the future, Dan might not disburse the money in the same way. Second, in Dan's system, everyone is deserving of some reward. Some children get more than others, but everyone gets something. As noble a solution as this may be, it does not deal with the question of the legitimacy of claims for reward. It does not consider the possibility that some of the children may not deserve any reward.

By the end of middle childhood, children are usually able to resolve the second limitation but not the first. They accept the fact that some claims are irrelevant to a situation and therefore should not be considered. They might express sadness at having to reject a claim, but they would nevertheless argue that it was the right thing to do. These children are not completely "heartless," however. One girl, in explaining why a poor child shouldn't get extra money just because she was poor, said that "it really wouldn't be enough money to make her not poor."

The evolution of social justice concepts during middle childhood reflects three relatively unique conceptions of fairness. Initially, fairness means that everyone should get the same reward. Then, it means that everyone should get something because everyone has some basis for a claim. Finally, fairness seems to mean that everyone who has a relevant basis for a claim should get a fair hearing and on the basis of that hearing should get some appropriate reward. At each step children are better able to separate the relevant from the irrelevant and to acknowledge multiple bases for equality. They are also more aware of individuality. The ability to put these new concepts into a broader social context must, however, wait for adolescence.

Construction of the Concept of Authority

To the preschooler, authority is represented almost exclusively by the parent. Other adults, such as a babysitter or a police officer acquire their authority by having it conferred upon them by the parent. In middle childhood, children encounter authority figures that are separate from the parent. Most typically, they are teachers, but these authority figures may also be principals and other school personnel, scout leaders, camp counselors, merchants, and friend's parents. Children must learn the appropriate response to each, the way dealing with one affects dealings with others, and the way these new relationships influence their relationship to their parents. At the same time, they begin to be aware of the differences between relations of equals and relations of unequals. The result of all this differentiation and integration is that parents are no longer viewed as omnipotent and omniscient, and as a result children develop a more realistic perception of their relationships to their parents.

Just as peer relations during the early part of middle childhood are reciprocal, so are parent-child relationships in the view of seven- and eight-year-olds. However, there is one important difference in the nature of the reciprocity in these two kinds of relationships. Reciprocity among peers implies equality; reciprocity between parent and child is complementary (Youniss, 1980). **Complementary reciprocity** means that the parent and the child each have certain rights and obligations within the relationship but that, unlike peers, one cannot take the role of the other. Parents have the right to be obeyed because they have the responsibility to care for their children. Eight-year-old Karen expresses this view very clearly when she is asked why it is important to do what her mother says:

> Because if it is something that you can't do she might help you. She is the one that can help you the most. (Why is she the one that can help you like that?) Because she is older and can do a lot of things that you can't. (Damon, 1977, p. 191)

This developing rationale for obedience in complementary relationships not only affects the child's view of the parent; it also extends to all hierarchical situations, even team sports. There is, for example, much more ambivalence in the feelings of ten-year-olds who aren't starting members of a team than there is in the feelings of eight-year-olds in the same situation. The eight-year-olds want to have as much playing time as the other children have. They don't seem to care that their presence on the field may actually hurt their team's chance of winning. The ten-year-olds also want to play, but added to that desire is the desire to be on a winning team. The conflict comes when the ten-year-olds realize that being on a winning team may mean less playing time. This conflict is rarely resolved by ten-year-olds. It does, however, clearly reflect their growing understanding of the nature and even legitimacy of complementary relations. Unfortunately, these developments don't help the parent of the "understanding" child sitting alone in the dugout to feel any less sad about the child's not getting to play.

By the time children reach the age of nine or ten, a fundamental shift in their conception of complementary relations begins to occur. The shift is influenced by a number of things. As children continue to grow, they begin to see themselves as "catching up" to their parents. The children are becoming competent in a number of skills. They can, for example, play many board games as well as their parents can: they may even win regularly at checkers. Although their math and reading skills are not equal to their parents', the fact that they can now do the same types of things that their parents can do leads them to feel a sense of greater equality. Finally, their continued involvement in peer groups and their increasing independence away from home both seem to gradually culminate in the view that a parent is a "fundamentally equal person with equal rights but with different amounts of training and experience" (Damon, 1977, p. 193). Consider nine-year-old Lisa's explanation as to why a girl named Michelle should do what her mother has told her.

> That would be the best thing to do. (Why do you think so?) Her mother knows best. (Why do you think her mother knows best?) Her mother went through all these times when she was a kid. (What difference does that make?) So now she knows what to tell Michelle to do, what's best for her and stuff. She knows because she's been through it. (Damon, 1977, p. 194)

Lisa's rationale for obeying a mother is very different from Karen's. In referring to the fact that Michelle's mother was once a kid herself, Lisa seems to sense the basis for the complementary nature of the parent-child relationship. A parent is simply someone who happened to be a child before you. (This discovery probably coincides with the cognitive gains in the understanding of time and chronological age.) It is the awareness that the parent once occupied the same role the child now occupies that prompts this new awareness of the reasons why parents should be obeyed. Once children realize that parents should be obeyed because they have learned things that the children have not, it is only a short step to the realization that there must also be times when the parent need not be obeyed—that is, in situations where the parent has had no experience. Many of the parent-child conflicts of adolescence hinge on this issue—the adolescent's perception of the parent's experience or inexperience in a particular area. In fact, in a rapidly changing technological society such as ours, it is not hard to imagine many situations in which the child is considerably more competent than the parent (a circumstance that may not be to the benefit of the parent or the school-age child).

By the end of middle childhood, the parent-child relationship changes in such a way that parents no longer rely on coercive measures as the primary means of maintaining their authority. Negotiations and contracts become the preferred methods of resolving differences. Certainly, there are still times when consensus or even discussion is impossible, but the new patterns are nevertheless established. They are patterns that recognize the experience and the rights of both the parent and the child. By learning to define respect for authority in terms of experience and competence, children have a basis for understanding the limits of authority. The eight-year-old thinks that teachers must always be obeyed; the twelve-year-old knows that their authority stops at the end of the playground. Across the street from the school, the teacher is no longer the teacher but simply another adult.

Conceptions about the Larger Society

Middle childhood is a time of growing interest in the larger world. Children begin to read the newspaper, at least the sports or entertainment sections. They may ask questions about current events. They begin to wonder what, if any, consequences might result from events far removed from their immediate experience (events such as war, economic crisis, and nuclear accident). This new interest reflects a number of developments. First, the development of concrete operations makes it possible for children to see themselves within a larger context. Second, they are acquiring the academic skills such as reading that make possible the gathering of information. Third, their increasing independence and competence allow them more opportunity to explore their communities and as a result to become

The preadolescent's growing sense of competence and feelings of equality with parents result in a shift in parent-child relationships. Toward the end of middle childhood, negotiation becomes a more important component of parent-child interaction.

more curious about their workings. This section surveys children's developing understanding of one aspect of this larger world: the political system.

When children are between six and eight years old, their understanding of government is best described as personalized, vague, and fragmented. In a sense, it is reminiscent of the preschooler's early sense of sex-role identification. That is, they seem to know some of the words but don't really understand what the words mean. A seven-year-old, when asked the duties of the president replies,

> Well, sometimes he . . . well, you know, a lot of times when people go away he'll say goodbye and he's on programs and they do work. (What kind of work does he do?) Studying, like things they got to do, what's happening and the weather and all that. (Greenstein, 1965, p. 34)

An eight-year-old's description of the White House shows a similar superficial level of understanding:

> Well, I heard that the White House is real big, and it's got a lot of rooms. There's a gray room, a green room, a gold room and the President lives in the White House. It's real big and nice. It is big and has two chairs, and if the king and queen want to come and visit, they sit down on the chairs. (Hess & Torney, 1967, p. 15)

For children of this age, the president is a nice person who tells people what to do, makes the law, and gives people jobs. He is the source of money (although few at this age understand taxes), and he pays for things. He is also a "he." Few eight-year-olds are yet able to go beyond their concrete experience to realize that he could just as easily be a she. Few understand anything about the legislative or judicial branches of government or the relationship between federal, state, and local governments (Furth, 1980). They may have learned specifics about the branches of government in school, but they don't yet seem able to comprehend their relationships to each other. It is not surprising that their conceptions of the president are similar to the preschooler's understanding of the all-powerful and all-knowing parent.

By the time most children are eleven, they no longer equate government with the president. They know of the three branches of government, and they can list a number of specific functions of government, but they still show little understanding of the relationship between the levels or the branches of government. They think, for example, that senators help the president rather than serving as an independent branch (Hess & Torney, 1967). They understand that the role of the mayor is parallel to the role of the president (each is head of a unit of government) and are aware of some of the more obvious functions of city government such as snow removal, street repair, and park maintenance. They may have less understanding of the less visible services such as social welfare, health clinics, and housing inspections. They understand that governments get their money from collecting taxes and that people can be elected to various offices, but they really don't understand the idea that government is the servant of the people. Government is still seen in a more or less parental role.

What is perhaps most interesting about this early development of political understanding is how closely it parallels the understanding of parents as authority figures and how clearly it points out the constructive nature of the child's understanding of social institutions. This pattern is also evident in the child's understanding of other social institutions. Seven-year-olds, for example, think that receiving change is actually getting extra money. They don't understand it in relation to the purchase price and the amount tendered. Nine-year-olds have little understanding of what merchants do with the money they receive, and many eleven-year-olds, even though they know that merchants use their money to buy new goods, still think that merchants get their personal money from some other source; in other words, they have little understanding of profit (Furth, 1980). In fact, it is not until adolescence that children are able to put the pieces of the economic or political systems together in an integrated fashion. Although these systems can be viewed as reciprocal and complementary just as peer relations and parent-child relations can, their relative abstractness and their relative distance from everyday experiences limit children's ability to understand their nature. Although school-age children are very much into their world, their understanding of that world is still rather fragmentary.

The Self-Concept in Middle Childhood

The development of concrete operational skills, the immersion of children into the peer world, the school system, and the community, and the corresponding changes in children's relationships to their parents all combine to produce a fundamental shift in children's conception of themselves. The "I" of early childhood becomes in middle childhood the "I with respect to you" (Youniss, 1980). The transition takes place in three steps.

The first step serves as the bridge from early childhood. By age six, children understand that people interpret situations differently and that people have different thoughts and different feelings. They also believe that people's perceptions are an accurate reflection of the objective facts of a situation, that is, that what people say is in fact what they are thinking. This set of beliefs is reflected in the fact that first and second graders' self-evaluations are based primarily on their assessment of their effort and accomplishment. From the children's point of view, the assessments are objective, and they are absolute; the children also think that since people's perceptions are copies of some objective reality, others must see them in exactly the same way as they see themselves (Ruble, Boggiano, Feldman, & Loebl, 1980). They don't yet understand how it is possible to have conflicting viewpoints about the same thing. In essence, they don't yet realize that people's perceptions are not carbon copies of situations. This rootedness in an objective reality causes children who are six to eight years old to be trusting, naive, and eternally optimistic (Selman, 1976; 1980).

This sense of certainty is also evident at this age in children's understanding of sex-role identity. A boy knows he is male; a girl knows she is a female, and both now know that a change of hairstyle or clothing won't change these basic facts. This certain and (at least to the seven-year-old) unambiguous view of males and females probably precipitates the relatively self-imposed sex segregation typical of much of middle childhood. Boys do boy things because that's what boys do; girls do girl things because that's what girls do.

By the time most children turn eight, their clear and certain views of themselves and their world begin to break down, as they come to realize an

The self-imposed sex segregation characteristic of middle childhood is probably the result of strong sex-role identification.

interesting fact: "I" can also be thought of in terms of "he" or "she." In other words, in the same way that I can talk about my teacher or my favorite baseball players, they can talk about me. For the first time, children are capable of limited self-reflection. This ability to view themselves in the third person and thus to look into themselves allows children to realize that the public self and the private self are different—a notion quite contrary to the six-year-old's view.

Although these new powers of self-reflection only function within the concrete operational world of real experiences, within that world, children now begin to understand that what they think and what they say need not be the same, that others can only see the public self and not the private self, and that they can influence another person's perception of them by what they say. It is not surprising that many children between the ages of eight and ten become increasingly concerned with their outward physical appearance (that is, with groomed hair and neat dress if not necessarily brushed teeth and regular showers or baths) and, unfortunately, come to use lying the way some adults use it, as a deliberate distortion of a known truth.

It is important to note that although ten-year-olds are capable of self-reflection, it is probably not their major preoccupation. They are still too eager to discover the world to spend much time reflecting on their discoveries or their place in that world. Self-reflection does not become the dominant mode until adolescence.

When children are between the ages of ten and twelve, their understanding of self represents a further clarification of their self-reflective abilities. There is a further differentiation between the public and private self. The private self is viewed as one's personality; the public self, as one's behavior. The latter is considered easily modified; the former, relatively enduring (Selman, 1980). This shift in the basis for self-definition is evident in these two excerpts from interviews from Montemayor and Eisen (1977). Notice the difference in the use of concrete descriptive versus abstract reflective conceptualization.

My name is Bruce C. I have brown eyes. I have brown hair. I have brown eyebrows. I'm nine years old. I LOVE! Sports. I have seven people in my family. I have great! eye sight. I have lots of friends. I live on 1923 Pinecrest Dr. I'm going on 10 in September. I'm a boy. I have a uncle that is almost 7 feet tall. My school is Pinecrest. My teacher is Mrs. V. I play Hockey! I'm almost the smartest boy in the class. I LOVE! food. I love fresh air. I LOVE School. (p. 317)

My name is A. I'm a human being. I'm a girl. I'm a truthful person. I'm not pretty. I do so-so in my studies. I'm a very good cellist. I'm a very good pianist. I'm a little bit tall for my age. I like several boys. I like several girls. I'm old-fashioned. I play tennis. I am a very good swimmer. I try to be helpful. I'm always ready to be friends with anybody. Mostly I'm good, but I lose my temper. I'm not well-liked by some girls and boys. I don't know if I'm liked by boys or not. (pp. 317-318)

The increasing breadth of the preadolescent's experiences coupled with an increasing flexibility of thought (in other words, the ability to distinguish actions from objects) allows the formation of a **generalized other,** that is, a generally accepted standard or perspective against which individual views and actions can be weighed (Mead, 1934). Preadolescents are beginning to see themselves not solely in the context of a specific situation or immediate reference group; they are beginning to judge themselves within a broader, more general context. This broadening of perspective is often "self-defeating." You may be the best ball player on the block or the smartest kid in your class but probably not in the whole school and certainly not in the whole world. The doubts, uncertainties, and criticalness of adolescence begin to become apparent as twelve-year-olds realize the sometimes growing disparity between their private and public selves and as their broadening explorations provide for an ever-expanding frame of reference (Elkind, 1978).

THE CONTEXT OF DEVELOPMENT DURING MIDDLE CHILDHOOD

The family continues to be the primary influence on development during middle childhood. However, it is no longer the sole influence. The peer group and the school system also exert considerable influence during middle childhood, as does television. The way each of these factors exerts its influence reflects its specific nature and its degree of compatibility with the other factors. The child who, for example, experiences great continuity among home, school, and peer-group experiences has a very different developmental experience from that of the child who finds significant differences in the values, demands, and attitudes of the three.

The Influence of Parents

Parents continue to be the dominant influence on children's development during middle childhood. Although the basic dimensions of parenting—control, communication, maturity demands, and nurturance—remain the same as they were during the preschool years, school-age children's new levels of ability alter the manner in which parents express these four aspects of the parent role.

Parenting Strategies

It should not be surprising that the same patterns that characterize good parenting during infancy and early childhood are equally salient during middle childhood. Good parents maintain a positive view of themselves as adults, establish a reasonable balance between the limits they set and the autonomy they allow, and set developmentally appropriate expectations for the child.

Coopersmith (1967) has noted that families of children with high self-esteem establish extensive sets of rules and zealously enforce them. This makes it possible for parents to establish authority, define environments, and provide standards against which children can judge their actions. However, enforcement of the rules is noncoercive, and the rights and opinions of the children are sought. Because there is latitude in the exercise of control, concessions are granted when differences exist; and children are permitted to enter into discussions, thereby learning to be self-assertive. Although Coopersmith's characterization of effective parenting is based on a series of studies investigating the antecedents of self-esteem in ten- to twelve-year-old boys, it is remarkably similar to Baumrind's (1973) description of authoritative parenting styles found during early childhood.

Like the parents of socially competent preschoolers, the parents of children with high self-esteem demonstrated emotional maturity, self-reliance, and a positive self-image. These parents provided for their children both a model of successful adult behavior and a safe, structured, responsive, nurturant home environment. Coopersmith has described these parents as individuals who recognize the significance of child rearing and believe that they can cope with the increasing duties and responsibilities it entails.

These **parenting techniques** result in other benefits during middle childhood besides the fostering of high self-esteem. They also appear to have a positive influence on the development of moral reasoning and behavior (Saltzstein, 1976; Shaffer & Brody, 1981). These techniques, known as **inductive techniques,** correlate positively with such examples of moral behavior in children as resistance to temptation, admission of guilt, acceptance of responsibility, and willingness to share with others.

These correlations exist because of the nature of the parent-child communication that characterizes inductive techniques. First, parents provide a set of standards that children can use to judge their future actions and intentions. Second, since they emphasize the consequences of children's actions not only in terms of the children but also in terms of other people, they help children see themselves in terms of other people. In this way, they foster the growth of empathy. Third, in discussing problems that children have dealt with unwisely, they suggest alternative strategies that the children can use in dealing with similar problems in the future as well as in making amends for their actions. Fourth, the nonhostile, supportive atmosphere in which these discussions ideally take place makes it relatively easy for parent and child to discuss how each feels about specific incidents (Shaffer & Brody, 1981). It is important to remember that such "well-tempered" discussions are more the ideal than the norm, but they are most often found in families using inductive techniques.

The two other control techniques that parents often use with school-age children, power assertion and love withdrawal (Saltzstein, 1976), may be equally effective in dealing with the immediate situation but often have undesirable side effects and usually have fewer long-term benefits.

Power assertion techniques work because parents are powerful. They control resources and rewards; they can withhold children's allowances or make them wash the dishes every day for the next two weeks or send them to their room or administer any number of equally nasty punishments. The problem is that children are provided little information about the right ways to deal with situations. They simply learn that their solution is unacceptable. Because they do not receive the information necessary to develop self-control and self-direction, children remain dependent on other people. Further, the anger and resentment these power encounters almost always generate separate children from their parents, who should be an important source of experience in dealing with problems. Instead of teaching children how to behave, these punitive techniques provide them with an undesirable model of adult behavior. No wonder there is a strong positive relationship between aggressive behavior in childhood and the use of power techniques in adulthood.

Love withdrawal techniques regulate children's behavior by appealing to the bond that exists between parent and child. These control techniques remind us that feelings of guilt and the development of conscience are an integral part of most people's development. There are few of us who haven't done things that we knew would make our parents feel bad if they found out; there are few of us whose parents at one time or another haven't said "how could you do this to us?" Since love withdrawal techniques foster a sense of conscience, they are an important element of the parent-child relationship. In the extreme or as a steady diet, however, they provide little guidance as to alternative ways of behaving, and they can produce considerable anxiety in adolescence and adulthood about self-worth.

Many a parent has attempted to use the sheer force of parental power to persuade a child to clean up his or her room. While this technique is often effective in the short run, less coercive techniques provide more positive results from the standpoint of future development.

Shifts in the Parent-Child Balance

Inductive techniques become increasingly appropriate and common as children pass through middle childhood. This progression reflects both the increasing competence of the children and their greater immersion into the world independent of their parents.

The increasing competence of the school-age child affects the entire range of the parent-child relationship. There are more shared interests and activities. Parent and child do more things together and talk about more things. Parents generally enjoy their school-age children. The children are usually reasonably well behaved and do not present the management problems of the preschooler or the young adolescent. School-age children are not very likely to make significant challenges to the parent's authority and, as will become important in adolescence, don't yet make the parents reflect on their own aging process. In a "left-handed" sense, this may be the ideal age group to parent. They can play a pretty good game of tennis but hardly ever win.

Competence is not the only factor that changes the nature of the parent-child relationship. Because children now spend a large part of their day away from their parents, strategies that worked solely because a parent was present to make them work are no longer useful. Two new strategies become important. The first strategy—that of allowing children more autonomy in decision making—acknowledges that since part of the child's world is relatively independent of the parents, children ought to have a degree of freedom in dealing with that world as best they can. What position to try out for on the team, which shirt to wear to school, which friend to call, and whether to join the French Club or the Science Club are the types of decisions that increasingly become the privilege of the child. With the privilege, of course, also comes the responsibility—a coupling that isn't always appreciated in middle childhood and, as a result, is a common source of conflict.

The second strategy is to teach children responsibility for their actions. Since parents are no longer physically able to protect their children, they begin to stress helping children understand why certain things must always be done and why certain things must never be done. Always look both ways before crossing the street; don't get into cars with strangers; let someone know if you're going to a friend's house after school instead of coming home, and so on. Parents worry about their school-age children's being on their own. But most acknowledge that they cannot and should not stop it from happening.

Techniques parents use with children always have both short-range and long-range consequences. They always have some impact on the immediate situation, the one for which they are usually intended. They also have indirect effects in the sense that they make future events more or less likely to occur. Beginning in middle childhood, both the short-range and long-range impacts of parenting techniques increasingly reflect children's interpretations of their parents actions. As children realize that it is possible for two people to view the same objective situation differently and that their behavior is often a reflection of their views, they begin to consider the motives behind people's actions. In particular, they begin to wonder about the intentions behind parents' actions or words.

When children conclude that their parent's actions represent a genuine concern for the children's welfare, they usually report a favorable attitude toward the parent. This was true of the high self-esteem children studied by Coopersmith (1967). Even though these high self-esteem children didn't necessarily like or even agree with many of the limits and demands placed on them by their parents, they nevertheless felt valued by their parents. It was this sense of being valued by their parents that helped create their sense of high self-esteem.

This perception of parental intentions affects not only the likelihood that children will comply with their parent's demands and requests (the short-term consequence) but, equally important, also influences children's perceptions of themselves as they develop beyond middle childhood. Since each stage of development involves a reanalysis of self and the factors that influence the self, it is quite possible that children may change their perceptions of parental motivations, in either direction, as new events provide the base for a new perspective. In particular, this continuing reevaluation process may be a necessary prerequisite to the development of their own ability to parent; that is, there may be more to developing a parenting philosophy than saying, "I'll never do to my kids what my parents did to me."

The Influence of Peers

Peers play a more significant role in the lives of school-age children than they do in the lives of preschoolers. Peers serve as constituents of a distinct peer culture and as the context for an elementary definition of sex-role identity.

Reciprocity of Relationships

As important as the role of the parent is in influencing the course of development, it has definite limits. Its influence is limited to activities that are typically done when the relationship is unequal in status, competence, and authority, that is, when the relationship is best described as complementary rather than reciprocal (Sullivan, 1953). As Youniss (1980) points out,

> parents believe that their knowledge of society is worth passing on. They want their children to be successful in this society and work to help them achieve it. By acting with this motive, parents cannot be but controlling and, in the process, establish the fact of interpersonal dependency of action. Therein lies the basis for reciprocity by complement in which children understand the exchange of their own conformity for the parent's approval. (Youniss, 1980, p. 273–4)

The development of peer relations during middle childhood provides children with a radically different perspective for viewing interpersonal relations. Through the establishment and maintenance of close friendships during this time period, they develop such traits as interpersonal sensitivity, appreciation of personhood, and mutual understanding. These traits become the basis for love and intimate relations during adulthood (Youniss, 1980).

During the early years of middle childhood, children attempt to make peer relations conform to the complementary character of the parent-child

BOX 5-1
THE UNIVERSALITY OF CHILDHOOD

Everytime you see an ambulance
hold your collar until you see a
four-legged animal.

Tread on lines your mother's kind;
Tread on squares your mother swears.

Sound familiar? You probably said phrases quite like these on your way to school or to a friend's house. I remember being very careful not to step on a crack just in case it really might break my mother's back (at least on the days I wasn't mad at my mother). These two quotes don't come from American children, however. They were gathered, along with thousands of others, by Iona and Peter Opie (1959; 1969) from children in the British Isles over the past forty years. What is perhaps even more surprising than the fact that these rhymes and games can span oceans is the fact that they also seem to be able to span centuries.

Finger games such as odds and evens and rock, paper, scissors—games often used to decide who goes first—date back to ancient Rome. The practice of having the finder in hide-and-seek turn around with his or her eyes closed three times was a common feature of children's games in the seventeenth century. The idea of having a "free zone" or, as I remember calling it, "safety" in games of tag dates back to the age of chivalry.

See if this rhyme sounds familiar.

Eachie, peachie, pear, plum
When does your birthday come?
(child gives the date such as
December 14)
1,2,3,4,5,6,7,8,9,10,11,12,13,14,
D-E-C-E-M-B-E-R. You are out.

You and I would both no doubt use a very similar sounding rhyme, or "dip," as the British call it, to decide who would be captain of the baseball team.

The Opies' ethnographic research strategy involved both observing and interviewing children between the ages of six and twelve in a variety of locations throughout the British Isles. Such a procedure allowed them to document regional variations on common themes and to show how, as a rhyme or game moved from one region to another or from one time period to another, slight variations in the wording gradually appeared. These variations never distorted the intent or even rhyme of a practice, only its wording or specific detail. The

Sources: Opie & Opie, 1959; 1969.

relation. Children are certain of their views, inform their peers of this fact, and assume that their peers will defer (just as they themselves usually end up doing in relation to their parents). Of course, their peers are thinking the same thing, and an impasse is quickly reached. Each child may then refer to the omnipotence of their respective parents, but this simply leads to another impasse. What children are confronting is a form of relationship in which one member need not necessarily defer to the other. No matter how strong one child's father, mother, sister, or brother is, the other child has one just as strong or stronger. The impasse is broken only when the

The nature of children's play appears to be universal, both geographically and historically.

Opies have even found that this common history of childhood games is not restricted to English-speaking children. Similar patterns, often involving the very same phrases, are common in all cultures. Capture the flag, a game played by many children, dates back at least to the early 1800s. Virtually the same game is called *guerra francese* in Italy and *kawat-kawat* in the Philippines.

Ethnographic research is a very useful strategy in documenting the enduring everyday aspects of culture and their socializing impact on children. The longevity of these games is a reflection of the importance of the oral traditions of a culture. The Opies believe that the common lore of childhood provides one means through which generations can establish a sense of continuity with each other.

two children implicitly, if not explicitly, realize that rather than one's deferring to the other, each must give a little. Through this process of concession, friendships develop (Youniss, 1980).

Once children understand the basic notions of compromise and cooperation, they are able to use these skills in many areas. They learn to negotiate, debate, and argue (Hartup, 1970). They learn to give and to take and, as a result, gradually come to understand, at a concrete level, that social structures are not given but are mutually agreed upon resolutions of conflicting viewpoints (Fine, 1981).

Like those in early childhood, peer relations in early middle childhood tend to be based on forced deference. Failure at these attempts eventually leads to the adoption of the negotiation techniques characteristic of peer relations in late middle childhood.

These "high-level negotiations" help school-age children understand the importance of considering the other person's point of view in trying to convince him or her that they should, for example, play at one house rather than the other, Monopoly rather than checkers, or hopscotch rather than jump rope, or in trying to resolve any other differences of opinion between the two (Fine, 1981). Through this process, children come to better understand others as unique individuals as well as the relationship between personality style and effective communication. It is not surprising that those children who are considered the most popular by their peers are usually the ones best able to tailor their message to suit the listener (Gottman, Gonso, & Rasmussen, 1975). They are children who have become skilled at pubic relations. In fact, in adolescence, many of their conflicts with their parents result from their tendency to view themselves as the equals of their parents and, as a result, to expect their parents to deal with them in a reciprocal rather than a complementary fashion.

Sex-Role Patterns

One of the most striking characteristics of peer relations during middle childhood is that they are almost always segregated by sex. This pattern gradually emerges toward the end of early childhood, becomes dominant during middle childhood, and begins to decline in adolescence. However, even in adolescence, many heterosexual contacts are occasioned not by consideration for the person of the opposite sex but by consideration of the significance the contact will have within the same-sex peer group.

There are a number of differences in the play and activity patterns of boys and girls during middle childhood. Although the differences are well documented (Lever, 1976; Schofield, 1981; Waldrop & Halverson, 1975), their origins and long-term significance are still a matter of debate (Dweck, 1981; Gilligan, 1982; Maccoby & Jacklin, 1974). The long-term question is particularly important. If the sex-role problems typically encountered during young adulthood (see Chapters 7 and 8) have their origins in the play patterns of school children, then some form of intervention would seem appropriate. However, if these play patterns are a necessary step in the sex-role sequence, then the problems of young adults are best dealt with in young adulthood.

One of the more extensive observations of boy and girl play patterns has been reported by Lever (1976; 1978). She notes that boys play outside more than girls do and that boys play in larger groups. Both patterns reflect the types of activities each sex prefers. School-age boys' play tends more toward sports, running and tag games, and other large motor activities such as bike riding. Girls' play patterns are more restricted in area (jumping rope and hopscotch) or more social-interactive in nature (dolls, make-believe, board games). Boys' games are more competitive and formally structured. They are more likely to involve team play and have a defined end point (that is, a winner and a loser). Perhaps because they are more structured, they tend to last longer. When disputes arise (Is the player out or safe?), the structure provides a framework for resolving the differences. If one side or the other doesn't prevail, the structure makes possible a "replay" or "cheaters' proof" as the ultimate arbitrator. Since girls' games involve less structure, a framework for resolution is less available. Further, since girls' games tend to be more social-interactive, conflict is more likely to be focused on personality than on procedure. Because girls are less willing than boys are to jeopardize their relations through argument, conflict is more likely to end their games.

Perhaps because boys' games are more structured, the age range associated with them tends to be greater and the expectations set for participants more uniform. If the game is baseball, everyone is expected to be able to hit, catch, and throw—even someone's six-year-old brother out in right field. Since greater social skills are required to structure and maintain much of girls' play, the age range for such activity tends to be more restricted—a pattern just the opposite of the one for boys.

In this fashion, Lever believes different peer cultures emerge. Girls engage primarily in small-group games in small spaces, games that allow them to practice and refine social rules and roles directly. Boys engage in larger-group games that are more physically active and wide ranging. These games tend to have a more extensive set of explicit rules within which the participants work (or play) toward an explicitly defined goal. They are also more competitive. The play patterns of girls emphasize close personal relationships rather than task completion.

The games of boys are seen as fostering the ability to act within and upon a larger system: to articulate long-range goals and to work actively toward achieving them. This broader perspective in the games of boys places considerably less emphasis on personal relationships than the games of girls do.

Out of these same-sex play patterns develop the close personal, even intimate friendships of the preadolescent years. Unlike the parent-child relations of middle childhood which can often be described from the child's point of view by the question "What should I do to get what I want?" these first true peer friendships are more likely to be defined by another question, "What can I do for my special friend?"

Although it is unclear to what extent these same-sex middle childhood friendships facilitate or hinder the development of intimate relations in adulthood, it is clear that middle childhood is the first time that this central pattern of adult relations becomes evident in the life-span. As Maccoby (1980) notes, the real issue may be the extent to which subsequent life experiences move children beyond these stereotypic and limiting views of appropriate male and female behavior.

The Influence of Television

Television continues to be an important influence on children during the school years. Viewing time gradually increases from $2\frac{1}{2}$ hours per day at age seven to as many as 4 hours per day at age twelve (Liebert, Sprafkin, & Davidson, 1982). The increase in viewing time is of great significance. It means that children are giving less time to other activities and that they are increasing their viewing during an age period when they are most vulnerable to television's message. Based on their research, Dorr and Kovaric (1980) conclude that television violence seems to be capable of affecting all children, regardless of their sex, age, social class, ethnic background, personality characteristics, or usual levels of aggression. While boys and girls are equally likely to be influenced by television violence, boys' behavior tends to be more aggressive than that of girls. Similarly, delinquents and other children whose normal behavior has been categorized as aggressive tend to become more aggressive than children in their obvious comparison groups become following exposure to television violence. Boys also exceed girls and children from working class backgrounds exceed children from middle-class backgrounds in terms of the actual viewing of and preference for this type of programming. Dorr and Kovaric also tentatively conclude that "middle-aged" children (children between the ages of approximately eight and twelve) are somewhat more likely to be affected by violence on television than are either younger or older children.

Almost all of the research on the association between the viewing of violent television shows and subsequent child behavior finds that, at least temporarily, children who view these programs use more aggressive behavior in their story telling and play. The fact that the relationship is certain only for a relatively short period has fueled the debate about television's potential for a more generalized, long-lasting impact. It is possible that the impact is only short-term and that for the most part it reflects the arousal value of such programs. The behavior of people pulling out of a parking lot after viewing an auto race provides evidence of such an arousal pattern. On the other hand, since there is also evidence supporting the position that television has a long-term effect (Lefkowitz, Eron, Walder, & Huesmann, 1977), a steady diet of violent television programs probably does influence a child's general predisposition toward certain types of aggressive behavior.

It is also important to point out that television's potential impact is not necessarily negative. Young school-age children are more likely than they normally would be to offer help to another person after watching similar actions on television, and they are more likely to participate cooperatively in a problem-solving task after watching programs such as "The Waltons" in which the entire Walton family is involved in a similar role (Liebert, Sprafkin, & Davidson, 1982). Unfortunately, there are more shows depicting violent acts than there are depicting positive social ones on most stations.

Television does more than simply influence the probability of various behaviors during middle childhood. It provides children information about the world at a time when they are becoming interested in it. The problem, as noted in Chapter 4, is that television does not often portray the world as it is. White males are overrepresented both in number and in positions of authority; other groups are underrepresented. Although preschoolers may not be much influenced by these distortions, the evidence is clear that older children are.

Young adolescents who are heavy television viewers are more likely to see the world as it is portrayed on television than as it actually is (Gerbner, Gross, Signorielli, Morgan, & Jackson-Beech, 1979). For example, heavy television viewers are three times more likely than other people are to believe that on an average day police officers are required to draw their guns five or more times (the average TV portrayal) rather than less than once per day (the actual frequency). When asked whether 3 percent or 10 percent of the people in the United States are involved in some kind of violent incident in a year, heavy viewers are more likely to report the larger (TV) percentage than to report the smaller (actual) percentage.

There is also evidence (Wartella, Wackman, Ward, Shamir, & Alexander, 1979) that school-age children are particularly susceptible to television advertising. Whereas preschoolers often don't realize that advertising is designed to get people to buy something, school-age children do. However, they have not yet had sufficient experience with products to realize that they rarely work as well as advertisements say they do or are as exciting as they appear to be on television. This experience-based skepticism does not become evident until early adolescence. As a result, school-age children are a particularly exploitable audience for television advertising.

This seemingly strong impact of television on school-age children is probably a reflection of their information-processing skills (Collins, 1979). Six- and seven-year-olds are often unable to integrate the various actions and subplots that constitute a program. They may not be able to predict what the next scene in a program is likely to be or to identify a particular scene as a turning point in the story. Their limited comprehension is also evident in their inability to describe a program fully to someone who has not seen it. Although they are often able to remember and accurately describe the separate incidents in the program, they are usually less able to correctly sequence the separate incidents or to accurately infer the intent of each incident. This limited information-processing ability may be of little consequence when children are watching programs designed for them (which usually have very direct story lines), but it can be a more serious problem as school-age children watch more and more adult-oriented programming (it boggles the mind to consider what an eight-year-old might think is happening in a "soap opera"). Gradually, over the course of middle childhood, children learn to integrate the various story lines in a program and to understand the relationship between a character's actions, the motives for these actions, and the circumstances surrounding them.

The Influence of School

The school represents a significant new context of development during middle childhood. Although for an increasing number of children, the elementary school is not the first educational experience (many children attend preschool and day-care programs), the philosophy of early childhood education is, in most instances, so distinct from that encountered at the elementary level as to make the latter a new experience. As Elkind and Lyke (1975) note, most early education programs still utilize a traditional child-centered approach. This informal, self-selected learning approach is in sharp contrast to the structured learning approach used in most elementary school systems. As a result, children from early education programs enter public kindergarten as autonomous creatures and run headlong into a classroom setting that is geared for programmed learning.

BOX 5-2
TELEVISION VIOLENCE AND CHILDREN

We live in a society in which portrayals of violence are commonplace, considerably more commonplace than what occurs in everyday life. Television programming is frequently cited as the most common culprit in this misrepresentation. More than three-quarters of all prime-time shows on any of the major networks contain incidents of violence. These acts of violence are almost always associated with the taking or maintaining of power or dominance over others. Males are the most frequent aggressors.

Does this steady diet of violent television programming make a difference? Is there any carryover to other situations? If there is carryover, who is responsible for doing something about it? The evidence, now collected over a period of thirty years, clearly and consistently demonstrates a relationship between television viewing patterns and behavior. The more children watch violent and aggressive events on television, the more likely they are to be aggressive in their dealings with their peers and parents and the less likely they are to be popular with their peers. There is now even evidence that these children have somewhat lower school

Research has indicated that television violence has a negative effect on children, regardless of the individual characteristics of the children.

Sources: Eron, 1982; Liebert, Neale, & Davidson, 1973. Liebert, Sprafkin, & Davidson, 1982.

Children spend a great deal of time in school: seven thousand hours through the elementary grades (Jackson, 1968). How does it affect them? First, and most obviously, school is a place where they acquire a great deal of factual information about the world as well as a number of skills that allow them to process this information and acquire more. They learn how to read, write in both block letters and cursive, spell, construct words into sentences, sentences into paragraphs, and paragraphs into compositions, add, subtract, multiply, divide, work with fractions, and not to run in the halls. In addition they learn 1001 facts about science, history, geography,

achievement test scores. The associations between television viewing and behavior are evident from both a short-term and long-term perspective, in some studies holding up over a period of ten years.

If the evidence is so damning, why are such programs still on the air? There are a number of reasons. First, because there are other factors that correlate with aggressive behavior in children, the television industry contends that the associations are spurious, that is, that other factors really account for the associations. In part, the industry is correct. The children who watch a lot of violent television programming are more likely than other children are to have parents who favor physical punishment as a disciplinary technique and are dissatisfied with their children's accomplishments. Thus, the industry asks, "Why not focus on parents and leave programmers alone?"

One reason that programmers need to be held more accountable is that the associations are evident, although to a lesser degree, even in families that do not model such inappropriate parenting strategies. A second reason is that even though the influence of television on a particular child may not be as great as that of the parent, given the millions of children who are exposed to these shows, the scope of its influence is vast.

How can programmers be best held accountable? Industry monitoring has not proved a viable solution. When it has been tried on Saturday morning shows, the programs did become less violent for a time but then gradually became more violent again. Given the highly competitive commercial nature of television, it is unreasonable to expect any network to take any action that might reduce its ratings. The fact of the matter is that violent programs are on the air largely because people like to watch them. Ultimately, the only long-range solution that offers the hope of permanence is to determine the factors that directly and indirectly influence children's and adult's viewing preferences and act upon them.

literature, and a host of other subjects. By the time most children have completed seventh grade (usually by age twelve or thirteen), they have acquired most of the skills and information they will probably ever need to function on a day-to-day basis as an adult.

Although the acquisition of these skills is of great importance in determining the course of future development, the indirect teachings of the school system may, in fact, be of even greater importance. Schools provide children their first exposure to independent, standardized evaluations, the workings of a bureaucratic system, and our values as a nation.

Introduction to a Bureaucratic Society

"Public failure in the classroom is probably one of the most destructive shocks to the individual self-esteem that life can deliver" (Grambs, 1978, p. 102). Presumably, praise would have the opposite effect. In either case, children entering the school system encounter a form of evaluation that is totally different from any of their previous experience. The evaluations are systematic, they are objective, and they are the same for each child. Each child must learn the same spelling words, correctly punctuate the same paragraphs, and solve the same math problems. Children receive reports of their progress and are often grouped according to their demonstrated ability. The teacher may attempt to soften the burden of evaluation by referring to the ability groups as the robins, bluebirds, and cardinals or the roses, daisies, and petunias rather than as the bright, average, and slow children, but any third grader will tell you that the robins are the smart kids and the bluebirds are the kids who can't read yet. Curricula designed to be self-paced make status less evident, but even in these programs each child has a good idea what level every other child is working at.

Objective evaluation is not necessarily a bad process. Like most other techniques used to influence development in the life-span, its consequences depend on the way it's done. Most elementary-age children like school, work hard, and want to get good grades. Most are eager to be part of the adult world, and grades provide them their first measure of status in it. It is one of the bittersweet aspects of parenting to have children bring home a project receiving only an average grade, assure them that you think it's very good, and have them reply, "you have to say that, you're my parent."

Our society is bureaucratic and technological. It is highly scheduled and relatively depersonalized. It is product oriented. Its concern is often the "bottom line." These statements are also true of the school. According to Ritchie and Koller (1964), the school system presents children with their first lesson in bureaucracy. Children find themselves basically on their own and surrounded by strangers. Their lives are now governed by rules and routines, and individual whims and moods are not as likely to be tolerated as they were at home. Similarly, less allowance is made for the children's inadequacies and failures.

Ritchie and Koller's description of the school is probably less true for the early elementary grades than it is for the others. There is a much greater interactive quality in the early grades (Brophy & Evertson, 1981). Through third or fourth grade, a teacher's reaction to a student is as much a reflection of the teacher's perception of the student's personality, motivation, and degree of self-control as it is a reflection of the student's demonstrated ability. Students who do average work but who seem to the teacher to be trying their best are more likely to receive positive comments about their work than are students working at the same level who seem more able but less motivated. Teachers respond most negatively toward students who are disruptive but seem to be able to control their behavior. Disruptive students who seem "immature" to their teachers are usually treated more tolerantly. Teachers particularly value children who seem bright, quiet, well-motivated, and helpful.

Success in bureaucratic school systems is often a function of how well a child has learned to make "legitimate" demands rather than spontaneous, personal ones (White, 1977). Children who ask for extensions before papers are due are probably more likely to meet with success than are children who ask for a second chance once the papers have been graded and returned. Children who save their extra questions for after class rather than—often unintentionally—monopolizing class time are more likely to get their questions answered and are less likely to be scorned by their classmates. Preadolescents and young adolescents often view these strategies as "selling out," but in fact they are the ways in which individuals see that their personal needs are met within any complementary social system. Clearly, children's early school experiences are as much a reflection of the qualities they bring to the classroom as they are of the nature of that classroom.

His School/Her School

The truth of the preceding sentence is perhaps nowhere more evident than in the different experiences that girls and boys seem to have in school. Girls are more attentive to teachers, more likely to initiate contacts, more willing to conform to class rules, more motivated to complete class assignments, and more successful in their efforts than boys are. Boys are more disruptive in the classroom, tend to receive more negative evaluations, and in general do not do as well on exams and assignments (Brophy & Evertson, 1981). But, in spite of all this, boys tend to overestimate their chances of success on exams, girls tend to underestimate their chances of success, and the achievement advantage enjoyed by girls is usually gone by junior high or middle school (Dweck, Davidson, Nelson, & Enna, 1978).

There are a number of factors that contribute to these differing patterns of experiences in school. The initial advantage enjoyed by girls through grade three or four probably reflects the greater emphasis placed on self-control in the socialization of girls, their slightly faster rate of maturation (which makes self-control more possible), and the importance apparently placed on self-control by most primary grade teachers. If these assumptions are correct, why do boys develop a more optimistic view of their academic abilities than girls do, and why does the achievement advantage enjoyed by girls fade by junior high? Part of the explanation lies in the fact that the well-socialized behavior patterns that were initially an advantage become a liability as a greater emphasis is placed on achievement in the middle grades. A second factor concerns the way boys and girls interpret teacher feedback. Dweck (1978) and her colleagues have noted that boys are much more likely to consider negative teacher comments to be reflections of teachers' attitudes rather than indications of the level of their own work. Girls are more likely to believe that teacher comments reflect their ability. Therefore, even though girls receive considerably less negative feedback than boys do, over time it may have a more pronounced influence on girls' expections for success and therefore on their degree of effort.

Dweck's findings parallel those of Lever with respect to children's games. In both instances, socialization patterns have a greater inhibiting effect on the behavior of girls than they do on the behavior of boys. The potential impact of these different socialization patterns will become evident in the next few chapters.

Transmitting Social Values

The schools have a third indirect function. They serve as the inculcators of our national values and heritage by passing on the beliefs, skills, and values that have been established over time. They teach us that Washington never told a lie; that if you want to be President, you might consider being born in a log cabin; and that America is the land of the free and the home of the brave. By transmitting values that are more broadly accepted than the values most children learn in the home, schools serve as a conservative socializing agency—an agency that presumably provides children with the resources they will need to function as adults in later years (Elkin & Handel, 1972).

Much of the controversy surrounding education in the last thirty years has focused on the nature of the values that are transmitted through the school. The portrayal of women, racial and ethnic minorities, and individuals with handicapping conditions in our textbooks has been a constant focus of debate. In many instances, textbooks provide the same distorted view that television provides. Affirmative action employment principles have been demanded not only so that all adults may have an equal opportunity for advancement within the school system but also so that all students will come to learn that within a society that views itself as a meritocracy, gender and color are not relevant indicators of ability.

Debates concerning bilingual education, family life, and sex-education programs often pit the values of families against those of the school and the larger community. Perhaps the most basic debate is whether the school posture should be conservative and reflect the traditional social

In addition to imparting knowledge, schools help to transmit the basic value system of society to the next generation.

order or progressive and prepare students for the anticipated world of tomorrow. Research indicates that the children in traditional school systems have a different worldview from that of children in progressive systems. According to Minuchin, Biber, Shapiro, and Zimiles (1969), children from more traditional environments seem to be oriented toward the established goals of society and the successful fulfillment of the conventional roles that are relevant to their sex and family status. Children from more modern environments, on the other hand, are relatively free from conventional images and roles, more sure of their own sense of self, and tend to live more in the immediacy of childhood.

THE SIGNIFICANCE OF DEVELOPMENT DURING MIDDLE CHILDHOOD

It should be clear from the reading of this chapter that many important developmental events occur during middle childhood. This section considers the significance of these events. In the first half of this section, middle childhood will be contrasted with the stages that precede it and the stages that follow it. Like any developmental stage, middle childhood is a time both for the consolidation of skills initiated earlier in the life-span and for the emergence of new skills whose impact will become more pronounced as development continues. In the second half of this section, the implication of the relative impacts of home, school, and peers on the course of development during middle childhood will be considered.

Patterns of Development

The notion of pattern implies repetition. It implies that some element is encountered and then, at a later time or place, reencountered. With respect to development, it suggests that certain topics and tasks appear and then reappear. With respect to the constructivist aspect of development, it suggests that understanding is a relative rather than an absolute quality. We don't understand so much as we understand at a particular level. A consideration of the developmental events of infancy and early and middle childhood provides a particularly good illustration of this constructivist view of development.

The developments of middle childhood are primarily a refinement and consolidation of those skills first evident during early childhood. When considered together, early and middle childhood form a developmental sequence that parallels that of infancy. As Piaget and Inhelder (1969) note,

> what is the most striking about this long period of preparation for and formation of concrete operations is the functional unity (within each subperiod) that binds cognitive, playful, affective, social, and moral reactions into a whole. Indeed, if we consider the preoperatory subperiod between two and seven or eight with the subperiod or completion between seven or eight and eleven or twelve, we see the unfolding of a long, integrated process that may be characterized as a transition from subjective centering in all areas to a decentering that is at once cognitive, social, and moral. This process

is all the more remarkable in that it reproduces and develops on a large scale at the level of thought what has already taken place on a small scale at the sensori-motor level. (p. 128)

In other words, the developments of the preoperational period of early childhood parallel those of the first half of the sensorimotor stage, and the developments of the concrete operational period of middle childhood parallel those of the last half of the sensorimotor stage. In both instances, the transition is marked by a major reorganization in children's understanding of the nature of their environment and, correspondingly, of their place in that environment. In the sensorimotor period, the milestone is the development of object permanence; in the preoperational/concrete operational sequence, it is the development of conservation (Langer, 1969).

Each of these milestones serves as an anchor for the child. Each helps children gain a foothold in understanding the nature of their physical and social environments by helping them appreciate the invariant nature of certain aspects of their world. In the case of object permanence, the invariant fact is that objects exist. Their existence is independent of our awareness of them. In the case of the conservation task, the invariant fact is that certain properties of objects are independent of our actions on them. Quantity remains invariant irrespective of the way objects are arranged or rearranged and irrespective of the shape of the containers liquids are poured into. Each of these milestones forces a shift in orientation because each in its own way at its own level, informs children that to be effective they must adjust their behavior to meet the characteristics of the world.

Development during the sensorimotor period is pragmatic and action oriented. Infants' knowledge of their environment is defined in terms of their direct actions on objects. This knowledge is initially limited to their own bodies and then gradually extends to objects in the environment. When coordinations are evident, they are coordinations of actions (looking and reaching); when goal-directed actions are evident, they consist of two actions, one of which is subordinate to the other (removing the blanket in order to reach the ball). Even when the first mental representations become evident around age two, they are limited to those situations in which infants have direct contact with an object (Fischer, 1980).

Development during early and middle childhood reflects children's ability to represent real objects and experiences. The coordinations of mental representations that eventually produce the reversible mental operations typical of the preadolescent follow the same sequence, but on a more complex level, that those of the sensorimotor period follow. Class inclusion at the level of representation, for example, is a parallel development to the means-ends differentiation of the sensorimotor period. What is initially understood at the level of direct action (the removal of the blanket is understood in terms of the larger goal of obtaining the toy) is reconstructed at the level of representation (dogs and cats are each understood in terms of the larger category of animals). And what is understood at the level of representation will be reconstructed during the period of formal operations in adolescence at the level of abstract thought. Each reconstruction in children's understanding of objects, space, time, and causal relationships changes the equilibrium that exists between them and their environments. In particular, each shifts the balance further away from reaction and toward action and from perception toward inference, giving children an increasingly broader perspective from which to interpret specific events. The system of reversible concrete mental operations evident by age eleven or twelve gives children very effective information-processing strategies for dealing with the events typical of middle childhood. It is in this sense that I suggested at the beginning of the chapter that twelve-year-olds, within their world of middle childhood, have their "act more together" than most of us ever will within our adult worlds. Bear in mind, however, that this "togetherness" is limited to the world of middle childhood. Their level of development is not yet adequate for dealing with adult issues of marriage, occupation, or parenthood.

In retrospect, middle childhood is the culmination of a long process of reconstructions that began during early childhood and translated onto the plane of representation what was initially limited to the plane of action (Kegan, 1982). In prospect, middle childhood is the first point in the life-span at which long-term predictability is possible (Kagan & Moss, 1962) and adult forms of interpersonal relations are clearly evident.

Long-Term Predictability

There are a number of reasons why developments in middle childhood make long-term predictability possible. Perhaps most central is the systematic, rule-governed nature of cognitive functioning during middle childhood. By providing a mechanism for coordinating and integrating discrete experiences, the cognitive operations of middle childhood make possible a degree of continuity previously lacking. These cognitive skills also allow children to see themselves in relationship to the larger society outside of the home. What usually follows from this awareness is a desire to become part of the larger society.

Children's entry into this larger society, most evident in their entry into the bureaucratic school system, also imposes upon them a degree of external continuity. It is a very common experience for children to move hundreds of miles, start a new school, and find themselves using the same book they were using in the old school, and perhaps even working on the same unit. Children become plugged into the system and like the rest of us are exposed both to its positive and negative aspects. This "standardizing" quality of the school system is equally evident in the fact that most children maintain approximately the same class standing throughout their entire school careers. With few exceptions, children who do well in the early grades continue to do well in later grades, and unfortunately, children who do poorly maintain an equal degree of continuity. What is particularly noteworthy about this degree of continuity is that it only holds within the school system (McClelland, 1973). Cross grade level achievement standings are quite high, but they are not very predictive of seemingly corresponding measures of status in adulthood such as salary, employment history, and occupation. In other words, this high level of continuity in achievement has more to do with the structure of schools than with the ability of students.

The highly sex-typed behavior of boys and girls during middle childhood also reflects the external continuity maintained by social forces. Although these behavior patterns need not remain the same during adulthood (Weitz, 1977), the fact that they often do, especially for men, further illustrates the powerful influence of social values on the maintenance of particular behavior patterns. The combination of children's more stable cognitive processes and their immersion into the broader society causes middle childhood to be the time when long-term predictability is first possible.

Interpersonal Relationships

The patterns of reciprocal and complementary interpersonal relations first evident during middle childhood continue to characterize interpersonal relations during the adult years. Although the level of the relationships does not remain the same, the pattern holds constant. The complementary pattern of child-adult relations is reconstructed during adulthood into the adult's willingness to respect legitimate authority such as the government. This adult complementary relation is assumed to have the same advantage (security) and the same disadvantage (lack of mutuality) that is typical of the child-adult relation (Youniss, 1980). On the other hand, the peer relations of middle childhood initiate the pattern of reciprocity that evolves during adolescence and young adulthood into a mature concept of intimacy and mutuality. It is from this sense of mutuality that a true adult morality evolves. This morality views the self in relation to others and is based on cooperation rather than competition (Youniss, 1980).

HUMAN DEVELOPMENT AND HUMAN SERVICES

In contrast to the world of early childhood, the world of the school-age child is increasingly one independent of parents. Children spend most of their day away from home, in school or playing with friends. Parents are less aware of and are often less able to influence their children's lives than they were when the children were preschoolers. Because of these developmental changes, attempts to influence the course of children's development can no longer be filtered solely through the parents but now must also focus on the other contexts in which school-age children function.

The school is the primary intervention agent during middle childhood and, appropriately enough, a great deal of research has been done concerning the way schools influence children. The findings of these studies have been quite consistent (Coleman, 1966; Jencks, 1972; Rutter, 1983). Those aspects of schools that might appear to exert a significant influence on children's learning—the age and quality of the school, the number of books in the library, the availability of laboratories and special equipment such as computers—have remarkably little influence on children during middle childhood. It is true that better schools are most commonly found in more affluent communities, and it is equally true that the children attending these schools score higher on most achievement tests than children attending less desirable schools do. However, when the influence of the school facility itself is separated from the family backgrounds of the children and the nature of the influence of peers on each other, the school falls a distant third in relative importance. As Coleman (1966) noted in one of the earlier surveys of the quality of American education:

> Taking all these results together, one implication stands out above all; that schools bring little influence to bear on a child's achievement that is independent of his background and general social context; and that this very lack of an independent effect means that the inequalities imposed on children by their home, neighborhood, and peer environment are carried along to become the inequalities with

which they confront adult life at the end of school. For equality of educational opportunity through the schools must imply a strong effect of schools that is independent of the child's immediate social environment, and that strong independent effect is not present in American schools. (p. 325)

Coleman's conclusion, which has been repeatedly confirmed (see Rutter, 1983), is based on comparisons of the average abilities of children in different schools. It does not mean that schools do not influence children—obviously they do. Rather it means that their relative degree of influence on the child seems less than that of the family—even though children are now spending less time in the home—and that the often extreme differences in the quality of schools are not reflected in the ability levels of the children. However, a closer look at the impact of the school on the child, shows that the range of differences between children in the same school is as much as four times greater than the average differences between children in different schools. In many cases the range, even within a single classroom, is quite great. In other words, the impact of the school on the child is a very individual matter, reflecting the nature of the child's home environment, the personality and skill of the classroom teacher, and the educational expectations and aspirations of the child's peer group (Johnson, 1980). The taxpayer (another form of human service provider) would do well to consider these findings the next time a bond issue to finance education is on the ballot. The quality of the building is of considerably less value than the quality of the relationships within that building.

The academic culture of the home (Garbarino, 1982) refers to the extent to which parents foster the intellectual and social skills that facilitate school learning, the extent to which they themselves model these behaviors, and the extent to which they value and support the school system in general and their child's classroom teacher in particular. The academic culture of the home continues to be a powerful influence on the child's achievement throughout the school years and as such is an important point of intervention in the child's development. Activities that strengthen the stability of the family almost invariably improve the achievement level of the child.

A second point of intervention is the classroom. Teachers are as prone to both implicit and explicit biases as the rest of us. They show clear preferences as to the types of children they enjoy working with, and their response to a child's efforts depends partly on their perceptions of that child's motivation and competence (Brophy & Evertson, 1981; Good, 1980). Given the normal chaos of an elementary classroom of anywhere from twenty to thirty-five children, teachers may be unaware that they are willing to wait longer for some children to answer than for others, that some children are rarely called on while others are called on frequently, and that some children receive a disproportionate share of negative comments. In many cases, intervention may only need to consist of someone's pointing these patterns out to teachers. The crucial element is that teachers become aware of the nature of their behavior toward specific children and of the way their behavior influences both the child's level of functioning and the level of expectations the child has about learning. This is especially true in the early school years when children spend most of their day with one teacher and when they are less able to separate task-relevant comments from more general interactions.

The spheres of home, school, and peer exert interactive influences on children. Each serves a unique function. The functioning of each is enhanced when there is continuity between them. The evidence is clear that educational attainment is enhanced when the school, peer group, and home function compatibly (Minuchin, Biber, Shapiro, & Zimiles, 1969). Equally clear is the fact that a mature adult personality requires both the complementary experiences found in parent-child relationships and the reciprocal experiences of peer relationships (Youniss, 1980). Each provides children a unique set of opportunities and experiences, and it is only when children view each from the perspective of the other that they come to appreciate what each has to offer.

In the next chapter, we turn to adolescence, a period in the life-span that is always described in extremes. It presents a marked contrast to the relative tranquility of middle childhood. But it is not a negative stage. It has its share of confrontation and rebellion, but its central feature is self-awareness. It usually begins with puberty. Puberty is a time when you see yourself as too "__________." (You can fill in the blank with any word; they are probably all true when you are thirteen.)

SUMMARY

THE COURSE OF DEVELOPMENT

1. Physical status and motor skills play a significant role in determining peer-group status during middle childhood.
2. One of the major accomplishments of the concrete operational period is the discovery of the invariance of relationships. This discovery is reflected in the various demonstrations of conservation. Other significant advances include an increase in self-regulatory capacity, increased awareness of metacognitive processes, and a greater ability to deal with events in context.
3. The ability to use concrete operational logic is restricted to circumstances that are real and can be experienced.
4. Peer relations during middle childhood still tend to be situation specific. However, they are now more likely to reflect psychological qualities as well as physical ones, and they now tend to be less egocentric in nature.
5. Fairness is no longer equated with personal desire in middle childhood. A sense of equity and a consideration of need also become part of the decision-making process.
6. By the end of middle childhood, parents in particular and authority figures in general are seen as having legitimacy largely in terms of experience, training, and responsibility.
7. School-age children become increasingly involved in the larger community although their understanding of its workings is often limited.
8. Children's concept of self becomes increasingly relative over middle childhood. School-age children become increasingly aware that they can influence the way others think of them.

THE CONTEXT OF DEVELOPMENT

1. Socially responsible behavior in school-age children is most common when parents have a positive view of themselves as adults, establish a developmentally appropriate balance between the limits they set and the autonomy they allow, and have appropriate expectations for mature behavior.
2. Inductive child-rearing strategies are associated not only with socially responsible behavior but also with internalization of moral standards and cognitive competence.
3. Peer activities during middle childhood tend to be highly sex typed and sex segregated.
4. The frequent viewing of adult-oriented violent television programs is associated with an increase in aggressive behavior, especially among children who are highly aggressive initially.
5. The school experience introduces children to standardized, objective forms of evaluation.

THE SIGNIFICANCE OF DEVELOPMENT DURING MIDDLE CHILDHOOD

1. The period of middle childhood is the first point in the life-span at which long-term predictions of adult behavior patterns are reasonably accurate.

HUMAN DEVELOPMENT AND HUMAN SERVICES

1. The provision of effective human services during middle childhood must involve effective coordination among the home, the school and, in many cases, the peer group.

KEY TERMS AND CONCEPTS

Erikson's Psychosocial Stage of Industry

COGNITIVE DEVELOPMENT

Piaget's Stage of Concrete Operational Thought
Conservation Experiment
Kendler and Kendler's Reversal–Nonreversal Shift Task
Awareness of Transformations
Inference
Reversibility
- Identity
- Compensation
- Negation

Decentration
Metamemory Skills
Metacognitive Skills

SOCIAL DEVELOPMENT

Reciprocal Relationships
Complementary Relationships
Generalized Other
Parenting Techniques
- Inductive Techniques
- Power Assertion Techniques
- Love Withdrawal Techniques

SUGGESTED READINGS

The educational and social experiences of the school-age child are often different from those of the preschooler. Educational experiences are more product oriented; social experiences, more sex typed. Elkind provides a good review of the issues involved in the education of children, and Lever provides some very intriguing insights into the origins of sex typing.

Elkind, D. (1976). *Child development and education.* New York: Oxford University Press.

Lever, J. (1978). Sex differences in the complexity of children's play and games. *American Sociological Review, 43,* 471–483.

Like preschoolers, school-age children spend long hours in front of the television. Unlike preschoolers, however, their viewing habits are more likely to include adult-oriented programs. The title of the Lefkowitz book suggests one possible outcome of this pattern.

Lefkowitz, M. M., Eron, L. D., Walder, L. O., & Huesmann, L. R. (1977). *Growing up to be violent.* New York: Pergamon Press.

The world of school-age children may be broader than that of preschoolers, but their understanding of the world is often quite limited. The books by Damon and Furth provide a clear demonstration of this fact.

Damon, W. (1977). *The social world of the child.* San Francisco: Jossey-Bass.

Furth, H. G. (1980). *The world of grown-ups: Children's conceptions of society.* New York: Elsevier.

Both parents and peers play significant roles in the socialization experiences of school-age children. The book by Coopersmith emphasizes the parent's role; the ones by Selman and Youniss, the distinct contributions of parent and peer.

Coopersmith, S. (1967). *The antecedents of self-esteem.* San Francisco: Freeman.

Selman, R. L. (1980). *The growth of interpersonal understanding.* New York: Academic Press.

Youniss, J. (1980). *Parents and peers in social development.* Chicago: University of Chicago Press.

6
ADOLESCENCE

CHAPTER OUTLINE

Perhaps the best way to begin Chapter 6 is with the definitions of puberty and adolescence. **Puberty** is the physical maturing of the individual that results in adult physical and physiological stature and reproductive capability. Under all but the most extreme conditions, puberty is an inevitable characteristic of our species; its sequence has remained unchanged since at least the beginning of recorded history (Tanner, 1972). **Adolescence** is a period of the life-span during which physically mature individuals retain an essentially childlike role in society. Adolescence is not universal. It is not an inevitable characteristic of our species. Instead it is a by-product of the social and industrial changes in technological societies over the past 150 years (Kett, 1977). There was no adolescence prior to the Civil War. And it has only been within the past thirty or forty years that adolescence has become a universal experience for teenagers in ours and other industrial societies.

The industrial revolution set in motion a series of changes that are still evolving today. It first reduced and eventually ended the economic and productive role of the family: the modern nuclear family is emotional and affective in nature. It centralized production and hastened the development of urban areas. It decreased the value of unskilled labor and increased the importance of educated workers. These changes did not occur at once. Early factories actually made much use of cheap, unskilled child labor. But as the workplace became more sophisticated and as the competition for industrial jobs became greater, children and teenagers were increasingly excluded from the workplace, and laws were enacted to protect them from being exploited in such settings (Elder, 1975). One consequence of these changes was that individuals entered a state of prolonged childhood, during which they remained economically dependent on their families. They were required by law to stay in school. They became less valued for their present skills, more for their potential skills.

A second set of events, occurring at the same time, had an opposite effect. It set these young people farther apart from the children whose world they were to retain. Adolescent girls today usually reach their full adult height and weight by age fifteen. For boys, the same status is reached by age seventeen (Tanner, 1972). One hundred and fifty years ago, adults did not reach their full height and weight until their mid-twenties. Today, by age fourteen, 92 percent of the girls in this country have experienced **menarche,** the beginning of menstruation. The earliest data available in this country show that in 1900, only 50 percent of all fourteen-year-old girls had experienced menarche (Garn, 1980). Data from European countries are even more startling (see Figure 6-1). In Sweden, in 1885 the 50 percent point (the median) was not reached until almost age sixteen. In 1890, in Germany, the median was not reached until after age sixteen. And in Norway in 1840 (these are the earliest reliable data available), the median was not reached until after age seventeen. In general, there has been a decrease of approximately three to four months per decade in the onset of menstruation. The trends in height and weight and menarche reflect the continuing improvement in nutrition and public health that has occurred over the past 150 years in the industrial countries of the world (Roche, 1979).

The consequences of the industrial revolution (protracted childhood) and improved public health (a reduced age of onset of puberty) have combined to create the stage of adolescence.

While adolescents retain an essentially childlike role in our society, they are not children. They are both physiologically and cognitively distinct from children. The basic conflict of freedom versus responsibility acquires greater significance during adolescence than it had during middle childhood. The peer group becomes increasingly influential (Elkind, 1978). Adolescence is also a time when children express their first serious concerns about their future adult roles.

Just as it is useful to distinguish the developmental patterns of early and middle childhood, it is useful to distinguish early adolescence from late adolescence. The central event of early adolescence (approximately twelve to fifteen years of age) is puberty. Much of the behavior of early adolescents reflects their attempts to adjust to the physical and physiological consequences of puberty. Peer groups remain relatively sex segregated. Heterosexual activity tends to be somewhat exploitive, serving as a device for confirming social status within the same-sex peer group (Gagnon, 1972).

The central event of late adolescence (approximately sixteen to nineteen years of age) is preparation for the future. Those who are college bound suddenly become very serious about their studies. The pursuit of high grades often becomes all-encompassing. Those who choose not to pursue a college education focus on vocational or commercial preparatory programs. Those who see little relation between their education and their

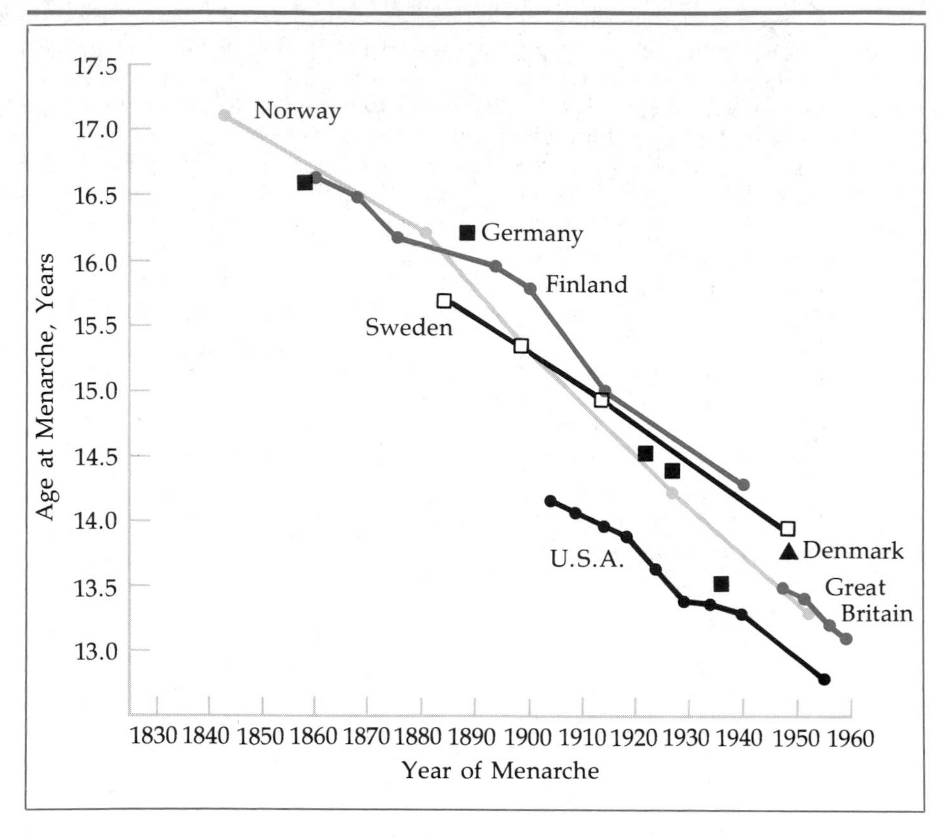

FIGURE 6-1
Over the last 150 years, the median age at menarche in the United States and Europe has decreased from a high of slightly over 17 years to a low of slightly less than 13 years (from Tanner, 1972).

future do what little they need to graduate or, when the law allows, leave school. The peer group is less influential during late adolescence. Heterosexual activity is more frequent and less exploitive. Although marriage is not usually an overriding concern of late adolescence, the increase in heterosexual activity leads adolescents to think of themselves and others in terms of marriage and husband-wife relationships.

With all of these things happening during both early and late adolescence, it is small wonder that Erikson (1968; 1980) considers the establishment of a sense of identity to be the major concern of adolescence. According to Erikson, the circumstances of the **psychosocial stage of identity formation** prompt adolescents to reflect on their own development. These circumstances are the development of sexual maturity, the cognitive capacity of introspective thought and hypothetical-deductive reasoning, and the increasing number of social demands for adolescents to choose a vocation and adult life-style.

But in puberty and adolescence all sameness and continuities relied on earlier are questioned again because of a rapidity of body growth which equals that of early childhood and because of the entirely new addition of physical genital maturity. The growing and devel-

> oping young people, faced with this physiological revolution within them, are now primarily concerned with attempts at consolidating their social roles. They are sometimes morbidly, often curiously, preoccupied with what they appear to be in the eyes of others as compared with what they feel they are, and with the question of how to connect the earlier cultivated roles and skills with the ideal prototypes of the day. (Erikson, 1980, p. 94)

Adolescents search for comfortable identities by "trying out" new roles or ways of behaving. The search can take any one of a number of forms. There may be a sudden and drastic shift in interests or activities. Parents may come home to find that their fourteen-year-old's room has been purged of once favorite childhood objects. Athletics, dance, and music lessons may now be seen as "kids' stuff." A new, more adult wardrobe may be demanded. Behavior toward parents, siblings, or peers may change abruptly. There may be a sudden and total (and usually temporary) rejection of family values. The adolescent may adopt an identity that is the mirror image of the parent. Parents are "dumb," religion is a "farce," school is a "joke," and capitalists have about as much "humanity" as Attila the Hun. There may be a strong commitment to a vocation, then suddenly to a new vocation. In Erikson's (1980) term, the period is one of **psychosocial moratorium,** allowing for a time of **role experimentation** without the limiting consequences of commitments.

This period is often a stressful time for both parent and adolescent. Behavior tends to be more extreme and variable than it was during middle childhood. Periods of concentration are rarely longer than thirty minutes. As is true of preschoolers, early adolescents seem to be creatures of the moment (Larson, Csikszentmihalyi, & Graef, 1980). Young adolescents may experience a great deal of inner turmoil (Rutter, 1976). Their parents may be ready to ship them to Siberia.

By late adolescence, most adolescents have defined an initial identity and may be allowed to return from exile. Initial commitments concerning future educational and vocational plans are made, a degree of harmony

Counterculture movements such as the punk rock craze are extreme manifestations of the natural adolescent search for identity. As distressing as this behavior is to parents and family, it seldom has long-term negative consequences.

exists between parents and adolescents, and most have established a relatively comfortable role within their peer group.

In adolescence as in all of Erikson's stages, the danger of resolution lies in the possibility that the resolution will be extreme. One extreme resolution during adolescence, a failure to establish any identity, creates what Erikson calls **identity diffusion.** Adolescents who fail to establish an identity fail to find a comfortable role. As a result, they are caught in limbo. Their ability to make plans for the future is constrained since such plans require a conception of self. The balance between satisfying present needs and satisfying future needs is tipped in favor of the former. In most cases, adolescents simply need more time to form an initial identity. In a few, the problem is more deep-seated and requires some form of intervention.

The other extreme is overidentification. In some instances, adolescents adopt the values and conventions of a particular group without really considering how they fit in. The group may have a religious or political quality. It may engage in delinquent acts or other forms of behavior (such as drug use or indiscriminate sexual activity) that are unacceptable to most people. In other instances, adolescents may take a "leap into adulthood." They may assume all the mannerisms, dress, and activity patterns of the adult culture. This pattern is most likely to be found among early maturing males, among those who have experienced continuous pressure to pursue a particular future, and among those who are prematurely required (often due to divorce or the death of a parent) to assume adult roles in the family. In all three cases, adolescents assume an identity without reflecting on the appropriateness of that identity. In some instances, this **identity foreclosure** creates no problem. The adolescent maintains the role into adulthood and is happy. In other instances, reflection only occurs after the individual is well into the role. The third year medical student who remembers always wanting to be a physician suddenly wonders if this is really what he or she wants and begins to question how and when the decision was made. No matter what the outcome of this questioning, the timing of the reflection (following rather than preceding the decision) creates additional stress for the adolescent or young adult.

It is important to bear in mind that adolescence involves the formation of an initial, not final, sense of identity. An initial sense of identity provides a first direction for the future. As that future is pursued, new events such as marriage, parenthood, or career change require adults to

Because of their strong need to belong and establish an identity, some adolescents overidentify with a particular group without understanding the full ramifications of group membership. This overidentification can be particularly damaging when the groups have delinquent or nontraditional religious or political overtones.

modify or redefine their sense of identity. In other words, identity formation is a lifelong process. Only the first step is taken during adolescence.

THE COURSE OF DEVELOPMENT DURING ADOLESCENCE

Adolescence is a time of both advancement and change in the course of development. The events of puberty transform children's bodies into those of adults. Advancements in cognition allow adolescents to deal not only with what has been experienced but also with what can be imagined. Within the social realm, adolescents demonstrate increasing skill in directing the course of their own development.

Physical Development

There is considerable uniformity in the sequence of physical developments that takes place in both males and females during puberty. The age of onset and the intervals between the various components, however, are highly variable. This variability of onset and the asynchronous nature of puberty foster the various anxieties typically accompanying puberty (Erikson, 1980). Even though parents rightly assure their children that they will soon stop growing, soon start growing, soon gain weight, or soon lose weight, the egocentric nature of young adolescents leads them to believe that they will be the one thirteen-year-old who never stops growing or never starts. Adolescents often find themselves having to learn to live in what seems like a new body. The following description of the plight of adolescent boys is typical of this period of development.

> Since most of his increased height is due to growth in the legs, many a boy finds himself equipped with pedal extremities that get him across a room rather faster than he expected, causing him to overrun his objective; they also tend to get tangled in furniture. His arms, having grown four or five inches in length, also contribute to his miscalculations of distance and lead him into a long series of minor tragedies, from knocking over his water glass because his hand reached it too soon to throwing a forward pass six feet over the receiver's head, because his elongated arm automatically produced far greater leverage than he has been accustomed to. (Cole & Hall, 1970, p. 22)

There are three major components to puberty (Katchadourian, 1977). The first involves changes in height, weight, body proportion, and body composition. The second involves accelerated development of the circulatory and respiratory systems, and the third involves the development of the primary and secondary sexual characteristics. As a result of these changes, adolescents acquire the physical stature and physiological functioning of adults, are capable of reproduction, and demonstrate greater sexual dimorphism than younger children do. That is, in adolescents, there are greater differences in appearance and functioning between males and females.

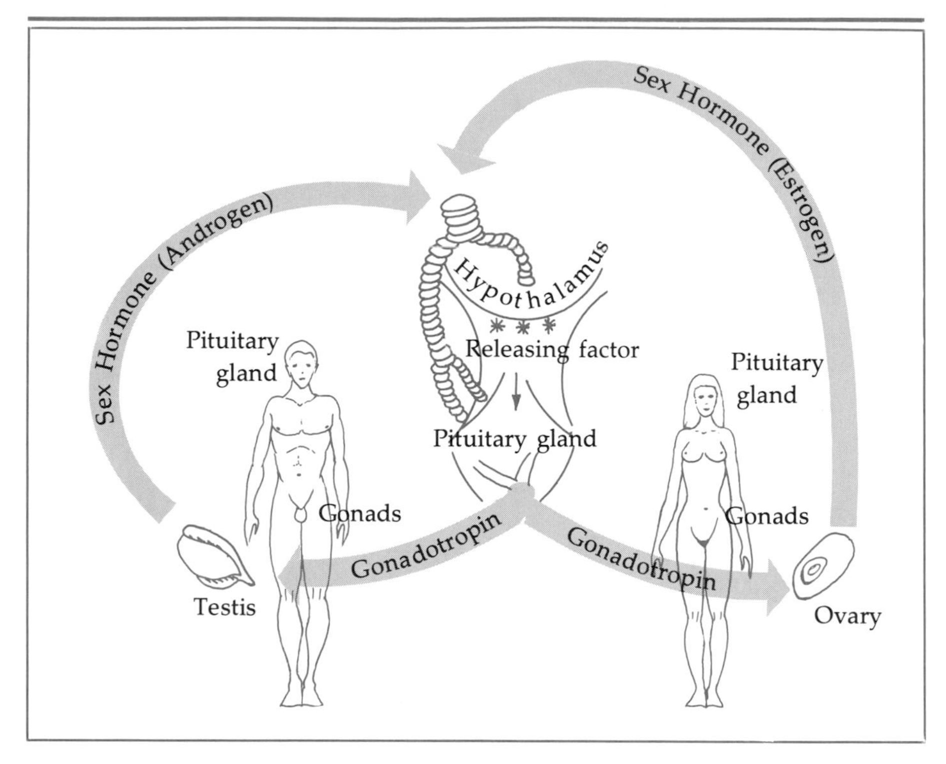

FIGURE 6-2
The onset of puberty is in part regulated by changes in the equilibrium of the interactive loop formed by the hypothalamus, pituitary gland, and the gonads.

The mechanisms that regulate the onset of puberty are still not completely defined (Peterson & Taylor, 1980; Sommer, 1978; Tanner, 1972). They involve the hypothalamus in the brain, the pituitary gland located at the base of the brain, and the gonads (testes in the male and ovaries in the female). Even before puberty, the testes and the ovaries produce small amounts of the sex hormones. The production of these hormones is regulated by the pituitary gland, which in turn is regulated by the hypothalamus. The hypothalamus is regulated by the amount of sex hormone circulating in the bloodstream (see Figure 6-2). Puberty begins when the equilibrium of this loop is altered.

Puberty Sequence

Figure 6-3 depicts some of the major events of puberty in males and females. The length of each horizontal line indicates the amount of time required for that characteristic to change from its immature state to its mature state. Breast development, for example, takes approximately four years (age 11 to age 15), and growth of the penis takes approximately two years (age 12.5 to age 14.5). The numbers under the ends of the horizontal

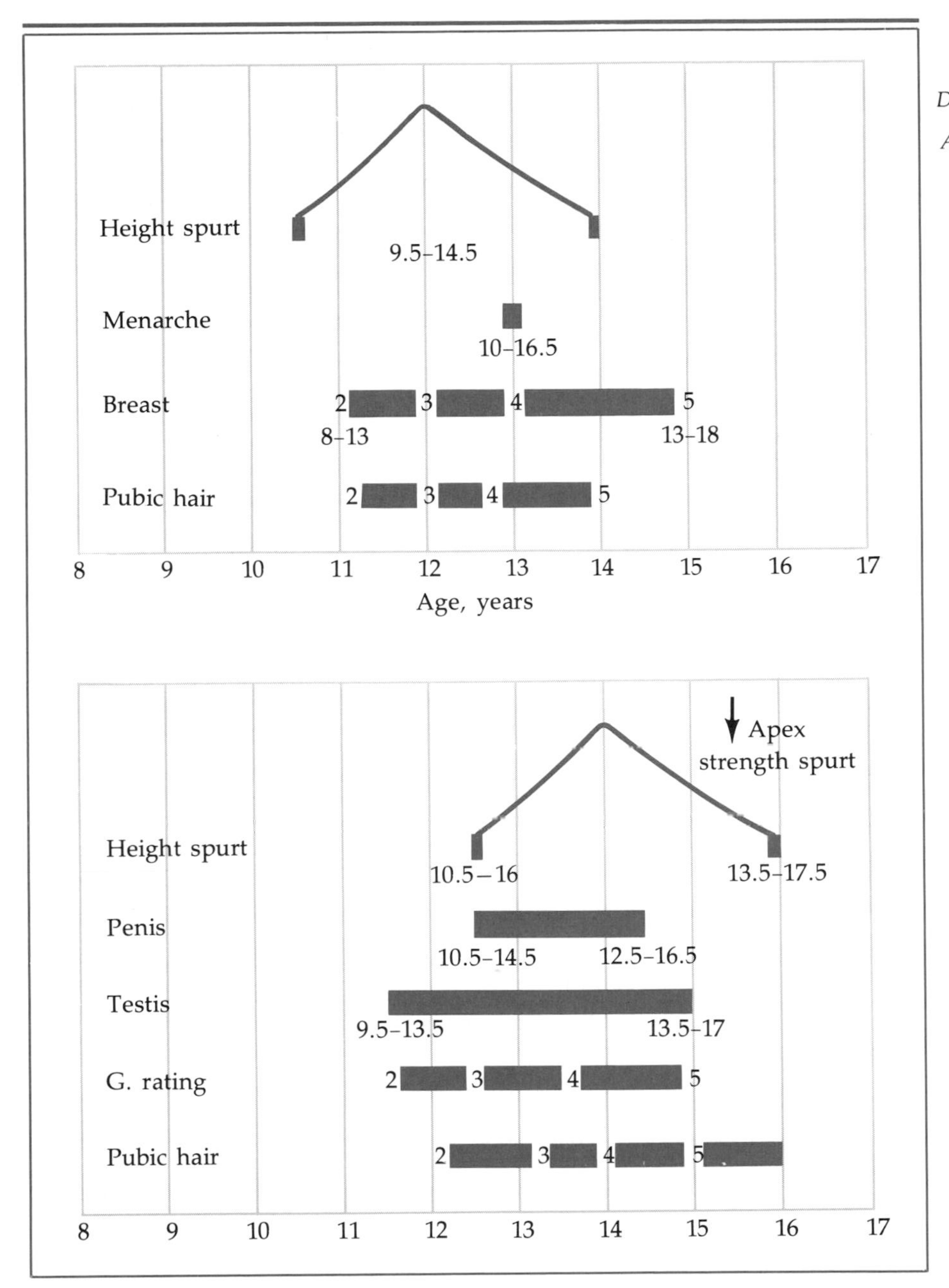

FIGURE 6-3
These two diagrams present the age range at onset and completion (indicated by the numbers under the horizontal bars) and the amount of time it normally takes (indicated by the length of the bars) for various puberty-related characteristics to reach their mature states in males and females. The numbers on the bars represent the degree of maturation completed by that age, while the G. rating refers to the texture and color of the scrotum (from Marshall and Tanner, 1970).

lines represent the normal range of onset and completion, respectively, for that particular characteristic (data on pubic hair are unavailable). For example, the growth spurt in females may typically begin as early as age ten or as late as age twelve; adult stature may be reached as early as age fourteen or as late as age sixteen. Onset of growth in males may come as early as age ten or as late as age sixteen; adult stature may be reached as early as age thirteen or as late as age nineteen. The significance of this great variability, particularly for males, will be discussed later in the section. For now, it is sufficient to note that it is quite common to find in adolescents of the same chronological age some who have not entered puberty, some who are in the middle of it, and some who have already completed the sequence of events.

The growth spurt reaches its peak around age twelve in females and age fourteen in males (see Figure 6-4). During this time, females may grow as much as six to eleven centimeters and males as much as seven to twelve centimeters (Tanner, 1972). The gain in weight accompanying growth reflects an increase in the size of the skeleton, muscles, internal organs, and body fat (Katchadourian, 1977). As a result of these changes, males emerge not only taller and heavier than females but also with greater muscle strength, increased pulmonary and cardiac capacity, and less subcutaneous fat.

Although the average age for breast development in females is slightly less than one year after the onset of accelerated growth (see Figure 6-3), the greater variability in the onset of breast development often means that the development of the breast buds is the first external indication of puberty. Once the breast buds have developed, the mammary glands make lactation possible; adipose tissue gives the breasts their adult shape (Katchadourian, 1977).

The growth of pubic hair, usually beginning around age eleven in females, is accompanied by the maturing of the external female genitalia and the growth and thickening of the muscular walls of the vagina and the uterus. Physiological changes in the inner lining of the uterus make it receptive to the future implantation of a fertilized egg. At the same time, the chemical balance of the vagina becomes increasingly acidic. This acidity adversely affects the survival of sperm traveling through the vagina, and as a result only the most robust sperm are able to attempt a union with the egg.

Menarche occurs relatively late in the puberty sequence. It begins approximately two years after the beginning of breast development. Early menstrual cycles are often irregular and anovulatory (unaccompanied by the shedding of an egg). It may be as long as one to two years following menarche before conception is possible (Tanner, 1972). Menarche also indicates that the peak of the growth spurt has passed. On the average, females grow approximately six additional centimeters following menarche. By age sixteen, most females have completed puberty and have achieved the physical stature and physiological functioning of adults.

The onset of puberty in males is marked by the growth of the testes, scrotum, and related reproductive structures (prostate gland, epididymis, vas deferens, ejaculatory duct, and the urethra). The growth of the penis begins approximately one year later, and one year following that (at approximately thirteen to fourteen years of age), ejaculation is first possible (Katchadourian, 1978). The first growth of pubic hair in the male and the onset of the growth spurt begin at about the same time as the growth of the penis.

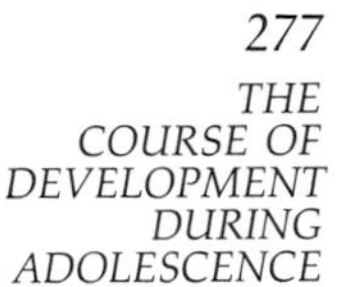

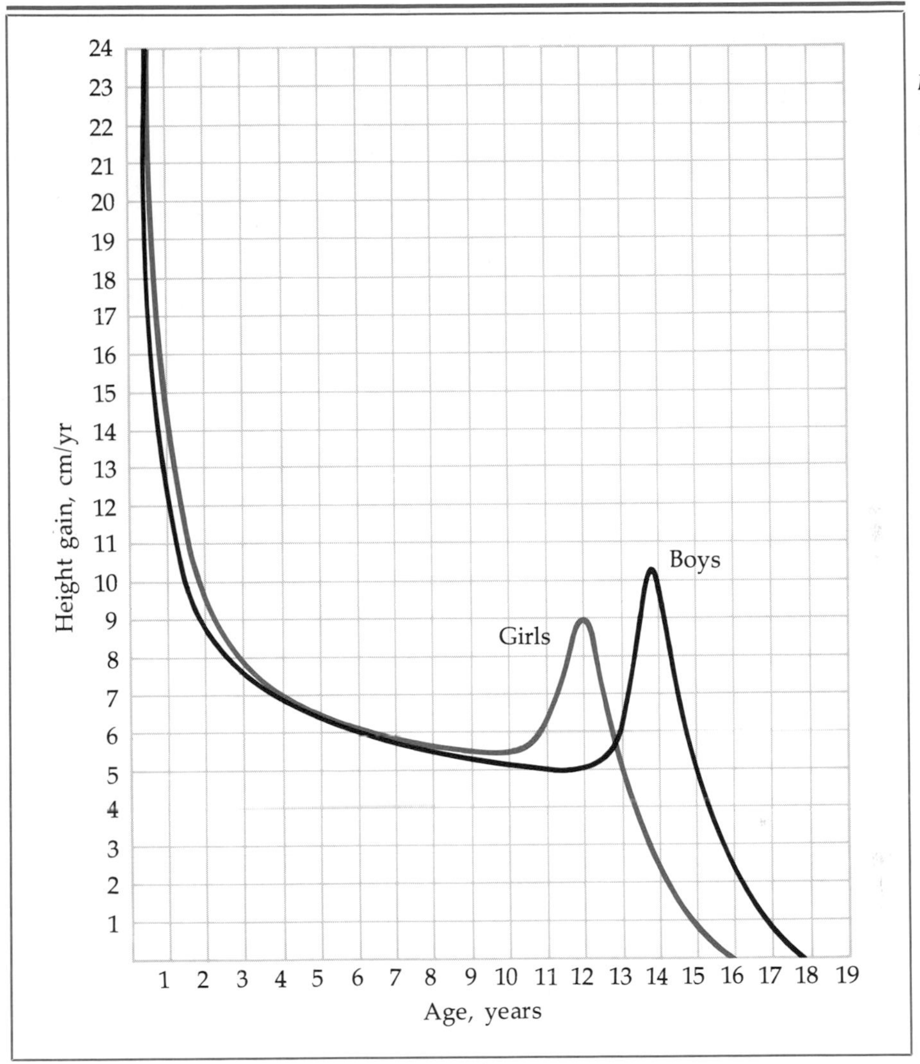

FIGURE 6-4
The growth patterns for males and females differ during puberty. Females reach their growth peak around age twelve, while males do not reach their peak until age fourteen (from Tanner, Whitehouse, and Takaishi, 1966).

The growth of face and body hair begins around age fourteen. Associated with the maturation of the hair follicles is the growth of the sebaceous glands in the skin. Although the sebaceous glands are present in both males and females, the increase in androgen production in males during puberty causes these glands to increase their secretions, and when the secretions block the openings of the hair follicles, acne results. (Acne is reported in 70 percent of pubescent males but only in 50 percent of pubescent females.) The deepening of the voice is the last aspect of puberty in males. This event usually takes place around age sixteen.

BOX 6-1
ASK BETH

For many years, Elizabeth Winship has been answering letters from adolescents in her newspaper column, "Ask Beth." Selected letters have also been reprinted in her book *Ask Beth: You Can't Ask Your Mother* (Winship, 1972). These letters have concerned the various aspects of puberty and adolescence that often prove troublesome. The content of these letters offers a personal glimpse into the transition to adolescence as seen through the eyes of young adolescents.

Many an adolescent boy has feared that he will never stop growing.

Changes in height, weight, and body proportion are almost always an embarrassment to adolescents, no matter what the rate or nature of the changes. They are a frequent topic of letters also. Here are two good examples:

> I'm a boy of seventeen and I'm so afraid I'm never going to grow. I can't stand it, Everyone just says, "There, there you will catch up soon." But I should have started to grow at least two years ago. Do you think I am going to be a midget? Is there anything that can make me start to grow?
>
> *Scared*

> Help! I'm growing endlessly. I am already the tallest girl in the whole elementary school. I'm taller than everyone in my family except my father. I have heard there is a medical treatment that can stop growth. Tell me quickly before I go through the ceiling.
>
> *Alice in Wonderland*

Not all of the problems of puberty are obvious to others. Consider these two rather direct letters:

> Is there anything a girl can do to herself to stop her period?
>
> *Inconvenient*

Source: Winship, 1972.

The Influence of Puberty on Adolescent Identity

The events of puberty have a major influence on young adolescents' image of themselves and their status in the peer group. Although the size of the breasts and the length of the penis are unrelated to their functioning, myths about such relationships are common in this age group (as well as in adult populations), and in a status-conscious society such as ours, devia-

> How long is a penis supposed to be?
>
> *Wondering*

Changes in physical appearance invariably lead to uncertainties about social status. Consider the plight of this group:

> There's a girl in school who dresses as if she were going to sing in a nightclub. She wears passionate purple eye shadow and false eyelashes two feet long. Now that low necks are in style, she wears hers down to where you can practically see what she had for breakfast. We think she looks cheap but the boys don't. They are all over her like a tent. It makes us mad! Should we copy her sexy clothes, or what?
>
> *The Plain Janes*

These same changes also lead to uncertainties about the self.

> Everyone says that if you want to get ahead, you have to have a "good personality." How can I have a good personality when I don't even know what my personality is? I'm not sure it has developed yet.
>
> *Steve*

Sometimes the changes that accompany adolescence appear more evident to the adolescent than to the parent:

> My parents are protective beyond belief. I can't even cross the yard without a lecture on half a dozen things to look out for. I realize they do this because they love me, but I'm strangling. I'm fourteen and my parents still insist on kissing me good-night and of all things, tucking me into bed! Don't you think a good night kiss is enough?
>
> *Hates Tucking*

> Why do my parents have a fit every time I get asked for a date? I am seventeen and have been dating on and off for a year with different boys. My parents almost always let me go, but first of all there has to be an international conference about who the boy is and where we're going. Is this normal?
>
> *Elsie*

The letters from these adolescents provide vivid examples of the concerns and uncertainties, the desire for personal autonomy, the need for recognition and approval from adults, and the egocentrisms that are unique to adolescence. From a more adult perspective, these letters have a naive quality about them. They are the stuff that prompts bittersweet memories at class reunions. But from the perspective of the adolescents who write them, they are the stuff of everyday events—a fact that adults who work with adolescents should never forget.

tions in body type from some idealized norm may be very traumatic for the individual. In fact, physical status is a significant correlate of popularity for both boys and girls throughout adolescence (Mattison, 1975).

Physical stature continues to have a major influence on peer-group status, heterosexual attraction, and self-acceptance well into the adult years. The adolescent and young adult seem destined to repeat the sequence of developments in peer-group status that was typical of middle childhood. Eventually having a nice personality again counts for something, but unfortunately it doesn't seem to count for much during adolescence.

Puberty influences other things besides peer-group status. It also requires adolescents to think of themselves as sexually mature individuals. The significance of this sexual maturity is considerably different for males and females. Puberty is associated with a much greater increase in sexual activity in males than in females. Masturbation is the primary form of sexual activity in early adolescence. Males are four times more likely to report frequent masturbation (twice a week or more) than females are. Females are six times more likely to report no masturbation during early adolescence (Gagnon, 1972). It is important to note that the differences are based on self-report. Given societies' generally greater tolerance of masturbation in males, it is quite likely that these percentages underestimate the figures for females and perhaps even overestimate the figures for males.

Of greater significance than the frequency of masturbation in males and females is the fact that males are considerably less likely to view sexual activity in an interpersonal context, while females are considerably more likely to do so. In particular, for young adolescent females, sexual maturity is often couched in the language of marriage (Gagnon, 1972; Stefanko, 1984).

The association of sexual maturity with marriage is replaced during late adolescence and young adulthood by an association between sexual maturity and romantic love. During early adolescence, females report relatively little sexual activity. The increase in sexual activity in males is in the form either of masturbation or of peer-induced heterosexual pursuits. As exploitive and awkward as these early sexual activities may be, they nevertheless serve one useful purpose. Gagnon (1972) believes that these early heterosexual involvements may increase male's emotional investment in male-female relationships, even though young males nevertheless often profess a rather cynical and nonchalant attitude toward such involvements. As Gagnon notes, "we often become what we thought we were only pretending" (p. 247).

Early and Late Maturers

One of the most significant aspects of development during puberty is the relationship between the age of onset of puberty and both short-term and long-term psychological development. Figure 6-5 shows the increments for growth in height for early and late maturing males and females. Early maturing males report getting along better with their parents at age seventeen than late maturing males report doing (Mussen & Jones, 1957). They are also judged more attractive by peers, show more attention to grooming, are more relaxed, and are less affected in their behavior (Jones, 1957). Jones (1957) has also reported that late maturing males are more expressive than early maturing males but that their "small boy eagerness . . . [is] also associated with greater tenseness and more affected attention-seeking behavior" (p. 116). Early maturing males are more likely to be involved in extracurricular activities than late maturers are and are more likely to be elected class leaders (Peshkin, 1967). Given the choice, it would certainly seem to one's advantage to be an early maturing male during the high school years.

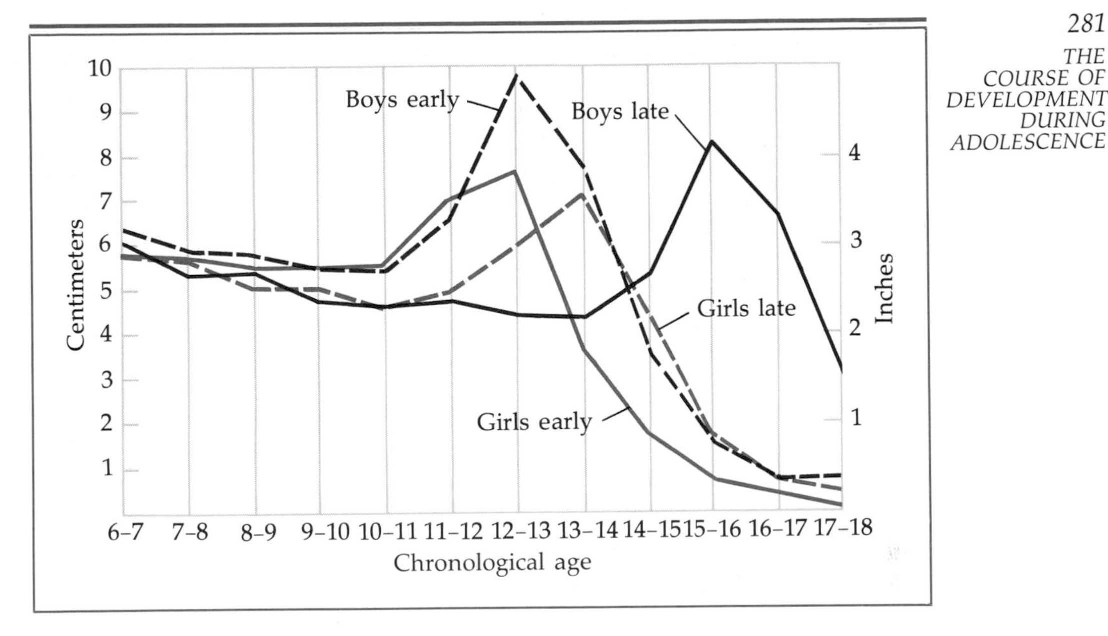

FIGURE 6-5
Increments of growth vary for early and late maturing males and females (from Shuttleworth, 1939).

There is no parallel pattern for females. Although early maturing females report an initial social disadvantage of being taller than everyone else in their class, (Jones & Mussen, 1958) the disadvantage is short-lived, and no other consequences have been found. The age of onset of puberty probably does not have as much influence on the psychological development of females as it does on the psychological development of males because there is more overlap in the growth patterns of early and late maturing females. The range for female peak growth is only one year, and both early and late maturers complete their growth at about the same time. Males show a much more drastic contrast. Most early maturing males have completed the majority of their total growth before the late maturers have even begun theirs. Late maturers grow to be as tall as early maturers, but as is evident in Figure 6-6, their rate of growth is quite different.

Follow-up studies of early and late maturing males show that the initial advantage enjoyed by early maturing males becomes somewhat "dubious" later on (Livson & Peshkin, 1980). In one such study, the early maturer was more likely to have married by age thirty and, if married, was more likely to have children; by age thirty, the early maturer was also likely to earn a higher salary than the late maturer earned (Peshkin, 1967). The early maturers in another study prided themselves, as adults, on being objective and rational; they considered themselves conventional and were judged by others as wanting to make a good impression (Jones, 1965). But early maturers have also been considered to be more rigid in attitude and more moralistic than late maturers are (Matteson, 1975). Late maturers as adults have been viewed as showing a greater tolerance of ambiguity and individual idiosyncrasies (Jones, 1965).

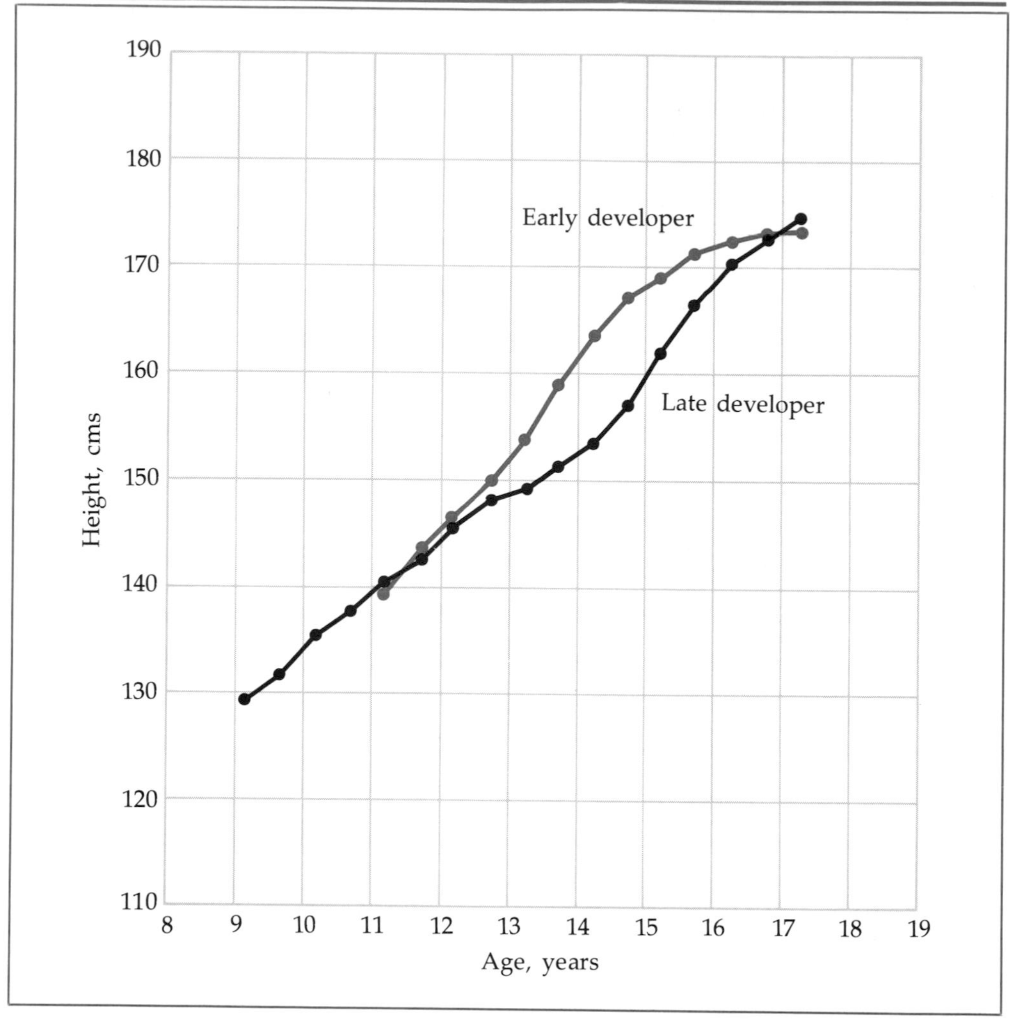

FIGURE 6-6
While late maturing males eventually grow as tall as early maturing males, their rate of growth differs (from Tanner, 1961).

The long-term consequences of early maturing in males seems to be a more conventional and conforming life-style (Peshkin, 1967). In other words, a psychological consequence of making early commitments seems to be a fairly rigid adherence to these early success patterns. The early maturing males of the studies just discussed resemble Erikson's identity foreclosure individuals. The late maturing males, having fewer adult demands made of them, enjoyed a longer period of role experimentation. A long-term consequence of this prolonged period of experimentation was greater flexibility of roles in adulthood (Jones, 1957).

The studies of early and late maturing males clearly indicate the influence of rate of development on the course of development. The two groups of males did not differ before puberty, nor did they differ physically or physiologically after puberty. But because the rate of development was faster for one group than for the other, the two patterns of adult development appear markedly different. These patterns are all the more startling when you consider that cognitive development during adolescence is unaffected by the age of onset of puberty. Cognitive development is the topic of the next section.

Cognitive Development

Adolescence marks the emergence of the fourth stage of cognitive development—the **stage of formal operational thought.** For Piaget (Inhelder & Piaget, 1958), formal operational thought represents the most advanced level of cognitive functioning. This is not to suggest that the adolescent's understanding of the world is as mature as it will ever become. Rather it means that adolescents have a mature set of cognitive competencies to use in understanding their world. They will spend the rest of their lives using these competencies to understanding that world better.

What exactly do formal operations allow adolescents to do? In a phrase, they allow adolescents to discover the world of the hypothetical. They allow adolescents to use their cognitive skills not simply on real or concrete experiences, as they did during the previous stage, but on any and all experiences, real or imagined. They allow adolescents to deal with the abstract as well as the concrete, to use **hypothetical-deductive reasoning,** to begin the development of a personal value system, to use propositional logic, and to exercise **introspection**—that is, to think about themselves not only as they see themselves, but also as they would like to see themselves.

The Unique Nature of Formal Operations

In comparison with the transitions between earlier stages of development, the transition from concrete operational thought to formal operational thought is unique in two ways. First, it does not involve a complete change of cognitive structures. Instead of being replaced by another, more mature set of cognitive structures, the cognitive structures of concrete operations become embedded within a larger set of formal operational structures. The achievements associated with concrete operations such as classification and seriation and reversible mental operations continue within the formal operational period. However, they are interpreted from a different perspective. Formal operations are not restricted to the real and experienced world. Real experiences are now appreciated simply as examples of what could be. In Piaget's words, "thinking takes wings" (1967, p. 63). Once adolescents realize that the real is only an example of the possible, their entire perspective undergoes a radical shift. The pragmatic, conservative child becomes the skeptical, idealistic adolescent.

The second unique aspect of the transition from concrete operational thought to formal operational thought is that it does not appear to be a universal achievement (Cowan, 1978; Flavell, 1977; Neimark, 1975). It is rare to find an organically intact ten-year-old who is unable to solve an elementary conservation problem; it is not uncommon to find in a population of adults as many as 30 to 40 percent who are unable to solve some of the various problems used to demonstrate formal operational thought. The development of cognitive skills requires the continuing maturation of the nervous system and the appropriate environmental circumstances (that is, circumstances of education and occupation).

Since concrete operations provide all the competencies necessary for dealing effectively with most adult tasks and since few adults find themselves in situations that frequently require the skills of abstract hypothetical-deductive reasoning, it is not surprising that these skills are not universally demonstrated. Piaget (1972) has suggested that formal operations may be evident only in certain aspects of an individuals's behavior. That is, individuals whose work requires the use of formal operational skills may not necessarily generalize those skills to other aspects of their lives. The day-to-day pressures of parenting, for example, may make it very difficult for parents to remove themselves enough from the situation to gain an abstract perspective. An abstract view of parenting may only be possible after the children are grown and the parents are free to reflect back on their parenting role. More generally, formal operational thought may only be possible when we can dissociate ourselves from the immediate situation and then view it from a broader perspective.

Tasks Demonstrating Formal Operational Logic

A review of the ways in which children and adolescents attempt to solve two Piagetian tasks used to mark the transition from concrete operations to formal operations should help clarify the nature and significance of this advancement. The first task concerns the ability to predict whether an object will float or sink and the ability to formulate a general law explaining why some objects float and others sink. The second task concerns children's and adolescents' ability to use deductive reasoning in re-creating a previously demonstrated outcome—specifically, in combining seemingly identical liquids to re-create a particular colored solution.

The behavior of objects in water. The ability to successfully predict and explain why some objects float and others sink requires an understanding of the concepts of density and specific gravity. Neither concept can be understood without the use of formal operations. Therefore, the ability to conserve volume, which develops around age eleven or twelve, is one of the earliest indicators of the transition from concrete to formal operations. The actual task is very simple. The subjects are shown a number of disparate objects and are asked to classify them according to their ability to float on water. They are then asked to explain the basis of their classifications. Once this is done, the subjects are given buckets of water so that they can actually test their predictions. Finally, they are asked to summarize their observations and develop a law of flotation (Inhelder & Piaget, 1958).

Piaget found that children attempt to solve the floating task in one of three ways. Young school-age children are not very accurate at predicting which objects will float and which will sink. They seem to be operating from an implicit theory that objects that are heavy sink and those that are light float. When an object in the water does not behave as they predicted it would, they are most likely to explain the outcome in terms of some special property of the particular object rather than to reconsider their implicit theory. Heavy objects that float are explained by the fact that they are "too heavy to sink." Other contradictions are explained in terms of color, shape, or material ("metal always sinks"). One five-year-old, finding her prediction that a piece of wood would sink contradicted, pushed the wood to the bottom of the tank and tried to "correct" its thinking by saying to the wood, "You want to stay down, silly" (Inhelder & Piaget, 1958, p. 22).

School-age children (age eight to eleven or twelve) are aware of the contradiction between their former implicit theory that large objects sink and small ones float and the actual behavior of the objects; that is, they realize that an explanation based on the absolute weight of the object is incorrect. However, like the five-year-olds who know that both the length and the width of the beaker must have something to do with the amount of water in it, ten-year-olds can't yet put the pieces of the puzzle together. They have an elementary notion of density, that is, the relationship between the weight and volume of an object. They have an implicit understanding that the amount of water in the tank has something to do with the likelihood of an object's floating or sinking. But these understandings are still tied to specific objects and empirical observations and therefore can't be generalized to new ones. Such piecemeal awareness is evident in the responses of one nine-year-old.

> (The big block of wood?) It will go under, there is too much water for it to stay on top. (And the needle?) They are lighter. If the wood were the same size as the needle, it would be lighter. (And the cover?) It's iron, that's not too heavy and there is enough water to carry it. (The cover is placed in the tank, it sinks, and the child is asked for an explanation.) That's because the water got inside. (The wooden block is placed in the water, it floats and the child is asked for an explanation.) Because it's wood that is wide enough not to sink. (If it were a cube?) I think that it would go under. (Inhelder & Piaget, 1958, p. 33)

Each object generates a prediction. Wood sinks if it's larger than a needle. Iron isn't too heavy so the cover will float. Needles are light so they float. The behavior of each object generates an explanation. The water got inside the iron; the wood is wide enough not to sink. What is missing still is any evidence that the child realizes that there are certain basic principles that regulate the sinking and floating of objects irrespective of the specific objects involved. As long as children are unaware of the logic of a general principle, contradictory explanations are simply not yet appreciated as being contradictory. In other words, the sense of necessity that is so evident in children's understanding of real events has not yet broadened to include their understanding of abstract concepts.

By age twelve or thirteen, young adolescents can correctly predict and explain the behavior of the objects in the water. A twelve-year-old explaining her prediction that a piece of silver will sink and a tree float even though the tree weighs more than the silver says, "You take the quantity of water for the size of the object, you take the same amount of water. (Can you prove that?) Yes, with a bottle of water. If it were the same quantity of cork, it would float because the cork is less heavy than the same quantity of water" (Inhelder & Piaget, 1958, p. 38). Her use of the phrase "if it were" indicates that her thought is taking flight. She is able to go beyond the specifics of the example and to formulate a concept of density. She is also able to conceptualize a volume of water that is different from the amount in the bucket but equal to the volume of the object placed in the tank. What is so significant about this conceptualization is that it requires the adolescent to create a situation (a hypothetical amount of water) that has no empirical correlate. Having done this, she is able to formulate a general principle that she can use to generate specific and consistent predictions. This ability moves her from the plane of representation to the plane of abstract thought.

Combination of liquids. This advancement in cognitive ability, specifically the development of hypothetical-deductive reasoning, is also demonstrated in Piaget's combination of liquids task. Adolescents using formal operational logic approach the task first with a hypothesis, then they generate all of the logically possible deductions or outcomes consistent with that hypothesis, and finally they systematically test these deductions to determine if the hypothesis is consistent with the empirical evidence.

The procedure is quite simple. Adolescents are shown four bottles of colorless liquids. The contents of the bottles are (1) dilute sulfuric acid, (2) oxygenated water, (3) tap water, and (4) thiosulfate. Two glasses of colorless liquid are then placed in front of the children, and a few drops of (5) potassium iodide are added to each glass. When the drops are added to the first glass (which unknown to the children contains an equal mixture of dilute sulfuric acid and oxygenated water), the solution turns yellow. When the drops are added to the second glass containing only tap water (also unknown to the children), the solution remains colorless. The adolescents are then asked to produce the yellow solution in a new glass. They may use any or all of the four bottles in combination with the dropper of potassium iodide. The task measures the extent to which adolescents are able to generate and systematically test all of the logically possible combinations of liquids as well as to draw conclusions concerning the role of each solution in the formulation of the yellow liquid.

The combination of 1, 2, and 5 produces the yellow solution. The presence or absence of 3 has no effect on the outcome, but the addition of 4 to the yellow solution bleaches it and returns it to a colorless state. Concrete operational children are often unable to discover the solution. The younger ones usually try only two liquids at a time; the older ones sometimes try three. Sometimes they re-create the yellow solution; sometimes they then discover that a new element makes it disappear. Their actions appear random and unsystematic. Although they are sometimes able to produce the color, they are never able to explain the role of each of the five chemicals in its formation.

Formal operational adolescents approach the problem differently. They are interested in re-creating the yellow solution, but they also seem interested in all the possible combinations. In some instances, they may even make a list of all the possible combinations before trying any of them. Such a list is presented in Table 6-1. It contains fourteen possible combinations because adolescents already know that potassium iodide must be present since they saw it added during the demonstration.

After making a list or at least mentally determining the possible combinations, adolescents are likely to attempt each combination and note the results. In the example given in Table 6-1, the fifth combination produces the yellow color. Whereas preoperational children would probably stop at this point (they would not be working from a list but might achieve the desired result on the fifth try by chance), adolescents who consider the empirical from the perspective of the hypothetical realize that other combinations might also produce the yellow color. They therefore attempt the remaining nine combinations. Their search becomes exhaustive, and when it is finished, they are able to determine the role each chemical plays in forming a yellow solution.

TABLE 6-1 Possible Combinations in Piaget's Combination of Liquids Task

COMBINATION	SOLUTION COLOR	COMBINATION	SOLUTION COLOR
1 + 5	colorless	2 + 3 + 5	colorless
2 + 5	colorless	2 + 4 + 5	colorless
3 + 5	colorless	1 + 2 + 3 + 5	YELLOW
4 + 5	colorless	1 + 2 + 4 + 5	colorless
1 + 2 + 5	YELLOW	1 + 3 + 4 + 5	colorless
1 + 3 + 5	colorless	2 + 3 + 4 + 5	colorless
1 + 4 + 5	colorless	1 + 2 + 3 + 4 + 5	colorless

Elements key: (1) dilute sulfuric acid; (2) oxygenated water; (3) tap water; (4) thiosulphate; (5) potassium iodide.

The real significance of formal operations becomes apparent when, having completed the fourteen combinations, adolescents are asked to determine the role of each element. They now go a step beyond merely observing the consequences of each action on the liquids. They reason about the products of their reasoning (what Piaget calls second-order operations). They are able to conclude that no single element combined with potassium iodide produced the yellow solution, that liquids 1 and 2 must both be present, that liquid 3 has no influence on the yellow color, and that liquid 4 is the bleaching agent.

The introspective nature of adolescence, especially early adolescence, is another example of these second-order operations. Introspective adolescents think about themselves thinking about something or someone.

Egocentrism in Early Formal Operational Thought

These two Piagetian tasks provide a demonstration of the abstract, hypothetical-deductive, proportional, higher-order thinking possible during adolescence. Most real-life experiences are not nearly so neatly packaged as these two experiments, however. In most real-life situations, adolescents are unable to identify all the relevant dimensions of the problem, may not be able to empirically test each possible outcome, and in fact may not have a very clear idea of the question that needs to be answered. Their logic alone is insufficient for successful resolution of many of the situations they encounter. This fact doesn't stop adolescents from spending long hours thinking and talking about their futures, but it does indicate that the practicality and probability of these hypothetical futures may be limited. Adolescents still lack the wisdom that seems to come only through many years of experience in the real world. In fact, the formal operational logic of adolescents often produces some very illogical, very egocentric conclusions.

Each stage of cognitive development is initially accompanied by a form of egocentric thought reflective of that cognitive stage. The egocentrism is due to an initial failure to distinguish between one's actions or thoughts and the actions and thoughts of others, or between one's actions and the objects that are acted upon (Inhelder & Piaget, 1958). At the sensorimotor level, infants are initially unable to separate their actions from the objects or people toward which they are directed. It is as if they believe that their actions regulate the very existence of those objects and people. As the concept of object permanence develops, they become able to place themselves in a spatially and causally organized field composed of permanent objects and other persons similar to themselves.

The transition to preoperational thought produces a second form of egocentrism, which is not resolved until the emergence of concrete operations. Once again egocentrism takes the form of an inability to differentiate, only now this lack of differentiation is representational rather than sensorimotor. Preoperational children are unable to distinguish between their own point of view and that of others. However, once these children reach the stage of concrete operations, they are able to determine the relationship between classes, relations, and numbers and to interact with other individuals within a cooperative framework. In parallel fashion, the emergence of formal operations in adolescence is accompanied by its own form of egocentrism. When young adolescents begin to think about their role in society, they are unable to separate their own point of view from that of the group as a whole (Inhelder & Piaget, 1958).

The egocentrism of adolescence has three unique qualities (Elkind, 1978). The first such quality results from a failure to distinguish transient from abiding thoughts. Adolescents are seriously embarrassed by personal experiences such as spilling milk in the school lunchroom or falling down the stairs in the hallway or making the last out in a big game because they think these experiences are extremely significant. What they fail to realize is that the experiences seem considerably less significant to other people. To embarrassed adolescents, the experiences seem abiding; to other people, they are at most transient. No doubt many conversations at class reunions end abruptly when one person says, "Remember when I . . . ," and another replies "No."

Elkind suggests that the second quality of adolescent egocentrism is a failure to distinguish the subjective from the objective. Adolescents are very taken with their new powers of self-reflection. It seems hard for them to appreciate the fact that others are not nearly as interested in them as they are.

The self-consciousness of the adolescent is a reflection of this failure to differentiate the subjective from the objective. You only need to visit a place frequented by young teens and to watch them almost constantly looking around while they are standing in small groups to appreciate the impact of their "imaginary audience." The obvious conclusion is that if other adolescents are also self-conscious, they can't be watching but in fact think too that they are being watched, but somehow this conclusion doesn't yet exert much of an influence.

The imaginary audience can be a very powerful motivator. Elkind suggests that it helps explain the self-improvement efforts common to young adolescents as well as seemingly senseless acts of vandalism. In both instances, adolescents believe that their status in a particular peer group will rise as this imagined audience becomes aware of their actions. They erroneously assume that others will be as impressed with their actions as they are. This assumption may also be a component of adolescent suicides. Adolescents who attempt suicide often comment that a primary

The preoccupation with self-improvement that is characteristic of many adolescents is due in part to their egocentric belief that they are constantly being looked at.

motive was the imagined reaction of those who would have been left behind.

The third form of adolescent egocentrism builds on the second. It results from a failure to distinguish the universal from the particular. If this imagined audience is always watching, then there must be some reason for their continuing interest. What is it? Young adolescents conclude that the source of the interest is the special feelings and thoughts that they are experiencing for the first time. In effect, Elkind (1978) believes these young adolescents create what he calls a personal fable about themselves.

The young adolescents are convinced that they are each the only one to have ever experienced the agony and ecstasy of first love, the embarrassment of acne, and the awkwardness of puberty. Elkind believes that the personal fable, like the imaginary audience, can serve as a powerful motivator. The perception of specialness can serve as a buffer against the minor setbacks of everyday life and as a reason to try to excel at some particular skill.

Elkind (1978) warns, however, that the personal fable can lead to negative as well as positive consequences. The sense of uniqueness leads some adolescents to conclude that what might happen to others will not happen to them. Others may not be able to drink and drive, but they can. Others may not be able to use drugs without becoming addicted, but they can. Others may become pregnant if they fail to use effective contraception, but they won't.

Adolescent egocentrism is not resolved through the emergence of new cognitive structures. It is resolved through continuing experiences in the adult world. Sooner or later adolescents realize that a temporary embarrassment is just that, that other adolescents are too busy worrying about being watched to watch others, and that maybe at least one other person had a first love like theirs (but it certainly couldn't have been their mother or father). Actually all three forms of egocentrism are still present in adulthood and serve the same positive and negative roles that they serve during adolescence. The range and depth of their influence may decrease, but they never completely disappear, a fact reflected in a statement made by the playwright and humorist William Saroyan shortly before his death: "I always knew that everyone had to die sometime, but I always thought an exception would be made in my case."

Social Development

The biological and cognitive changes that accompany adolescence result in significant changes in the way in which adolescents view themselves, in the way they relate to their peers and adults (particularly their parents), and in the ways in which they perceive events in their environment. As a result of puberty, adolescents acquire the physical stature and physiological functioning of adults. Consequently, they come to see themselves as occupying a position of equality with adults and increasingly demand the privileges and independence of adulthood and the opportunities to prove that they are no longer children. Because of changes in their cognitive abilities, adolescents come to view their interpersonal relationships—both present and future—in a different light. All of these changes have a profound effect on adolescents' self-concept.

Physical and Cognitive Influences

The cognitive developments of adolescence have a profound impact on adolescents' social development. The emergence of abstract thought allows adolescents to view individuals and events not simply as isolated elements having a limited, immediate impact but as parts of a larger, broader social system. Adolescents begin to understand such abstract concepts as society and government and the way individual effort fits into the scheme of things.

The use of hypothetical-deductive reasoning allows adolescents to see the actual as merely one example of the possible. In early adolescence, this skill often results in overly critical evaluations of parents, peers, siblings, and just about anyone else adolescents happen to think about. They compare the real ("our house is a dump") to the ideal ("we should be living in a beautiful house with a beautiful garden and a swimming pool, etc., etc., etc.,"), and the real always comes out second best.

These emerging cognitive skills also make possible the viewing of the present in terms of the future—certainly one of the most significant advances occurring during this period. Adolescents begin to formulate a life plan and to view present events as consistent or inconsistent with it. They view school from both a present and a future perspective and take less desired courses because they need them for college or a future job. Similarly, adolescents debate how much of the salary from their part-time job should be spent on records and how much should be put in their savings account. They begin to think about their friends as the kind of people they would like to know forever, as people they would consider marrying, or as the kind of people that will help them achieve their future plans. They begin to view their family as a possible model for a family of their own someday. These advancements are not apparent during early adolescence. In fact, just the opposite seems more typical. But by late adolescence, these perspectives become more consistently evident.

Adolescents' higher-order thinking allows them the mixed blessing of introspection. Increasingly they focus on self-improvement and speculate about the future. As much as one-third of the young adolescent's day is spent in daydreaming and fantasizing (Larson & Csikszentmihalyi, 1978). Young adolescents seriously attempt to integrate the diverse elements of their life—to develop some sort of ideology and moral code. For the first time, they try consciously and deliberately to form a unique personality—

to achieve some personal sense of wholeness. These attempts are not always successful or even well integrated. They rarely provide a clear guide for present and future behavior. They are likely to be frequently changed. But they are attempts at deliberate life planning.

Adolescence is not the first stage in which deliberate, intentional, integrated behavior is evident. Such behavior is evident in infants, and it is possible to talk about the personality of a child of any age. But during earlier stages, coherence is more evident to the observer than to the child. It is the observer who explains the infant's development of the object concept by first saying, "it is as if." The phrase "as if" is no longer needed in adolescence. The role adolescents take in constructing an understanding of their environment becomes deliberately intentional (premeditated, if you will) and increasingly the result of self-reflection. This is not to say that infants or children are any less actively involved than adolescents are in influencing the course of their development or that the thought processes of adolescents are always accurate. Rather, it is to emphasize that the nature of this active involvement changes in adolescence.

The impact of these cognitive changes is perhaps most apparent in adolescents' interpretation of the consequences of puberty. They begin to become aware of themselves as sexually mature individuals and to establish relationships consistent with this perception. One of the major desires of adolescents is to establish close, personal, intimate relations; one of the major fears, that these relations will be used in an exploitive fashion.

This fear of exploitation is more common in female adolescents, a fact that suggests their active conception of sex-role identity differs from that of males. Even though the double standard for sexual behavior has declined dramatically over the past thirty years (Zelnick & Kanter, 1977) and female

The desire for the establishment of intimate relationships is often accompanied by the fear of sexual exploitation, particularly in the case of adolescent females.

adolescents are now as likely to be sexually active as males (Stefanko, 1984), they are nevertheless still more likely than males to feel the need to define their current sexual activity within the context of close, personal relationships.

The Self-Concept in Adolescence

The single defining element of the self-concept in adolescence is a quest for personal integration, or a sense of personal identity. This emphasis on personal integration is not evident at earlier stages of the life-span and during the adult years typically acquires equal importance only during mid-life and old age. The quest for identity is evident in the efforts of adolescents to integrate actions, thoughts, and words in an attempt to establish continuity among the various aspects of their lives. Adolescents' frequent noting of "phoneyness" in others and their denial of "phoneyness" in self are reflections of this strong desire for integration.

One of the more prominent features of adolescents' attempts to establish a sense of personal identity is the development of a moral code—a set of principles that adolescents use to guide individual actions and to judge the behavior of themselves and others. This moral code differs from those of earlier years in that it is abstract rather than situation specific and its standards are internal rather than external.

The evolution of this moral code follows a sequence that has been studied by Kohlberg (1964; 1976; Colby, Kohlberg, Gibbs, & Lieberman, 1983; Kohlberg and Gilligan, 1972). The sequence involves three levels of moral reasoning. The preconventional level is most typical of children; the conventional, of preadolescents and young adolescents; and the postconventional, when present, of older adolescents and adults.

The behavior of children at the preconventional level is a reflection of the reactions of others to them. These children are particularly sensitive to being labeled good or bad; they seem motivated primarily through the promise of reward and the threat of punishment. The behavior of those at the conventional level reflects an internalization of rules and norms. Behaving in ways consistent with family and culture is now seen not so much as a way to avoid punishment or gain reward but rather as something desirable in its own right. The feelings of reward and punishment now come from within the individual. Postconventional moral reasoning develops in those who sense a potential or even real conflict between behavior that is consistent with a socially sanctioned value system and behavior that is consistent with universal conceptions of the rights of people. The history of race relations in this country and elsewhere provides a vivid example of just such a conflict. Postconventional individuals justify acts of civil disobedience, saying that they are necessitated by universal principles of justice and human rights, principles and rights that no social group has the right to negate.

Moral level is defined in terms of the reasons an individual gives to justify a particular action rather than in terms of the action itself. Preconventional children might base a decision not to steal on the likelihood of being caught; conventional preadolescents might argue that since stealing is against the law, it is wrong and should not be done; postconventional adolescents might argue that even though stealing is against the law, there may be certain circumstances in which stealing would nevertheless be morally acceptable (for example, to save a life). These adolescents might acknowledge that if they broke the law, they might have to go to jail, but would add that they would still steal to save the life of another person. In other words, they believe there is a law or moral order that is higher than socially based laws.

Although development of moral principles moves from preconventional to postconventional, the movement is not inevitable. Most people, most of the time, use conventional moral reasoning to explain their everyday actions. Even those who demonstrate postconventional reasoning acknowledge that there is not always a perfect correspondence between their actions and their words. Further, even though the progression is equally evident in males and females (Walton, 1984), there is an active controversy concerning the relationship between sex-role socialization and the elements (as opposed to the levels) that contribute to the moral reasoning of men and women (Gilligan, 1980). This latter issue will be discussed in the next chapter.

Nevertheless, postconventional reasoning first becomes evident in adolescence and continues to develop within adult populations. Its presence indicates a serious attempt on the part of the adolescent to develop an internal set of standards that can be used to guide behavior in the various aspects of adolescent life. Perhaps some of the difficulty adolescents encounter in developing an initial sense of identity stems from their inability to consistently behave in the "lofty" manner their model ideology requires.

A second prominent factor influencing the adolescent self-concept is sex-role identification. How does the attainment of a sex-role identity influence adolescents' conceptions of themselves, present and future? To what extent are these conceptions consistent with adolescents' future planning, and to what extent are they consistent with society's expectations for male and female adolescents? Most reviews of adolescent development (Douvan & Adelson, 1966; Gallatin, 1975; Rogers, 1981) suggest that this aspect of self-concept formation follows different patterns in males and females.

The masculine sex-role is more narrowly defined than the feminine sex role is. It centers on vocation. Marriage, family, and home are assumed to be consistent with the primary component of vocation. Because the masculine sex role focuses on vocation, the issues of vocational development and achievement in adulthood tend to provoke a great deal of anxiety in men. The feeling of having great responsibilities and the feeling of having

Coeducational home economics and industrial arts classes may prove to have long-term effects on sex-role identification.

to prove one's worth continue through the early adult years (Pleck, 1983), perhaps explaining why men have shown less willingness than women to broaden their sex-role identity.

The situation is more complicated for women. The relationship of the adult roles of worker, intimate, and parent are not as well integrated for women as they are for men. In fact, in many instances, these roles are in conflict. For women favoring a traditional role, many decisions influencing definition of self must wait for marriage. Those favoring broader role definitions are faced with the problems of timing and sequencing in their decisions about marriage, occupation, and parenting.

It is uncertain how many adolescents, male or female, are aware of or concerned about the potential dilemmas of sex roles in the adult years (Walters & Walters, 1980). High school adolescents often have a poorly defined sense of the potential conflicts inherent in any sex-role conceptualization. Perhaps, for now, this is just as well. Research with middle-aged adults finds that those who have the more realistic perception of their present and future circumstances are those who tend to be the most depressed (Gurin & Brim, 1984).

A third factor influencing the self-concept of adolescents is their ability to accept their behavior as an expression of themselves. For example, probably the single best predictor of contraceptive use among sexually active adolescents is their ability to view themselves as sexually active individuals (Reiss, Banwart, & Foreman, 1975). This ability involves more than simply acknowledging that they have had sexual intercourse. It also involves acknowledging that they choose to be sexually active and that such activity is an acceptable form of adolescent behavior. Those who are unable to view themselves from such an objective perspective are less likely to use contraceptives than those who are (Hornick, Doran, & Crawford, 1979).

It is significant to note that Fox (1980) found that although few adolescent girls received sexual instruction from their mothers, those who did became sexually active later than other girls did and were more effective users of contraceptives. Fox believes that such discussions are important "not so much because of the factual information imparted as because such discussions can make the daughter's sexual behavior explicit and can encourage the daughter's awareness and acceptance of her own sexuality" (p. 25). It is also likely that such discussions help adolescents accept themselves as sexually active individuals because they make possible a greater degree of continuity between the family and peer domains of the adolescents' world. Such continuity-enhancing experiences may also help adolescents make decisions that take into account considerations other than the pressures of the moment. In this regard, it is worth noting that a frequent characteristic of adolescents who conceive children out of wedlock is a "reluctance to pass up the opportunity for physical intimacy in spite of the known risks of pregnancy" (Rogel, Zuehlke, Petersen, Tobin-Richards, & Shelton, 1980).

More generally, the acceptance of the self involves a realization of personal limits and the realities of the adult world. Young adolescents are optimistic. They believe that they can do just about anything or be just about anybody if they try hard enough. Late adolescents are more aware of the limits of individual effort and, as a result, are more likely to channel their efforts in ways that are consistent with their individual abilities and life situation (Selman, 1980). This doesn't mean that late adolescents merely accept the status quo. Rather it suggests that older adolescents have a more realistic understanding of what is required to pursue various courses in life and how likely they are to succeed in a particular course. This understanding becomes further refined as they deal with the tasks of the adult years.

THE CONTEXT OF DEVELOPMENT DURING ADOLESCENCE

Although the contexts of development during adolescence are the same as those of middle childhood, the nature and degree of their impacts are quite different. In particular, the peer group assumes a much larger role in the lives of adolescents than in the lives of school-age children. This greater influence does not come at the expense of the family or school, however. Rather it reflects the continued broadening of adolescents' environments. Of even greater contrast is the manner in which adolescents see themselves in relationship to their families, schools, and peers. Because adolescents view themselves as more actively involved in directing the course of their individual development, the influence of these three contexts of development is increasingly mediated by adolescents' interpretation and evaluations of their meaning and importance.

The Adolescent and the Family

Most adolescents continue to see their family as a primary source of support and advice. They generally hold the same attitudes and values as their parents on most moral, political, and social issues and consider them important sources of vocational guidance.

Parent-Adolescent Relations

Parent-child relations are generally more stressful during adolescence than during middle childhood. However, they are rarely as extreme as the unbridgeable generation gap that is frequently depicted in the popular media. The major source of conflict involves the increasing requests, or in some instances demands, of adolescents for greater autonomy and independence (Hill, 1980). Adolescents do not want to completely sever their ties to the family. Rather, they seem to want the freedom to make their own decisions while maintaining the assurance that the family will always be available as an emotional safe haven and a source of physical and monetary support (Elkind, 1978). The strength of adolescents' requests, the areas in which the requests are made, and the ease of the resolution of conflicts reflect the quality of the parent-child relationship previous to adolescence, as well as the willingness and ability of the parents to recognize their children's continuing development.

The tone of the parent-child relationship is usually well established before adolescence. Its nature may either hinder or facilitate the resolution of the issues of independence, autonomy, and self-management. Adolescents who were overprotected as children may feel overwhelmed when other people begin to expect them to show more self-direction. They may retreat back into the safety of the home, and they may be easily exploited

The desire for greater autonomy on the part of adolescents often makes parent-child interaction more stressful during adolescence.

by peers. Overindulged children may find these greater demands for self-reliance frustrating (Conger, 1972).

Parents who have been consistent in their parenting techniques, who have made age-appropriate demands on their children, who have recognized their children's legitimate place in the family structure, who have dealt with children in a rational, issue-oriented manner, and who have provided children appropriate and sufficient justifications for their actions are more likely to successfully cope with and resolve the issues of adolescence than are parents who have behaved in an inconsistent, punitive, arbitrary, and unresponsive manner (Baumrind, 1975). During adolescence, parents who demonstrate the former set of behaviors are better able to acknowledge the increasingly reciprocal nature of the parent-adolescent relationship, allow the adolescent a reasonable degree of autonomy and independence, use negotiation and compromise as the primary means of expressing parental authority, and recognize both the short-term and long-term value of providing adolescents with justification for parental action.

Parenting the adolescent is more difficult than parenting the school-age child, in part because the old techniques are no longer effective. It's hard to spank children who tower over you or to threaten not to take them to grandma's when they don't want to go in the first place. Parents need to elicit the cooperation of adolescents, and parents who have always believed that "because I say so" is sufficient justification may find themselves dealing with hostile, resentful, unresponsive children.

School-age children accept parental authority because they have a limited ability to conceptualize alternative paths of action, because they view parents as all-powerful, and because their notion of being grown-up involves behaving in ways desirable to parents (Selman, 1980). None of these conditions is true during adolescence. The formal operational nature of adolescent thought allows adolescents to conceive of many alternatives to the parent's recommended course of action. The increasingly reciprocal nature of the parent-adolescent relationship means that, in comparison with younger children, adolescents see themselves as more equal and therefore less subservient to parents. Certainly, being as tall as the parent also adds to this new perspective. In some areas, adolescents acknowledge the parent's wisdom by virtue of greater experience; in others, they perceive their own experiences as providing equal or greater knowledge. Further, conforming to adult demands is no longer necessarily consistent with the adolescent ideal self-image (Baumrind, 1975). For some, oppositional behavior is equated with independence and autonomy. As a result, a rational justification for parental requests and demands is no longer simply a desirable accompaniment to the exercise of parental power; it now becomes the source of that power (Hoffman, 1980).

To further compound the parent's dilemma, the stakes of the parent-child relationship are higher during adolescence than they were previously. The consequences of adolescent behavior are potentially more severe and long lasting than are the consequences of younger children's actions. Parents are confronted with the possibility that their adolescent will become pregnant or contribute to the pregnancy of another, abuse alcohol or drugs, smoke, drive recklessly, or behave in any number of other harmful ways. It does not matter that the parents are fairly sure that their child won't do these things, that they in fact have been good parents. The fact that these events are possible adds a sense of urgency and stress to the parents' situation.

Some parents believe that adolescence provides them with the last chance they will have to influence their child's development. They believe that in a few years, their child will leave home and they will no longer be able to correct perceived lacks in their child or make up to the child for what they view as inappropriate parenting in the past. This sense of urgency is often further heightened by the parents' awareness of their own

Because of their growing physical and emotional independence, adolescents are more likely than school-age children to engage in activities that are potentially destructive to their well-being.

aging (most are entering middle age). To compound the situation still further, the parent's parents may be reaching old age, and the increased dependency and frailty that often accompany old age become another indicator to the parents that time is running out. The parents feel that their generation is caught in the middle, a topic that is discussed more fully in Chapter 9.

It is somewhat ironic that so many parents worry about losing influence during adolescence. The evidence indicates that most adolescents like their parents, seek their advice, and accept the fact that they are as much to blame as their parents when conflicts exist (Conger, 1972). In fact, there is a remarkable similarity in the values held by adolescents and their parents. Most adolescents, like their parents, believe that competition encourages excellence—few would be supportive of abolishing the grading system in schools. Most subscribe to such enduring notions as the idea that hard work leads to success, that success is a worthy goal, that accomplishment is a reflection of personal character, and that compromise is an essential aspect of progress (Conger, 1972; Offer, Ostrov, & Howard, 1981).

Those issues that generate the most parent-adolescent conflict tend to be ones that affect the immediate life-style of the adolescent. Music preference, use of language, choice of friends, curfews, clothing styles, and use of family resources such as money or the car are the major points of conflict (Stefanko, 1984). As long as parents are able to realize that these high-conflict issues are very important to adolescents and have little lasting impact on or significance for adult development, unnecessary conflicts can be avoided.

Although adolescents today accept most of their parents' basic values, they often don't consider these values complete. They share with their parents the desire for material possessions but also want, for example, a work setting that provides personal fulfillment. They are less willing to "sell their soul to the devil" than they believe their parents are. They are more liberal and more socially conscious but at the same time want the "good life." They are less willing to defer gratification (Yankelovich, 1974).

These value patterns certainly reflect both the idealism and cynicism that is typical of adolescence, but they also seem to reflect a generational shift. Previous generations of adolescents were less likely than adolescents today to value both materialism and personal fulfillment, and they did not place such great weight on personal friendships. Contemporary adolescents seem more likely to feel entitled to these goals than previous generations of adolescents did. Contemporary adolescents believe they have a right to both material success and personal fulfillment. It is this sense of entitlement that highlights the shift in adolescent value patterns (Konopka, 1976).

Vocational Planning

Vocational planning becomes an increasingly important topic during adolescence. For many adolescents, it provides a blueprint for future life planning and educational goals. For others, it raises conflicts concerning personal identity and the integration of adult roles. The fact that male adolescents are more likely to be found in the first group while females are more likely to be found in the second emphasizes the increase in sex-role differentiation that takes place during adolescence and the continuing conflicts that women in our society face concerning the nature of their future roles. The family serves as the primary base for adolescent vocational planning.

Most school-age and even some preschool-age children can tell you what they want to be when they grow up. They may want to have the same jobs as their parents; follow in the footsteps of a famous sports hero, actor, or actress; or assume some role that seems exciting and special. These early occupational choices have a fantasy quality about them (Ginzberg, 1972). They aren't realistic, they don't take into consideration any special qualities necessary for the role, and they rarely ever come true. These initial choices do acknowledge, however, an awareness of one part of the adult role and a desire to identify with it.

Early adolescents typically make a tentative commitment to an occupation. They understand that different jobs require different skills, interests, and aptitudes. They understand that different jobs require different types and amounts of training. They begin to realize that there is a relationship between how well they do in school and the types of jobs that will be available to them (Ginzberg, 1972). Most expect to find jobs in professional or technical fields (Borow, 1976). The fact that these expectations rarely reflect an awareness of the state of the economy or a realistic sense of the amount of training or competence necessary for such positions is probably why they rarely are realized.

These early commitments are tentative for a number of reasons. First, most adolescents change their minds many times before finally deciding on an occupation. Second, the basis upon which these initial decisions are made is often very limited. Our age-segregated society prevents adolescents from gaining much firsthand knowledge about occupational roles (Bronfenbrenner, 1970; Steinberg, Greenberger, Jacobi, & Garduque, 1981; Stipek, 1981). Third, the vocational guidance programs in most high schools are very limited. Most vocational counselors either have too limited a knowledge of the current job market or too limited a knowledge about the students or both (Coleman, 1975). Fourth, adolescents' initial vocational choices probably are related more to the "trying on of roles" typical of early identity formation than to any serious consideration of future occupational roles.

The occupational choices of late adolescence tend to be more realistic (Borow, 1976; Ginzberg, 1972). Late adolescents are more likely to be aware that the job market reflects current social and economic factors, although they may not have a very clear image of the current status of these factors. They also have a clearer image of themselves as unique individuals, and they know that vocational decisions should in some way reflect their unique individual qualities. However, since their understanding of the actual job market is limited, they often don't have a very clear sense of the possibility of finding work compatible with their individual characteristics. The greater seriousness of the vocational thinking of later adolescents is reflected in their choice of courses in high school, in particular, the decision to pursue college prep or vocational training programs (Borow, 1976).

Families play both a direct and indirect role in helping adolescents develop vocational plans. They assist directly in the definition of occupational roles. This process takes place through discussion and through visits to occupational and educational settings. The indirect role begins long before adolescence. Families provide models of appropriate adult roles and foster the development of those personal qualities that probably, in the long run, are more influential in determining job choice and job satisfaction

While their final career goals may be far from settled, by high school most adolescents have decided whether to pursue college prep or vocational training programs.

than are discussions about the specific jobs available in our society. Female adolescents, for example, are more likely to pursue nontraditional work roles and to desire roles as parent, spouse, and worker if their mothers have a commitment to a career and if they view their mothers as competent in nonfamily roles (Hill, 1980).

Parents also model a variety of personal characteristics that are not directly related to specific occupations but rather to qualities that influence success in a variety of work settings. These characteristics include a sense of personal autonomy in meeting and handling everyday tasks, internal control over one's behavior, a personal desire to achieve, and a feeling of personal satisfaction in success (Borow, 1976). The more realistic quality of late adolescent vocational planning, as compared to earlier vocational planning, is reflected in the more deliberate attempts of late adolescents to integrate these indirect and direct influences into some coherent sense of present identity and future direction. It is probably at this time that female adolescents begin to realize that this integration process is more conflict prone for them than for male adolescents.

The two main worlds of the adult are home and work. Traditional views of male and female roles have defined the home as the domain of the female and work as the domain of the male. These definitions don't mean that women don't work or that men don't come home; rather they stress the fact that, traditionally, future planning for women has focused on family and marriage, and future planning for men has focused on work. These traditional views give the male role dominance over the female role. The father gives his daughter to his about-to-be son-in-law at the altar; the daughter then follows the husband from job to job and acquires her new name and status through his place in society. Erikson's early writings present an extreme version of this pattern through the argument that women cannot complete the process of identity formation until they marry (Erikson, 1950). That is, they acquire an identity by adopting that of their husband (a view Erikson tempered in later writings). For adolescent women who favor such a traditional role, their future course is set. They seem unaware or unconcerned that this traditional role is often reported to be personally unfulfilling (Bernard, 1982). When confronted with the evidence, they are likely to argue that for them it will be different (Rubin, 1976). And for some, but probably not most, their prediction will hold true.

Adolescent women who favor nontraditional roles are faced with integrating the worlds of home and work in their life planning and their personal identities. Although adolescent males must also integrate home and work, the evidence shows that the process produces less conflict for men. Adolescent females are more likely to experience conflict and ambiguity over the integration of career and marriage, over the incorporation of strong achievement motivations into their conception of appropriate sex-role behavior, and over the compatibility of their value systems with those of adolescent males (Matteson, 1975). The conflict is perhaps first evident in the sudden and often dramatic decline in the math and science grades of many junior high and high school women (Mills, 1981).

The adolescent female soon finds that her male counterparts in high school and in college are considerably more liberated in principle than in fact (Komarovsky, 1973; 1976). These adolescent males are not opposed to their future wives working; in fact they would welcome the added income. They are not, however, very willing to increase their involvement in house and child-care routines. In other words, they still expect their wives to fulfill the more traditional roles of women. It is small wonder that some female adolescents inhibit their achievement motivations during high school and college (Horner, 1972; Houts & Entwisle, 1968) and abandon them in adulthood (Gallatin, 1975). The conflicts these views about sex roles create for men, for women, and for couples, as well as the degree to which they change during adulthood, are discussed in Chapters 7 and 8.

The Adolescent and the Peer Group

The peer group assumes greater influence on behavior during adolescence than it does during middle childhood. However, it usually does not do so at the expense of the family (Bowerman & Kinch, 1959; Condry & Siman, 1974; Seltzer, 1982). Rather, because of the greater autonomy granted the adolescent and because of the relatively age-segregated nature of adolescence, the peer group assumes influence over those issues that are unique to the adolescent peer culture. These issues tend to be immediate in nature and focus on dress, language, music, friendship patterns, use of money, and curfews. The family maintains its influence in such areas as future life plans, moral judgments, and basic values.

When the peer group's influence does extend to all aspects of adolescents' lives, it usually reflects a lack of attention and concern at home

BOX 6-2
HOW ADOLESCENTS SPEND THEIR TIME

Parents report a variety of changes in their children's behavior as they enter adolescence. One of the most common and, for some, troubling changes is an increase in the amount of time spent alone. Parents often wonder if something is bothering their children or if they have suddenly become unpopular with friends. They wonder if time spent alone is predictive of other, more disturbing developments. Questions about whether something is wrong are usually met with, "No, everything is fine. I just like spending time in my room." This response seldom quiets parents' fears. To them, spending time alone is a sign of loneliness and social alienation, undesirable qualities from an adult perspective.

Thanks to the work of Larson and Csikszentmihalyi (1978), two University of Chicago developmentalists, parents can now rest a little easier. In an effort to determine how adolescents spend their time, they asked a group of adolescents to fill out the self-report forms, five to seven times per day, between the hours of 8:00 A.M. and 11:00 P.M. To ensure that the reports would be filled out at random times during the interval, Larson and Csikszentmihalyi hit upon the ingenious solution of providing each adolescent with a pocket-sized paging device. Whenever the device beeped, they were to fill out the report. The report concerned what the adolescent was doing and with whom, his or her current emotional state, and the level of interest in the reported activity.

The research confirmed, in part, what parents already suspected: adolescents do spend a lot of time alone, approximately one-third of their day. However, even more time (50 percent) is spent with peers. The remaining 20 percent of the day is spent with parents.

According to the adolescents studied, time alone is generally taken up with daydreaming, thinking about themselves and the future, or just fantasizing, usually in the privacy of their own bedrooms. Strangely enough, even though it is

Source: Larson & Csikszentmihalyi, 1978.

rather than the attractiveness of the peer group (Conger, 1972). These peer-oriented adolescents tend to view themselves, their parents, and even their peers negatively. They report little guidance, affection, or support from parents. They are more likely to hold pessimistic views of their future than are adolescents who report that their parents show genuine concern about their welfare. One evidence of this lack of effective guidance and affection is that these adolescents are more likely to be involved in delinquent activities (Bronfenbrenner, 1970).

While adolescents do spend a lot of time alone, research has shown that the time is not nearly as much as parents fear.

usually voluntary, time alone seems to leave adolescents feeling lonely, sad, and self-conscious. To cloud the picture even more, those adolescents in the study who spent a lot of time alone, with the accompanying negative moods, had the highest overall mood ratings; in other words, they were the ones generally most happy and positive in outlook.

Larson and Csikszentmihalyi offer an interesting explanation for this seeming paradox. They suggest that time alone, even with its accompanying negative moods, may be a necessary component of the identity-formation process. In particular, the self-reflective processes associated with identity formation may proceed best when the adolescent is least self-conscious, that is when he or she is alone. In fact, those adolescents spending moderate amounts of time alone reported feeling the least alienated from others. Perhaps for these adolescents, time spent alone facilitated the self/other differentiation process, the core issue in the development of self-identity.

Early Adolescent Peer Groups

Early adolescent peer groups maintain the same pattern as the peer groups of middle childhood. Groups are self-segregated by sex, and most members feel a strong allegiance to their group. The peer group helps the young adolescent cope with the anxieties and uncertainties associated with puberty. It confirms the fact that everyone seems to be in the same boat (Seltzer, 1982). Male groups tend to be larger than female groups, and

Adolescent peer group influences tend to strongest in the areas of age-related behaviors. These Boy George look-alikes are most likely bowing to this type of pressure.

females usually have a few special friends within the peer group. Solidarity seems to be a primary concern of young adolescents, and they view the peer group as the primary mechanism for maintaining this closeness. Peer-group leaders are valued because they reflect the concerns of the group itself, rather than imposing their own will on it. They are seen as promoting the group's sense of community and bringing the group together "as a whole" (Selman, 1980).

The desire for closeness is also evident in the friendship patterns of early adolescence. Adolescents emphasize the establishment of close interpersonal bonds—truly getting to know other people rather than just doing things with them. Because of this shift in emphasis, adolescents become increasingly concerned with the psychological qualities of their peers. It now takes more than merely living next door to someone to develop a friendship. Conflict resolution is now understood to require some sort of working-through process. Each person has to feel satisfied (Selman, 1980). Because resolution requires some sort of negotiating process, close friendships are considered harder to maintain than they were during middle childhood but, at the same time, the resolution process is seen as having the potential to make the friendship even stronger. As one young adolescent puts it:

> Like you have known your friend so long and you love her so much, and then all of a sudden you are so mad at her you say, I could just kill you, and you still like each other because you have known each other for years and you have always been friends and you know in your mind you are going to be friends in a few seconds anyway. (as quoted in Selman, 1980, p. 112)

The occurrence of puberty adds an element to early adolescent peer groups that is absent in those of middle childhood. There is an increased interest in sexuality. Sexual topics are discussed more frequently, sexual activity is often prompted and supported by the peer group, and one's status in the peer group is increasingly a reflection of one's "professed" sexual knowledge and experience. This early sexuality takes very different forms in male and female adolescent groups.

In this period, the male and female are being trained in essentially opposite modalities, though there is more interconnection between boys and girls than in other social locations. The central themes for females are commitment to affect-laden relationships and the rhetoric of romantic love. It is not so much that their sexual development is inhibited, though it is policed, but that it exists as a vacuity and can only be expressed as an absence. The male themes are more complex. There is a growing and developing commitment to sexual acts, in part for themselves, in part as acts as they relate to other main patterns in male gender role training: autonomy, aggression, control, achievement, normative transgression. What most males, except for the middle class, are not learning are socioemotional skills that are directed toward females. Both through masturbation and heterosexual contact males can be committed to directly sexual activities, and in some part, to gratification from them; females, even when they report sexual arousal, do so in the haze of romantic as opposed to erotic illusions. (Gagnon, 1972, p. 249)

The young adolescent male defines his sexuality from a present perspective (Stefanko, 1984). Most male adolescents masturbate and report the experience as pleasurable (Gagnon, 1972). Their early heterosexual encounters are usually limited to kissing and petting above the waist (Diepold & Young, 1979). The exploitive character of these early sexual adventures is evident in the fact that they are often encouraged by the male peer group and the participant often reports no special regard for the girl (Gallatin, 1975).

Young adolescent females are more likely to associate sexuality with love and marriage, and therefore they are less likely to consider sexuality only from a present perspective. Fewer adolescent females masturbate and those who are involved in kissing and petting activities often report that they felt their partner was simply using them to gain status with his peer group (Chilman, 1979; Gallatin, 1975; Konopka, 1976). In general, sexual activity is less often viewed as a status symbol in most early adolescent female peer groups.

Late Adolescent Peer Groups

By the time adolescents are fifteen or sixteen, peer-group patterns begin to undergo a significant shift. The solidarity of the large peer group begins to decline. Perhaps as a result of having resolved the uncertainties of puberty and feeling some initial resolution of the identity conflict, adolescents feel less need for peer-group support. The group is still valued, but it is seen as serving different purposes by different members. Its cohesiveness decreases, and participation in it becomes increasingly casual (Matteson, 1975; Selman, 1980).

The value placed on friendship shows a similar decline in late adolescence. Perhaps as a reaction to the overdependence typical of early adolescent friendships, older adolescents begin to view close friendships as possible inhibitors of autonomous growth and development (Selman, 1980). The decline in the relative importance of same-sex friendships occurs at the same time as the increase in male-female friendships. The older adolescent seems unable to integrate the two friendship patterns and as a result tends to forsake one for the other. (This problem is not unique to adolescence. It continues to be a major problem during many of the adult years.)

During late adolescence, male-female friendships tend to replace the same-sex friendships common through childhood.

Sexual activity is an integral part of late adolescent peer relations. By age eighteen, approximately 60 to 70 percent of male and female adolescents report having had sexual intercourse at least once (Diepold & Young, 1979). Sexual intercourse is no longer viewed as morally or personally wrong or sinful. On the contrary, two-thirds of all adolescent males and females say that sexual activity (not necessarily intercourse) makes life more fun (Walters & Walters, 1980). Relatively few adolescents place great importance on marrying a virgin (Cvetkovich, Grote, Lieberman, & Miller, 1978). This shift in attitude is as evident in those who are not sexually active as in those who are. Premarital sexual activity is now generally accepted by adolescents if it is associated with a close personal relationship. Promiscuous and exploitive sexual activity is still disapproved of, however. The major concern is that the sexual activity take place within the broader context of a close interpersonal relationship. Perhaps the most significant result of this change in attitude is that the percentage of adolescent females who are nonvirgins is now approximately equal to the percentage of nonvirgin males (which has not changed significantly in the past forty years).

Research indicates that more late adolescent and young college-age women are now reporting their first sexual experience as pleasurable and that more are reporting orgasm during their first sexual experience (Matteson, 1975). All of these changes indicate a greater appreciation in both partners of the importance of the interpersonal context of sexual activity and a decrease in the view of sexual activity as a status symbol. At the same time it is important to note that those high school females engaging in relatively early coitus (at age fifteen or sixteen) are more likely than other females of their age to hold rather traditional sex-role values. The evidence suggests that they become sexually active at this time because they feel that boys expect them to and that if they don't they might lose their boyfriend (Chilman, 1979).

The transition in sexual activity patterns is occurring as part of a more general loosening of male-female interaction patterns during adolescence. This change is probably a reflection of the changes in adult relationships that have taken place over the past twenty to thirty years. In general these changes are characterized by a greater equality in male and female roles and a decrease in prescribed role behaviors. The change in adolescent behavior is perhaps most evident in the decline of formal dating patterns.

Dating traditionally had almost the quality of a well-defined ritual (McCabe, 1984). Although specific dating patterns differed in various segments of society, they always involved an acknowledged set of rules concerning what night a date was made, who made the date, where the date would take place, how each person would behave on the date, and how the couple's behavior would change toward each other as they continued to date. If Wednesday night was the night to call for a date, boys never called on Tuesday (they didn't want to appear too eager), preferably didn't have to call on Thursday (which would indicate that they didn't have any luck on Wednesday), would possibly be refused on Friday, and would almost certainly be refused on Saturday afternoon. Boys always called girls, picked them up at their home, and paid for the evening. Depending on the peer group, there were four or five places to go on a date. It was generally understood within the peer group what sexual advances could "properly" be made on what date. It was equally well understood that the male's role was to make the advances and the female's to decide when they had gone far enough. The following Monday was a busy day at school—everyone reported back to their friends about the events of the weekend. Tuesday was then used to digest the "intelligence reports" so that everyone would again be ready for Wednesday.

The main advantage to such a rigid pattern is its rigidity. It is clear and well defined, and if the participants adhere to its rules, their behavior falls well within the limits accepted by their peer group. The pattern's main advantage, however, is also its main disadvantage. It creates the same type of complementary relationship that exists between parent and child. Specifically, it limits the extent to which one member of the relationship can gain an understanding of the other member of the relationship. The developmental differences between the parent and the child explain why their relationship must have a complementary quality. No such parallel necessarily exists for male-female relations. Rather, the complementary nature of the relationship reflects culturally defined roles for men and women in our society. As the roles have become less rigid, so too have the interaction patterns between male and female adolescents.

Dating patterns are now becoming more reciprocal, although many of the old patterns are still evident. It is probably most accurate to say that attitudes concerning dating are changing faster than the actual practices (Rogers, 1981). Both males and females are increasingly assuming the same responsibilities and privileges formerly reserved exclusively for one or the other. For example, couples are more likely to share the expenses, especially if they have been dating for awhile. While first dates still seem fairly rigid, dating, in general, is becoming more spontaneous. Adolescents will gather at a favorite place and pair off as the evening goes on, or they may agree to meet somewhere rather than having the boy pick up the girl at her house. Sexual activity is more likely saved for a "special" person rather than for the second or third date, and a "live and let live" attitude is increasingly evident in dating patterns and more generally in peer relations. The consequence of this reciprocity is a greater appreciation for the perspective of the other person and, as a result, the possibility of a closer interpersonal relationship—the universal desire of most adolescents (Conger, 1972).

Like vocational choice, changing sex roles have different meanings for males and females. The dramatic increase in the sexual activity of adolescent females, as well as their increasing comfortableness in this new role, indicates a movement toward greater equality in relationships. Unfortunately, they often encounter males who don't share this sense of equality. Many adolescent males still feel that their loss of control and power in the dating situation is not compensated for by gains in reciprocity. Adolescent females are often forced to choose between relatively traditional dating patterns and not dating at all. Further, since adolescent activities are increasingly viewed from the perspective of the future, adolescent females are also forced to consider if the same bind awaits them when they marry and have a family. Again, they have less ability to form a clear picture of their future than adolescent males do. The fact that males' sense of clarity often proves an illusion is of little comfort to adolescent females and little concern to adolescent males.

The Adolescent and the School

Adolescents' school experiences are very different from those of middle childhood. The transition takes one of three paths, but the outcome is typically the same. In one path, children enter a four-year high school after completing eight years of elementary school. In a second, children experience grades seven and eight in a separate junior high; and in the third, grades seven, eight, and nine (or six, seven, and eight) in a middle school. In all instances, the class schedule becomes increasingly individualized. Different courses are taught by different instructors, and students gain greater freedom in the selection of courses. All students are required to take a minimum concentration in English, mathematics, science, and social studies. Beyond these courses, the electives chosen reflect individual interests and future career plans.

The school experience of early adolescents is often chaotic and minimally rewarding (Kagan, 1972). Teachers seem less sympathetic and concerned about individual problems than they were during elementary school. Moreover, the trials of puberty leave young adolescents self-conscious and vulnerable at the very time that the individualized class schedule makes them feel more evident to others. Adulthood still seems too far away to serve as an effective motivator for serious future planning or academic effort. Although approximately 25 percent of early adolescents seem highly motivated to get good grades, most seem more interested in the social aspects of adolescence and quickly learn to do what is necessary to survive with acceptable grades (Grambs, 1978).

The large, institutional nature of most secondary schools requires adolescents to cope with what are considered an unbelievable number of arbitrary, unnecessary, and inhumane rules, regulations, and procedures (Grambs, 1978). The problem of depersonalization has actually increased as secondary schools have continued to become larger. Although schools with more than five hundred students are better able than smaller schools are to offer specialized instruction, they have an adverse effect on marginally competent students, generally offer fewer opportunities for participation in extracurricular activities, are less likely to promote a sense of social responsibility among students, and are more prone to arbitrary regulations and procedures (Garbarino, 1980).

The sorting, selecting, and evaluating process that began in the primary schools continues during high school. In fact, it is one of the main functions of the high school. The top track is college bound. Academic excellence is valued because good grades are the primary basis for admission to good universities. Teachers enjoy working with college-bound students because they are usually highly motivated. Administrators value these students because they prove that the high school is doing its job.

Lower tracks are not college bound. Some students plan to attend vocational or technical programs after graduation, but many see high school as the end of their formal education. High grades are less valued because they are not a prerequisite for employment. Because these students are less grade conscious, they are less valued by most teachers and administrators. Because they are less valued, less attention is paid to them. What develops is a self-fulfilling prophecy. In fact, these students are not less motivated to learn, but rather see high grades as less relevant to their learning. Where high schools have instituted good vocational and technical programs and where high schools have developed work-study and apprentice type programs, the evidence suggests that these students do as well as any other students (Coleman, 1975; Rosenbaum, 1976).

The tracking system is not restricted to the academic aspects of high school; it also exists at the social level (Cohen, 1979). Social groups in the high school tend to reflect the academic tracking system. The academic group tends to be the highly motivated college-bound students. They have clearly defined career goals, are not typically involved in the extracurricular activities of the high school, and date less often than other students do. Although they are the best students, they are not highly valued by other students (Stefanko, 1984). They are viewed as working too hard and as being too motivated. The extracurricular group is also college bound, but it tends to be more present oriented than the academic group. This group is composed of athletes, elected class leaders, and those active in extracurricular activities. Whereas the academic group sees high school as an educational setting, the extracurricular group views it as a social setting. The extracurricular group's views of college reflect this same orientation. This group dates frequently and devotes a great deal of time to social activities.

The non-college-bound students form their own social group. It is usually separate from both of the college groups and is not defined in relation to the school. The social activities of this group may not differ from those of the extracurricular group except that they occur away from the school. There is also a small percentage of students who are loners. They don't appear to belong to any social group. They appear to have few friends and are rarely involved in extracurricular activities. They may or may not be college bound; they may or may not earn high grades. They are not easily characterized. Some appear to have resolved the issues of adolescence; others appear unaware of them. Some work, some hang around the house, and some are very involved in individual activities, projects, or hobbies.

Although high school does not exert as significant an impact on the adolescent as the family does, there is ample evidence (e.g., McDill & Rigsby, 1973; Rosenbaum, 1976; Rutter, Graham, Chadwick, & Yule, 1979) that the general climate of the school does exert an influence on the educational achievement and academic orientation of its students. A positive atmosphere results when teachers set high standards of achievement, show interest and involvement in student concerns, respect the school facility, prepare thoroughly for class, exhibit competence in class management, and grant responsibility to students. This positive atmosphere is further evident in the willingness of school administrators to acknowledge their faculty as competent and well motivated and to grant them a considerable measure of responsibility and self-direction. The beneficial effects of such a positive school environment are evident in such diverse indicators as high attendance, a low incidence of vandalism, high achievement test scores, and a high percentage of students aspiring to go to college or seek other forms of postsecondary education and training. Unfortunately, such supportive and motivating school environments are the exception rather than the rule.

The developmental events of adolescence mark the transition from childhood to adulthood. Although it would be hard to determine in some objective sense the relative significance of this transition in comparison with others in the life-span, it is probably safe to say that from the adolescent's perspective, this is the big one. Most twenty-year-olds describe themselves as adults (or at least young adults) and consider themselves light-years away not only from the eleven- or twelve-year-old but from the fifteen- and sixteen-year-old as well. They consider themselves grown-ups and wish to be treated as such. Perhaps the trials of the metamorphosis they have just completed prevent them from appreciating the fact that development is truly a lifelong process—one is probably never really "all grown up."

THE SIGNIFICANCE OF DEVELOPMENT DURING ADOLESCENCE

Adolescence marks the passage from childhood to adulthood. It is the period of the life-span during which the individual attains the physical, sexual, and physiological, as well as the legal, status of the adult. It marks the end of compulsory education and, more generally, of the time during which society views the individual as childlike and therefore in need of protection and special care.

In a sense, adolescence completes what can be thought of as the second phase of the life-span—the maturation phase. The first phase—the establishment phase—consists of the prenatal period. This first phase witnesses the differentiation of the various biological systems. The second phase, which begins at birth, witnesses the evolution of these biological systems from an immature to a mature state. The end of the second phase does not mark the completion of development. It only indicates that the nature of the developmental process after adolescence acquires a different flavor. In comparison with earlier stages of development, adult development is much less a reflection of the biological maturation of the individual and much more a reflection of the cultural demands of the society in which the individual lives. The adult, or third, phase of the life-span continues until the individual experiences the primary decline of the biological systems that accompanies advanced age. This decline characterizes the fourth phase of the life-span.

Although this four-phase system is useful in gaining a broad conceptualization of the life-span, it is important to bear in mind that culture is certainly a major influence on development during childhood and adolescence, and that the status of the biological systems remains a major influence on development after adolescence. The four-phase system primarily serves to emphasize the fact that the nature of the roles of culture and biology changes as development proceeds.

Patterns of Development

Throughout the entire life-span, individuals remain active participants in directing the course of their own development. In adolescence, the nature of this active involvement takes on a new quality. For the first time, individuals become introspective. As a result, they become much more effective as information processors. They gradually become better able to delay a response, to sift and sort through information, to evaluate information from a number of perspectives, and to consider both the potential long- and short-term consequences of the possible alternatives before acting. As a result, it may be more difficult to predict an adolescent's behavior from knowledge of the immediate situation than it is to predict the behavior of a child. And in case this fact is not evident to an adult, most adolescents are quite eager and willing to point it out. This increased capacity for self-reflection also explains a good deal of the desire for privacy first noticed in early adolescents ("I want to be alone with myself") and the anxieties and uncertainties that can accompany late adolescence as well as many of the adult years ("how can I be sure that I have made the right decision?").

These self-reflective skills make possible the emergence of a true self-concept. The evolution of the self begins during infancy. The differentiation of self from other and the awareness that the world exists independently of oneself is the first step. The second step occurs during early and middle childhood. It involves the perception that the self transcends time and space and specific circumstance. It involves such realizations as that a change of clothes doesn't change the person, that two people can be angry with each other and still be friends, and that even if you are some day taller than your mother, you will still be her "little" girl or boy.

With the emergence of self-reflection, the adolescent is able to ponder the significance of an identity, conceptualize many alternative selves, and decide which "self" to make public. In other words, the development of the self-concept in adolescence involves the differentiation of a public from a private self (Breger, 1974). It involves the ability to portray the self as the adolescent would like to be seen by others while at the same time "protecting" the private self from public scrutiny. What makes the close friendships of adolescence so special is that they are relationships in which the private self can be shown. The differentiation of the private self from the public self remains a major component of development throughout the life-span as does the search for safe relationships in which to share the private self. What becomes added in adulthood is a desire to integrate the private and public selves—a desire to have others see you as you see yourself.

Adolescence is unique not only in its emphasis on the emergence of a true self-concept but also because the issues of trust and autonomy are central to the adolescent's sense of identity. Neither trust nor autonomy are major issues during middle childhood. On the contrary, the school-age child seems quite eager to identify with society. But for the adolescent, the path to identity is independence—specifically from parents, more generally from the adult culture. It does not matter that this search for independence often takes the form initially of total conformity to a peer culture or that many of the values the adolescent adopts bear a striking resemblance to the parents'. What is important is the adolescent's perception that these values have been achieved independently of the parents.

For adolescents, the quest for autonomy is also a quest for trust. They are asking parents to trust that they will be able to make intelligent decisions on their own. The ability of parents to grant adolescents this trust is partly a reflection of how much trust they have in their parenting skills. Parents who feel they have done a good job raising their children are more likely to grant requests for autonomy than are those who do not. This doesn't mean that these parents will have no trouble going to sleep on a Saturday night when their seventeen-year-olds are out with the car. But it does mean that they are fairly certain that their adolescents will call if they are likely to be late, and the seventeen-year-olds, in turn, know that if a request to stay out later is made, it will probably be granted. In other words, the ease with which independence is granted the adolescent reflects not only the nature of the request or the manner in which it is made but the entire history of the parent-child relationship.

Adolescents want trust not only from their parents but also from other adults and peers. (Perhaps the most famous two words spoken during a date are "trust me.") Adolescents are asking others for the chance to define and abide by their own limits. The ability to know which way to go, how far to go, and how to stop when you get there serves as the basis for defining the boundaries of the self. This boundary testing is not unique to adolescents. A parallel pattern of behavior is typical of toddlers. In this sense the teenage years could be described as the second period of adolescence. In both instances, individuals seek to separate or differentiate themselves from others. And in both instances, this differentiation serves as a necessary prerequisite for the integration that follows.

Given the nature of adolescent development, it is not surprising that when compared to middle childhood, adolescence is a stage of relative disequilibrium (Larson, Csikszentmihalyi, & Graef, 1980). The rapid changes in the biological and cognitive domains, as well as the social changes that result, all contribute to a major shift in the way adolescents view themselves and others. Much of the behavior of adolescence can be thought of as an attempt to reestablish a sense of both personal and interpersonal equilibrium. The emotionality of adolescents and the relatively rapid shifts of mood, interest, or goals suggest that in tone adolescence more closely resembles early childhood than middle childhood.

Stage development is cumulative. One stage is not simply replaced by another. Rather, the accomplishments of one stage form the basis for the developments of the next. The concrete operational skills of middle childhood, for example, are not simply replaced by the formal operational skills of adolescence; these concrete operations are now viewed as only one instance of the total range of possibilities. In a similar fashion, the basic issue of adolescence is not resolved or forgotten with the advent of the adult years. The question of personal identity continues to be a major influence on the course of development throughout the entire life-span. It is reflected in the attempts of adults to establish and maintain a personal value system, a sense of integration among the various domains of their lives, and a feeling of trust and autonomy in their dealings with others. The developmental tasks of the adult years are certainly different from those of adolescence, and the wisdom that is acquired through experience provides adults a different perspective from which to view themselves, but the basic issue of personal identity continues to be a major factor in directing the course of development.

The dominant characteristic of the developmental tasks of adulthood is a focus on others. Adults are expected to make a contribution to society. They are expected to be productive, to defer personal gratification, and to tend to the needs of others. The roles of parent and worker encompass all of these qualities. Adolescence, on the other hand, is characterized by a focus on self. For most, adolescence is a time of psychological moratorium, a time when the adolescent is free to search for a sense of identity, free from the responsibilities that characterize the adult word. Within our culture, adolescence thus becomes a period of personal indulgence. In this characterization, adolescence and adulthood form the end points of a continuum. In the extreme, adolescence is a focus on self; adulthood, a focus on other. The adult in our society is continually faced with maintaining an acceptable balance between these two extremes. Each of the events typical of the adult years (such as marriage, parenting, work, retirement) requires that the balance be reestablished. Each new point of equilibrium redefines the sense of personal identity. In this sense, adolescence never ends. Rather, the developmental tasks of adolescence become integrated into the developmental tasks of adulthood.

HUMAN DEVELOPMENT AND HUMAN SERVICES

Providing services to adolescents is often a difficult task. In contrast to school-age children who consider cooperation with adults as a sign of being grown-up, adolescents tend to view maturity as independence from adults. As a result, adolescents often view the provision of direct services as intrusive and the acceptance of these services as a sign of dependence (Rutter, 1980). A second complicating factor is that many of the problems adolescents encounter are essentially the same problems that adults encounter (Gallatin, 1975). Substance abuse, crime, reckless driving, and sex-

With the increase in teen-age pregnancies, more attention needs to be paid to providing direct services that enable teen-age mothers to care for their children while continuing their own education and development.

ual promiscuity are all serious circumstances for both adolescents and adults. In other words, the problems of adolescence anticipate the problems of adults; and since we, as a society, have had limited success in dealing with these problems in our adult population, there is no reason to think that we would necessarily have greater success in an adolescent population.

A third contributing factor is that adolescents are neither fish nor fowl. They are not children, but they are not adults. They are like each but at the same time unlike each. This ambiguity is evident in the way our institutions treat adolescents. We raise and lower the drinking age as if changing their legal status will solve the problem of teenage alcoholism. We lower the age at which an adolescent may be tried as an adult whenever a fourteen- or fifteen-year-old commits a particularly heinous crime and raise it whenever a study comes out reporting on the abuse that adolescents suffer when incarcerated with adult offenders. We expel adolescents by the score from high schools when we decide that a get-tough policy is the way to treat vandalism, truancy, and disruption, and then we invite these same "incorrigibles" to participate in a faculty-student forum when active participation is seen as the answer. We tell adolescents that we will treat them as adults when they act like adults. They of course respond that they will act like adults when they are treated like adults.

Suggesting solutions to the problems of adolescent substance abuse, unplanned pregnancy, delinquency and crime, school dropout rates, and poor educational achievement is beyond the scope of this text. However, an understanding of developmental theory does provide one insight into the origin of these problems. Adolescence is probably the most age-segregated stage of the life-span. It is the period of the life-span in which our dealings with individuals seem most inconsistent with the gradual, continuous nature of the process of development. Adolescents live in a world that is separate from both childhood and adulthood. This world has its own subculture and its own frame of reference. There is a very strong positive relationship between the extent to which adolescents are immersed in this subculture and the likelihood of their experiencing some or all of the major problems typical of adolescence (Bronfenbrenner, 1970; Coleman, 1975; Rutter, 1980). According to Coleman,

> segregation and the shift of opportunities to youth have grown out of some of the most basic forces of 20th Century society. Family values no longer sanction, as they once did, the early commitment of youth to productive activity. Trade unions and professional organizations, for understandable reasons, fear large scale incursions by

> youth into their labor markets. Humanitarian sentiment opposes the exploitation of youth. Professionalization and bureaucratization have sharply narrowed the range of youth's contacts with adults outside leisure pursuits. The forces that have isolated young people and cut off certain options once available to them have not, thus, been necessarily mean or reactionary. Paradoxically, they have been, at least in original intent, enlightened and altruistic. Yet it seems equally clear that we have reached a point in history in which these forces are spinning out of control. What was once done to protect youth from manifest exploitation now serves to reinforce the "outsider" status of youth. Ideas and institutions that once served explicit and genuine needs, and in some cases, still do, have uncritically been extended to the point where they deprive youth of experience important to their growth and development. (Coleman, 1975, p. 130)

For Coleman, the secondary school is the prime example of an institution whose initially good intentions have spun out of control. The creation of middle schools or junior high schools has isolated young adolescents from older adolescents as well as from younger children. The trend toward large, comprehensive high schools has created impersonal bureaucratic settings—settings in which students feel little sense of belonging and in which the opportunities for personal contacts between students and teachers are limited. In an effort to solve some of these problems, the President's Advisory Panel, which Coleman chaired, recommended greater diversity in secondary education, a recommendation that continues to have merit today (Garbarino, 1980; Stefanko, 1984). Among the possible alternatives would be to encourage the development of smaller schools in an effort to foster greater teacher-student contact, a sense of belonging in the student body, and a greater opportunity for individual attention. This could be done in part by clustering a number of relatively autonomous school units on the same campus. In this way, access to special resources such as computers, athletic facilities, and laboratories could be provided while at the same time maintaining the relative intimacy of a small administrative and academic unit.

A second set of recommendations concerns the relationship between the school and the workplace. The transition between the two needs to be made less abrupt. A greater emphasis needs to be placed on career and vocational education. Students need to be made more aware of the vast range of occupations in our society. A greater emphasis also needs to be placed on individual vocational counseling. Yet neither of these efforts will prove worthwhile unless various sorts of work-study and flexible scheduling programs are developed to give students actual experience in the workplace (Steinberg, Greenberger, Jacobi, & Garduque, 1981). In addition to providing the adolescent valuable career guidance, these programs would reduce the segregation of adolescents from the adult community and, in the process, expand their perspectives and range of potential influences. Community service programs, patterned after the Peace Corps but not requiring a change of residence, could also be used to increase adolescents' involvement in the community and contact with adults. A dual minimum wage system (adolescents receiving less than adult workers) would also increase the likelihood of adolescents being incorporated into some aspects of the adult work force.

None of these recommendations is without its drawbacks. Increasing the percentage of adolescents in the work force may deprive adults of jobs. Work-study programs require industry to assume the educational role now associated with the public school systems. Spending more time in the workplace and less time in the classroom may hinder the development of abstract thought, a valuable skill in adulthood (Stipek, 1981).

Smaller schools may be viewed as running counter to the trend toward specialization within an academic discipline that has characterized the training of secondary school teachers over the past few decades. However, the problems that confront adolescents will not be resolved until the adult community acknowledges that these problems are often a by-product of the social and technological changes in the adult community. Those interventions that seem most promising are ecological in nature; that is, they deal with systems rather than individuals (Bronfenbrenner, 1979).

If age segregation is a primary contributor to the problems of adolescence, then one essential human service is the removal of age barriers—barriers that exclude adolescents from the workplace, that exclude them from necessary services such as family planning and sex education, and that exclude them from acquiring the sense of social responsibility they need to function effectively in the adult world. Broad-scale indirect ecological interventions seem the ones with the greatest potential for dealing with the problems of adolescents, but they are also the ones that will require the greatest cooperative effort among adults and between adolescents and adults.

At the close of this chapter, it is important to bear in mind that most adolescents, most of the time, enjoy their adolescent years. It is certainly true that adolescence involves a greater emotional upheaval than middle childhood, and a greater potential for serious problems. However, it is also true that adolescents possess greater social, cognitive, and physical capacities to deal with the period's demands.

SUMMARY

THE COURSE OF DEVELOPMENT

1. Puberty refers to the biological process of achieving physical and sexual maturity; adolescence, to a set of social circumstances typical of affluent, industrialized societies.
2. Adolescence represents the initial period of conscious identity formation.
3. Puberty follows a common sequence in both males and females. The rate and age at onset are highly variable, however.
4. Age at onset of puberty in males tends to have a relatively enduring impact through early adulthood.
5. The emergence of formal operational thought provides adolescents with the skills to deal with the abstract as well as the real in a systematic hypothetical-deductive fashion.
6. The discovery of the self in adolescence prompts a temporary increase in egocentric thought.
7. The single defining element of the self-concept in adolescence is a quest for personal integration.

THE CONTEXT OF DEVELOPMENT

1. Most adolescents share their parents' basic goals and values; they tend to differ most on more immediate, more peer-oriented topics such as dress and music.
2. Male-female sex-role differentiation becomes increasingly apparent during adolescence, especially in the areas of vocational planning and interpersonal relationships.
3. Peer-group patterns change over the adolescent period, from large same-sex groupings to more personal same-sex friendship pairs and more heterosexual groupings.

4. Adolescent's school experiences increasingly reflect evaluation and tracking for future educational and vocational goals.

THE SIGNIFICANCE OF DEVELOPMENT DURING ADOLESCENCE

1. The process of identity formation during adolescence is as much a search for trust as a search for autonomy.

HUMAN DEVELOPMENT AND HUMAN SERVICES

1. Providing services to adolescents is often a difficult task. They are less responsive to adult initiatives than school-age children are, and their problems are often those that adults are either unable or unwilling to contend with.

KEY TERMS AND CONCEPTS

PUBERTY

Puberty
Adolescence
Menarche
Erikson's Psychosocial Stage of Identity Formation
  Psychosocial Moratorium
  Role Experimentation
  Identity Diffusion
  Identity Foreclosure

COGNITION

Piaget's Stage of Formal Operational Thought
  Hypothetical-Deductive Reasoning
  Introspection

SUGGESTED READINGS

Three readings offer a clear understanding of the social forces that have created our modern conception of adolescence and the factors that maintain it.

Kagan, J. (1972). A conception of early adolescence. In J. Kagan & R. Coles (Eds.), *Twelve to Sixteen: Early adolescence.* New York: Norton.

Kett, J. F. (1977). *Rites of passage.* New York: Basic Books.

Rutter, M., Graham, P., Chadwick, O. F. D., & Yule, W. (1976). Adolescent turmoil: Fact or fiction? *Journal of Child Psychology and Psychiatry, 17,* 35–56.

The longitudinal research on the later development of early and late maturers is well summarized in these articles by Jones and Peshkin.

Jones, M. C. (1965). Psychological correlates of somatic development. *Child Development, 36,* 899–911.

Peshkin, H. (1967). Pubertal onset and ego functioning. *Journal of Abnormal Psychology, 72,* 1–15.

Did you ever wonder how researchers find out that adolescents spend so much of their time alone in their rooms if they in fact spend so much time alone in their rooms? This article provides one answer.

Larson, R., Csikszentmihalyi, M., & Graef, R. (1980). Mood variability and psychological adjustment of adolescents. *Journal of Youth and Adolescence, 9,* 469–491.

One of the best examples of the difficulties adults encounter in providing advice or even information to adolescents concerns the infrequent use of contraceptives among sexually active adolescents. Four articles offer some insight as to why this has been and continues to be a serious problem.

Chilman, C. S. (1979). *Adolescent sexuality in a changing American society.* DHEW Publication No. (NIH) 79–1426.

Cvetkovich, G., Grote, B., Lieberman, E. J., & Miller, W. (1978). Sex role development and teenage fertility related behavior. *Adolescence, 13,* 231–236.

Hornick, J. P., Doran, L., & Crawford, S. H. (1979). Premarital contraceptive usage among male and female adolescents. *Family Coordinator, 28,* 181–190.

Zelnick, M. & Kanter, J. F. (1977). Adolescent sexual behavior. *Family Planning Perspective, 9,* 55–73.

7
THE TRANSITION TO ADULTHOOD: AGES EIGHTEEN TO TWENTY-FIVE

CHAPTER OUTLINE

A LIFE-SPAN CONCEPTUALIZATION OF DEVELOPMENT DURING THE ADULT YEARS

Before considering the developmental events that occur to young adults between the ages of eighteen and twenty-five, I want to raise a more general issue. Specifically, to what extent is the nature of the developmental events of the adult years similar to the nature of the developmental events of the stages already presented? My reason for asking the question at this point in the text should be self-evident; the need to ask it at all may be less so. Two considerations lead me to ask the question.

The first concerns those who study the process of development. In the preface to this text, I mentioned that the increasing collaboration between scholars interested in different segments of the life-span is making possible, perhaps for the first time, an understanding of the factors that influence the course of development over as grand a scale as an entire life-span. As true as this statement is, it is also true that much developmental research has been and continues to be done from a narrower perspective. The study of development through adolescence has traditionally been the domain of the more psychologically oriented members of the social sciences. Therefore, emphasis has been placed on the study of individual development and the delineation of the underlying processes (both internal and external) that regulate individual behavior. Development during the adult years, on the other hand, has been the interest of the more sociologically oriented members of the social sciences. Consequently, interest has focused less on individual development and more on the nature of interpersonal relationships within social settings. Although the differences between the two groups are, to some extent, merely semantic—the psychologists study "parent-child relations" while the sociologists study "family dynamics"—they nevertheless create a chasm that needs to be bridged if the process of development from the broader perspective of the life-span is to be fully understood.

The second consideration is a logical outgrowth of the first. It concerns the definition of development itself. Consider the following quote from Bower (1979), a well-respected scholar primarily interested in infant development.

> To be sure, there are changes after twenty or so, but they are not changes resulting from gene expression, nor are they age-linked, nor, indeed, are they at all universal (with the exception of death). Anyway, any such changes are, I feel, at the most, the differentiated end of psychological functioning. As such, they are continuations of the abstract accomplishments of the growth phase of development. (p. 432)

Bower's argument reflects the view that developmental transitions through puberty are associated with significant changes, or maturations, in the biological structure and functioning of the individual and that (with the exception of death) these changes are not as apparent during the adult years. Bower is not suggesting that people don't change after the age of twenty, only that the changes are no longer developmental.

Now consider a second quote, this one from Neugarten (1968), an equally well-respected scholar primarily interested in adult development.

> What then are the salient issues of adulthood? At one level of generality it might be said that they are issues which relate to the individual's use of experience; his structuring of the social world in which he lives; his perception of time; the ways in which he deals with the major issues of love, work, time and death; the changes in self-concept and changes in identity as he faces the successive contingencies of marriage, parenthood, career advancement and decline, retirement, widowhood, illness and personal death. (p. 139)

Neugarten's definition of development is broader than Bower's. Rather than emphasizing the behavioral correlates of biological structure and functioning, she emphasizes the behavioral correlates of the social world. In fact, Neugarten's concept of development easily encompasses many of the developmental issues discussed in previous chapters. Bower and Neugarten are not contradicting each other so much as they each are emphasizing different factors influencing the process of development. From a life-span perspective, development is a lifelong process, influenced by both biological and social factors at all points. Even though the degree of difference between the twenty-year-old and the fifty-year-old may not be as great as that between the one-year-old and the twenty-year-old, the differences are nevertheless very evident. But as Neugarten and Bower are each suggesting, the nature of the interplay between biological and social factors during the adult years may be different from what it was at earlier times. What then is this shift in the interplay of biological and social factors?

Child Development—Adult Development

As already indicated, perhaps the most significant difference between adult development and child development involves the fact that adults are more directly affected by the **sociohistorical context** of their lives than children are (Riegel, 1972; 1975). That is, adults are more directly influenced by the economic, social, and political conditions of their world. This does not mean that children are unaffected by these events, but rather that children's experience of these events is filtered through, buffered by, and represented by adults.

Because adults are more directly affected by the sociohistorical context of their lives, they are more likely to consider this context when making judgments about their development. Specifically, judgments about the rate and course of development increasingly reflect the **social clock** of the culture rather than the **biological clock** of the species (Neugarten, Moore, & Lowe, 1965; Neugarten & Datan, 1973). Table 7-1, based on research by Neugarten, Moore, and Lowe (1965), illustrates the pervasiveness of these culturally based social clocks. What is remarkable about the data is that, although the age range for each category could be considerably different from what is given, a very substantial number of both men and women agree as to the appropriateness of these time intervals for the various life events. Norms such as these serve as the criteria individuals use to decide if their behavior is "on course." They help individuals decide how to act and provide expectations as to how other people will respond to them. In general, they provide a guide to the appropriateness of behavior.

The guidance provided by the social clocks of adulthood is considerably less precise than the guidance provided by the biological ones of childhood. As a result, development is considerably more individualistic and considerably less age specific during adulthood than it was during the previous stages (Whitbourne & Weinstock, 1979). Put another way, chronological age is a significantly more useful piece of information for predicting the behavior of a two-year-old or a seven-year-old or even a twelve-year-old than it is for predicting the behavior of a twenty-eight-year-old or a forty-two-year-old or a sixty-seven-year-old.

A second distinction between adulthood and the years that precede it concerns the role of experience in regulating behavior. During adulthood, past experience assumes an increasingly important role in an individual's decision-making process. For this reason, the logic inherent in concrete and formal operations becomes integrated with life experiences to form a more comprehensive basis for decision making—ideally resulting in the individual's developing a degree of wisdom (Clayton & Birren, 1980). In

TABLE 7-1 Consensus in a Middle-Class Middle-Aged Sample Regarding Various Age-Related Characteristics

	AGE RANGE DESIGNATED AS APPROPRIATE OR EXPECTED	*PERCENT WHO CONCUR*	
		MEN (N-50)	*WOMEN* (N-43)
Best age for a man to marry	20–25	80	90
Best age for a woman to marry	19–24	85	90
When most people should become grandparents	45–50	84	79
Best age for most people to finish school and go to work	20–22	86	82
When most men should be settled on a career	24–26	74	64
When most men hold their top jobs	45–50	71	58
When most people should be ready to retire	60–85	83	86
A young man	18–22	84	83
A middle-aged man	40–50	86	75
An old man	65–75	75	57
A young woman	18–24	89	88
A middle aged woman	40–50	87	77
An old woman	60–75	83	87
When a man has the most responsibilities	35–50	79	75
When a man accomplishes most	40–50	82	71
The prime of life for a man	35–50	86	80
When a woman has the most responsibilities	25–40	93	91
When a woman accomplishes the most	30–45	94	92
A good-looking woman	20–35	92	82

Note: Reprinted, by permission of The University of Chicago Press, from B.L. Neugarten, J.W. Moore, & J.C. Lowe, "Age Norms, Age Constraints, and Adult Socialization," *American Journal of Sociology, 70,* (1965), 712.

other words, the formal operational skills first evident during adolescence become integrated into other actions and become a means rather than an end in themselves. More generally, this change means that adults are able to assume a more conscious and deliberate role in directing the course of their development than children or adolescents are.

As a result, evaluations of developmental status become increasingly internal and relative. They reflect not so much the actual status or accomplishments of any individual at any particular point in the life-span but, rather, the way individuals see their present status in relationship to some idealized future (Campbell, 1981; Levinson, 1980). This relative perspective helps explain why it is not uncommon to find adults who have accomplished much in their lifetimes and have received considerable rewards for these accomplishments yet still feel dissatisfied or even unsuccessful.

The Three Careers of the Adult Years

Because the factors regulating development during adulthood are somewhat distinct from those regulating childhood and adolescence, a different conceptualization (or, more to the point, organization) is needed in the chapters that follow. This section elaborates this conceptualization. It is based on the three developmental careers that most adults in our culture encounter during their adult years—the careers of work, intimacy, and parenthood. The timing of the transitions within each career affects the other careers as well. Depending on the social ecology of the culture, it can either hinder or facilitate development and, in the process, the acquiring of a sense of satisfaction with life and an integrated sense of personal identity.

Virtually all adults in all cultures, at one time or another, find themselves involved in these three careers. Within our culture, over 95 percent marry. Although marriage is not the only means of establishing a sense of intimacy, it is clearly the most typical. Of those that are biologically able, over 90 percent conceive at least one child. Many of those who are not able to conceive children adopt them. Although it sometimes appears that an increasing number of adults have decided to forgo marriage or parenthood or both, the evidence indicates that what is actually happening is that more people are marrying later and having their first child later (Hoffman, 1977). Likewise, virtually all of us work. The nature of the work varies greatly as do the working conditions, the level of prerequisite training, and the nature of the intrinsic and extrinsic rewards, but at one time or another during our adult years, we work in or outside the home or in both places.

One way to gain some insight into the three careers of adulthood is to consider them in relationship to Erikson's (1980) psychosocial stages. Although Erikson associated each stage of development with a particular psychosocial task, he also suggested that, in one form or another, all of the tasks are present at all stages of the life-span (Erikson, 1980). This observation appears to be particularly true during the adult years. The desire of almost all adults to feel a sense of "wholeness" within themselves and within their environments represents the continuing need to establish a sense of identity. The three adult careers of work, intimacy, and parenthood are associated, respectively, with the psychosocial tasks of industry, intimacy, and generativity. All three maintain their importance throughout the adult years. Similarly, although the notion of satisfaction with one's life is the defining characteristic of Erikson's last psychosocial task, acquiring a sense of ego integrity, satisfaction begins to be important well before old age. The tasks of industry and identity have been presented in earlier chapters and, at this point, need no further discussion. However, the other three do need an introduction.

Mastering the **psychosocial stage of intimacy** is, for Erikson, the primary task of young adulthood. A sense of intimacy presupposes a healthy initial resolution of the adolescent's identity question. Intimacy involves a fusing of identities, both sexual and nonsexual. Although most adults associate intimacy with sexuality and with marriage, the pairings are not necessarily inevitable. Rather, the associations are to a large degree a reflection of our sociohistorical context. According to Erikson (1968), sexuality is only one facet of intimacy. In the broader sense, intimacy involves a concern for others that equals the concern for self. The failure to acquire a sense of intimacy leads to a sense of isolation. Erikson implies that individuals whose sense of identity is "shaky" are particularly vulnerable to isolation since their focus remains on defining the ways in which they are unique.

The **psychosocial stage of generativity** is expressed in a concern for the next generation. Parenting is, of course, the most common form of this expression, but it need not be the only one. In fact, Erikson notes that merely conceiving a child is no indication of having achieved a sense of generativity. Rather, it is in the desire to care for the child, to truly parent the child, that generativity is achieved. Any activity that is motivated by this concern for the next generation is an expression of generativity. Erikson sees generativity as being valuable, not only in terms of maintaining

According to Erikson, developing a sense of intimacy is the most important task of young adulthood.

the species, but also in terms of helping adults expand their perspectives beyond themselves (the task of intimacy also performs this function) and their own time. Such an expansion of perspective helps adults better appreciate the meaning of the events in their lives. An inability to foster a sense of generativity results in a sense of stagnation.

The **psychosocial stage of ego integrity** is associated with advanced age. It is primarily a task of accepting the past as the past. For Erikson, the extent to which individuals can accept their lives as they have lived them, that is, the extent to which they can accept the past as the past, is an important component of their ability to accept death as the inevitable end of life. A failure to achieve a sense of ego integrity leads to a sense of despair—a feeling that there is no more time to try and undo or change the past. What follow, Erikson believes, are feelings of disgust and displeasure with particular institutions and people and, ultimately, with the self.

The psychosocial tasks of intimacy and generativity emphasize a focus on others. They reflect one side of what is perhaps the central dilemma of the adult years—maintaining a balance between meeting one's own needs and meeting those of others (Elder & Rockwell, 1978; Gilligan, 1979). The two are not necessarily antithetical; parents, for example, often take great personal pride in the accomplishments of their children. But many adults tend to see self and other as being in conflict. Those considering marriage often fear "being tied down." Parents of young children frequently report "never being able to find time for themselves." Attempts to integrate the adult careers of intimacy and work produce similar conflicts. Each stage of adulthood presents a unique relationship among intimacy, work, and parenthood. Rather than reaching resolution one at a time and then being replaced by another task, the tasks of industry, intimacy, and generativity maintain a dynamic equilibrium across all the adult years.

Viewing old age as the time of looking back on one's life, of being "out of time" and acknowledging the necessity of accepting the past as the past, is certainly an important task of old age. In a broader sense, however, the process occurs throughout the adult years. It is only in fairy tales and the minds of children that a magic wand is waved and the past undone. The rest of us can perhaps compensate or make amends for the past, but we can't undo it. The drunk driver who kills a small child may spend the rest of his or her life doing volunteer work with children as a personal penance, but no number of hours of good deeds will bring the child back. The ability to acknowledge and accept this fact, which develops through the adult years, suggests that a sense of ego integrity is not exclusive to old age but, more generally, may serve throughout adulthood as the fulcrum that progressively enables us to balance the three careers.

Out of this "simmering psychosocial stew" emerge the data that adults use to define and redefine their sense of personal identity. These data provide them with both a factual statement of their position in terms of the tasks of industry, intimacy, and generativity and an evaluative statement that indicates how well they are accomplishing their life goals.

The Course of the Three Careers of the Adult Years

The last few years have witnessed an increased interest in documenting the sequences of events typical of the careers of work, marriage, and parenting. One of the major motivating factors behind this resurgence of interest is the continuing debate over the degree to which the patterns of adult development for men and women are similar. More specifically, is it possible to document a sequence of developmental events that is as representative of the lives of men as it is of the lives of women? Where differences are found, are they best understood as relatively superficial in importance, or are they more fundamental in nature, reflecting basic differences in the ways adult men and women attribute meaning to the events in their lives?

This section describes the findings of four researchers—Levinson, Gould, Frieze, and Gilligan—who represent a range of opinion on these questions. Since their findings are discussed throughout the next several chapters, only an overview of their work is presented here.

The Work of Levinson

Levinson and his colleagues (1978; Levinson, Darrow, Klein, Levinson, & McKee, 1976) documented five relatively distinct phases in the cross-sectional sample of forty men, aged thirty-five to forty-five, they studied. Their data were collected through in-depth interviews with each participant. Levinson restricted his sample to men because of the amount of information gathered from each participant. He felt that adding an equal sample of women would be beyond the capacities of his research group. In reporting his findings, he stated that even though women were not included in his sample, the findings were equally representative of their adult development—a conclusion much debated.

Phase I: Leaving the family. Levinson refers to the first stage, which marks the beginning of the transition to adulthood, as "leaving the family." According to Levinson, the period begins during the late teens and lasts until the early twenties, and during this time young adults first begin to separate themselves, both physically and psychologically, from their parents. The transition is neither abrupt nor total. Rather, it is a milestone in the ongoing parent-child differentiation process. Levinson notes that college students living on campus are physically independent from their parents but often maintain a high degree of financial dependence. On the other hand, those who look for work after high school are often able to gain a degree of financial independence but continue to live at home in what could be described as a "boarder" status. In either case, Levinson thinks that during this interval young adults achieve a significant increase in autonomy in some, if not all, areas and, for the first time, begin to think of themselves as adults.

Phase II: Getting into the adult world. The twenties are described by Levinson as a time of "getting into the adult world." He reports that during this phase young adults make "provisional commitments" to occupations, people, roles, and responsibilities. They try to fashion a life structure consistent with their self-concept.

Levinson believes that individuals' life structures play pivotal roles in their adult development. Life structures are composed of roles, memberships, interests, and goals. These structures reflect an adult's values, identities, and even fantasies. The life structure is often couched in the form of a "dream," which, for Levinson's male sample, was almost always defined in terms of occupation. However realistic or unrealistic the dream proves to be (and not surprisingly, it usually proves the latter) it does provide young adults with a sense of direction and a set of goals. In so doing, it enables them to make the transition from being provisionally committed (keeping their options open) to being truly committed to a course of action.

Phase III: Settling down. Levinson suggests that during the third phase, adults, who are at this time entering their thirties, "settle down." A sense of order or stability, or in other words, a sense of commitment to making the dream a reality, replaces the provisional commitments of the previous stage. Goals such as to be promoted twice during the next three years, to become shop manager, to have an office with a window, and to earn a large salary by age thirty-five or forty become the types of personal incentives that fuel their commitment.

Levinson identifies two components in the life structure of the fully committed adults of this phase that he believes subsequently cause them difficulty. One concerns their relationship to their family; the other, their relationship to their peers. In the process of trying to establish a well-ordered life-style, the one most compatible with making their dream come true, Levinson's sample invested considerably less time and energy in their marital and family relationships than in their occupations. To help justify the focus on work, these men created a personal myth, which stated that time spent away from family members was justified as being time spent for family members. In Levinson's opinion, difficulty arises when the myth is found to be just that—a myth.

The other difficulty concerns autonomy, in particular, that idea that the end of childhood and adolescence marks the end of the need for guidance, support, and direction from others. These men thought of themselves as highly autonomous and self-directed. They tended to deny the influence that individuals and the structure of the work setting had on their behavior. These "tribal influences" as Levinson calls them were nevertheless present, and as these adult men in their early thirties began to realize that they had come to rely on others to provide the support once provided by parents, they again were faced with the task of establishing a true sense of autonomy.

Phase IV: Becoming one's own man. By the late thirties, Levinson's men were increasingly aware of the extent to which they were following the lead of others and consciously began to establish their own identity—a process Levinson refers to as BOOM (Becoming One's Own Man). This

During their early thirties most individuals begin to seek stability in their lives. Central to this settling down is a serious commitment to career advancement.

process often involved distancing themselves from a trusted and valued mentor and therefore was at times a painful process. After resolving this conflict, Levinson's men no longer thought of themselves as "apprentice adults" but instead as true peers of their coworkers.

Phase V: The mid-life transition. Most dreams do not come true, and these men's were no exception to the rule. As Levinson's men entered their forties, they began to realize that they were about as close to the window as they were ever going to get, that promotions and salary increases were becoming fewer and smaller, and that younger men were being promoted over them. Such realities prompted a mid-life transition—a time of taking stock, maneuvering midcourse, and attending to the parts of the self that the personal myths had allowed to be temporarily put aside. Levinson noted a wide range in the ease of the mid-life transition but also found that, once it was completed, most of the men seemed to feel more "comfortable" with themselves. A new life structure was created for the middle years—one that was broader and less driven.

Levinson's work is important because it emphasizes the continuing development of a personal identity throughout the adult years and because it demonstrates that developmental patterns continue to occur and reoccur at different levels across the life-span. In particular, there is a striking parallel in the fact that these men had to experience the same differentiation process with regard to their mentors that they experienced twenty years earlier with regard to their parents. Levinson's findings depict a sequence that is very linear and goal directed and defined primarily in terms of occupational growth. Notwithstanding Levinson's claim, it is a sequence that appears to be a more valid description of the course of adult male development than it is of adult female development (Frieze, 1978; Gilligan, 1982).

The Work of Gould

A second stage theory of adulthood is that of Roger Gould (1972; 1978). His initial sample consisted of a group of sixteen- to sixty-year-old outpatients at the UCLA Outpatient Psychiatric Clinic. Information obtained through in-depth interviews served as the initial basis for defining a sequence of stages of adult development. The stage sequence was validated through the use of a questionnaire administered to a sample of over five hundred nonpatient adults. Unlike Levinson, Gould included both men and women in his sample. He believes that the sequence he documented is as representative of the lives of women as it is of the lives of men. Like Levinson, Gould reports that the late teens and early twenties are primarily devoted to leaving home and establishing a degree of independence from parents. In Gould's opinion, the desire to leave home initially remains at the level of rhetoric. There is no clear plan of action, and the future is still defined in rather vague and glorified terms. Gradually, however, plans become specified, a move is made, a car perhaps bought, a job obtained, and the future begins to take focus.

Gould's twenty- to thirty-year-olds had a much clearer vision of their future and a stronger commitment to their life plans than did Levinson's. Gould believes that they were engaged in the work of being adults. They felt themselves to be distinct from their parents and wanted to show their parents that they could make it on their own. They also felt a need to distance themselves from their adolescent selves. They expressed a conscious desire to act in a more tempered, adultlike manner. They clearly wanted others to view them as adults. As they entered their thirties, however, they became aware of the same uncertainties that Levinson's group reported at the end of their thirties. Gould (1972) describes the uncertainties with the following question: "What is this life all about now that I am doing what I am supposed to be doing?" (p. 37). The adults in Gould's sample had successfully completed most of the developmental tasks of the early adult years. They had married, were raising children, and had established themselves in their work. Yet having done what they believed was expected of them, they didn't experience the satisfactions they had expected.

Apparently, Gould's adults developed an awareness of the lack of symmetry between the hopes and dreams of their youth and the realities and limitations of their life situations. This "die is cast" perception continued to dominate their thought into the early forties. Initially, it was met with a degree of resignation and at times even depression, but these negative perceptions gradually gave way to an acceptance and "mellowing" as Gould's adults passed through their forties and into their fifties.

Gould heard the "what is life all about" question asked as frequently by women as by men. Therefore, he believes that the sequence he described is equally representative of the adult lives of men and women. It is quite possible, however, that even though the men and women in Gould's study perceived their lives in similar ways, the components being evaluated were different. Frieze's (1978) revision of Levinson's sequence in terms of the life events of adult women and Gilligan's (1982) study of moral reasoning in adult women give considerable weight to this possibility.

The Work of Frieze

Frieze (1978) was interested in determining how well the Levinson sequence really does represent the lives of adult women. By reviewing the literature on adult development, she found the pattern for adult women to be distinct from Levinson's in two respects. First, it focuses primarily on the home and family rather than on occupation. Second, it is more complicated. Frieze believes that adult women are less able to put some aspects of their adult lives on the "back burner" than Levinson's men were. Therefore, at all ages, women experience a much larger interplay between the three careers of work, intimacy, and family than men do. Further, this interplay is probably increasing as women (at least those who are college educated) seek to define themselves in all three domains.

Phase I: Leaving the family. According to Frieze, the pattern for women also begins with an emphasis on leaving home, on psychologically emancipating themselves from their parents and beginning to formulate an adult life plan. The primary component of the life plan is marriage. However, the ability to actualize this life plan is hindered by the fact that, unlike a career, marriage can't be accomplished alone. Therefore, many of the factors regulating the early adult years of women remain undefined until they marry. For example, a woman who wants to pursue both marriage and a career, but who gives priority to marriage, has to wait until marriage and (most typically) until her husband finds work before she can act upon her career plans. Although there is no reason why this need be so, this has been the pattern and, as the discussion on male views on marriage later in the chapter will make clearer, it is not likely to change in the near future.

Phase II: Entering the adult world. Frieze reports that during their twenties, most women marry and become mothers. Although marriage and career are usually seen as quite compatible and rewarding, the arrival of children places additional burdens on women. Women who wished to pursue all three roles reported that the task was difficult, especially when their children were of infant or preschool age. This conclusion is consistent with data reported by Pleck (1983). He found that fathers rarely assume a significant share of the home and family tasks, and as a result the women Pleck interviewed reported that, even though they found each role an important source of satisfaction, they still felt overburdened. Frieze believes that this frequently reported sense of being overburdened probably explains why many women choose to leave work when their children are young and hope to return when the children enter school. She also notes that such a strategy can seriously hinder job advancement and even reentry into the job market. In any event, women seem to feel more of a sense of tension between their different roles than men do.

Phase III: Entering the adult world (again). Frieze notes that by the time women reach their early thirties, their children have entered school and they often find themselves in a position similar to the one Levinson's men encountered in their twenties—entering the occupational world as a relatively unencumbered adult. She also notes that, for many women, the early thirties represent the first opportunity to define themselves, not in terms of others (that is, as someone's daughter, wife, or mother) but in terms of an independent task. Frieze finds it somewhat ironic that many women find themselves becoming deeply committed to a career at the very time that their husbands are having second thoughts about theirs.

The Work of Gilligan

While the major differences between Frieze's findings and the findings of Levinson involve the timing of events and the interplay between roles, Gilligan's research (1977; 1979; 1982) indicates that there may be more significant differences in the adult patterns of men and women than the tim-

For women who have stayed home with their children, the early thirties often mark a major transition. With their children now in school, many of these women enter or reenter the work force, often with a strong career commitment.

ing of life events. Her analysis of the distinct perceptions men and women have of the relationship between the three careers is seen as one of the most significant arguments supporting the hypothesis that there are significant differences in the developmental patterns of adult men and women.

Gilligan's work evolved from her research with Kohlberg (Kohlberg & Gilligan, 1972) on adolescents' resolutions of moral dilemmas. In the course of evaluating adolescents' explanations for their moral choices, she became increasingly impressed with the different strategies males and females used to explain their choices. Her observations have led her to examine the impact of male and female socialization patterns on adult development. Gilligan (1982) finds that the male pattern of adult development places great emphasis on individuation, autonomy, and goal direction. This pattern is well reflected in Levinson's sample. Levinson's men defined themselves almost exclusively in terms of their career or occupation and believed that their developmental tasks were to become independent first from their parents and then, in their thirties, from mentors and institutions. They did not realize until mid-life that this pattern might be based on unrealistic expectations and that it might cause poorly nurtured relationships. They came to this realization at mid-life only because their dream was not to be.

Gilligan believes that for males the pattern of adult development represents a continuation of the independence training they received during childhood. She also observes that for females the emphasis during childhood is less on independence and autonomy and more on attachment and affiliation. Therefore threats to attachment for women become just as anxiety provoking as threats to autonomy for men. Gilligan criticizes theories such as Levinson's. She says that by claiming the male sequence is equally appropriate for females, when in fact it isn't, these theories stack the deck against women. That is, they make autonomy, the very quality that female socialization patterns render most anxiety provoking, the cornerstone of mature adult development. As a result, for women to function effectively in the workplace, they must either "resocialize" themselves or find work settings that are not rigid and inflexible.

The consequences of these socialization patterns are reflected not only in the differing sources of anxiety for adult men and women but also in their decision-making strategies. Gilligan characterizes the male pattern as being abstract and analytic, the female as relative and contextual. By placing a greater emphasis on resolution, the males in Gilligan's sample make secondary what is primary to the females. In the extreme, the male perspective appears unfeeling ("let the chips fall where they may") and the female perspective uncertain and diffuse. How similar Gilligan's adults are to Lever's (1978) school children in Chapter 5.

The distinction between male and female patterns is the distinction between a relatively greater focus on self and a relatively greater focus on other. This distinction separates the developmental patterns of many men and women from early adulthood through middle age. It is not until middle age that these sex-role patterns become less differentiated and more integrated. That is, it is not until middle age that adult men and adult women each seek to establish a greater balance between their needs for autonomy on the one hand and their needs for affiliation on the other. At both stages the self/other balance is a significant element of the self-concept. What differs is the way the balance is approached.

The Timing of the Three Careers of the Adult Years

The work of Levinson, Gould, Frieze, and Gilligan emphasizes the sequencing of events during the adult years. A second way to consider the developmental patterns of the adult years is to examine the timing of the transitions between the various steps in the sequences that these authors have documented. The two factors—timing and sequence—are closely related but not identical. The age or point at which a particular step in the sequence is initiated (such as age at marriage, length of marriage before first birth, and spacing between children), the length of time that step is maintained, the factors that prompt a transition between steps in the sequence, and the ease with which the transition is accomplished are as important in understanding adult development as are the characteristics of the steps themselves. For example, pregnancy can occur before or after marriage, the birth of a third child can come three or fifteen years after the birth of the second, and a career change can be undertaken at thirty or fifty years of age. In all three instances, the events are the same, but because they may stand in different relationships to each other, their meaning and consequences are often quite different (Elder & Rockwell, 1979).

The social clocks (Neugarten & Datan, 1973) of the adult years provide adults criteria against which to judge the timeliness of change and the reasonableness of expectations. They provide adults a basis for judging whether a promotion is really a sign that the boss likes their work or merely an indication that the boss thinks it's easier to keep them than to train someone else for the job. These social clocks also provide adults a framework for planning the future. For example, by giving adults a sense of the time when they might reasonably expect to buy a new home, they provide them with a basis for deciding how much of each paycheck to put into a down-payment savings account.

Not all of the timing of adult events is based on social clocks. Biological clocks also continue to exert an important influence during the adult years. The death of a spouse is always traumatic but is perhaps more so if it occurs "before his or her time." Showing signs of age such as balding or graying hair before they are age typical can send many people to the hair-coloring shelf or the toupee shop.

Brim and Ryff (1980) have developed a typology of life events that sorts possible life events in terms of their degree of association with chronological age, their likelihood of occurrence, and their typicality (see Table 7-2). Events such as those in category 1 represent the normative events of the life-span. They are events that are usually correlated with chronological age, have a high probability of occurrence, and are likely to occur in the lives of many people. At the other extreme are the events in category 8. These events bear little relationship to chronological age, are not very likely to occur, and when they do happen, affect few individuals (although Brim and Ryff note that most people are likely to experience at least one category 8 event sometime in their life). Not surprisingly, we are best prepared to deal with category 1 events. We receive the most anticipatory socialization for events such as the transition to parenthood, marriage, and recovery from a heart attack; we have the empathy and support of cohorts experiencing the same events; and since these are events experienced by many,

TABLE 7-2 *Life Events Typology*

CORRELATION WITH AGE	EXPERIENCED BY MANY: HIGH PROBABILITY OF OCCURRENCE	EXPERIENCED BY MANY: LOW PROBABILITY OF OCCURRENCE
Strong	1 Marriage Starting to work Retirement Entering school Woman giving birth to first child Bar Mitzvah First walking Heart attack Birth of sibling	3 Military service draft Polio epidemic
Weak	2 Death of a father Death of a husband Male testosterone decline "Topping out" in work career Children's marriages Accidental pregnancy	4 War Great Depression Plague Earthquake Migration from South
	EXPERIENCED BY FEW: HIGH PROBABILITY OF OCCURRENCE	**EXPERIENCED BY FEW: LOW PROBABILITY OF OCCURRENCE**
Strong	5 Heirs coming into a large estate Accession to empty throne at 18	7 Spinal bifida First class of women at Yale Pro football injury Child's failure at school Teenage unpopularity
Weak	6 Son succeeding father in family business	8 Loss of limb in auto accident Death of daughter Being raped Winning a lottery Embezzlement First black woman lawyer in South Blacklisted in Hollywood in 1940's Work disability Being fired Cured of alcoholism Changing occupations Grown children returning home to live

Note: Reprinted, by permission, from O.G. Brim & C.D. Ryff, "On the Properties of Life Events," in P.B. Baltes & O.G. Brim (Eds.), *Life Span Development and Behavior*, Vol. 3 (New York: Academic Press, 1980), 375.

we are able to make use of relatively formal social support networks to help us cope with their stresses.

We are considerably less prepared for events in the other categories because we don't know when they are going to happen or even if they are going to happen at all. Adjustment to these events is considerably more difficult. Since they are unexpected, little preparation is likely. And for those that are relatively unique, little support is available. Losing a home in a fire, as far as the actual loss is concerned, is no different from having it blown away in a hurricane. However, in the second instance, the loss was probably one of many, and in such cases, not only are various formal support systems available (such as the Red Cross or Government Disaster Relief Agencies), but also having to share the terrible loss with others somehow seems to lessen the burden of each.

It is important to note that the placement of items in Table 7-2 is itself a reflection of the sociohistorical context. For example, the last item in category 8—grown children returning home to live—has become somewhat more common in recent years as a by-product of the increasing divorce rate. Many divorced adults find that two apart can't live as cheaply as two together and that a return to their parent's house offers at least a temporary resolution of their financial dilemmas.

The balance of this chapter will examine the developmental patterns discussed in this introduction as they apply during the age period of 18 to 25. The three chapters that follow will do the same for the age intervals of 25 to 40, 40 to 65, and 65 and older. Because of the added significance of sociohistorical factors for the course of development during adulthood, the sections on the course of development and the context of development will be merged in Chapters 7 through 10.

THE COURSE AND CONTEXT OF DEVELOPMENT DURING THE TRANSITION YEARS

During the eighteen to twenty-five interval, the primary developmental events are completion of education, entrance into the work force, and marriage. The order in which the three events occur can have a significant influence on the course of development during much of the adult years. For example, the age at which individuals complete their schooling usually defines the age at which they enter the labor force and the age at which they marry (Otto, 1979). And the age at which they marry usually defines their fertility patterns and therefore the size of the financial burdens they place upon themselves.

High school graduation serves as one of the clearer points of divergence between social classes in our society. The probability that people will continue on to college correlates strongly with their social class. The more economically privileged their background, the more likely it is that they will continue their education beyond high school. The impact of the decision about future education is cumulative. Continuation of education usually delays the decision to marry, which in turn reduces the length of the childbearing period. The likelihood that college attenders will have large families is further reduced by the attitude and value changes often resulting from continued education beyond high school. College graduates are likely to have higher-paying jobs than non–college graduates and to think of their jobs as careers rather than simply work. Moreover, college-educated adults are more likely to be two-income families (Otto, 1979). However, conclusions about the advantages of higher education must be tempered by the fact that, although tangible material rewards are associated with years of school, personal judgments of life satisfaction are often not (Campbell, 1981; Flanagan, 1980; Trent & Medsker, 1968).

The timing of the three events—school completion, the beginning of work, and marriage—has changed over the past seventy-five years. For people born in 1910, the average age at school completion was seventeen, the average age for starting work was eighteen, and the average age for marriage was not until twenty-six. Those born in 1950 stayed in school longer (the average age of completion was at this time almost twenty), began work at about the same time or even slightly before they completed school, and married at the age of twenty-three (Hogan, 1981). Since the percentage of high school graduates currently enrolled in college is not expected to increase significantly beyond the current 50 to 60 percent, there is little reason to expect a change in the average age at completion of schooling. There is, however, likely to be an increase in the average age for first marriage. This increase is primarily a reflection of the increasing number of college graduates who are choosing to delay marriage until the middle or late twenties, sometimes even into the early thirties.

These historical trends are largely a reflection of political and economic factors. Economic conditions determine the length of time individuals are able to remain in school. When economic and political factors are positive, the pressure on people to make decisions concerning future life plans is reduced, and the rate of transition is lessened. Further, such comparatively ideal conditions provide enough support so that even when individuals marry while in college or work part-time while completing their education, the additional stresses are not overwhelming. When economic circumstances are less favorable, educational plans may be foreclosed, and job options may be lessened. If anything does increase during such hard times, it is probably debt. In general, favorable economic and political conditions provide individuals adequate time to assimilate the various changes that are occurring in their lives.

Economic and political conditions can be easily disrupted, however. For example, the mobilization of young men into the armed forces during wartime has historically been one of the greatest disruptive factors faced by men undergoing the transition to adulthood (Hogan, 1981). Mobilization interrupts the continuation of schooling and often forces individuals to make sudden decisions concerning marriage. Further, the individual's sense of developmental continuity is broken. Plans people make before entering the armed services may prove unrealistic or no longer possible upon their return home. Further, the stresses of military service may make adjustment to civilian life difficult.

Transition Patterns of College Youth

Between 50 percent and 60 percent of all high school graduates go directly on to college. But as many as half of those who do go on do not graduate within the typical four-year period. Whether or not they graduate, college has a pronounced impact on all those who attend. The influence is reflected in changes in attitudes and values, in career preparation and opportunity, in dating and marriage patterns, and in perception of self as a unique individual.

Academic ability and family background are the strongest predictors of college attendance (Davies & Kandel, 1981; Trent & Medsker, 1968). Family background influences the likelihood of college attendance in a number of ways. The most obvious is financial. As college tuition continues to rise, it becomes increasingly more difficult for individuals to finance their own education through part-time work. Second, parents, in particular, and the broader social network, in general, provide role models for achievement and career expectations. Many who attend college report that, even though it may never have been stated explicitly in their home, they somehow always knew that their parents expected them to go on to college (Trent & Medsker, 1968).

The decision to stay in college until graduation seems primarily a matter of motivation. Neither the high school grades nor the family socioeconomic status of people who go on to college is predictive of their graduation. Rather, those who graduate report that they never seriously considered not finishing school. Graduates do report that their parents remain an important source of support and encouragement during college, especially when academic problems are encountered.

The Influence of College on Attitudes and Values

One of the most consistently reported findings on the effects of college attendance is that changes in attitudes and values occur during the college years. On the average, students become less authoritarian, less dogmatic in their views, less ethnocentric, more interested in world events, less religious, more hedonistic in their life-style, more tolerant of others, and less traditional in their sex-role orientations (Astin, 1977; Feldman & Newcomb, 1969; Freedman, 1967). The fact that these differences are "on the average" is important to keep in mind. It implies that these changes are more evident in some students than in others. It is also important to realize that these general trends are also evident, though to a lesser degree, in non-college-attending youth. In other words, these changes in attitudes and values reflect the life experiences typically encountered during the transition years. What seems to make the college environment (both in and out of the classroom) particularly supportive of these changes is that students are exposed to a wider range of attitudes, values, and behaviors there than they were previously and find a greater emphasis placed on independent, autonomous thought and action.

This "opening up" process is equally evident in changes in career plans during college. Over 50 percent of all students change their major at least once. Almost half of all entering students choose business, engineering, or education as their major. A high percentage of males choose business or engineering while a high percentage of females choose education. The fact that the choice of majors by first year students is so heavily sex typed and so oriented toward traditional disciplines no doubt is a reflection of earlier sex-role socialization experiences. Because students learn of new career opportunities and become more independent in their decision making, their second choice of majors is likely to be broader and somewhat less sex typed. Nevertheless, the shifts for males most often tend to be into business or prelaw; those for women, into nursing and the social sciences

BOX 7-1
THE CLASS OF '84

For the past twenty years, entering first year students at many two- and four-year colleges and universities across the country have completed a questionnaire developed by Alexander Astin and his colleagues (1974; 1984) for the American Council on Education. A look at some of the findings reveals how entering students see themselves and the world around them. If we compare the views of the class of '84 with those of the class of '74, it is possible to provide a measure of the degree of social change that has occurred in this country over the past decade.

Probably the most drastic change in the views of students has come in the area of future goals. Especially dramatic are the reversal in their views concerning the relative importance of developing a philosophy of life and being very well-off financially and the increasing amount of value they place on recognition by colleagues (see Table A). The percentages are also interesting because of the potential conflicts they reveal. Many students may find that the value they place on raising a family may conflict with financial success and professional recognition.

This shift in future goals has brought about a change in choice of major, particularly in the case of women. The traditional female majors of education and nursing have lost ground, while business, computer science, law, and engineering have all gained in female enrollments (see Table B).

Another interesting change is in the area of political orientation. Students in 1984 were less likely to label themselves liberal and more likely to call themselves middle-of-the-road or conservative than were students in 1974 (see Table C). This was true for both men and women.

TABLE A Important Goals

	MALES 1974	*MALES 1984*	*FEMALES 1974*	*FEMALES 1984*
Recognition in field	54	66	43	63
Influence social values	25	30	30	35
Raising a family	53	67	57	69
Being very well-off financially	54	75	36	67
Developing a philosophy of life	57	44	65	44

TABLE B Future Career Goals

	MALES 1974	MALES 1984	FEMALES 1974	FEMALES 1984
Business	18	17	15	19
Engineering	9	18	1	3
Medicine	7	5	5	4
Law	5	5	3	4
Computers	2	7	1	5
Nursing	2	.2	13	8
Teaching	5	2	16	9
Undecided	12	10	5	12

TABLE C Political Orientation

	MALES 1974	MALES 1984	FEMALES 1974	FEMALES 1984
Far left (%)	2	2	2	2
Liberal	29	19	27	21
Middle-of-the-road	51	54	58	61
Conservative	16	23	12	16
Far right	1	2	1	1

Some things don't seem to change, however. For both men and women, in 1974 and in 1984, the single most important reason for applying to a particular school was its academic reputation. Average high school grades also do not seem to have changed much over the past decade. Women's grades were higher than men's in 1974, and they still are today (see Table D).

Interestingly enough, students' attitudes don't seem to have changed much over the past ten years either. With the exception of feelings about the legalization of marijuana, most of the differences are more a reflection of sex than they are of cohort (see Table E). We will need to wait until the data are in for the class of '94 to see if these social and economic trends are to continue.

(continued)

BOX 7-1 (CONTINUED) THE CLASS OF '84

TABLE D Average High School Grades

	MALES 1974	MALES 1984	FEMALES 1974	FEMALES 1984
A	17	15	23	23
B	56	56	62	59
C	26	27	15	16

TABLE E Student Attitudes

	MALES 1974	MALES 1984	FEMALES 1974	FEMALES 1984
Gov't not controlling pollution	80	57	84	80
Male/female equal pay for equal work	88	87	95	97
Woman's place is in the home	39	30	19	15
Should live together before marriage	74	50	39	40
Sex okay if partners like each other	60	63	30	32
Should legalize marijuana	49	26	43	20

Sources: Astin, Green, Korn, & Maier, 1984; Astin, King, Light, & Richardson, 1974.

(Astin, 1977). For women, the other "career" option is to be a homemaker. The reason for the quotation marks is that it is difficult to determine the extent to which this choice reflects a genuine desire to be a homemaker or a sense of resignation to the idea that this might be the most realistic option available. To an extent, this issue is true of all career choices, but as was discussed in the last chapter on sex-role development and as will be discussed later in this chapter, the evidence suggests that the dilemma is greater for women than for men.

Because of the concentrated exposure to a wide range of experiences and viewpoints, college is often the most politically active period of an individual's life. However, while the level of involvement may decrease in later years, the underlying values and attitudes usually persist.

The Influence of College on Reflective Judgment

The liberalization of attitudes and values, as well as the broadening of career choices during the college years, reflects a basic shift in what Perry (1970) labels the "forms" in which students construct their experiences. The progression defined by Perry, as well as a similar one by King (Kitchener & King, 1981; Strange & King, 1981), involves **three forms of experience construction.** The first, most typical of first year students, demonstrates what Perry calls a **dualistic world view.** These students believe that, in essence, there are right and wrong answers to all questions and that these answers are knowable. Such students feel very frustrated in class when their questions to the professor are tossed back at them with the statement "What do *you* think about it?"

Both Perry and King find that this rather absolute view gives way to something quite opposite—a **relativistic world view.** Perhaps as a result of always being asked what *they* think, students begin to believe that their view is as valid as that of the professors, that all viewpoints are equally correct and that even if there is a true answer to a question, it is probably not knowable. It would seem likely that the hedonistic life-styles explored by many students are a reflection of this form of thinking.

Gradually, toward senior year, students increasingly demonstrate a third form of experience construction. Perry refers to it as an **evolving of commitments.** Ideas and opinions are no longer considered valid because of their relation to absolute knowledge or because of the individual's right to have them, but rather they are judged according to the soundness of the foundation upon which they are based. Students at this third level would, for example, place as much emphasis on gathering evidence to support a conclusion as they would on the conclusion itself.

The sequence described by Perry and by King (see Table 7-3 for the complete King sequence) is reminiscent of the other developmental sequences described in the text. The hedonistically oriented middle phase (sometimes involving behaviors quite worrisome to parents) reflects the dismantling of the earlier, more rigid form but offers only the earliest glimmer of the more rooted final form. The evolution of identity during the college years, discussed next, also follows a similar pattern.

The Influence of College on Identity Formation

Factors promoting identity change during college. The identity established during high school continues to undergo revision and expansion during the college years. Three factors support this continuing progression. First, college students, even those living at home, are considerably less influenced by their parents in day-to-day activities than they were during high school (Sullivan & Sullivan, 1980). Although parents continue to serve as an important source of support and knowledge (Klos & Paddock, 1978; Tokuno, 1983), gradually, during the college years, students come to think of school as home. They become less likely to return home during the summers or on semester breaks and sometimes even feel like visitors when they are at home. In general they report feeling less close to their parents over the course of the college years (Moore & Hotch, 1981). This distancing process appears somewhat restricted to these transition years. As adults move through their twenties, they report closer ties to their parents, especially when they themselves become parents (Aldous, 1978; Hill, Foote, Aldous, Carlson, & McDonald, 1970). In other words, the lessening of ties to parents during college seems very situation specific. It probably reflects a heightened peer influence brought about by the age-segregated nature of the college environment.

The second influence on identity during college is the diversity of peer contacts. The college environment presents a much greater range of life-styles, attitudes, values, and experiences than high school does. Further, this broadening of experiences comes at a time when individuals are developing a more relative and contextual perspective from which to view the world. It is not surprising that many students, especially those from relatively cloistered backgrounds, find the diversity of the college environment overwhelming at first.

The third shift in identity concerns the future-oriented component of the self-concept. For high school students, the topics of marriage and career tend to have a romantic tint to them. Perhaps as a result of the adolescent's burgeoning sense of omnipotence, all problems are seen as resolvable—to the satisfaction of everyone. This outlook is less typical of college students. Through gaining a better sense of their ability to pursue a career, of possibilities of obtaining work in different professions, and of both the assets and liabilities inherent in any career choice, they begin to sense the problems inherent in juggling the adult roles of intimate, parent, and worker. The awareness that the future is getting "closer" is also heightened by the awareness that some of their high school classmates have already entered the job market or married or are both working and married. The impact of such realizations is a frequent component of transition periods during the adult years. When people learn that their peers are already dealing with issues that seem far in the future to them, they are often prompted to reevaluate their own lives. In fact, this type of realization is seen as one of the major causes of the mid-life review. Few things are more likely to cause a forty-five-year-old man to realize that he is not as young as he might like to think than hearing about a same-age colleague dying from a sudden heart attack.

The identity formation process. Keniston (1975) distinguishes the college phase of identity formation—a period he calls "youth"—from the high school phase by noting that while in high school, adolescents are concerned with defining who they are, in college, youth are concerned with determining how well they fit in. That is, young adults are increasingly aware of the potential disparity between the emerging self and the social order.

The work of Marcia and his colleagues (1966; 1980; Marcia & Friedman, 1970; Schenkel & Marcia, 1972) represents one of the most thorough efforts to document the ways in which young adults between the ages of eighteen and twenty-five attempt to resolve the disparity between their sense of self and their awareness of the expectations of society. His views on identity formation are based on interview data collected from college students. Students were asked to respond to a number of open-ended questions concerning their attitudes, values, and goals on the topics of occupation and political and religious ideologies. The responses were scored using a system based on Erikson's (1968) conception of the identity formation process. Marcia's sample was limited to males and to these two specific aspects of an individual's sense of self. The significance of these two restrictions will be dealt with in the next section.

Marcia was able to distinguish four unique response patterns or, as he calls them, **identity statuses.** Status was determined by evaluating the responses in terms of two distinct qualities. The first concerned the degree of resolution evident in the response. The second concerned the degree of conflict or crisis evident in the resolution process (see Table 7-4). Consistent with Erikson (1968), Marcia believes that the greater the degree of resolution, the more aware individuals are of their similarities to others and their unique strengths and weaknesses. The greater the degree of their awareness of self, the greater is the probability of their establishing mutually satisfying relations with others. Also consistent with Erikson, Marcia sees the resolution process as inevitably involving a degree of personal crisis and conflict. Given these criteria, those in the **identity achievement** status represent the highest level of identity formation and those in the **identity diffusion** status represent the lowest level of identity formation. Marcia is less clear on the relative desirability of the other two statuses. From his longitudinal data, he does report that those individuals reaching identity achievement status did so by moving from the **psychosocial moratorium** status. Therefore, moratorium would seem to be a necessary prerequisite to achievement.

TABLE 7-3 *The Development of Reflective Judgment*

STAGE	A) ASSUMPTIONS ABOUT REALITY	B) ASSUMPTIONS ABOUT KNOWLEDGE	C) CONCEPTS OF JUSTIFICATION
1	There is an objective reality which exists as the individual sees it. Reality and knowledge about reality are identical and known absolutely through the individual's perceptions.	Knowledge exists absolutely. One's own views and those of authorities are assumed to correspond to each other and to absolute knowledge. Knowledge is gained through the individual's perceptions and prior teaching.	Beliefs simply exist; they are not derived and nee not be explained. Differences in opinion a not perceived, and justification is therefore unnecessary.
2	There is an objective reality which is knowable and known by someone.	Absolute knowledge exists, but it may not be immediately available to the individual. It is, however, available to legitimate authorities.	Beliefs either exist or are based on the absolute knowledge of a legitimat authority.
3	There is an objective reality, but it cannot always be immediately known, even to legitimate authorities. It is possible to attain knowledge about this reality, but our full knowledge of it is as yet incomplete and therefore uncertain.	Absolute knowledge in some areas, but in other's it is uncertain, at least temporarily. Even authorities may not have certain knowledge, and therefore cannot always be depended upon as sources of knowledge. Knowledge is manifest in evidence which is understood in a concrete, quantitative way such that a large accumulation of evidence will lead to absolute truth.	Beliefs either exist or are based on an accumulation of evidenc that leads to absolute knowledge. When such evidence is not available, individuals claim that while waiting for absolute knowledge to become available, people can temporarily believe whatever they choose to believe.
4	There is an objective reality, but it can never be known without uncertainty. Neither authorities, time, or money, nor a quantity of evidence, can be relied upon to ultimately lead to absolute knowledge.	Absolute knowledge is for practical reasons impossible to attain, and is therefore always uncertain. There are many possible answers to every question, but without certainty and a way to adjudicate between answers, there is no way to decide which one is correct, or even whether one is better than another. Knowledge is idiosyncratic to the individual.	Beliefs are justified with idiosyncratic knowledge claims and on idiosyncratic evaluations of data ("What is true is true for me, but not necessarily for anyone else.") The individual is the ultimate source and judge of his or her own truth.

Note: From "Intellectual Development and Its Relationship to Maturation during the College Years" C.C. Strange and P.M. King, 1981, *Journal of Applied Developmental Psychology, 2,* 288–89. Copyright 1981 by Ablex Publishing Company. Reprinted by permission.

'AGE	A) ASSUMPTIONS ABOUT REALITY	B) ASSUMPTIONS ABOUT KNOWLEDGE	C) CONCEPTS OF JUSTIFICATION
	An objective understanding of reality is not possible, since objective knowledge does not exist. Reality exists only subjectively and what is known of reality reflects a strictly personal knowledge. Since objective reality does not exist, an objective understanding of reality is not possible.	Knowledge is subjective. Knowledge claims are limited to subjective interpretations from a particular perspective based on the rules of inquiry and of evaluation compatible with that perspective.	Beliefs are justified with appropriate decision rules for a particular perspective or context, e.g., that a simpler scientific theory is better than a complex one.
	An objective understanding of reality is not possible, since our knowledge of reality is subject to our own perceptions and interpretations. However, some judgments about reality may be evaluated as more rational or based on stronger evidence than other judgments.	Objective knowledge is not possible to attain, because our knowledge is based on subjective perceptions and interpretations. Knowledge claims can be constructed through generalized principles of inquiry and by abstracting common elements across different perspectives. The knower must play an active role in the construction of such claims.	Beliefs are justified for a particular issue by using generalized rules of evidence and inquiry. However, since our understanding of reality is subjective, any such justification is limited to a particular case, time, or issue.
	There is an objective reality against which ideas and assumptions must ultimately be tested. Despite the fact that our knowledge of reality is subject to our own perceptions and interpretations, it is nevertheless, possible, through the process of critical inquiry and evaluation, to determine that some judgments about that reality are more correct than other judgments.	Objective knowledge is possible to attain. Knowledge is the outcome of the process of reasonable inquiry. The process of inquiry, however, may not always lead to correct claims about the nature of reality, since the process itself is fallible. Knowledge statements must be evaluated as more or less likely approximations to reality and must be open to the scrutiny and criticism of other rational people.	Beliefs reflect solutions that can be justified as most reasonable using general rules of inquiry or evaluation. Criteria for evaluation may vary from domain to domain (e.g., religion, literature, science), but the assumption that ideas, beliefs, etc. may be judged as better or worse approximations to reality remains constant.

TABLE 7-4 *Marcia's Identity Status Typology*

DEGREE OF RESOLUTION ACHIEVED	DEGREE OF CONFLICT OR CRISIS INVOLVED	IDENTITY STATUS
Hi	Hi	Identity achievement
Lo	Hi	Psychosocial moratorium
Hi	Lo	Identity foreclosure
Lo	Lo	Identity diffusion

Source: Marcia, 1980.

Marcia (1980) reports that college students at each of the four statuses differ not only in terms of their answers to the interview questions but also in a number of other ways. Identity achievers are described as the most self-accepting of the four groups. They demonstrate the least discrepancy when their own self-assessments are compared to those made of them by others. They tend to hold the strongest values and show the most willingness to act on the basis of those values. Not surprisingly, given Marcia's allegiance to Erikson's theory, achievers are seen as the only group capable of real intimacy with others. Finally, Marcia reports that identity achievers' fathers tend to be relaxed and encouraging rather than controlling.

Those in the psychosocial moratorium status are described as often uncertain and overly critical. Perhaps because of their desire but inability to reach some sense of personal identity, they are unwilling to accept advice or support from others. They tend to have a low sense of confidence in their ability to make choices and demonstrate high scores on tests of anxiety level.

Identity foreclosure reflects a very different pattern. In terms of resolution identity foreclosurers resemble identity achievers. However, achievers reached their status through a deliberate effort to choose a value structure; foreclosurers reached theirs through adopting the value structure of another person, often the parent. They place a high value on maintaining traditions. They avoid conflict and tend to be fairly rigid in their judgments. In general, they seem more influenced by the judgments of others than by their own personal evaluations.

Identity diffusers are also strongly influenced by the judgments of others but, having no commitments to any value structure, will readily change to meet the expectations of those they are currently impressed by. They demonstrate the least amount of basic trust and, no doubt as a result, are the least able to form close relationships.

It is important to bear in mind that these statuses are not permanent qualities of individuals but rather represent their present degree of resolution on topics that are significant in their lives. In other words, degree of resolution is likely to increase over time. Equally important is the fact that status is defined in terms of a particular aspect of self-identity. It is quite possible to identify a student who has reached identity achievement on a topic such as occupation, is experiencing moratorium with respect to marriage, and is both uncertain and unconcerned about a far removed topic such as retirement.

Sex differences in identity formation during college. Marcia, like Levinson, limited his samples to males. For this reason, the same questions that were asked about Levinson's conclusions have been asked about Marcia's. In particular, others have asked if the findings are as representative of the identity formation process in college women as they are of the identity formation process in college men. The answer appears to be no. The identity formation process differs in three fundamental ways. These differences concern the core component of self-identity, the role of crisis in the identity formation process, and the visions men and women hold about their futures.

The core component of the identity process for college males is occupational and ideological competence (Hodgson & Fischer, 1979; Marcia, 1980). This sense of competence is defined in terms of career and reflects an internal set of beliefs about the individual's future role in society. The core component of the identity process for college women focuses on relationships (Hodgson & Fischer, 1979). The primary emphasis is on acquiring interpersonal competence and a set of internal beliefs that allow women to resolve the issue of getting along with others in ways that are satisfying both to themselves and to those important to them. Neither the core components of male identity nor the core components of female identity emerge for the first time during the college years. Rather they reflect the cumulative impact of the socialization patterns of girls and boys in our culture. They are evident in the different ways in which parents respond to infant girls and infant boys and in the ways in which girls and boys in middle childhood resolve disputes during games. They are equally evident in the different ways in which male and female adolescents approach sexual activity. What is now added to the process is a greater awareness of the distinct nature of these central components, or in Keniston's (1975) view, a greater concern with integrating these distinct components into society.

Male and female college students differ not only in terms of the core element of identity. The research suggests that they may also differ in terms of the ease of reaching identity achievement status. It is through crisis and conflict, either internal or external, that individuals clarify their beliefs, differentiate themselves from others, and become autonomous, mature adults. Forged and tempered are words that come to mind to describe the result of this process. But as was noted by Gilligan (1982) and others earlier in the chapter, our socialization patterns are more supportive of males entering into conflict than of females. Because conflict places relationships in peril, it may be more anxiety producing for females who value nurturance and affiliation than for males who value autonomy and independence. For this reason, it is possible that if moratorium is a necessary prerequisite to identity achievement, the step may be more willingly pursued by males in our culture than by females (Ginsburg & Orlofsky, 1981).

As already mentioned, identity achievement status is not a global state. It is achieved in terms of each of the components of an individual's self-structure. Is the relationship between conflict and other aspects of the self-structure the same as the relationship between conflict and the core component? In other words, do college women have more difficulty achieving resolution on all aspects of the self-structure? The research says no. Although the core component of identity for males is instrumental competence and for females interpersonal competence, these are not the only components of personal identity. Each sex must also deal with the other, and the manner in which young adult men and women do so suggests that the problem of dealing with conflict may be just as great for males as for females—but in different areas.

Identity foreclosure, the "backdoor" to achieving a stable identity structure, is considerably more common in college women in the areas of occupational choice and ideology but considerably more common in males in the areas of sexual identity and interpersonal relations (Waterman & Nevin, 1977). The difficulties males have in dealing with the interpersonal domain are evident in the fact that they have fewer same-sex friendships than women do and in their heterosexual behavior patterns. Fischer (1981) finds that male same-sex friendships are considerably less likely than female same-sex friendships are to involve self-disclosure, sharing of confidences, and requests for emotional support. Since males see such behaviors as inconsistent with their sex role, they avoid them. In the same fashion that women's difficulty in dealing with the core component of male identity creates problems for them in developing an instrumental identity, men's uninvolved style in same-sex friendships may impair their ability to achieve a sense of interpersonal competence.

Because the social role expectations for adult men and women continue to be distinct in our culture and because identity formation in the areas of instrumental and interpersonal competence takes different forms for college men and women, their visions of self in the future are also distinct. For men, these visions tend to be narrow in scope and long-term in focus; for women, they tend to be the opposite (Angrist, 1975). These patterns reflect the relative amounts of importance placed on the instrumental and interpersonal domains for men and women and the extent to which college men and women are aware of the interplay of the three adult careers across the life-span. In particular, they reflect the greater involvement of women than of men in the parent career. As becomes increasingly apparent during young and middle adulthood, these visions of the future each carry major liabilities. Men, for example, are considerably less likely to foresee conflict between the three adult careers than are women (Johnson & Jaccard, 1980). As a result, most of the research on mid-life transitions (see Brim, 1976) finds that men have a more difficult time resolving the issues encountered in the transition. Women, fully aware of the greater potential conflicts between the three careers and more likely to be influenced by the circumstances of others (such as spouse and children) than men are, often feel frustrated in their attempts to implement their visions of the future (Josselson, 1973).

Testing the waters. As life plans are formed, they are, in Keniston's (1975) view, tested within the broad social context. One of the most important and immediate of these "wary probes" concerns the compatibility of women's life plans with those of men, and vice versa. For an increasing number of college students, the probe finds less than a perfect match. In particular, over the last two to three decades, the sex-role patterns and ideology of college women have changed considerably faster than those of college men (as will be discussed later, this change is less true of the noncollege population). Further, because of these changes the two patterns now often conflict (Huston-Stein & Higgins-Trenk, 1978). Specifically, women's concept of their sex role has changed to emphasize a greater balance between the instrumental and interpersonal domains of identity.

Although there has only been a very small increase in the number of college women who forgo marriage and parenthood, there has been a significant increase in the number of college women wishing to combine these

The close friendships that many young women form during college help them to establish stable identity structures.

roles with that of a career (Parelius, 1975). The problem is that a significant number of the college women in this second category feel that men are less interested in marrying a woman who wants to combine home and career than they are in marrying one who favors a more traditional sex-role orientation. Table 7-5 shows that the percentage of female college students who agree with feminist positions changed significantly between 1969 and 1973. It also shows the percentage of female college students in each of these years who believed that men would want to marry a woman favoring such a position. If the perceptions of these women are correct, they are faced with the prospect of having to either significantly alter many of their basic values to make them more compatible with men's, somehow alter male value structures to make them more compatible with theirs, delay marriage until a compatible mate can be found, or consider the possibility of not marrying at all. Irrespective of which of these options is actually taken and even irrespective of the outcome, the realization that the value structure they have crafted has the potential to create significant problems for them must prove a considerable source of anxiety and stress. As one woman put it, "I'm pretty much what I would like to be but I am not what I think men want me to be" (Rappaport, Payne, & Steinmann, 1970).

How accurate are the perceptions of college women? Have the changes in sex roles been restricted to women? The answer is mixed and is perhaps best depicted in the following description by Komarovsky (1976) of one of the male college seniors she interviewed concerning their changing sex-role values.

> He "disagreed" [author's quotes] with thirteen out of the sixteen propositions asserting sex differences. He endorsed enthusiastically the proposal to counsel qualified high school girls to "train for occupations now held by men". In fact, he termed "sick" the current practice of influencing girls to enter only feminine occupations. But describing his preferences for the future, he wanted his wife to stay at home to rear his children. In the interview, in answer to our standard question: "List three most favorable qualities of your mother as a mother in relation to you," he said: "She was completely devoted to me. She was a very bright and a very intelligent woman and she stayed home till I was grown. I mean she didn't go to work till I was in high school; she then took a job till three o'clock so that she could be home to cook dinner." He hoped his future wife would be as good a mother to their children. (pp. 23–24)

TABLE 7-5 Percent Giving Feminist Responses and Percent Believing That Men Would Want to Marry a Feminist, 1969 and 1973

	FEMINIST RESPONSES		BELIEF THAT MEN WOULD WANT TO MARRY A FEMINIST	
DESCRIPTIVE ITEM	1969	1973	1969	1973
Work and Finances	%	%	%	%
Believes that a wife's career is of equal importance to her husband's.	49	81	20	31
Believes that both spouses should contribute equally to the financial support of the family.	37	65	27	45
Intends to work all her adult life.	29	60	15	32
Division of Labor in the Home				
Does not expect to do all the household tasks herself*	56	83	18	24
Expects her husband to help with the housework.	47	77	14	28
Expects her husband to do 50% of all household and childrearing tasks.	17	43	10	10
Marital and Maternal Role Supremacy				
Does not think the most important thing for a woman is to be a good wife and mother.*	31	62	8	9
Would marry only if it did not interfere with her career.	10	22	8	13
Would forego children if they would interfere with her career.	17	28	7	8
	(N = 147)	(N = 200)	(N = 147)	(N = 200)

Note: Reprinted from A.P. Parelius, "Emerging Sex Role Attitudes, Expectations and Strains among College Women," *Journal of Marriage and the Family, 37* (1975), 147. Copyrighted 1975 by the National Council on Family Relations, 1910 West County Road B, Suite 147, St. Paul, Minnesota 55113. Reprinted by permission.

*These questions were actually phrased in the affirmative; a negative answer was considered to be a feminist response.

Komarovsky believes that this student is not exceptional. Rather, he holds a set of attitudes and values that are fairly typical of her sample of male college seniors. For the most part, these students endorsed and supported most of the broad social and legal changes that have been designed to bring a greater degree of equality into the life structures of men and women. However, their liberation seems to have stopped at the front door.

A similar pattern is noted by Almquist (1974). In this study, 75 percent of the males reported that they did not want their wives to work while their children were still young. The contradictory values noted by Komarovsky were also evident in this study. Most of these men wanted their wives to be their intellectual equal. However, only a small percentage saw the need for their wives to pursue a career. Rather, most felt that occasional jobs were more appropriate.

It would be wrong to conclude that college men are really "pigs in sheep's clothing" (to twist a phrase a bit). Wrong for two reasons. First, as in all other developmental dimensions, in attitudes about sex roles variability is the rule rather than the exception. Although the evidence suggests that the values reported by Komarovsky and Almquist are held by a large percentage of male college students, they are, nevertheless, not held by all. (It is equally true that, although an increasingly larger percentage of college women profess feminist values, a significant number still hold and value the more traditional patterns.) Second, Komarovsky believes that these men favor traditional roles for their wives because they are locked into a dualistic view of sex roles. In the same sense that Perry (1970) found that first year college students used an either/or pattern of reasoning in approaching general issues, these men seem to show the same kind of reasoning with regard to a more personal and immediate issue. This kind of reasoning reflects the general developmental pattern that has been noted elsewhere in the text. Namely, people are able to develop a relative rather than an absolute perspective sooner for issues that they are less personally involved in.

There is clearly a developmental quality to views on the appropriate roles of adult men and women. The view presented by Komarovsky's male college student may be as much a reflection of the importance he places on believing that he is capable of adequately providing for his family as it is a reflection of his feelings about his future wife's employment. Indeed, most young adult males welcome the added income provided by a working wife. Unfortunately, this change in attitude is still not often accompanied by an increased willingness or desire to be more involved in household routines and child care (Pleck, 1983).

Long-term predictions of social change suggest that the present dilemma facing college men and women as they attempt to integrate their value structures may be, from a broad historical perspective, a temporary condition. The lack of accord between male and female college students' views about sex roles may indicate that the relative transition rates for men and women from a traditional (complementary) to an egalitarian (reciprocal) perspective are different. Women may be changing their sex-role orientation at a faster rate than men. This seems likely for two reasons. First, male roles have historically been more valued in our culture than female roles. Since they are seen as more desirable, the change for women is in a "positive" direction; for men, it is in a "negative" direction. If we want to accelerate the rate of change for males, we need to change the values attached to male and female roles. Second, the socialization patterns of males may leave them less flexible about adopting alternative role expectations. Because they have been reared believing that they will someday bear the responsibility of providing for the needs of others, they may be less able or willing to tinker with the procedure.

There is, however, evidence of change. Both male and female children reared in homes where both parents work have less traditional sex-role values and are more supportive of egalitarian roles, in their personal lives as well as in the general society (Huston-Stein & Higgins-Trenk, 1978; Meier, 1972). Further, these children are more likely to choose less sex-typed majors in college. The fact that these findings are from middle-class samples, as well as the fact that these changing sex roles are more evident in college populations, stresses the role social class and education play in career opportunities and life-style during the adult years.

Transition Patterns of Noncollege Youth

If approximately half of all high school graduates go on to college, then the other half must not. What happens to them? Do they experience a transition period parallel to the one experienced by college students or do they simply skip the transition and "leap" into adulthood? The limited evidence suggests that there is a transition phase but that it is probably shorter for noncollege youth. The evidence is limited because the noncollege population is less accessible to researchers, many of whom are college faculty members. However, research by Rubin (1976) on low-income working-class families gives some indication of the nature of this transition phase, at least at the lower end of the working-class continuum:

> "When I was young"—the phrase spoken by a twenty-seven year old sales clerk, mother of three, married ten years. "When I was young"—a phrase used repeatedly by working-class women and men well under thirty; a phrase that surprised my middle-class professional ears, accustomed as they are to hearing students, patients, children of friends, my own children—all of like age—struggling with problems of identity, reluctant to step into adulthood, still defining themselves in terms of youth and the youth culture. Such an extended psycho-social moratorium and noncommittant crisis of identity is, it would seem, a luxury that belongs to an economy and a culture that can afford to permit its young the privileges of adulthood without the responsibilities. For the working-class youth, those privileges—including separate domiciles and sexual relations outside marriage that are accepted and legitimated by the community—come only with marriage. For them, this is yet another of the hidden injuries of class. For them, there is no time for concern about the issues of their own growth and development that so preoccupy the college-educated middle-class youth in this era; no time to wonder who they are, how they can differentiate themselves from parents, how they can stand as separate autonomous selves. Instead, early marriage and parenthood combine to catapult them, ready or not, into adult responsibilities. (pp. 72–73)

Changes in Attitudes and Values

The changes in attitudes and values among college students that began in the 1960s are also evident in surveys of noncollege populations (Yankelovich, 1974). Within the entire eighteen to twenty-five-year-old age group there has been a liberalization of attitudes and behavior concerning sexual activity, a decrease in the view that religion is a source of moral guidance, a decrease in automatic respect for established authority, and an increase in the amount of importance associated with self-fulfillment as opposed to the amount of importance associated with self-sacrifice.

Because they often go from high school to motherhood without the break for college, working-class women seem to experience a shorter transition period between childhood and adulthood than do their college-educated counterparts.

Yankelovich (1974) sees these changes as the legacy of our involvement in the Vietnam War. The emerging pattern represents an attempt to find conventional careers that are both personally satisfying and financially rewarding enough to enable individuals to pursue full lives outside of work. But as much as this desire has become part of the noncollege population's value structure, the reality is that these individuals have fewer opportunities than college educated men and women to find work that makes such a synthesis of private and professional lives possible. And they know it. Most believe that they will be able to find work that is adequate in terms of income and job security; few feel that they will be able to find work that has the potential for self-expression and self-fulfillment (Osterman, 1980).

Although the general trend in the noncollege population is toward the same value structure that the college population holds—toward this new synthesis—the shift in value structure is more evident for men than for women. There is a greater degree of similarity between the attitudes and values of college and noncollege men than there is between the attitudes and values of college and noncollege women. According to Yankelovich (1974), noncollege women see work as a way to make "ends meet" rather than as a source of personal fulfillment. Further, most look forward to marriage as the opportunity to stop working—a dream not realized in the majority of cases. These women place great value on the role

Many young working-class women see a job as a means to make ends meet and a stopgap until marriage. Unfortunately, the reality of the situation is often very different.

of wife and mother and profess little support for the feminist positions supported by college women. Since the job opportunities available to non-college-educated women tend to offer the least possibility of adequate income or job security, not to mention self-expression and self-fulfillment (West & Newton, 1983), their adherence to traditional values seems quite understandable.

The fact that college-educated youth have a greater chance of finding Yankelovich's "new synthesis" than the noncollege population does reflects the types of jobs available to each. The college population is more likely to be hired for jobs that are comparatively free of close supervision, that involve a degree of personal decision making, and that are relatively complex in nature. On the other hand, the jobs available to the noncollege population are generally characterized by close supervision, little opportunity for self-direction, and relatively simple tasks (Kohn, 1977). It is important, however, not to overemphasize the degree of difference between the kinds of jobs available to the two groups. It is certainly not correct to think that all jobs available to college graduates involve great opportunity for personal decision making and self-fulfillment and that those for the noncollege population are the most menial and dehumanizing. However, the jobs of the college population increasingly take on the quality of careers, whereas the jobs of the noncollege group remain merely work, and the gap between the two kinds of jobs continues to widen throughout the adult years. A career offers the potential for intrinsic as well as extrinsic rewards;

work usually only offers the latter. Therefore, one of the consequences of a college education is that it leads to job opportunities that are different in nature from the job opportunities open to the noncollege population, and these opportunities define the role and meaning that work acquires in the life structure of adults.

Early Work Histories

The actual work histories of noncollege youth follow a two-step sequence (Freeman & Wise, 1982; Osterman, 1980). Jobs taken immediately after high school graduation are seen primarily as a means to obtain enough money to buy things that are wanted—a car, stereo system, clothes, entertainment, and so forth. These early jobs are selected without much thought to future plans, and "on-the-job" performance is often erratic. Working hard, making a good impression on the foreman or supervisor, and being considered for promotion don't seem to be considered important. Attendance records at work are sometimes poor, and since there are always a number of these "dead-end" jobs available, few young workers are reluctant to quit if they feel like it, hang out for a few weeks until their money runs out, and then find another job. The fact that many of the noncollege youth between eighteen and twenty-two who feel this way about work continue to live at home helps make this life-style possible. Further, even though many continue to live at home, most find their income gives them more independence from their parents than they had during high school. In many cases, they gradually acquire the status of a boarder, often paying a nominal rent (West & Newton, 1983).

Sometime during the early twenties, these employment patterns change. There is a greater emphasis placed on finding a good job, one that pays well and offers some degree of job security and a future (Osterman, 1980). There is a corresponding change in job performance. Attendance becomes more regular; work performance, more satisfactory. Jobs are held for increasingly longer periods, and when a job does end, there is a much greater effort made to find another one. Often, this shift in job behavior corresponds to the development of a steady relationship and plans to marry. The timing of this shift is an important predictor of future job performance. If the shift is made by the middle or late twenties, the rather haphazard first phase of work performance is not predictive of later patterns. If the shift does not occur by this period, the probability that it will ever occur steadily decreases (Freeman & Wise, 1982).

What kinds of jobs are available? For men, the most common kinds of jobs are in manufacturing, construction, and the service industries. For women, the most common kinds of jobs are in clerical work and sales (Fox & Hesse-Biber, 1984). These jobs tend to be more sex segregated than the jobs available to college graduates are. For the most part, these jobs are found rather than chosen. Their availability is primarily a reflection of national economic conditions and the nature of local industries. One of the major distinctions between college and noncollege youth is that the former are more likely to choose a job; the latter, to find one (West & Newton, 1983).

The transition to a more stable job pattern is usually valued by men as a positive step, a sign that they are finally assuming their place within the adult community. Although some are unhappy with their specific working

During their early twenties, most working-class men begin to place greater emphasis on finding and keeping a good job.

conditions, they enjoy the fact that they have a steady job. Women seem more ambivalent about the transition. Since adulthood for them is primarily defined in terms of marriage and motherhood, the taking on of more responsible jobs may be associated with difficulties in finding a husband (West & Newton, 1983).

Marriage

Over 90 percent of the adults in this society marry at least once during their adult lives. For most of these people, the first marriage usually takes place while they are between eighteen and twenty-five years of age. Over the last two decades however, there has been a gradual trend, at least among the college population to delay marriage into the middle or even late twenties. This trend reflects the desire of an increasing number of college-educated adults to establish a foothold in the workplace or complete their postgraduate education before considering marriage. It probably also reflects a greater willingness on the part of other college-educated adults to give serious consideration to the option of remaining single. It also reflects a heightened anxiety generated by the significant increase in the divorce rate over the past fifteen years.

The Decision to Marry

What are the social pressures that are so effective in getting people to marry? Certainly the first must be the simple fact that marriage is a virtually universal characteristic of adult life; indeed, it is one of the elements that confers adult status. Second, there are a variety of economic and legal benefits to marriage. There are tax breaks for married couples, inheritance rights, employee fringe benefits, social security benefits, joint property arrangements, and other similar practical advantages. Although it's unlikely that many people marry because one has a family dental plan fringe benefit and the other a number of cavities, considerations such as these tip the scale just that much farther towards marriage.

There are a variety of religious, moral, and social sanctions favoring marriage. All of the major Western religions place great emphasis on marriage as one element of being a religious person. Society is much more accepting of married than nonmarried adults and is particularly critical of adults who parent without first marrying.

Finally, there are the values of companionship and love. Historically, these emotional aspects of marriage have taken a backseat to the more "practical" considerations reflected in the giving of a dowry and the like, but as marriage in our culture has continued to become a decision between two people, the special feeling of being in love and the comfort of being with a special person have become very strong incentives to marry.

Finding a Mate: College Youth

As is true of age at entry into the work force, length of education is a significant predictor of whom a person will marry and, increasingly, of the type of marital relationship the person and the spouse will establish. Propinquity and homogamy (being in the same place at the same time and having similar backgrounds, that is) seem to be the two major factors in early mate selection (Reiss, 1980), and the college environment is very good at fostering both.

Initial contacts. The college environment brings together large numbers of people of similar age, provides a variety of official and unofficial channels through which they can meet, provides them with ample free time to do the meeting, and lets the meeting take place, usually in the absence of parental or other adult supervision (Gagnon & Greenblat, 1978). It is true that the social patterns found at a small, rural school with a strong religious orientation are different from those found at a large state university enrolling thousands of students. The social patterns at a large state university are, in turn, different from those at a commuter college where most of the students continue to live at home or in apartments scattered throughout the city. But in all three cases, the very fact that people are meeting other people within the college community means that their potential field for future mates has been reduced by half—that is, they are now most likely to meet people from the group of high school graduates going on to college (and approximately one-half of all high school graduates go on to college). Within the college community, the potential field is further reduced by such factors as living arrangements (dormitory versus fraternity/sorority), family background, and future career plans (students considering postgraduate education are usually not interested in developing serious relationships). These factors don't predict whom a person will marry, but they do define the type of person he or she is most likely to marry, if for no other reason than that they influence the likelihood of the person's meeting different types of people (Murstein, 1976; 1980; Reiss, 1980).

These "sorting" factors continue to operate after college since type of work and, to a lesser extent, the single adult life-style correlate with college attendance. In fact, only about 15 percent of all college students marry

BOX 7-2
DELAYING MARRIAGE

Three events combine to define the character of the transition years: completion of schooling, entry into the work force, and marriage. These three events are important for other reasons as well as the behavior changes required by each. Of more long-range significance, the timing and sequencing of the three events set in motion a series of events that essentially defines the character of the adult years often even into the middle years.

The timing and sequencing of the three transition events are best understood as cohort factors. In particular, they reflect the political, social, and economic circumstances that form the context for a particular generation's developmental experiences. One of the most significant of these cohort factors has been the impact of the women's movement. There is a tendency today to downplay this impact, to argue that the current cohort of young adult women is less interested in and concerned about issues that made previous cohorts feel very militant. It may be true that there are fewer marches and headlines about women's issues, but the data indicate that the women's movement is having a very clear impact on the lives of the current cohort of young adult women. Specifically, there has been a drastic shift upwards in the predicted age at first marriage for the current population of college-educated women in their early twenties.

This shift has been noted in longitudinal studies of the developmental patterns of young women, and perhaps of even greater importance than the shift itself is the variable that best predicts it—the answer to the question, "What kind of work would you like to be doing at age thirty-five?" The more likely the respondent is to say that she wants to pursue a career in addition to her roles as parent and spouse, the more likely she is to delay marriage, on the average, about two years. Interestingly enough, present work plans are not a predictor, only future work plans. That is, more and more young women are beginning to believe that life decisions such as the decision to marry must be compatible with individual goals, especially those concerning

Source: Cherlin, 1980.

while in college, and by graduation only about a third of the graduates are involved in relationships that result in marriage. Another third are involved in serious relationships but do not yet define them in terms of future marriage, and the remaining third report no serious relationships. For almost all college attenders, however, the experience prompts the beginning of a process that will most likely result in their finding a person they want to marry and in the decision actually to marry. As will be discussed later, the sequence is similar but not identical for noncollege youth.

Although most people eventually find someone to marry, the evidence indicates that few start out looking for a mate in a very systematic or

careers. Historically, the opposite pattern has been the norm.

What are the consequences of this shift? It will probably result in smaller families, and it may reduce the divorce rate. Both factors correlate with age at marriage, income, and education. It may well change the "standard operating procedures" in the workplace. The evidence indicates that tolerance and appreciation of others usually follow integration. Not all the changes may be positive. The fact that this shift in age at first marriage is only evident in the college population suggests that social class differences may actually increase. The percentage of non–high school graduates answering the question about work at thirty-five has shown just the opposite pattern: they are now more likely to want only to be housewives than members of previous cohorts were.

In the not-so-distant past, many people—including women—viewed college attendance as a means to find a suitable husband. Today, the justification for female college attendance is much more likely to be career related.

The cohort shifts associated with the women's movement provide strong evidence for the continuing cumulative impact of events within the sociohistorical context. Perhaps some day it will be possible to say that its cumulative impact, as well as that of other disenfranchised groups in society, has been so great that it doesn't even need to continue to exist as a distinct movement.

deliberate manner (Adams, 1979; Clayton, 1975; Murstein, 1976; 1980; Reiss, 1980; Ryder, Kafka, & Olson, 1971). First meetings are often chancy affairs. A friend of a friend of a friend has a friend in town for the weekend. You can't seem to concentrate on your reading and decide to get a cup of coffee at the student union. Your roommate drags you to a student mixer you would prefer not to attend. You get to class late and the only seat left happens to be next to this person. Not all first encounters are this random, but most couples report that their first meetings were usually uneventful. Apparently, fate doesn't play much of a role: few couples report love at first sight. Whether or not the two will plan to meet again usually depends

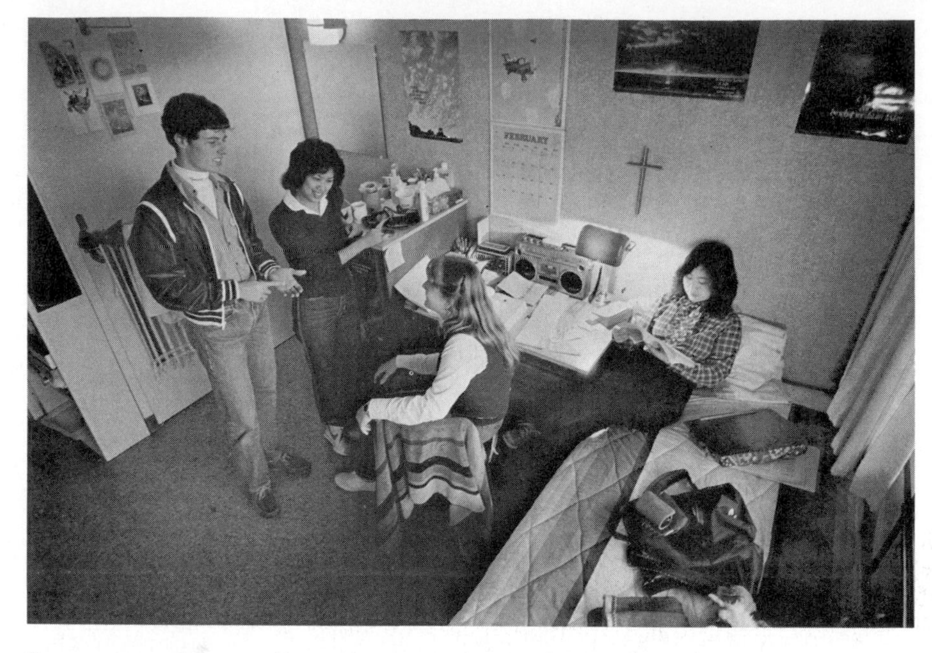

For many students, college often provides the first opportunity to meet informally with the opposite sex away from the watchful eyes of their parents or other concerned adults.

on physical appearance, similarity of interests and behaviors, and what, if anything, each knows about the other from friends.

What often follows from the first encounter is a series of dates or shared experiences as part of a larger group (such as a social group or special interest club). On some campuses, these meetings tend to resemble a structured dating pattern; on others, a more informal structure is the norm. One person mentions to the other that he will be at the library, the other mentions that she will be too, "why don't we get a beer afterwards?" "If I finish my human development reading, fine." And so it goes. There is, as yet, no real effort to have frequent contacts. Each may be seeing other people, even involved in a more serious relationship. Each may like the other and think the other fun to be with, but perhaps, like the three-month-old, they show little unhappiness about separation and little anxiety over the fear of loss. Circumstance and peer social patterns are still as instrumental as mutual attraction in maintaining the relationship.

Establishing relationships. If the relationship continues, it begins to take on a new quality. Conversations that previously focused on classes, mutual friends, campus events, previous experiences, and the like now gradually shift toward attitudes and values, fears and concerns, and hopes and goals (Clayton, 1975). Both people gradually expose their inner selves. They gradually test to see whether the other person can be trusted with being told anything more intimate or substantial than their favorite foods or love for playing football. This transition represents a first step in the development of what Erikson (1968) calls a sense of intimacy, or a sense of mutuality. The couple seems to be trying to determine, at a more personal level than before, how much they have in common in terms of values, attitudes, and goals. The transition is probably not very intentional or premeditated. Only in retrospect would the demarcation be evident. But this process of self-disclosure and rapport building provides each with independent evidence of the correctness (in terms of the agreement of the other) of his or her attitudes, values, feelings, and beliefs (Adams, 1979). It also identifies the other as a person who will both listen to and understand the private self that is being revealed (Clayton, 1975). With this transition, the couple begins to establish bonds. The relationship is no longer merely held together by circumstance or mutual friends. It can now survive a "change of major" or the like. It now exists in part because the other person has taken on a special quality and value.

In the traditional sequence, the two people continue the value-clarification, rapport-building process, at some point decide that they are in love and wish to marry, and become formally or informally engaged. They then become involved in the whirl of events involved in planning a wedding, establishing a home, and learning to live as a married couple. In many cases, this pattern still proves the norm, but over the past twenty years, particularly among the college population, an additional step has been added to the sequence.

Committed relationships. This new step is the establishment of what, for want of a better term, can be called a committed relationship. It is usually an exclusive relationship. Often the two people are sexually intimate, and in many cases, they set up some sort of joint living arrangement (Macklin, 1972; Peplau, Rubin, & Hill, 1977; Reiss, 1980; Scanzoni & Scanzoni, 1981). What is unique about this new step is that it is primarily present rather than future oriented. Couples who take this step do not see it as a new form of engagement. Likewise, those who actually live together do not see their arrangement as a trial marriage. Rather, committed relationships have been made possible by changes in society's views on premarital sexual activity; by the availability, effectiveness, and convenience of modern contraceptives; and by the virtual elimination on many campuses of curfews, dorm visitation restrictions, restrictions on off-campus living, and, more generally, the view that the college must act in loco parentis in regard to its students. From the couple's standpoint committed relationships offer a haven from the loneliness of large universities, an intimate context for sexual activity, and a chance to further the identity process into the area of intimate interpersonal relationships (Davidoff, 1977).

Committed relationships can take any one of a number of forms. In about 25 percent of the cases, the two people live together for six or more months—usually one moves in with the other rather than the two jointly looking for housing. In some cases, the two people maintain separate housing but live together on weekends. Although food and entertainment expenses are usually shared, money is rarely pooled into a joint checking account. Within the campus community, the living arrangements are usually public, and friends confer upon the two people the status of "going together" or "living together" or some other term that recognizes their special relationship.

Parents are apparently another matter, however. Some people report that their parents are unaware that they are living with someone and would be hurt if they found out. A few even go to very elaborate lengths to prevent their parents from knowing. Friends are enlisted as accomplices, and rooms are rented to establish a separate mailing address. Even when parents are aware of the arrangement, some people feel uncomfortable about bringing the other person home for a weekend, perhaps fearing conflict over sleeping arrangements. The ambivalent status of these relationships probably contributes to the fact that so few do become more permanent. Only 20 percent of the couples who live together eventually marry (Peplau, Rubin, & Hill, 1977).

Nevertheless, committed relationships, often involving cohabitation, are viewed favorably by many college students. In addition to the 25 percent who establish such relationships, an additional 50 percent report that they would like to if they met the right person. And those who do have such relationships report a deeper understanding of themselves, a broader knowledge of what is involved in maintaining a relationship, an increase in self-confidence, and greater emotional maturity as a result of the experience (Macklin, 1972).

Since these relationships are present rather than future oriented, most eventually end before marriage. In this sense, they are probably more akin to going steady than to being married. They end for a variety of reasons, both practical and interpersonal. Some people always fear that their parents might find out and for some, the constant juggling of living arrangements gets to be a hassle. In many instances, even effective use of contraceptives does not eliminate the fear of pregnancy when people don't want to become pregnant. Finally, some people report feelings of being trapped in a relationship, sexually exploited, or isolated from other friends. In effect, committed couples face the same basic dilemma that confronts young married couples—namely, to achieve a sense of emotional security without losing a sense of personal freedom.

Commitment to marriage. Feelings of love are not necessarily a part of committed relationships. They are, however, an integral part of relationships that become future oriented and are seen as potentially resulting in marriage. Poets certainly do a better job of defining love than we more scientifically oriented souls, but we do our best. The very notion of "falling" in love reflects the romantic aspect of the emotion. It is certainly a time of grand emotions, rose-colored glasses, and the like. It is certainly a state of ecstasy (which, by the way, comes from the Greek word meaning deranged). It is the emotion that prompts wonderful visions of the future and the glossing over of more practical matters. Companionship love, in comparison, is a more restrained emotion and perhaps more predictive of the future health of a relationship. It is an emotion of deep affection and attachment, of caring about another as much as you care about yourself.

Men appear to fall in love earlier in relationships than women do. Women, on the other hand, fall "farther" (Kanin, Davidson, & Scheck, 1970). The two patterns reflect the often different perceptions men and women have about the role of intimate relationships in their lives, as well as about the effect these relationships have on a sense of personal identity. It is often at the point of falling in love that the two people may begin to decide how much each is willing to compromise individual goals and plans. This process is similar to what occurred earlier in the relationship, but this time the emphasis is on the future.

Once the couple acknowledges to themselves, as well as to others, that they are considering marriage, a new set of forces begins to act upon the relationship. One involves the extent to which parents and peers either support or oppose the match. Parents may be relatively unconcerned about the people their child dates, and if they aren't aware that their child is living with another person, they can hardly have much to say about it. Peers may feel that a couple is not well suited, that one is simply using the other, but they may be reluctant to raise the issue if they believe that the relationship will probably end on its own. However, once the couple announces their intention to marry or even if one mentions to parents or peers that they are considering marriage, concerns are much more likely to be voiced or support given.

Other forces also begin to be felt. The couple becomes identified as a "couple." Neither person is invited to a party alone any longer; the two are invited together. Other people become aware of the couple's new status and perhaps treat them a little differently. Announcements are made in the local paper. Each person is brought home to meet the other's family—grandparents, aunts, uncles, cousins, and so forth. Wedding arrangements begin. Housing and furnishings are discussed and explored. Down payments are made. All of these activities serve to foster the transition during which the two people come to think of themselves as part of a couple, or to use "us" or "we" rather than "I".

As the two people jointly deal with each activity, they become more aware of their mutual dependence (Clayton, 1975). They each have the other to turn to when a parent wants to invite "just a few more old friends" to the wedding. At the same time, this growing interdependence can be quite frightening. Fears about what actually is being given up and doubts about whether this is really the right decision are very common among engaged couples (Gagnon & Greenblat, 1978). Both feel that they are no longer in control of what is happening to them, that they couldn't get out of the relationship now even if they wanted to (Adams, 1979). Given the nature of these fears, many people are unable to talk about them with their partner. Often, an older sibling, trusted friend, or parent provides the necessary support and assurance that the feelings are normal and temporary.

The pattern of couple formation described in this section is most representative of couples that form while still in the campus community. Moreover, even though the pattern is typical for them, there is considerable variability at each of the points in the sequence. A good deal is known about this sequence simply because it's accessible to investigators who, for the most part, are college faculty members. Less is known about the patterns typical of noncollege youth or college graduates who have made no commitments at the time of graduation. What is known suggests the pattern is similar but with a few important differences.

Finding a Mate: Postcollege Singles

Graduation from college is one of those critical incidents that prompt people to make important decisions such as to marry. Quite a few couples report that, although they would probably have decided to marry anyway, the timing of the decision was primarily due to the fact that they were graduating at the end of the year (Gagnon & Greenblat, 1978).

Those who graduate uncommitted enter a singles world that gives them less support in meeting other people than the college environment did. Work and family become more important sources of finding new people. The singles bar replaces the library or student union as the place to meet new people. Old friends from high school, recently married class-

The desire to meet eligible partners makes the after-work gathering at the local bar a ritual for many postcollege singles.

mates, even neighbors come to serve as matchmakers. The life-style maintains much of a collegiate flavor. Going out and doing exciting and new activities are still valued. However, when more serious relationships are formed, they move more rapidly into exclusive relations than they did during college (Gagnon & Greenblat, 1978).

Four factors distinguish the postcollege mate selection process from the college mate selection process. One is that family and workplace become important resources for social contacts. Second, all settings, including the family and workplace, provide much less support for the making and maintaining of contacts than the campus setting did. Third, because these settings provide less support, there appears to be a less playful and more serious or deliberate tone to social contacts when they are made. Knowing that you're going to be in the same class with a person every Monday, Wednesday, and Friday at three for the entire semester gives you a certain leeway that isn't available in most other settings. Fourth, because the contacts are more serious and because there is less of a social structure to support them, the potential for exploitive relationships to develop is somewhat greater.

Finding a Mate: Noncollege Youth

What about the noncollege population? Again, the basic sequence seems to be the same (Simon, Gagnon, & Buff, 1972), but those factors that do distinguish the sequence for noncollege youth from the college pattern indicate that, although the sequences may be similar, the meaning and timing attached to the events are often quite different.

Because home and work play a larger part in noncollege mate selection, the process probably takes on the same degree of deliberateness that is typical of the postcollege population. Therefore, the informalness and the playfulness that are typical of college populations are less typical in the noncollege population.

For the college population, marriage and leaving home are usually separate matters. For the noncollege population, they are more often the same issue. In situations where there is a great deal of parent-adolescent conflict, the motivation to marry may be wanting to get out of the house as much as it is wanting to be with a person loved.

Because sex-role differentiation appears greater in the noncollege population, the men and women in this population are more likely to differ in their concepts of marriage than the men and women in the college population are. Since more activities are sex typed, shared activities are fewer. Since work plays a less important role in the lives of women, a larger part of them is invested in the desire to marry and the role of wife and mother. And, since the sexual double standard remains a more prominent characteristic of this population, sexual activity, especially for women, is more often couched within the context of a future marriage.

Further, since contraceptive use is less frequent in the noncollege population, a larger percentage of women are pregnant at the time of their marriage (Chilman, 1979). The fact that the average length of time in Rubin's (1976) sample between marriage and the birth of the first child was nine months suggests that either many of those marriages would not have occurred if there was no pregnancy or, at the very least, would have occurred later in the relationship. Neither circumstance represents a very good way to start a marriage.

The Establishment Phase

Sooner or later, irrespective of the particular circumstances surrounding the event, whether on top of a mountain or in a very proper church, with one set of words or another, each person says "I do." And these fateful words announce to the assembled congregation that a formal commitment is being made between two people to live together as a couple. Once the cake is cut, toasts made, and bouquet thrown, the two people must begin the process of learning to live as a couple.

It has become an axiom within the field of human development that people tend to be more vulnerable and impressionable during periods of major transitions than during periods of relative stability (Bloom, 1964; Golan, 1981). For this reason, the range of reactions to situations often increases during major transition periods. Situations that at other times elicit a mild reaction may elicit a more extreme one. The **establishment phase of marriage**—the period from the wedding to the birth of the first child—provides a good example. Some researchers see it as the most positive phase of the entire marriage career (Rollins & Galligan, 1978), a time relatively free of debt and a time of high communication between partners. But the establishment phase also has a higher rate of divorce than other periods in the marriage (Aldous, 1978). This wide range of potential reactions reflects the nature of the tasks encountered during this period, as well as the difficulties sometimes encountered in their resolution.

The developmental tasks of the establishment phase serve to help the husband and wife shift their orientation from thinking of themselves as relatively autonomous individuals to thinking of themselves as members of an interdependent pair (Golan, 1981). Couples certainly differ in the ways in which they make the transition and in the degree of interdependence they eventually establish, but in all cases, the "bottom line" is that both people must begin to consider the consequences of their actions, not only in terms of themselves, but also in terms of their partner.

The various tasks can be grouped into three categories. The first involves the relationship of the two people to each other; the second, the couple's relationship to other people and events; and the third, the balance between the first two tasks (Clayton, 1975).

Learning to live with another person. Learning to live with another person on an intimate day-to-day basis can prove a trying task. Although establishing a mutually satisfying pattern of sexual activity is a large part of this task, intimate living involves a great deal more. In the course of growing up, people acquire a variety of idiosyncrasies. Most are no doubt based on observations of others, and for the most part, they are behaviors that receive little reflection. They are certainly not the stuff of discussions between couples who are trying to decide if they can marry and live together successfully. But toothpaste tubes left uncapped, toilet seats left up, dirty socks under the bed, jelly jars left on the kitchen table, stockings hanging from the shower bar, mud on the carpet, and the like can prove more annoying than most of us like to admit and, for this reason, become one of the elements that must be dealt with as two people learn to share the same space.

More generally, the couple must begin to develop mutually satisfying styles of decision making, conflict resolution, and communication. How are purchases going to be made? Are purchases going to be charged, or is the money going to be saved first so that the price can be paid in full? What is going to be bought? Where are they going to live? Who is going to do the cooking and the shopping? How much money is to be saved and how much spent on entertainment? The list is probably endless, but as more and more decisions are made, patterns of decision making gradually emerge.

Conflict is an inherent part of all intimate relationships. It is evident in successful marriages, as well as in unsuccessful ones. What distinguishes successful from unsuccessful marriages is how couples learn to manage this conflict. Does one person always defer to the other? Does violence ever become part of the resolution process? How able is the couple to focus on the issue and not engage in personal insult and derogation? To what extent does conflict spill over into other aspects of the relationship? How long does conflict fester before it is dealt with and, once it is resolved, how long does it take the couple to reestablish a sense of normalcy? Consider the difference in the couples studied by Aldous (1978):

> There were definite differences, however, in the conflict sequences of the more happily married couples as contrasted with those having trouble getting along. The happily married couples tended to resort to rejecting and coercive statements much less than other couples. They preferred instead to provide information about the issue causing the disagreement or to suggest a solution. The husbands of the happier couples were also more likely than other husbands or even their own wives to initiate peace-making statements. In contrast, the statements of the discordant couples spiraled quickly into power struggles in which neither one listened to the other and in which both resorted to destructive personal attacks. In contrast, the happier couples who engaged in conflict were sensitive to each other's feelings and more concerned about their interpersonal relationship remaining positive than the outcome of any conflict interchange. (p. 147)

Another element of an intimate relationship is the extent to which couples share their thoughts and feelings with each other. Again, couples can differ greatly in this regard, but during the first few years, expectations gradually form concerning the extent to which each uses the other as a source of emotional support, as a sounding board, as a source of advice, and, more generally, as the person who understands him or her best.

A perverse pattern of violence and reconciliation becomes an enduring feature of some relationships.

Styles of intimate living, styles of conflict resolution and communication, styles of decision making, and divisions of labor serve to define the two people's relationship to each other. An equally important element of the establishment phase is the way the couple defines their relationship beyond each other.

Learning to deal with family and peers. One important element of this second category of tasks concerns the nature of the relationship the couple establishes with each set of parents and, more broadly, with each partner's extended family. Family relationships not only involve social contacts but also serve as important sources of information and advice and, in many cases, as equally important sources of money and other tangible goods and services (Hill, 1970). In addition, since marriage redefines the parent-child relationship, both partners must establish a new balance between themselves and their mother and father.

A parallel process occurs with friends. Even in cases where the couple held most friendships in common before they married, there is usually a period of redefinition. When old friends were not held in common, each partner must meet the other's and establish some sort of friendship patterns with them. In either instance, one of the most common issues is the extent to which each partner will maintain friendships independently of the other. For example, what happens if one person does not like a close friend of the other? If one is used to going out with the girls or the boys two nights a week, will the pattern continue? And, often more difficult to deal with, how will opposite-sex friendships be integrated into the relationship? Although in-law jokes are legion in our culture, the evidence (Ryder,

Kafka, & Olson, 1971) indicates that peers often present a greater integration problem during the early years than families do.

Establishing a balance. The third category of tasks, which is concerned with establishing a balance between the tasks of the first two categories, involves the extent to which the two people redefine themselves in terms of the new relationship and the extent to which they continue to see themselves as before. One way to judge the new balance is to note the extent to which each person is willing to and actually does alter old patterns. Another is to determine the extent to which each expects the other to accommodate himself or herself to previously established patterns. How much time and energy are actually put into the relationship? How much, for example, are reserved for career? Once again, couples vary greatly, and resolution of this issue involves establishing mutually satisfying patterns that allow the marriage to grow rather than modeling any one particular pattern. However, as will be discussed shortly, on this issue husbands and wives often soon find that, in Bernard's (1982) words, "her marriage" and "his marriage" are quite different.

Correlates of Marital Satisfaction

What are some of the factors that determine how easily the developmental tasks of the establishment phase are accomplished? One set of factors concerns the events surrounding courtship and marriage. In particular, the age of the couple at marriage and the length of their courtship each correlate positively with ease of resolution (Aldous, 1978). The younger the two people are at marriage (at least within the age range of this chapter) and the shorter the courtship, the more likely one or both of them are to see the marriage as an escape from a home life viewed as oppressive. Further, since age at marriage correlates positively with years of schooling completed as well as with the interval between marriage and the birth of the first child, the younger the couple, the fewer resources they have and the lower their potential is to obtain additional resources. The result of this set of circumstances is that the teenage couple often is forced to make do with minimal resources or to ask help from parents, the very people they were trying to get away from in the first place. In either case, limited resources and continued conflict with parents increase the likelihood of conflict between husband and wife and decrease the likelihood that it will be easily dealt with.

A second set of factors concerns the expectations about married life that each person brings to the marriage. These expectations may reflect the married patterns of other people, usually parents and close friends, but may also reflect the media, as well as whatever hopes and dreams the two people cherish. In the population of low-income families that Rubin (1976) studied, the importance of hopes and dreams was very evident.

> Thus, despite the fact that the models of marriage they see before them don't look like their cherished myths, their alternatives are often so slim and so terrible—a job they hate, more years under the oppressive parental roof—that working-class girls tend to blind themselves to the realities and cling to the fantasies with extraordinary tenacity. (p. 41)

Lederer and Jackson (1968), through their work in counseling married couples, noted that problems could often be traced back to expectations

When teenage couples use marriage as a means to escape an unhappy home life, they often find that they have created a worse situation.

that proved unrealistic and incorrect. In particular, they found that many people saw marriage as a cure for a personal sense of loneliness, that many people felt that the presence of conflict within a relationship was a sign that the relationship was not healthy, and that many people also felt that unsatisfactory sexual relationships were a cause of bad marriages.

A third set of factors affecting the accomplishment of developmental tasks concerns the extent to which previous experiences serve to foster or hinder their resolution. It would seem reasonable to assume that those who have had sexual experiences, particularly with the person they later married, would have already begun to resolve some of the life-style issues. Further, it would seem that people who have cohabitated, especially if their partner became their spouse, would again have resolved some of the life-style issues. The evidence (Macklin, 1974; Peplau, Rubin, & Hill, 1977), however, does not present a strong case for this argument. To date, there is little evidence that either prior sexual or prior cohabitation experiences correlate with success in marriage, at least in terms of rate of divorce. There are two possible explanations for this lack of relationship. One is that, since most people who are sexually active before marriage or cohabitate do not conceptualize these activities in terms of future marriage plans or patterns, they may not transfer the insights gained from these experiences to an experience they view from a very different perspective, that is, marriage.

A second interpretation is that there are insights gained from premarital activities that do transfer to the experience of early marital adjustment but that other factors offset the advantages provided by these insights. In particular, there has been an increasing emphasis over the past thirty years on the idea that couples should establish their own unique identity rather than adopting more typical patterns of married life (Aldous, 1978). Couples choosing this route are faced not only with the necessity of resolving a set of developmental tasks but also with the problem of first determining which tasks are relevant to the style of married life they wish to pursue. Raush, Barry, and Hertel (1974) have noted that as divisions of labor and relationship patterns have become more relaxed and more individual, there has been more anxiety associated with the development of acceptable life-style and communication patterns.

Those couples who opt for traditional patterns establish a relationship that gives most of the decision making and resource control to the husband. The division of labor is relatively clear-cut, with little overlap in roles and responsibilities. The husband assumes responsibility for providing money for the family; the wife assumes responsibility for maintaining the home and eventually for child rearing (Clayton, 1975). This pattern has been the most common among married couples, but cultural shifts, especially within the more educated segments of the population, have significantly modified this balance. In particular, as more and more women enter the work force and as more who do so view their work not solely in terms of a second income for the family but also as a source of personal growth and fulfillment, couple relationships are becoming more like partnerships (Scanzoni & Scanzoni, 1981).

In most dual-earner relationships, the husband still earns the majority of the family income, the wife's contribution is usually between 20 percent and 40 percent. In only 10 percent of the cases does the wife contribute one-half or more of the family income. But even when the wife's contribution is less than half, the fact that she is also bringing money into the family, that she conceptualizes her work in the same way her husband does, and that household routines and responsibilities must often be altered to accommodate both adults' working is resulting in an equalizing of power within relationships and a gradual breaking down of the traditional divisions of labor in many young married couples (Scanzoni & Scanzoni, 1981). It is still too early to tell whether these cultural trends in couple relations will continue to move in the direction of greater equality of roles and responsibilities or stay with this "senior partner/junior partner" pattern. It is also too early to tell if this shift will become equally common among couples from less privileged backgrounds. What is clear is that as long as husbands think these shifts benefit their wives more than them, the rate of change will continue to be gradual (Scanzoni & Fox, 1980).

Earlier in this chapter, in the section on identity formation in young adults between eighteen and twenty-five years of age, mention was made of the imbalance, reported more and more frequently, between the expectations that men and women have about the way they will live their early adult years and their beliefs about the expectations of those they might marry. In particular, many college-educated women who wish to pursue adult roles both in and out of the house and who wish a marriage that works as an equal partnership believe that most college males do not share these goals. This dilemma may explain the trend toward delaying marriage within the college population (Cherlin, 1980). But the vast majority of this population does marry, and unless women manage to find husbands whose expectations about married life are similar to their own, they are faced with the task of changing one set of expectations or the other. The evidence indicates that the former course of action is more difficult, and thus the latter, more common.

Wives are much more likely to report making adjustments in their marriage than their husbands are (Bernard, 1982). Because interpersonal competence is a more central element in the self-concept of women than in the self-concept of men, women seem more willing to make concessions in order to make the marriage work. In cases where wives report that their husbands actually have made more concessions than they have, it is not uncommon to find the husband reporting that he is unaware of making any (Bernard, 1982).

Ironically, even though all of the evidence supports Bernard's argument that women have a greater sense of investment in marriage than men do and that men appear less aware of any adjustments they make in their marriage, on virtually every measure of longevity, mental and physical health, and personal life satisfaction, married men do better than non-married men (Barry, 1970). Or perhaps this fact is not so ironic. Perhaps it simply indicates that marriage serves as a much better support system for men than it does for women—a circumstance that continues to seriously affect the ease of resolution of the many developmental tasks of the adult years for both men and women.

THE SIGNIFICANCE OF THE TRANSITION YEARS IN THE LIFE-SPAN

Adolescence marks the passage from childhood to adulthood. By the age of eighteen, most people and, by the age of twenty-five, certainly all people have achieved the legal and biological status of an adult, if not necessarily an adult's social status. For this reason, the interval from age eighteen to age twenty-five provides the first opportunity for observation of the way individuals function given their new social status and given the comparatively unique social structure of the adult years. The new social status reflects the decline of the age-stratified system that characterizes development in our culture through adolescence, and the changing social structure reflects the increasing importance of the "social clock" of adulthood in regulating the rate and course of development (Neugarten & Datan, 1973).

All this suggests that during the first eighteen years of life a gradual shift occurs in the regulatory roles that biological and social forces play in individual development. By the early twenties, almost all people have achieved full biological maturity and, by virtue of being considered legally adult, are freed from such age-related restrictions as laws requiring school attendance and status offender laws that impose penalties for actions such as curfew violations, running away from home, and truancy. Likewise by the early twenties, almost all people are no longer subject to protective custody laws. The acquisition of adult status provides them with a greater degree of autonomy and responsibility for their lives and therefore marks a significant shift in the pattern of development.

Patterns of Development

From the instant of conception, chronological age serves as an important index against which to measure and observe the course of development. The significant units gradually increase from hours to days to months to years as a person moves from the prenatal period through infancy and childhood to adolescence, but during this interval chronological age continues to serve as the single most useful predictor, or correlate, of developmental status. A child's chronological age provides a better basis for making predictions about that child than any other single piece of information. Notice that the ability to predict does not necessarily imply the ability to explain. Nor does it suggest that chronological age is the only factor that predicts level of development during childhood. Sex and social class, to mention just two, are also significant factors. However, the fact that chronological age is such a useful index suggests that, whatever the regulatory mechanisms, they must somehow operate in an orderly, sequential fashion that correlates with chronological age.

The psychological/biological/social matrix is a lifelong phenomenon, but the interplay between the elements changes over the course of the life-span. The eighteen to twenty-five age period marks a shift in the roles biological and social forces play relative to each other in determining the course, rate, and variability of development. During the adult years, variability between individuals seems more a reflection of life circumstance and less a reflection of biological status.

There are two important corollaries to this pattern shift. The first is that, since the rate of cultural evolution (measured in units as small as decades) is faster than that for biological evolution (measured in units as large as eons), successive generations, or cohorts, of adults can be potentially more distinct from each other than successive cohorts of children can be. Further, the degree of variability between individuals of the same cohort can increase with chronological age. The implications of this increased variability during the adult years will be discussed in the following section.

The second corollary to this pattern shift is that the effects of the social forces that serve to categorize individuals within the culture become more evident during the transition period. In particular, the effects of social class and sex-role socialization become increasingly evident during the transition years. To say so is not, however, in any way to suggest that these factors have no influence on the course of development earlier in the life-span. Nothing could be further from the truth. But the age-stratified social and biological patterns through much of childhood and adolescence tend to restrict the full expression of social class and sex-role socialization. If, for example, the law says that everyone under age sixteen is to attend school, then sex and social class become secondary to chronological age in determining school attendance. Once chronological age is removed as a determinant of behavior, other factors gain in prominence. Social class, for example, is a significant predictor of college attendance, the effect of which is cumulative over the adult years. In particular, college attendance predicts a wide variety of future life conditions including income, occupation, choice of mate, number of children, and many attitudes and value judgments.

Sex-Role Transitions

The transition years witness a significant evolution in men's and women's sense of themselves as unique, autonomous individuals. There is a marked improvement in the ability to view themselves within a social context. This new competence is evident in Keniston's (1975) distinction of adolescence from youth. Whereas adolescence is characterized by a form of reflection that idealizes self, youth is seen (particularly college youth) as a time when the idealized image must be integrated into the individual's social world.

While far from absolute, social class and sex-role socialization influence developmental choices. When these influences are extreme, they severely limit the choice of adult lifestyles. In the case of this young woman, they have lead her to view herself as little more than a desirable sex object.

Again, when compared to the introspective thoughts of the adolescent, the introspective thoughts of youth seem more practical and reality oriented (Bocknek, 1980). This increasing practicality is not surprising. The decisions made during the transition period set the course of development for much of the next ten to fifteen years (Hogan, 1981).

At the same time that these new experiences and decisions set the course of development for the near future, they also make possible a new perspective on old patterns. In particular, the development of intimate relationships often prompts reflection on the parent-child relationship. Volpe (1981) finds that young adults are better able than adolescents to differentiate the interactional patterns that characterize parent-child relationships from those that characterize peer relationships. The former are understood to be unilateral and complementary; the latter, mutual and reciprocal (refer to the middle childhood chapter for a fuller discussion of the difference between complementary and reciprocal patterns).

The ability to distinguish the two patterns may have important implications for the establishment of adult relationships. In particular, it may be a necessary prerequisite for the ability to understand and accept the parent's role within the complementary parent-child relationship. Many children and even adolescents vow never to treat their children as their parents treated them. Except under circumstances in which parents abused and neglected their children, one important development before the assumption of the parenting role in adulthood may be the realization that the vows of childhood don't always stand the test of time.

If this logic holds true, the nature of the experiences that facilitate the ability to distinguish between reciprocal and complementary interaction patterns becomes an important developmental issue. One likely source of this ability is the sex-role socialization patterns of men and women. Given the arguments of Gilligan (1982) and Bernard (1982) concerning male and female socialization patterns, the relatively greater task orientation of male social interaction patterns might render the distinction between reciprocal and complementary relationships less clear to males. That is, if male peer relations more closely resemble the hierarchical structure of complementary parent-child relations than female peer relations do, males may be somewhat handicapped in their ability to move toward the establishment of a true sense of intimacy and mutuality. If this speculation proves correct, males may be more likely to associate the marital relationship with a unilateral pattern while females may associate it with a pattern emphasizing mutuality. If this description is accurate, it would help explain why Bernard (1982) makes the distinction between "his marriage" and "her marriage."

The one element that characterizes all of the major developmental events of the transition years is that due to the changing nature of social forces as well as the continuing maturation of self-reflective skills, individuals begin to become sensitive to the dialectics of adult life. A **dialectic perspective** (Riegel, 1973; 1975) recognizes the inherent contradictions that are present in many of the roles that adults play. Further, it emphasizes the fact that most adult social roles are intertwined. Marriage, for example, offers commitment but also restriction of individual freedom. There is no way to have one without the other. The awareness that most adult social roles consist of a variety of sometimes contradictory components first makes its appearance during these transition years. The acceptance of this notion, however, often takes longer. Indeed, the mid-life dilemmas discussed by Levinson (1978) and Gould (1978) are, to a large extent, the acknowledgement of the dialectics of adult social roles.

HUMAN DEVELOPMENT AND HUMAN SERVICES

The transition years are a period when people find themselves making decisions that have consequences that are longer-lasting and more significant than those of earlier decisions, when people are particularly vulnerable to broad exosystem events such as economic and political policies, and when they must learn to shift their frame of reference away from the age-graded system that has been such an integral part of their lives. Human service providers can play a significant role in all three domains.

The Timing of Decisions

Most people complete their education during the transition years, and the time during the interval at which this event occurs has consequences that last for many years to follow (Hogan, 1981). This statement is true largely because of the way our society is structured. Specifically, age at completion of education correlates with age at marriage, age at first parenting, number of children, and age at entry into the labor force. These various correlates of age at completion of education set in motion a series of events that essentially define the patterns of life during much of the young adult years.

Two implications follow from the research findings on the timing of decisions during the transition years. First, those who are in a position to influence decisions concerning the completion of education—teachers, school counselors, parents, peers—should recommend that young adults stay in school. School does not necessarily mean college, however. When the two are equated, when staying in school means going on to college, many young adults who have neither the means nor the desire to attend college are inadvertently cast aside. Staying in school has a broader meaning than going to college. In the broadest sense, it means continuing to do something that will delay marriage, parenting, and entry into the job market because the evidence indicates that for young adults between the ages of eighteen and twenty-five such a delay is beneficial. In this broader sense, schooling may mean vocational and technical training, a two-year college program, military training, or some sort of apprentice program. Again, it may mean anything that further prepares young adults for successful entry into and stability in the labor market—that is, anything that either makes them more employable or prevents them from becoming less employable.

The second implication of these research findings is that efforts also need to be made to reduce the association between age at completion of education and other transition events. This may seem to contradict the first implication, but it really doesn't. Rather it reflects the limited impact that those in positions to convince people to continue their education have on such decisions. For example, it hasn't been many years in some communities since pregnancy meant leaving school, and few of the adolescent girls who found themselves in such a situation ever returned to complete their education. Rather they tried to establish an adult life having a child but no high school degree, having expenses but few job skills, and having needs for emotional and practical support without a partner to help meet those needs. In most communities, the situation today is better. The correlation between pregnancy and probability of graduation has been lowered because of interventions. These kinds of interventions need to be expanded. Those in control of educational programs need to become ever more sensitive to the way their policies and procedures intentionally or unintentionally influence the enrollment patterns of young adults who are married, who are parents, or who are working full- or part-time. To a degree, this change is already taking place as the declining birthrate has forced educators to learn to function in a buyers' market. It needs to happen more, however, even in a sellers' market.

Vulnerability during the Transition Years

In a sense this section really only needs four words to convey its message—last hired, first fired. These four words convey in very direct fashion how vulnerable those entering the adult years are to events at the level of the exosystem (Bronfenbrenner, 1977). As discussed in Chapter 1, the exosystem is a level of the social ecology on which individuals have no direct impact, while decisions made at the level of the exosystem have direct and significant impact on them. The economic and political decisions made at the level of the exosystem—through government action, through the

Few, if any, people have absolute job security. However, young people between the ages of eighteen and twenty-five are particularly vulnerable to unemployment, since they have the least job experience.

workings of the marketplace—affect all people, but they especially affect those in the transition years. Young adults between the ages of eighteen and twenty-five are particularly vulnerable because they have the least leverage in the adult system. They can't gain employment if the economy isn't offering employment, and they are least able to keep employment if the economy is slowing. Because they have the meagerest of employment histories, they are the least eligible for compensation and benefit programs. They are also the most likely to have to go to war, certainly the ultimate vulnerability.

Economists and politicians and others who function at the level of the exosystem probably don't think of themselves as human service providers, but they are. Their actions have profound and long-lasting impact on the lives of all people.

Decline of the Age-Graded System

Albrecht and Gift (1975) note that the transition to adulthood involves four changes in socialization patterns. First, it involves an increase in the number of social roles that individuals assume. The lives of adults are simply more complicated than the lives of children. Second, it involves a decrease in what Albrecht and Gift refer to as boundary conditions. Adults don't get promoted to a new job every June. If they get promoted at all, their promotion is a more individual than typical experience. Third, the transition to adulthood involves a decrease in the directness of reinforcement, that is, in the clarity and frequency of information informing people of their status and competence. And fourth, during this period, role definitions become more obscure and individualized. There are simply more ways to be an adult than a child.

These four observations are significant because young adults in the transition years who continue to think that developmental timetables are uniform and age related, in other words, who do not realize that these timetables change, are creating expectations that can produce unnecessary stress and inappropriate negative evaluations.

How are interventions possible? What roles can human services play? Two seem reasonable, one direct, one indirect. Acting directly, human services can provide information, advice, counseling, and education. Such information and advice should help young adults in transition appreciate the differences between the socialization patterns of adulthood and those of childhood. The direct intervention, however, is a difficult one. People who are about to embark on a new venture, irrespective of their stage in life, are often the ones least responsive to advice and warnings. Thus, there may be distinct advantages to indirect interventions.

Indirect interventions can be of two types. One might involve nothing more than modeling, thereby making it possible for young adults to see that adult life patterns are more highly individualized and less age-related than those of childhood. A second indirect strategy involves the examination of policies and procedures to determine if they portray a view of adulthood that is more uniform than in fact it is. Employment policies, for example, that place restrictions based on age not only discriminate against those directly affected by those policies but also convey the impression to others that age is a significant correlate of ability during the adult years, an impression that is not supported by the evidence.

Reading this chapter must give you the impression that the lives of adults are very different from the lives of children, that in terms of sense of self, sex-role identity, interpersonal relationships, and cognitive competence, adults' perceptions of their present lives are very different from their perceptions of their childhood and even adolescent years. This is certainly the impression most of the adults give in the next chapter.

SUMMARY

THE COURSE OF ADULT DEVELOPMENT

1. Differences between adults are more likely a reflection of sociohistorical than biological factors.
2. The three careers of the adult years are intimate, parent, and worker.
3. Research indicates that adult development follows a sequential pattern, regulated largely by social expectations. Research also indicates that the timetables and components of this sequence are probably different for men and women.
4. Adults assume a more conscious and deliberate role in directing the course of their own development than children and adolescents do.

THE TRANSITION YEARS

1. The timing and sequencing of school completion, marriage, and entry into the work force correlate with social class and define much of the character of development through the early adult years.
2. The college experience tends to make students less dogmatic, more relative in their thinking, and more tolerant of diversity.
3. Males have more difficulty defining a personal identity in terms of interpersonal relationships; females have more difficulty defining a personal identity in terms of vocational choice. The difficulties reflect sex-role socialization experiences relative to the three careers of the adult years.
4. Noncollege youth share the goals and aspirations of college youth but are less optimistic about the likelihood of achievement.
5. Since similarity of background and frequency of contact are significant correlates of mate selection, educational history serves as a major predictor of marriage partners.
6. Cohabitation is seen as providing a sense of intimacy without forcing people to confront the choice of a long-term commitment.
7. The developmental tasks of the establishment phase of marriage serve to help the husband and wife shift their orientation from thinking of themselves as relatively autonomous individuals to thinking of themselves as members of an interdependent pair.
8. Age at marriage, social class background, and similarity of experiences correlate positively with marital satisfaction. These correlations reflect the likelihood that husband and wife will fulfill material and interpersonal expectations.

THE SIGNIFICANCE OF THE TRANSITION YEARS

1. The differing orientations of males and females concerning personal identity and interpersonal relations may reflect the degree to which each conceptualizes adult relationships as complementary or reciprocal.

HUMAN DEVELOPMENT AND HUMAN SERVICES

1. Young adults trying to establish a "beachhead" in the adult world are particularly vulnerable to social, political, and economic factors that influence the nature of transition experiences and the availability of adult roles.

KEY TERMS AND CONCEPTS

THE TRANSITION TO ADULTHOOD

Sociohistorical Context
Social Clock
Biological Clock
Erikson's Psychosocial Stage of Intimacy
Erikson's Psychosocial Stage of Generativity
Erikson's Psychosocial Stage of Ego Integrity

THE TRANSITION YEARS

Perry's Three Forms of Experience Construction
- Dualistic World View
- Relativistic World View
- Evolving of Commitments

Marcia's Sequence of Identity Statuses
- Identity Achievement
- Identity Diffusion
- Psychosocial Moratorium
- Identity Foreclosure

Establishment Phase of Marriage
Dialectic Perspective

SUGGESTED READINGS

One of the major issues raised in this chapter is the degree of similarity in the developmental patterns of adult men and women. These four authors provide contrasting opinions.

Frieze, I. H. (1978). *Women and sex roles.* New York: Norton.

Gilligan, C. (1982). *In a different voice.* Cambridge: Harvard University Press.

Gould, R. L. (1978). *Transformations.* New York: Simon & Schuster.

Levinson, D. J. (1978). *The seasons of a man's life.* New York: Knopf.

Self-concept, sex-role identity, and the establishment of intimate relationships are important and interrelated issues during the transition years. These four articles provide a good understanding of these issues.

Ginsburg, S. D., & Orlofsky, J. L. (1981). Ego identity status, ego development and locus of control in college women. *Journal of Youth and Adolescence, 10,* 297–309.

Hodgson, J. W., & Fischer, J. L. (1978). Sex differences in identity and intimacy development in college youth. *Journal of Youth and Adolescence, 8,* 37–51.

Marcia, J. E., & Friedman, M. L. (1970). Ego identity status in college women. *Journal of Personality, 38,* 249–263.

Peplau, L. A., Rubin, Z., & Hill, C. T. (1977). Sexual intimacy in dating relationships. *Journal of Social Issues, 33,* 86–109.

For most adults in our society, marriage is the most significant social relationship of the adult years. The survey reported by Campbell supports this notion. The fact that this most important social relationship is often seen as less satisfying by wives than by husbands is discussed in Bernard's book.

Bernard, J. (1982). *The future of marriage* (1982 ed.). New Haven: Yale University Press.

Campbell, A. (1981). *The sense of well-being in America.* New York: McGraw-Hill.

8

THE EARLY YEARS OF ADULTHOOD

AGES TWENTY-FIVE TO FORTY

CHAPTER OUTLINE

If a word can capture the character of an entire period of the life-span, then the word for the early years of adulthood must certainly be "busy." Adults between the ages of approximately twenty-five and forty often find themselves more involved in more distinct endeavors than people are at any other period of the life-span. Not only are these roles distinct, but they may often be in conflict. Marital roles conflict with parenting roles. Husbands' needs conflict with those of wives. Pressures from home conflict with those from work. Many adults find themselves in a difficult position as they try to meet the needs of their aging parents while, at the same time, trying to meet the needs of their growing children.

Given that this phase of the life-span is so busy and full of conflicting pressures, it might seem that adults would view it negatively. But they usually don't. Even while fully acknowledging that their roles during this period are demanding, most adults still report that this time is personally satisfying. Parenthood is seen as bringing a greater division of labor for husband and wife, but the interdependence of functions and the sharing of common goals are reported as bringing them closer together. Children are viewed both as a restriction on individual freedom and as a source of enormous satisfaction (Hoffman & Manis, 1978). Perhaps these judgments reflect adults' increasing acknowledgement of the dialectic character of the adult years.

In a sense, the early years of adulthood are reminiscent of middle childhood. Both are times of great industriousness, during which people try to master a number of socially defined roles. The nine-year-old may be very unhappy about not being a starter on the team but thinks it's fair that the better players should be the ones who play the most; similarly those whose marriages end in divorce nevertheless still value the institution of marriage (and, in fact, usually remarry). In Piaget's terminology, both middle childhood and the early years of adulthood are periods that emphasize accommodation, and both are preceded by periods that emphasize assimilation. Early childhood and adolescence are remarkably narcissistic (in the best sense of the word) periods of the life-span. Both are periods in which the focus is on self. However, the analogy between middle childhood and the early years of adulthood is not perfect. The school-age child accommodates to a world constructed in terms of "or statements"; the adult, to one constructed in terms of "and statements." That is, experiences are no longer seen as good or bad. Instead they come to be understood as good and bad. It is to an investigation of the adult's world that we now turn.

THE COURSE AND CONTEXT OF DEVELOPMENT DURING THE EARLY ADULT YEARS

The age interval of twenty-five to forty typically overlaps two periods of the family life cycle (Aldous, 1978). The first, known as the expanding phase, begins with the birth of the first child and ends with the birth of the last child. The second, the stable period, begins with the birth of the last child and ends when the first child leaves home. Because most adults marry and most parent, the family life cycle provides a useful frame of reference for viewing the development of individuals within a family structure. In particular, it emphasizes the significance that the parent role has for individual development during the early adult years.

Over the past two to three decades, however, a number of significant changes have been occuring in the patterns of early adult development. The consequence of these changes has been an increasing variability in family forms within all of the family life cycle periods, as well as an increasing variability in the chronological ages associated with the boundaries of each period. As a result of these changes, the present cohort of young adults, although dealing with the same developmental tasks that their parents had to deal with, will do so within a very different social context.

What are these changes? The first is the increase in the number of high school graduates going on to college. Although this number, currently estimated at 50 to 60 percent, is not expected to continue rising (Masnick & Bane, 1980), it represents a significant increase over the numbers recorded for past decades. Perhaps of even greater significance is the fact that the percentage of women entering college now equals that of men.

The second change, no doubt in part a result of the first, is that first marriages are now being delayed until the mid or late twenties (Cherlin, 1980). In 1960, almost 75 percent of all twenty-four-year-olds had married at least once; by 1977, the percentage had declined to slightly more than half. Not only are many people delaying marriage, an increasing number are choosing not to marry. Norton (1983) estimates that this number has increased from 4 to 8 percent.

Two other changes have correlated with the trend toward delayed marriage: a trend toward delay in the birth of the first child following marriage and a decrease in the preferred number of children. In the early 1960s the first child was born, on the average, fourteen months after the wedding; by 1980, this interval had increased to twenty-four months (Wilkie, 1981). Further, the birth rate has declined to the point where most couples are now simply replacing themselves, that is, having two children (Norton, 1983). These changes in the timing of marriage and fertility patterns have been most pronounced in the college-educated segment of the population, but the trends are also becoming evident in the noncollege segment.

The changes in family planning patterns, delayed parenthood and reduced family size, clearly reflect the influence of the birth control pills

Increasing numbers of people are choosing to delay first marriages until their mid to late twenties.

first made generally available in this country in the early 1960s. Increased control of fertility has also influenced the timing of marriage and has distinguished decisions concerning sexual activity from those concerning marriage and decisions concerning marriage from those concerning parenthood (Hoffman, 1977). As families have gained more effective control over fertility, they have found themselves in a better position to attempt to integrate their family and work roles more effectively. This has been especially true for women. More than half of all married women work, an almost 100 percent increase from the 1950s (Masnick & Bane, 1980). The percentage is projected to climb to over three quarters of all married women by the end of the century. Of these numbers, an increasingly large percentage is coming from families with young children. Between 1960 and 1980, the percentage of married mothers working full- or part-time increased from 15 percent to 27 percent for those with a child under three, from 24 percent to 38 percent for those with a preschool-age child, and from 34 percent to 50 percent for those with children in school.

The second cohort change among married couples, certainly the one receiving the greatest amount of publicity, is the increase in the divorce rate. Although the rate of increase has slowed, it is estimated that from 30 to 40 percent of all marriages will eventually end in divorce (Norton, 1983). The corresponding figure for new marriages in the 1950s was 20 percent; in the 1960s, it was 25 percent. Even though most of these divorces occur early in the marriage (before children) and even though most who divorce eventually remarry (in three years on the average for men, six for women), this change has increased the number of single-parent households and the percentage of children who will spend part of their childhoods living in a single-parent family. Norton (1983) estimates that this latter figure will climb to as much as 50 percent by the end of the century.

Whatever the causes of the increase in the divorce rate, it is not a disenchantment with the institution of marriage: most divorced people do remarry. Marriage has not lost its pivotal role in the lives of adults. What has changed is the ability of couples to maintain a marriage. It is now both legally easier to divorce and more socially acceptable than it was in the past; but these factors seem more likely to be effects of the change in rate than to be its causes. Causes cited range from the increased pressures of modern life to the alleged self-centeredness of young adults (Yankelovich, 1981) to an increased emphasis on the emotional as opposed to instrumental aspects of a relationship (Scanzoni & Scanzoni, 1981) to the increasing length of the life-span and the lessening economic dependence of women on men.

Why have these cohort changes occurred? There is no consensus as to the specific reason for these changes, but a number of contributing factors have been identified. Changes in our economy over the past thirty years have increased the availability of employment, particularly of jobs requiring advanced education. This change has had more impact on women than on men. At the same time that jobs have become more available, the rising cost of goods and services has required more married couples to become dual earners in order to maintain a desired standard of living. Further, the meaning of work has changed. Primarily among college-educated men and women, work is seen not only as a means of earning money, but also increasingly as a potential source of personal satisfaction and development (Iglehart, 1979).

The bulk of this chapter will discuss the impact of these cohort changes on development during the early adult years. For now it will suffice to note that these changes have influenced the patterns of child rearing, have altered the meaning of the careers of the adult years (work, intimacy, and parenthood), and have significantly influenced the roles of adult men and women in our culture. It is important to note that the roles of adult women have been more substantially affected than the roles of adult men—a fact that is becoming an increasingly significant element in the evolution of adult relationships.

The chapter is organized around the three careers of the early adult years: marriage, parenthood, and work. Since the three are not independent, a good part of the discussion of each of the three will focus on its integration with the other two. Following these sections, a separate section will be presented on divorce. The discussion will concern the factors that contribute to the dissolution of a marriage and the marriage's aftermath both for the couple and for others affected by the divorce. The chapter will conclude with a consideration of the place of the early adult years within the total life-span and the implications of the developmental events of this period of the life-span for the provision of human services.

Socioeconomic status and sex role are the two major qualifiers of the developmental events of this period. They influence the opportunities that are available to people, their potential sources of life satisfaction, and the role expectations they hold for themselves, as well as for others. They clearly influence the marriage patterns couples evolve during the early years of adulthood.

The Marriage Career

Marriage is the single most important influence on the course of development during the entire adult period. Based upon over twenty years of questionnaire surveys involving adults of all ages and socioeconomic backgrounds, Campbell (1981) believes that to the extent that it can be separated from other influences, marriage offers both the greatest sources of satisfaction and the greatest sources of dissatisfaction.

> The basic source of social support among adult Americans is marriage. The need for human relationships can undoubtedly be met in various ways, and countless individuals live what they regard as very satisfactory lives outside of marriage, but, on the average, no part of the unmarried population—never married, separated, divorced, widowed—described itself as happy and contented with life as that part which is presently married. This may seem curious in light of the probability that a sizeable fraction of those married will eventually terminate their marriages in divorce, but the evidence is consistent and substantial: married people see their lives more positively than unmarried people. Despite the fact that attitudes toward marriage are changing in this country, especially among young people, the marriage pattern continues to contribute something uniquely important to the feelings of well-being of the average man and woman. (pp. 226–227)

Marriage Patterns

The relationships that husbands and wives establish with each other can be conceptualized in terms of three relatively distinct dimensions (Scanzoni & Scanzoni, 1981). First, the relationship can be defined in terms of its task orientation. Those relationships that are most concerned with earning power, child-rearing tasks, household maintenance, and the like place great emphasis on the instrumental aspects of a relationship. Those that are more concerned with companionship, with providing the partner emotional support, sexual gratification, and a degree of empathy place great emphasis on the expressive side of a relationship.

The second dimension concerns the division of labor in a relationship. Relationships that are complementary in nature have a rather complete division of labor. The roles and responsibilities of one spouse are distinct from those of the other. There is relatively little overlap in their activities. Relationships that are reciprocal in nature are just the opposite. One spouse is as likely to be found doing the laundry as the other. The terms complementary and reciprocal should have a familiar ring. They have also been used earlier in the text to distinguish the structure of the parent-child relationship from that of peer relationships. The two kinds of relationships are different in the degree of equality and distinctiveness that exists between participants.

The third dimension involves the distribution of power in a relationship. The distribution of power is reflected in a couple's decision-making processes. How closely, for example, does the final decision reflect the original positions of each spouse? Power is not a unitary dimension. Its distribution might depend on the topic in question. In a traditional marriage, most decisions concerning children are in the domain of the wife; those concerning such things as when to buy a new car are in the domain of the husband. More generally, power is related to the rights that each spouse feels entitled to. The rights of one spouse then become the responsibilities of the other (Scanzoni & Scanzoni, 1981).

Historically, when wives were legally considered the property of their husbands, all power was vested in the husband. He had virtually all the rights, and she had all the responsibility to see that those rights were fulfilled. It was her responsibility to care for her husband's needs, to maintain his household, to bear and rear his children, to reflect favorably on his status, and to obey him in all matters. The evolution of marriage patterns over the past century has been away from this owner/property pattern and toward patterns favoring a greater balance in all three dimensions. The three **patterns of spouse interaction** that the Scanzonis note as typical of today's culture differ in the degree to which they favor equality in the relationship. The Scanzonis define the three patterns as the head/complement pattern, the senior partner/junior partner pattern, and the equal partner/equal partner pattern.

The head/complement pattern. The **head/complement pattern** is the traditional marriage pattern. The division of labor is rather complete, and therefore the relationship is complementary in nature. The husband's responsibilities for the most part concern providing for the financial welfare of the family; the wife's concern raising children and maintaining the home. Decision-making power remains vested primarily in the husband. The couple may discuss the wisdom of a change of jobs or a company move, but the final decision is usually his. As complement, the wife is supposed to accompany her husband through the life course he defines. The corporate wife who accompanies her husband from one location to another as he moves up in the corporation is the classic example. Her life-style, friendship patterns, and opportunities for independent development outside of the family setting are largely defined by her husband's career advancement.

So far, the head/complement pattern may not sound very different from the owner/property pattern. With the exception of a change in wives' legal status, their roles seem very similar to those of a century ago. In a sense, this is true, but the Scanzonis find that what distinguishes the two is a greater emphasis in the head/complement pattern on the expressive component of the relationship. The head/complement pattern places a greater emphasis on companionship than the owner/property pattern does. Husbands and wives are seen as friends and lovers as well as providers and housekeepers. They are sources of emotional security as well as sources of economic and material security. As the Scanzonis put it, "They were expected to enjoy one another as persons, to find pleasure in one another's company, to take one another into confidence and share problems and triumphs, to go places together and to do things together" (Scanzoni & Scanzoni, 1981, p. 318).

The senior partner/junior partner pattern. The second type of relationship, the **senior partner/junior partner pattern,** is increasingly coming to be the norm, especially among college-educated couples. Compared to the head/complement pattern, this pattern is characterized by a greater degree of overlap in division of labor and a more equitable sharing of power. It is like the head/complement pattern, however, in the balance it maintains between instrumental and expressive aspects of the relationship. This pattern has evolved largely as a result of the increasing numbers of women entering the labor force. As a result of their employment and the income it generates, wives in this kind of relationship acquire a larger share of the decision-making power than wives in head/complement relationships have. However, both the husband and wife view her employment as secondary to her still primary responsibility as wife-homemaker-mother. Most major decisions are still more consistent with his occupational status, and her employment patterns are largely defined by what is possible given her child-care and household responsibilities. Moreover, since there are often times when the wife's primary responsibilities prevent any employment, many couples find themselves during the early adult years shifting back and forth between a head/complement pattern and a senior partner/junior partner pattern.

The equal partner/equal partner pattern. The third pattern, the **equal partner/equal partner pattern,** is still quite rare in our society, although it has received a great deal of attention in the popular media as perhaps the marriage of the future.

The equal partner/equal partner pattern is significantly different from the others in three ways. First, the husband and wife are equal in power, status, rights, responsibilities, and all other matters. It is a true reciprocal relationship. Neither owns the "football," and therefore neither can arbitrarily define the rules. Negotiation, compromise, and consensus become the only decision-making strategies consistent with a relationship of equals. Second, the decision to parent is no longer seen as an inevitable consequence of the decision to marry but rather as a separate decision. Further, the decision to parent has a direct impact not only on the wife but also on the husband. Like all other roles in the equal partner relationship, parenthood is viewed as a joint venture, one affecting the career patterns of the husband as well as the wife. Third, roles are no longer linked to the sex of the spouse but are now seen as potentially appropriate for either.

The sharing of all rights, responsibilities, and statuses is the premise of the equal partner/equal partner marriage pattern. The actual practice of this pattern is still very rare in our society.

The couple's job histories and friendship patterns could as likely reflect her occupational development as his, or they could reflect some mutually agreed upon combination of factors. As in the other two patterns, both the instrumental and expressive aspects of the relationship are considered important to the quality of the relationship.

Marital Satisfaction

Patterns of relationship. The quest for the answer to the question "What makes for a happy marriage?" is probably timeless. And it seems remarkably elusive. There certainly doesn't appear to be a magic formula. We do know that most adults are generally satisfied with their marriages (Campbell, 1981) although this generalization is more true for men than for women (Rhyne, 1981). We also know that the various objective measures of marital status, such as income or occupation, or even marriage pattern, do not show very high correlations with marital satisfaction (Spanier & Lewis, 1980). Satisfaction appears to be a highly subjective phenomenon, based mostly on a match between expectations and accomplishments (Houseknecht & Macke, 1981). In other words, there is nothing inherent in any of the three marriage patterns that guarantees either satisfaction or dissatisfaction. On the other hand, the fact that the least satisfied group of married adults are full-time housewives (Bernard, 1982) raises the possibility that the three patterns may place different ceilings on the degree of happiness or satisfaction that is possible—an issue that will be discussed more fully later in the chapter. Given the subjective nature of marital satisfaction, it becomes important to determine the factors that contribute to the construction of expectations about marriage.

Figure 8-1 depicts the variety of factors that have been shown to influence marital satisfaction and in turn marital stability (Spanier & Lewis, 1980). Three sets of variables are seen by Spanier and Lewis as determining marital quality. The first concerns the qualities each individual brings to the relationship; the second, the degree to which each spouse is satisfied with his or her life-style; and the third, the extent to which each is satisfied with the quality of the relationship. Although the variables within each set are not independent (the way each is evaluated is in part a reflection of the evaluation made of the others), they nevertheless reflect the various elements that go into a marital relationship. Spanier and Lewis (1980) suggest that since each variable can have a positive or a negative influence on the relationship, there are a very large number of ways that relationships can be defined and yet still result in the same degree of satisfaction or the same quality. Further, different marriage patterns emphasize different sets of variables. Traditional relationships place greater emphasis on the life-style variables than equal partner relationships. Equal partner relationships place more emphasis than traditional relationships on spousal interaction variables.

You may have noticed that although there is a variable labeled "satisfaction with wife's work," there is no corresponding variable labeled "satisfaction with husband's work." This omission is not an error. Rather it reflects the relative imbalance typical of many marital relationships, in particular, the fact that the man has historically been the more powerful spouse and therefore the one whose evaluation sets the tone of the relationship. Perhaps a replication of this figure some years from now will reveal a more balanced picture.

The differentiation of variables contributing to marital quality provided by Spanier and Lewis (1980) is useful in identifying the variety of factors that influence the quality of a relationship. However, the individual subjective perceptions of the couple for the most part reflect the degree to which the two people feel a sense of equity or fairness in the relationship, a sense that each is doing what is expected by the other, and the extent to which each is appreciative of the other's efforts (Houseknecht & Macke, 1981; Rhyne, 1981; Schafer & Keith, 1981; White, 1983). Within the traditional pattern, the primary expectation for the husband is that he provide an adequate degree of economic stability; for the wife, that she be a competent mother and housekeeper. As long as each performs consistently with the expectations of the other and is cognizant and appreciative of the efforts of the other, the relationship is satisfying to both spouses. Equal partner relationships have a very different set of expectations but just the same potential for satisfaction. Both kinds of relationships have a relatively clearly defined set of expectations for each spouse. In dual-earner relationships where the husband is still the primary wage earner (the senior partner/junior partner pattern), the role overload experienced by many women and the shifts in role that occur as women move in and out of the labor market may make expectations less clear and therefore may make it more

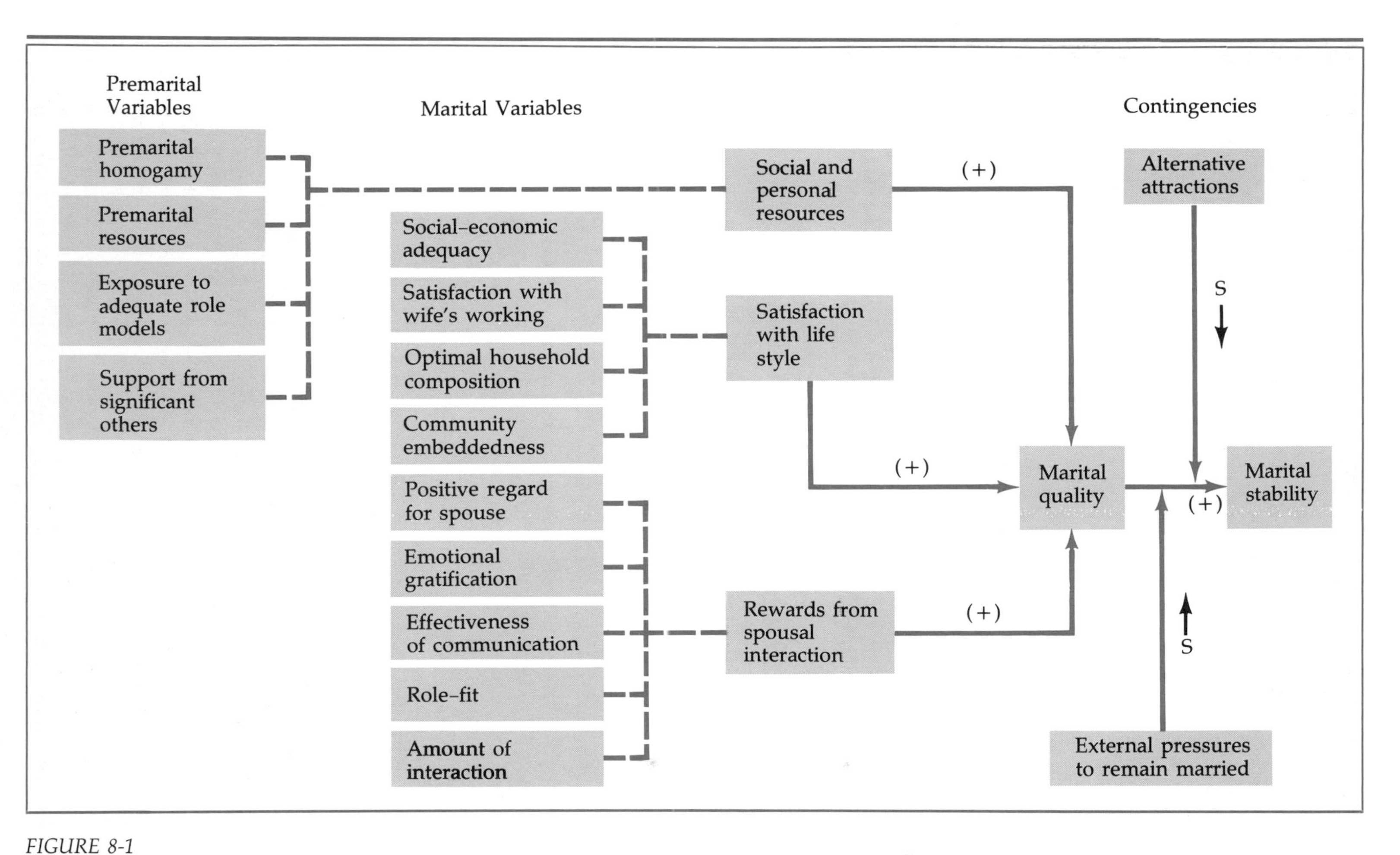

FIGURE 8-1
The theory of marital quality (from Spanier & Lewis, 1980).

difficult for each spouse to be appreciative of the efforts of the other. The more effective the communication between the spouses, that is, the better able they are to express their frustrations and their feelings of not being appreciated as well as their positive regard for each other, the less likely this is to be a problem (White, 1983). Unfortunately, during the early adult years, most of these couples have young children, a circumstance that often limits time alone together and, for that reason, effective communication.

Social class. As indicated at the beginning of this section, socioeconomic variables such as occupation, income, and education show only moderate correlations with marital satisfaction (Campbell, 1981). In general, the higher the social status of the couple, the more positively they view their marriage, but again, the correlation is only moderate. The couple that earns five times more than another is not very likely to be five times happier (five times richer, yes; five times happier, no). The positive correlation, however, does reflect the fact that the higher the status, the easier it is to fulfill expectations concerning economic issues. For example, in non–college-educated populations where marital patterns are more likely to be traditional, the economic responsibility is placed on the husband. Since his limited educational background restricts the range of jobs available to him, he may find himself unable to adequately fulfill this responsibility. His lack of satisfaction in his own efforts may be further heightened if his wife needs to find work to supplement the family income and thus to assume a role inconsistent with her expectations for marriage (Rubin, 1976). In this instance, both spouses are dissatisfied with the marriage because they are neither fulfilling their own expectations nor meeting those of the other.

Such dissatisfaction is less of a problem in well-educated families. If the relationship is traditional, the husband is in a better position to find a well-paying job. If the relationship is more equal, the wife is more likely to include the role of worker as part of her self-concept. However, the research (Campbell, 1981) also indicates that as well-educated couples become more economically secure, the importance attached to economic security as a measure of marital satisfaction decreases. As couples become more affluent, the interpersonal aspects of their relationship become increasingly significant as a measure of marital satisfaction. This shift in the nature of the expectations, no doubt, explains why the correlations between satisfaction and social status are only moderate. Even though economic expectations are fulfilled, the additional psychological ones may be less easily realized.

His marriage–her marriage. The other generalization made at the beginning of this section was that, irrespective of social status, men usually report greater satisfaction with their marriages than do women. Although this finding is consistently reported in the literature (Campbell, 1981; Rhyne, 1981; Verhoff, Douvan, & Kulka, 1981; White, 1983), its interpretation is not clear. The full discussion of this issue must wait until we consider how the careers of parent and worker overlay the career of intimate during the early adult years. For the moment, a few points should be kept in mind relevant to this sex difference. First, women seem to place greater emphasis on the companionship aspects of relationships than men do (Rhyne, 1981). Second, husbands are more likely to see their role responsibilities as equitable than their wives are (Schafer & Keith, 1981). Third, over the past twenty to thirty years, there has been a greater shift in the adult roles and expectations of women than of men. And fourth, this shift in the roles and expectations of women has not been paralleled by any shift in the role behaviors of men. Husbands of working wives don't spend much more time in household and child-care tasks than husbands of nonworking wives do (Pleck, 1983).

Before you draw what might seem an obvious conclusion from these four points, consider the fact that some surveys (Robinson, 1977; Yogev, 1981) find that a substantial minority of working women report that they would not like their husbands to be more involved in household and child-care tasks. In spite of the fact that most of these women feel overburdened by their multiple roles, they still don't want their husbands to assume many of these responsibilities. Certainly in a few cases, the husband's help would prove more a liability than an asset, but more to the point, these findings suggest that whatever the ultimate advantages of moving away from traditional marriage patterns, the immediate cost may be the role insecurity that many women feel and that at least some of them deal with by holding on to more traditional roles while, at the same time, trying to expand into newer ones.

The Parent Career

Most married couples, even those pursuing an equal partner arrangement, eventually parent. The major question is not "if" but "when." Most couples answer the question by considering their financial stability, the quality of their relationship, and their sense of readiness for taking on the responsibility of parenthood (May, 1982). In fact, the transition to parenthood is seen by most adults, especially adult women, as the major transition event during the early adult years (Rossi, 1968). As farsighted and rational as this strategy may sound, it is important to bear in mind that as many as a third of all teenage marriages involve a pregnancy (Chilman, 1979) and that even among young adult couples who use some form of family planning and who eventually hope to parent, the pregnancy is "unanticipated" or "ill-timed" in as many as one-third of the cases (Entwisle & Doering, 1981).

The Transition to Parenthood

Most couples report a good deal of stress associated with becoming new parents (Belsky, Spanier, & Rovine, 1983; Miller & Sollie, 1980; Wandersman, 1978). Even couples that have a good understanding of the parenting role and its repercussions on other aspects of adult life still report that coping with the constant dependency of a young infant is both emotionally satisfying and emotionally draining.

Time—or lack of it—is the most commonly reported problem. In interviews with new parents, LaRossa (1983) found that both mothers and fathers reported sleeping less, watching television less, having less time to talk to each other, and having less time to be sexually intimate. They even reported having less time to go to the bathroom! Paradoxically, LaRossa also found these new parents reporting more frequent periods of boredom. There is only so long a period of time that even the most involved parents can find the experience of watching their newborn take a nap emotionally satisfying.

BOX 8-1
JUST WAIT UNTIL YOU BECOME A PARENT

The number of times that a parent, perhaps in a state of exasperation, says to a child "just wait until you become a parent, then you'll see what it's like" must be many across the child-rearing years; equally often must the child reply, "I'll treat my kids different than you treat me." What in fact does happen when the day of reckoning finally arrives? Who proves the better fortune-teller?

A study by Lucy Fischer (1981) suggests that they both may be right. Fischer was interested in the way the transition to marriage as well as the transition to parenthood affected the mother-daughter relationship. Fischer wanted to determine if the two role transitions changed the perspective from which each person viewed the other and, if so, if there was a corresponding change in their interaction patterns. Fischer obtained her information through structured interviews with young women who were single, married without children, and married with children and their mothers. The daughters were in their twenties and all lived in small towns; the mothers ranged in age from the mid-forties to the early sixties. Half of the mothers lived in the same community as their daughters, half were scattered across the country.

Fischer found that although both marriage and parenthood affected the mother-daughter relationship, each did so in a different way. For the mothers, the marriage of the daughter proved a major turning point. For the daughters, the turning point in their view of their mothers came about through their own transition to parenthood. Fischer believes that mothers of single daughters are more likely to see their daughters as focused on their own needs, to be in effect rather egocentric. The daughter's transition to marriage leads the mother to reevaluate her view. She sees the daughter's decision to marry as a desire to become more responsible and to concern herself with the needs of others. In other words, she now sees her daughter as more mature and therefore deserving of a different form of relationship.

The parent transition changes the daughter's view of her mother. The

Source: Fischer, 1981.

Coupled with the loss of personal time is the anxiety frequently felt over parenting competence. Very young infants can seem rather fragile, unresponsive creatures. They don't necessarily give parents many clues as to the adequacy of their parenting skill. Further, given the variety of sometimes conflicting advice provided by child guidance books, "helpful" friends, and family, new parents often feel great uncertainty about whether to let their infants cry or not, whether to feed them more or less, when to pick them up and when to put them down, when to hold them, when not to hold them, and so forth. Many of these uncertainties gradually resolve themselves as infants, by the third month, become more lively

Marriage often changes the relationship between mothers and daughters.

daughters reported that they now were in a position to appreciate why mothers do what they do. They also reported finding themselves, often unintentionally, treating their children as they remember being treated. The daughters did not always like recognizing the similarity in parenting styles, but in all cases they reported a greater appreciation of their mothers' efforts and a more positive view of their mothers. So powerful is the effect that many daughters even felt that when their children reached adolescence (none were older than three at the time of the study), they would deal with them as they had been dealt with. These changing perspectives were also reflected in changes in behavior. Daughters were in much more frequent contact with their mothers after they became parents than they had been before. Maybe mother knows best after all.

and responsive and, in so doing, provide the sensitive parent more feedback about their parenting techniques (Lamb & Easterbrooks, 1981; Wandersman, 1978).

Incorporating the Parent Role into the Marriage

Changing roles. In adjusting to parenthood, people have more to cope with than the new demands made on their time and their uncertainties about ministering to the needs of an infant. They must also acquire a new perception of themselves as an adult and adjust to the changes in the

Loss of time for each other is often the hardest adjustment for new parents to make. Special effort has to be made to compensate for changes in preparental interaction patterns.

couple relationship that invariably accompany the transition to parenthood.

Couples find that coping with the new demands on their time and energy often results in a greater division of labor between husband and wife, one invariably more sex typed in its character (Gutmann, 1975). They also report that their relationship loses some of its romantic or spontaneous quality, becoming more instrumental and practical (Hoffman & Manis, 1978).

The relative ease or difficulty of these transitions often depends on the pattern of the relationship (Polonko, Scanzoni, & Teachman, 1982). In traditional arrangements where the husband has already been successfully fulfilling his economic role for a few years, the wife may welcome the instrumental nature of her parental role as a way to fulfill her end of the agreement (Feldman, 1971). As long as the husband and the wife continue to be appreciative of each other's efforts, marital satisfaction often increases in such traditional families. In more egalitarian patterns, the greater role imbalance often accompanying parenthood—largely reflecting the greater involvement of the wife in the parent role—can create additional stress in the relationship.

In general, appreciation of the role of the other spouse is not always easy at this time, irrespective of marital pattern. Behavior patterns are in transition, time and energy seem in short supply, expectations are yet to be clearly defined, and it is difficult to determine whether or not different tasks are equitable (LaRossa & LaRossa, 1981). Are a flat tire, a traffic jam on the highway, and two canceled appointments equivalent to an infant with diarrhea, a broken washing machine, and the cat's knocking over a gallon of milk? It is not surprising that many people report feeling less appreciated and less loved by their spouse than they did before the birth of their child (Campbell, 1980; Wandersman, 1978).

Changing sources of satisfaction. The relationship between the onset of parenthood and a decline in self-reported level of marital satisfaction is a frequent finding in the research literature. Beginning with the birth of the first child, there is a gradual decline in marital satisfaction which does not begin to reverse itself until children reach adolescence and does not end until the children are old enough to leave home (Aldous, 1978; Bernard, 1982; Hoffman & Manis, 1978; Rollins & Galligan, 1978). At all points in this U-shaped function, the degree of dissatisfaction is greater for mothers than for fathers. It seems most intense for all mothers when their children are infants and preschoolers, that is, when they are most in need of constant care. It seems most intense for fathers when their children are entering adolescence and the financial strains of the cost of a college education are encountered. For families in which the children are not likely to go to college, fathers encounter the greatest strain when their children are young and their own earning power is relatively low.

These research findings are disturbing and even depressing, especially to people about to assume the roles of intimate and parent. Perhaps we ought to place a label on contraceptives that says, "Warning, The Surgeon General Has Determined That Failure To Effectively Use This Product May Result In A Decrease In Marital Satisfaction." Actually, the findings aren't quite as bad as they may seem, especially when examined more closely.

First, most people over this entire period continue to report that they are happy with their marriage. It is true that they are not as satisfied with their marital relationship as they were before they became parents, but the change is better understood as relative than as absolute; in other words, the rating may have gone from "very happy" to "usually happy" rather than to "miserable." Second, couples seem to maintain the same relative degree of marital satisfaction with respect to other couples. The most satisfied childless couples become the most satisfied couples with children; the least satisfied childless couples become the least satisfied couples with children. Third, as Campbell (1981) notes, couples do not report a decline of satisfaction with all aspects of life during the early adult years but only with their marital relationship. Other aspects of life, work for example, may be going extremely well. Morever, the decline in marital satisfaction doesn't mean that spouses stop loving or even liking each other. Rather, it means that they are dissatisfied with the ways in which they have come to relate to each other during these years (Schiavo, 1976). Fourth, it doesn't mean that parents are unhappy with their decision to parent or that parents see their children as the direct cause of the changes in their marital relationship. Most parents enjoy their children and the experience of being parents and are glad they made the decision they did.

What then are the causes of this shift in marital satisfaction and what, if anything, can be done about it? Because children are in need of adult supervision and because they place additional financial burdens on the family, most couples find that to cope with these added demands, their roles must become more specialized. Usually the husband assumes the primary, if not total responsibility for meeting the economic needs of the family (even to the point of taking a second job) and the wife takes responsibility for the supervision of the child or children and maintaining the household. Even in the increasingly common situation in which the wife continues to work, at least part-time, she is still most likely to assume the primary responsibility for child care. Even when children are involved in day-care programs, the wife is more likely to assume the responsibility for choosing the setting, monitoring the child's involvement in the program, providing or arranging transportation, and staying home from work when the child is ill.

Three changes in the couple's relationship result from this greater division of labor (Feldman, 1971; Goth-Owens, Stollak, Messe, Peshkess, & Watts, 1982). First, the distribution of power in the relationship becomes increasingly unequal. Husbands typically acquire a greater share. This is not an inevitable consequence of becoming parents, but to the extent that family decision-making power is associated with sources of income (almost all family decisions involve the spending of money), husbands, by virtue of being the primary wage earners, gain power. Second, there is a decrease in the number of shared activities experienced by the couple. Since companionship is one of the prime components of the expressive side of the marital relationship, a decline in the frequency of these experiences is usually viewed negatively. Even when couples do get time together, especially when children are young, they are often too tired to make full use of the opportunity. This shift is reported for both sexual and nonsexual activities (Schiavo, 1976). Third, the decline in shared experiences, coupled with the greater division of labor, makes it more difficult for each spouse to be aware of the activities, frustrations, and rewards of the other's daily routines. As a result, couples become less empathetic toward each other, and each person comes to feel less appreciated and understood.

Coping with the demands of parenthood. Couples deal with the multiple demands of the early adult years in a number of different ways. Those that use a little forethought often choose either to delay parenthood or to remain childless. As mentioned earlier in the chapter, the first option is becoming increasingly popular among college-educated adults who favor nontraditional marriage patterns. Delaying parenthood allows both partners additional time to establish themselves in their respective careers and to earn sufficient income to help defray the costs of rearing a child. It also provides the couple more opportunity to enjoy the companionship aspects of their relationship. Couples who delay parenthood typically pursue an equal partner marriage pattern, which usually changes to an interrupted career pattern for the wife when the couple become parents. In some instances, husbands also assume a share of the child-care responsibilities but, as will be discussed later in the chapter, this shift is still the exception rather than the rule (Pleck, 1983)—a situation that becomes a common source of stress among these couples.

The voluntary decision to remain childless is found primarily in equal partner relationships in which both spouses are highly educated and have strong career orientations (Houseknecht, 1979). Although this pattern has become more common and will probably continue to increase in popularity, it is unlikely that it will ever represent more than a small minority of

While voluntary childlessness has increased along with the level of career commitment among women, most women still choose to combine careers with parenthood. For some women, however, starting a family becomes a race between career advancement and their biological clock.

the young adult population. Among couples that pursue this option are found both some of the highest ratings of marital satisfaction reported for this age group (Houseknecht, 1979) and some of the highest divorce rates. There appears to be nothing middle-of-the-road about voluntarily childless equal partner relationships.

Another option, and probably the most common one, is to revise one's expectations concerning the sources of satisfaction in a relationship so that they are more consistent with the realities of the situation. The reality is that most people find themselves adopting relatively traditional marriage patterns, especially women. Their role invariably changes more than their husbands' does with the birth of a child. A great deal of ambivalence is felt by women who find themselves in this position. On the one hand, most women, irrespective of educational background, wish to parent and still place greater importance on their parenting role than on their occupational role. On the other, assuming the primary parenting role usually involves leaving the work force or finding a part-time job, and either decision can adversely affect career development. The shifts in decision-making patterns and the decrease in the number of activities shared with the husband are usually viewed negatively (Bernard, 1982). To a degree, the extent to which the husband is successfully fulfilling his role as provider and the extent to which he is cognizant and appreciative of the wife's parenting efforts affect the wife's ability to resolve some of her ambivalent feelings. There is a positive relationship between the degree of marital satisfaction reported by wives and the degree to which they feel their husbands are supportive of their parenting efforts (Lamb, Frodi, Chase-Lansdale, & Owen, 1978). However, some still find themselves confronting the fact that there appears to be no way to successfully accomplish all of the tasks they have set out for themselves.

Perhaps because couples are becoming increasingly aware of the conflicting demands of the early adult years, as well as the fact that it is the wife who typically makes the most drastic changes in behavior, a small but increasing number of couples are opting for the father to assume a greater role in child-care and household tasks. This is certainly true in equal partner arrangements and in couples favoring the senior partner/junior partner pattern who don't wish to assume more traditional roles. Results from national surveys (Pleck, 1983; Robinson, 1977) indicate that this shift is still not very common. Most fathers appear relatively uninvolved in the primary care of their children and in the performance of household routines. However, this lack of involvement may be, for some men, an expression of uncertainty about assuming a new role rather than an indication of their lack of desire to do so. There is relatively little direct evidence to support this conclusion, but there are a number of findings that seem consistent with it.

First, men seem to be as captivated by their young infants as their wives are (Greenberg & Morris, 1974; Parke, Power, Tinsley, & Hymel, 1980). The term **engrossment** has been used by Greenberg and Morris to describe the father's fascination with his newborn. The engrossed father takes great pleasure in looking at and touching his newborn. He is able to describe the infant in great detail, often judging the child perfect. His sense of self-esteem is enhanced by virtue of becoming a parent, and he feels a close attachment to the infant. Although Parke finds that fathers are just as adept as mothers are in sensing their infant's needs during the first few days, they are less likely than mothers to bathe, diaper, or feed their child. Over the course of the infant and preschool years, fathers are considerably less involved in child-care than mothers, and when they are involved, they usually restrict their activities to playful, noncaretaking situations.

It is quite possible that many fathers who would like to be more involved in the care of their children find themselves confronting the same problems that are confronted by women who wish to develop a professional, as well as parental role. Namely, they find themselves anxious about assuming a role that is not consistent with their primary socialization experiences, and, intentionally or unintentionally, they find those who have already assumed that role somewhat ambivalent about sharing it with "outsiders" (Robinson, 1977; Yogev, 1981). As Gilligan (1982) was reported as saying in the last chapter, the primary socialization experiences of most males in our culture is toward autonomy; for females, it is toward affiliation. It is inevitable that some anxiety accompany the assumption of roles more closely identified with the other sex than with a person's own. In fact, the difficulty may be even greater for fathers since the parent role is less valued in our culture than the role of worker (Bronfenbrenner, 1979; Lein, 1979).

Although more fathers are becoming involved in the primary care of their children, in most families, this responsibility still falls more heavily on the mother. The resulting division of labor tends to distance husband from wife. Each is less able to empathize with the role of the other, each finds the equity equation more difficult to balance, and each seems less satisfied with the interpersonal aspects of their marital relationship. What happens to these patterns when the role of worker, as well as the role of parent, is overlaid onto the role of spouse?

The Work Career

As a culture, we use the word "revolution" a lot. We use it to herald some new product or some new perspective on the order of things or some alleged social phenomenon. Usually the light of day shows the event to be far short of revolutionary. Something rather revolutionary has been occurring, however, with respect to the role of the worker during the early adult years. Specifically, the percentage of women, even those with children under five, who are working, who choose to work for reasons other than income alone, and who anticipate working all of their adult lives has risen dramatically during the past thirty years. The change has been so dramatic that, by the end of the century, it is estimated that 75 percent of all adult women, irrespective of their parenting role, will be working full- or part-time. Further, it is estimated that, by the end of this decade, our traditional conception of the family (in which the father is the sole breadwinner and the mother stays home caring for home and family) will account for less than 20 percent of all married couples (Kamerman, 1980).

The consequences of this social revolution are as yet far from clear. But what is clear is that these consequences are being measured in terms of the other careers of spouse and parent and the various social, political, and economic institutions that influence our lives. This section will concern itself with the implications of this change in terms of the spouse and parent roles. (The broader social implications will be discussed in the concluding two sections to this chapter.) One of the most obvious implications is that the life structure of women is now more parallel to that of men; that is, each now functions within all three roles. The parallelism may be more apparent than real, however. Because of the fact that most working women also maintain their roles as the primary parent, the interplay of the three careers of the early adult years has a different character for women than it has for men (Geerken & Gove, 1983; Pleck, 1983; Scanzoni & Fox, 1980).

The Changing Nature of the Work Force

Why has there been such a dramatic increase in the percentage of working women? There are four interrelated reasons (Burke & Weir, 1976; Hoffman & Nye, 1974). The first concerns the growth of our national economy over the past thirty years. In spite of periods of economic recession, the overall pattern has been one of growth, and therefore, the number of jobs available to women, as well as to men, has increased. Further, many of these jobs have been in the health- and service-related sectors of the economy, as well as in the hi-tech areas—all settings that women find accessible and appealing.

A second factor has been the increasing percentage of women going to college and earning a degree. The increase in the rate of employment has been faster for the college population than for non–college-educated women (Kamerman, 1980). A college degree gives women access to jobs that are considered desirable and that were previously unavailable. College not only provides access to better jobs. It also fosters a set of attitudes, values, and self-images that are consonant with a career. This shift in attitudes and values coincides with the more general reevaluation of the role of women in the culture that the women's movement has inspired and championed.

The third and fourth reasons have to do with changes in family life. Because of decreasing family size and greater health and longevity of older adults, more women find themselves with more time to pursue work (Hoffman, 1977). Their traditional role of caring for family members has not lessened in importance. Rather, the number of those in need of care has declined. At the same time, the variety of labor-saving devices that have become the necessities of our modern lives gives women even more

BOX 8-2
WORKING

Studs Terkel (1974) could easily be called the chronicler of the average man and woman. Like the Opies who studied the everyday activities of children, Terkel records the everyday experiences of average adults. Unlike the Opies, Terkel is as much the activist as the historian. Through the words of others, Terkel vividly shows how "surviving the day is triumph enough for the walking wounded among the great many of us" (p. xiii). Consider the day of Will Robinson, a forty-seven-year-old city bus driver in Chicago.

"Then we have people get on the bus and pay their fare just like any other passenger, but all the time they're a spotter, see? They're watching everything that goes on. If there's anything you do wrong, two or three days later you're called into the office. I was called in about a year ago. We have the fare boxes. As the people drop the money there is a little lever there, and you're supposed to continuously hit the lever so that the money can go down into the bottom. I was called in. Some spotter on the bus said I didn't make the money go down—which was very erroneous. I'd forced a habit of just steady hitting this all the time. There's a little door that lets the money go through. It's spring loaded. Once so much money gets

free time. However, washing machines cost more than scrub boards, dishwashers more than dishrags, and frozen foods more than home-prepared foods. In other words, these new necessities are expensive, and many women find themselves working to pay for these items. In sum, the variety of social and economic changes that have occurred have provided women with both the opportunity and the necessity to earn additional income and to define themselves in additional ways.

in there the weight'll make it open anyway and it'll fall down. There's nothing you can do about it anyway. Once the money goes down all you can do is see it.

"They will report if any passengers are getting by without paying. They check up on the transfers that you issue—if you give someone a transfer with too much time on it, or if you accept a transfer that's too late. A spotter will get on the bus and give you a transfer that's too late, purposely, to see if you'll observe it.

"Then you have the supervisors on the street. They're in automobiles. If you're running a minute ahead of time, they write you up and you're called into the office. Sometimes they can really upset you. They'll stop you at a certain point. Some of them have the habit of wanting to bawl you out right there on the street. That's one of the most upsetting parts of it.

"If you're running hot, ahead of time, there're afraid you're gonna miss some passengers. If I go out there and run three or four minutes hot, then the guy in back of me gets all the passengers. You got a guy in front of you two or three minutes ahead, you gotta carry the whole street. It's pretty tough.

"They call these checkpoints. On my run I have three, four checkpoints between one terminal to the other. You'll never know when they'll be there. Most of 'em are in little station wagons. If you come late to a checkpoint, there isn't much they can do about it. They allow you time for being late, with traffic conditions. But they say there is no excuse for running ahead of time. They'll suspend you for a day or two, whatever the whims of the superintendent. He's the guy who has the say in the garage. If he decides to suspend you for a week, you lose a week's work. If you're caught running ahead of time, within about six months you'll get whatever he feels he wants to give you" (p. 202–203).

For Terkel and others who write about "violence to the spirit," the pen is indeed mightier than the sword.

Work in the Lives of Adult Women

Most women do not see their role as workers as a means of fulfilling the parent and spouse roles or as something to substitute for them. Rather, it is viewed as an additional but equally important and potentially satisfying aspect of their identity as an adult. Because most women who are working maintain primary and sometimes even exclusive child-care and household

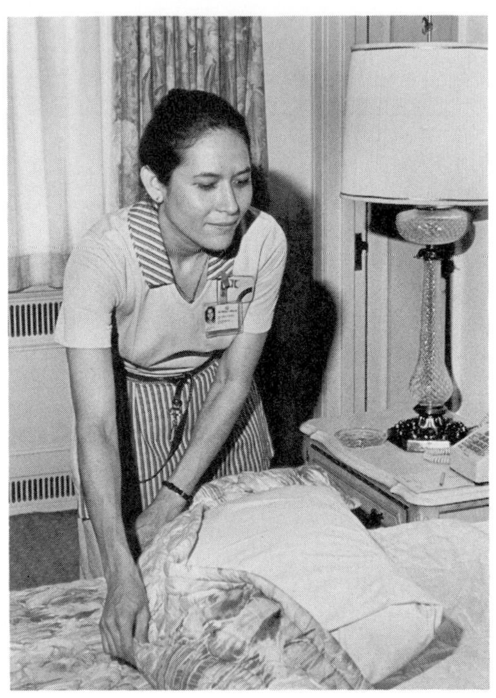

The women's movement has helped to open many better-paying areas of employment to women. Notable among these areas is the hi-tech industry. However, the lower-paying service-related industries still account for the majority of female employment.

responsibilities (Geerken & Gove, 1983; Hill & Stafford, 1980; Kamerman, 1980; Pleck, 1983), they invariably find themselves in a bind for time to perform their various responsibilities. As a result, they often have relatively little time for leisure activities.

The two typical daily schedules in Table 8-1 provide vivid evidence of the time bind that many working women report. The first schedule is for a woman working as a nurse. She has two young children, and her husband is a full-time graduate student. They share many of the child-care responsibilities. The second schedule is that of a secretary. She also is married and has two young children. Her husband works as a salesman, sometimes having to work at night. The two women were part of an in-depth interview study by Kamerman (1980) on the consequences of being a working mother. You should bear in mind that these schedules reflect days when all the pieces fit together. That is, on these days, no one had a flat tire, no child had an earache, the bus didn't run late, and the washing machine didn't break. (Schedules such as these become considerably more flexible once children reach school age.) As difficult and hectic as these schedules appear, they must nevertheless be seen as a reasonable price to pay for working. Indeed, approximately four times more women who are not working wish they were than vice versa (Geerken & Gove, 1983).

These schedules reflect the fact that women are undergoing a **role expansion process** (Yogev, 1981). They are attempting to broaden their range of competence and sources of satisfaction without abandoning traditional functions, obligations, and sources of satisfaction. The difficulties

TABLE 8-1 *Two Schedules Typical of Working Women*

FIRST WOMAN'S DAY	
5:00 A.M.	Get up, shower, brush teeth.
5:30	Make coffee; prepare baby's breakfast.
5:45	Dress.
6:15	Change, dress, and feed baby.
6:45	Leave for work, bringing baby to sister's along the way.
7:00	Husband awakens four-year-old and helps him get washed and dressed.
7:30	Husband prepares breakfast for son and himself, and they eat together.
8:15	Husband and son put on heavy jackets and walk to school.
Noon	Husband picks up son at school and brings him to sister's house and then leaves for school.
4:00 P.M.	Arrive home from work and go to get children. Visit with sister for a half-hour or so.
5:00	Play with children.
5:45	Prepare dinner.
6:15	Feed baby.
6:45	Dinner for husband, self, and four-year-old.
7:15	Make lunches for next day while husband cleans up.
7:45	Baby gotten ready for bed. Parents play with baby.
8:00	Husband puts baby to bed.
8:30	Put older child to bed.
9:00	Parents have coffee and relax.
10:00	In bed.
SECOND WOMAN'S DAY	
6:00 A.M.	Out of bed and shower.
6:15	Wake up baby, wash and dress her and wake up Sharon.
6:30	Give baby breakfast.
6:45	Put baby in crib and prepare Sharon's breakfast.
7:00	Get dressed.
7:15	Collect baby's clothes for sitter, check Sharon.
7:20	Leave house.
7:25	Drop baby off at sitter's.
7:30	Take Sharon to mother's (grandmother's).
8:00	Begin work.
8:30	(Grandfather takes Sharon to school when he leaves for work.)

TABLE 8-1 *Two Schedules Typical of Working Women (Cont'd)*

SECOND WOMAN'S DAY

5:00 P.M.	Leave work.
5:15	Catch bus for home.
5:40	Get off bus into car to pick up children.
5:45	Reach parents' house to get Sharon. Visit.
6:30	Arrive home and start to prepare dinner. (Husband already home.)
7:00	Dinner.
7:30	Dishes (done together).
7:45	Play with children.
8:00	Get baby ready for bed.
8:15	Bathe Sharon while husband puts baby to bed.
8:30	Husband plays with Sharon. Straighten the house, do some laundry.
9:00	Sharon to bed (either parent).
9:30	Prepare baby's clothes and bottle for next day, finish laundry.
10:00	Watch TV with husband; talk.
11:30	To bed.

Note: Adapted with permission of The Free Press, a Division of Macmillan, Inc. from PARENTING IN AN UNRESPONSIVE SOCIETY: Managing Work and Family Time by Sheila B. Kamerman. Copyright © 1980 by The Free Press.

encountered in this endeavor reflect both the multiplicity of roles and the degree of support available to help women pursue their varied interests. It is worth noting that the title of the Kamerman (1980) book from which the two schedules are taken is *Parenting in an Unresponsive Society.*

Work in the Lives of Adult Men

For men, the role of work occupies a different place in their adult identities. First, it is a more central component of identity than it is for women, and second, it is viewed as the means through which a significant portion of the roles of parent and spouse are fulfilled (Levinson, 1978).

> A man's occupation is one of the primary factors determining his income, his prestige and his place in society. . . . Over a number of years a man chooses and forms an occupation. All men make one or more changes, some of them quite marked, within the original occupation or from one occupation to another. A man's occupation places him within a particular socio-economic level and work world. It exerts a powerful influence over the options available to him, the choices he makes among them, and his possibilities for advancement and satisfaction. His work world also influences the choices he

> makes in other spheres of life. . . . Occupation has important sources within the self and importance consequences for the self. It is often the primary medium in which a young man's dreams for the future are defined, and the vehicle he uses to pursue those dreams. At best, his occupation permits the fulfillment of basic values and life goals. At worst, a man's work life over the years is oppressive and corrupting, and contributes to a growing alienation from self, work and society. (Levinson, 1978, p. 45)

Two factors are paramount in men's assessment of their worker role. The first is the amount of money (and other compensations) they earn, and the second is the nature of the work and the work setting. Clearly, the more money a husband earns, the more he is fulfilling the provider aspect of his role. Factors that influence salary, such as education and occupation, are therefore valued by men at least to the extent that they predict income. On the other hand, a high-paying job may have a number of disadvantages such as the expected level of performance, long working hours, excessive travel, irregular schedules, company moves, and pressure from superiors. These factors are particularly relevant during the early adult years since most men are eager to establish themselves within their work setting.

With regard to men who don't bring a briefcase home at night or work late finishing reports for the management meeting the next day the

While many men assess their jobs on the bases of salary and compensation benefits, other men rate job satisfaction on the basis of the intrinsic value of the work performed.

common assumption is that they leave their jobs behind when they leave the workplace. This is only partially true. These men may not bring their work home with them, but they may well bring the effects of their work home with them (Kohn, 1977; Piotrkowski, 1979; Piotrkowski & Katz, 1982). A negative carryover from work to home is not unique to lower-income jobs, but the nature of these work settings does make it more common to them. In particular, men in lower-status jobs have a relatively small degree of power with which to influence the specifics of their individual jobs and limited mobility within the work setting.

Piotrkowski (1979) finds that one of the carryovers of such job settings is an interpersonal unavailability in the home. The husband comes home from work and simply wants to be left alone. He is often irritable with his children and his wife. Eventually a pattern develops in which family members distance themselves from the husband. A very different carryover pattern is found when the work setting is viewed as enjoyable, when the worker has a degree of control, when there is some opportunity for initiative and creative effort, when the work is seen as intrinsically rewarding, and when the worker's efforts are recognized and rewarded by others. In these cases, the husband is much more likely to be involved with his wife and children, both emotionally and physically.

The Interplay of the Three Careers

Most adults find the task of balancing the sometimes competing demands of parent, intimate, and worker both frustrating and rewarding. Managing, even for a short time, to successfully integrate the three careers produces a sense of personal satisfaction that may not be possible at earlier stages of the life-span. Confronting the sometimes seemingly irreconcilable conflicts among the three careers soon makes apparent to adults how unsupportive society can be of the developmental tasks of the early years of adulthood.

Maintaining a Sense of Equity

When both the husband and the wife work in a family, a different balance exists from the one that exists when only one spouse works. But in both cases, the issue is still a sense of equity. In both cases, husbands and wives judge the equity of their relationship by considering whether their functioning is consistent with their own expectations, as well as those of their partner, and by considering the degree to which their expectations are being fulfilled. In particular, the sense of equity is low when one spouse feels overburdened relative to the other or does not fulfill what is seen as a joint venture or decision (Kamerman, 1980). For example, little sense of equity would be felt in a situation where both husband and wife prefer a head/complement pattern but the wife is forced to find work to supplement the husband's limited earnings. Neither is meeting his or her own expectations or those of the partner. On the other hand, a dual-career couple, both successful at their jobs, are doing what they each want to do and what each wants the other to do, and therefore have a relationship that seems equitable to them. There are a large number of patterns that could be defined to demonstrate equitable and inequitable relationships. The important element is not the specific details of the roles but the extent to which those roles are consistent with people's expectations. In both examples, both husband and wife are working. In the first, the sense of equity is low; in the second, it is high.

Establishing a mutually satisfying sense of equity may prove particularly difficult when the husband feels ambivalent about his wife's working. The extra money is certainly welcome, but the greater equality of decision-making power is sometimes seen as a liability (Burke & Weir, 1976). Such husbands are often reluctant to significantly increase their household and child-care commitments, sometimes feeling resentful of the fact that their "social support system" is less effective than it was before. Some feel that their wife's employment is, in some way, detrimental to their children. They may even see the wife's worker role as a threat to their sense of masculinity. In these circumstances, it is not surprising that wives see their work as a sort of "sideline." They pursue their work or return to school in such a way that they continue to fulfill the role expectations of their husbands. In so doing, they resolve, or at least avoid, some of the spouse conflicts that arise because of their employment. The price they pay for this is usually less personal time and often less sleep. The satisfaction provided by the couple's relationship is also lessened.

When children are included in the family equation, the calculation of equity becomes even more difficult. It might seem logical that if women are assuming some of the responsibility for the economic well-being of the family, then men should show a corresponding shift in the amount of responsibility they assume for the house and child care. As has been mentioned a number of times already in this chapter, the evidence does not show such a corresponding shift (see Table 8-2). For whatever the reason, reluctance on the part of husbands or resistance on the part of wives, men remain relatively uninvolved in these tasks.

TABLE 8-2 Time Use (Hours/Day) of Employed Husbands and Wives

GROUP	HOUSEWORK HUSBAND	HOUSEWORK WIFE	CHILD-CARE HUSBAND	CHILD-CARE WIFE	COMBINED HUSBAND	COMBINED WIFE
All couples	1.61	4.52	.24	.91	1.85	5.43
Wife employed	1.63	3.37	.24	.64	1.87	4.00
Wife not employed	1.59	5.60	.25	1.16	1.83	6.76
Youngest child 0–5	1.35	4.68	.44	1.86	1.79	6.51
Wife employed	1.38	3.48	.43	1.20	1.81	4.69
Wife not employed	1.34	5.25	.44	2.16	1.78	7.40
Youngest child 6–17	1.75	4.73	.28	.81	2.03	5.55
Wife employed	1.74	3.71	.26	.52	2.00	4.24
Wife not employed	1.77	5.97	.29	1.16	2.06	7.14
No children in household	1.72	4.14	.03	.12	1.75	4.26
Wife employed	1.76	3.02	.02	.11	1.79	3.12
Wife not employed	1.64	5.62	.02	.14	1.66	5.76

Note: Adapted, with permission of the publisher, from J.H. Pleck, "Husband's Paid Work and Family Roles: Current Research Issues," in H.Z. Lopata & J.H. Pleck (Eds.), *Research in the Interweaves of Social Roles: Family and Jobs,* Vol. 3, (Greenwich, CT: Jai Press, 1983), 259.

Maintaining Good Environments for Children

How does having a working mother affect children's development? A complete understanding of the answer to this question first requires an appreciation of the reason why the question is not phrased "How does having working parents affect children's development?" In other words, the issue really isn't the effects on children of having parents who work but rather the effects of having someone other than a parent, invariably the mother, provide direct care for them. The issue concerns the effects on children of some form of substitute child care or, when they are older, some degree of self-management. It also concerns the effects of reduced direct involvement, at least in terms of clock time, with the mother or the father on a day-to-day basis. The answer to the question is that researchers have found remarkably few differences between the development of children whose mothers work and of those whose mothers don't. There appears to be little relationship between maternal employment and school achievement (Hayes & Kamerman, 1983), between maternal employment and children's involvement in community activities (Rubin, 1983), and between maternal employment and children's compliance, affection, aggression, and anxiety (Henggeler & Borduin, 1981). Findings like these are based on overall comparisons, that is, comparisons between all working and nonworking mothers. When the research digs a little deeper, however, some differences do emerge.

The most important generalization to emerge, noted in almost all research in the area, is that the issue isn't whether or not mothers work but whether or not they are satisfied with their present life situation. The children of working mothers who are satisfied with their work, with their marriage, with the quality of the available child care, and with the other aspects of their lives do very well on the various measures of development that have been discussed in earlier chapters. The same statement is equally true of children of nonworking mothers who are satisfied with their life situations. Unfortunately for working and nonworking mothers who are dissatisfied with their life circumstances, their dissatisfactions also seem to transfer to their children's development (D'Amico, Haurin, & Mott, 1983; Etaugh, 1974; Hoffman & Nye, 1974).

A second finding from the research is that the actual amount of time working and nonworking mothers spend with their children is considerably more similar than their distinct schedules would lead one to expect (Hill & Stafford, 1980). Nonworking mothers do not spend their entire day with their children. Mothers and children may be in the same house together, but that doesn't mean they are in constant contact. On the other hand, working mothers usually set aside time when they are home (usually early evenings or the hours before bedtime) just for their children. The fact that this special time is more typically set aside by college-educated women (Hill & Stafford, 1980) probably reflects the work carryover issues that were discussed earlier (Piotrkowski & Katz, 1982). Although the actual number of contact hours between mother and child for working and nonworking women is not identical, the two numbers are similar enough to be seen as functionally equivalent.

There are other consequences for children of having mothers who work. Older children are more likely to be involved in child-care and household responsibilities (Hayes & Kamerman, 1983), and there is a slightly greater likelihood that school-age males will be involved in delinquent activities. Both of these findings are more likely to occur in less economically advantaged communities where quality child-care settings and attractive, well-supervised after-school activity programs are less frequently found.

Having a working mother seems to exert a significant influence on the sex-role development of girls. D'Amico, Haurin, and Mott (1983) report that in comparison with the daughters of nonworking women, the daughters of working women who viewed their work as successful and satisfying held more egalitarian views about the adult sex roles of men and women, had a greater desire to pursue a career when they became adults, and favored a greater range of both traditional and nontraditional careers. No such differences were found between the sons of working and nonworking women. It may be that adolescent and young adult men, feeling a strong cultural pressure to assume the provider role, are somewhat less able during this period to be responsive to diverse influences. If this is so, the impact on men of having a working mother may become more evident later in their lives as they become more comfortable with their adult tasks. On the other hand, the lack of relationship may reflect the fact that there remain greater cultural pressures on males to assume specific adult roles than there are on females. If so, this would help explain the reluctance of many men to become more involved in household and child-care responsibilities. In other words, as a result of the social changes in adult roles over the past thirty years, the range of socially acceptable adult roles for women has become greater than that for men.

This discussion of the interplay of the three careers of the early adult years highlights the fact that people experience this period of the life-span as both very busy and very satisfying. They may never have enough hours in the day to accomplish everything, but the satisfactions most adults report (Campbell, 1981; Robinson, 1977) indicate that this time is viewed very positively. Most adults feel a great sense of accomplishment in becoming fully participating members of the community of adults, in seeing their children grow, and in being able to provide adequately for themselves and their family. However, not all adults feel so successful. Some do not accomplish what they had hoped to accomplish; others feel little sense of accomplishment at all. This sense of failure is perhaps greatest when it is associated with the most pivotal aspect of adult life, marriage.

DIVORCE

It is estimated that approximately one-third of all first marriages will eventually end in divorce in this country (Jacobson, 1983). This rate is the highest in our history and is among the highest of those in all industrialized nations. Ironically, the high incidence of divorce is due to a set of circumstances that over the past century (especially the past twenty-five years) has made marriage at the same time both more important and less important in the lives of adult men and women.

Divorce is not a random event. Certain life circumstances are more predictive of divorce than others. Income and education are the two major ones. The higher a couple's income is, the less likely they are to divorce. Generally, the same statement is true concerning years of education (Price-Bonham & Balswick, 1980). However, as will be discussed later, these two generalizations are qualified when the wife has an independent source of income, most typically a job, or when she has adequate education to obtain a satisfactory income.

Age makes a difference. Teenage marriages have the highest divorce rate, especially if the woman is pregnant at the time of marriage. First marriages involving couples over thirty have a somewhat higher divorce rate than those involving couples in their middle to late twenties. Half of all divorces occur by the seventh year of marriage. However, one-quarter involve couples over the age of forty (Jacobson, 1983). Slightly more than half of all divorces involve children. It is now estimated that by age eighteen, one-third of all children in this country will experience their parents' divorce (Glick, 1979).

Changing Patterns of Divorce

Since the beginning of this century, there has been overall a gradual but steady increase in the rate of divorce. The rate of increase slowed during the economic depression of the 1930s, showed a steep but temporary increase following the end of World War II, and remained relatively stable during the 1950s. Beginning in the 1960s, the rate again increased sharply and now again seems to be leveling at a record rate of one out of three marriages. What accounts for the gradual increase over the past century and, in particular, the drastic change since the early 1960s?

The gradual overall increase seems associated with the same set of social forces that created our modern conception of adolescence—namely, the industrial revolution that decreased the economic role of the family, placed greater emphasis on individual effort, and fostered the growth of urban areas in this country (Weiss, 1975). This association of divorce rate with degree of urbanization has been noted in other countries and is believed to reflect a shift from the family to the individual as the basic economic unit. Bear in mind that this is a correlated association, not a causal one. All it indicates is that there are factors associated with an urban lifestyle that influence the rate of divorce. It does not explain whether these factors actually influence the likelihood that people will have a successful marriage or merely the likelihood that they will be willing to end an unsatisfying relationship. An examination of the factors influencing the post-1960s shift suggests that both explanations are relevant.

What seems to be happening is that the social forces that influence adult roles are changing. Specifically, they are changing in such a way that marriage no longer serves as the keystone of adult social roles. As the various aspects of adult life become more independent, the necessity of marriage (quite distinct from its value) has lessened (Furstenberg, 1982). Traditionally, marriage was associated with a move away from home, the beginning of sexual activity, the securing of a stable income, parenthood, and the establishment of an independent residence. Although in reality these activities may not have corresponded, it is nevertheless true that both through social value and legal practice, this "package deal" was viewed as the ideal norm. Over the past few decades, however, these activities have become increasingly independent. Adults living on their own before marriage are no longer considered to be behaving improperly. Premarital sexual relationships may not yet be condoned but are, at least, tolerated by most adults in our culture. Parenthood is no longer viewed as the necessary and immediate sequel to marriage. For a small number of adults, marriage is even seen as irrelevant to parenting. Appropriate roles for women are no longer restricted to home and children. In fact, many researchers (e.g., Furstenberg, 1982; Jacobson, 1983; Price-Bonham & Balswick, 1980; Scanzoni & Scanzoni, 1981; Weiss, 1975) see the increasing economic power of women as one of the most significant factors influencing the changing divorce rate. These researchers argue that, as women have gained economic independence both through greater opportunities for employment and through increased education, the necessity to remain in an unsatisfying relationship has lessened.

As social forces have changed, so too have the legal and social values that regulate marriage. Divorce no longer carries the social stigma it once did (Ahrons, 1980). Couples seeking divorce no longer must prove that one or the other in some way violated the marriage contract, and therefore they don't have to sue for damages. In most states, they may petition the court, in a no-fault fashion, to dissolve the legal bonds that they voluntarily accepted previously. Divorce is no longer a barrier to political office, hiring and promotion, or participation in civic and many religious activities, and it is no longer viewed as inevitably a negative influence on children's development (a point discussed more fully later). All of these changes have meant that adults are increasingly less likely to feel the necessity to get married or to remain married in order to participate in the various other aspects of adult life. For this reason, the meaning of marriage is changing from one described in terms of economic necessity and social responsibility to one that involves companionship, empathy, affection, and individual benefit (Weiss, 1975).

The meaning of marriage may be changing, but all the evidence indicates that its value to adults is not. Marriage remains a virtually universal aspect of adult life, for the never married and the previously married as well as for those who marry and whose marriages remain intact. If adults still value marriage, then why has the divorce rate increased? The increase reflects the need for adults to learn the coping strategies and negotiating skills necessary for maintaining an increasingly egalitarian, increasingly voluntary relationship, a relationship that no longer exists out of economic necessity alone (Price-Bonham & Balswick, 1980). Put another way, the drastic increase in the divorce rate over the past thirty years reflects a period of transition in the roles of adult men and women in our society. This transition is more noticeable in the more economically advantaged segments of the society, particularly in the college-educated population, but there is no reason to think that it will not eventually spread to other segments as well. As Norton and Glick (1976) put it,

> There seems little doubt that a basic transformation of the institution of marriage is underway and that many variables are influencing the direction of the change. This transformation appears to be predicated largely on a restructuring of the roles which men and women play within the traditional boundaries of marriage and family living. Some people can confront this type of change and adapt to it without much difficulty; others find that the process of adjustment is much more difficult and leads ultimately to marital conflict and disruption. Fundamental to the understanding of this change is a comprehension of the redefinition of expectations of individuals involved in a modern marriage.
>
> . . . Viewed in this manner, the high rate of divorce can be interpreted as an understandable pursuit of happiness. This does not necessarily mean that people are marrying and subsequently divorcing without care or concern, but rather that there exists a new awareness that a marriage which is subjectively viewed as not viable can be dissolved and—hopefully—replaced by a more nearly viable one. (p. 17)

The Experience of Divorce

As adults continue to acquire more effective ways of dealing with the changes in adult roles, the divorce rate should stabilize (there is some indication that this is already beginning to happen), perhaps even declining somewhat. But those adults for whom divorce becomes a necessity must deal with one of the most painful and stressful of all transitions in the life-span (Jacobson, 1983).

It is perhaps only when a marriage begins to unravel that the participants come to realize what a physical, social, and psychological investment they each have had in their relationship. The process is not a simple one. It involves a number of facets and follows a developmental course that, in some ways, resembles the separation distress experienced by infants and the loss through death of a spouse in later life.

The Six Facets of Divorce

The complexity of the process is reflected in Bohannan's (1971) observation that divorce involves six interdependent but identifiable issues. The **six facets of the divorce process** highlight the multiple bonds involved in a marital relationship. Bohannan refers to them as the emotional divorce, the legal divorce, the economic divorce, the coparental divorce, the community divorce, and the psychic divorce.

The **emotional divorce** concerns the failing relationship of the couple. It involves the falling out of love with the spouse, the decline in trust and intimacy, and the building of emotional barriers to protect oneself. Needless to say, it is an extremely difficult process, one frequently involving reversals, false hopes, and shattered dreams.

The **legal divorce** involves the actual termination of the marriage contract. By the time a couple has reached this step in the divorce process, their relationship has already come to an end, but as couples who live together before marriage report, the public commitment of the marriage ceremony in some way changes the nature of the relationship, and so too does the divorce decree. In both instances, the change in legal status prompts adults to think about themselves and others in new ways.

Couples amass many possessions in the course of a relationship. The worth of some is easily determined; others whose value is based on memories rather than economics are more difficult to appraise. In any case, all of the possessions of the couple must be divided. It appears rare that a couple can complete this facet, the **economic divorce,** without some degree of bitterness and resentment, without both feeling that they are not getting their fair share. In some cases, the division of property is contested in the courts. For example, in situations where the wife's employment made the continuation of her husband's education possible (for example, in medical school or law school), the courts have argued that the wife is entitled to some amount of money reflective of the future economic status of her husband, given the fact that, through her efforts, he was able to continue his education and therefore now enjoys continued economic well-being. Even under the most favorable economic circumstances, however, most couples confront the fact that two together can live more cheaply than two apart.

Divorce ends the marriage bond; it does not end the parent bond. But the **coparental divorce** facet does change the way these parental roles are maintained. About 40 percent of all divorces involve children under the age of eighteen. In the sense that children are property, they must also be "assigned." Custody must be awarded to one or the other spouse or to them jointly. Decisions have to be made about where the children will live, how each parent will spend time with them, how each will provide for child support, and how future decisions about the children's welfare will be made. In some cases, custody is contested and fought in the courts. Whatever the outcome, the couple's bitterness toward each other is often heightened, a problem that hinders their ability to continue to communicate as parents even though they are no longer husband and wife. The effect of divorce on children will be discussed in the following section, but for now, it is useful to know that one of the best predictors of a child's recovery from the parents' divorce is the parents' ability to deal reasonably with each other as parents after the divorce is finalized (Wallerstein, 1983).

Marriage is a public event. It announces to family, friends, and the community in general that a special relationship has been formed between two people. The **community divorce** tells this same public about the end of that relationship. Some people are told directly; others learn through the grapevine. Official documents, bank accounts, credit cards, and the telephone listing must be changed. Family and friends sometimes take sides or sometimes find themselves caught in the middle, with each spouse trying

Changes in divorce laws and judicial philosophy are resulting in a small but increasing number of fathers gaining full or joint custody of their children.

to gain their support and sympathy. The informal social support network of friends and family enjoyed by a couple is almost always less available and less supportive of singles.

Bohannan (1971) feels that, for many, the **psychic divorce** may prove the most difficult adjustment. It is the process through which people again come to think of themselves as single, independent adults. For a time, many react with feelings of loneliness and isolation, rather than seeing the situation as a new beginning. The fact that most divorced adults eventually remarry indicates how necessary this sense of coupleness seems to be.

The Process of Divorce

The various aspects of the divorce process described by Bohannan do not all occur at the same time. As mentioned previously, there seems to be a developmental course to the divorce process. The sequence can be divided into three phases (Golan, 1981). The first phase is bounded by the "beginning of the end" and the actual separation and decision to divorce. The second consists of the divorce and the initial period of readjustment. Since the readjustment invariably involves a period of disorganization and distress, the third period begins when the now single adult starts to put his or her life together again and ends when a new positive life-style has been formed. The process can be a long one. Postdivorce adjustment often takes as many as four or five years; the first phase, as long or longer (Kitson, 1982).

Phase I: The beginning of the end. When couples look back on a failed marriage, they tend to group the specific circumstances in one of three ways (Weiss, 1975). Some view the marriage as having been flawed from the start. Some say they were aware of the other's drinking or gambling or inability to keep a job or whatever but felt that the situation could be changed. Others argue that being young and being in love blinded them to qualities in the other that only began to become apparent as the marriage progressed. Finally, there are those who feel that they grew apart from their spouse. The relationship started well and continued to be satisfying for a time. At some point, however, each person realized that a basic conflict was developing between the stability and vitality of their marriage and the course they each individually wanted to pursue. Perhaps the husband wanted a job that required frequent moves, and his wife didn't want to be continually uprooted. Perhaps the wife wanted to finish her education and find a full-time job, while her husband felt that her place was in the home and his in the office. Whatever the specifics, it is common for the two partners to see the same events in very different ways and for the idea of divorce to be presented by one to the other rather than to emerge from joint discussion.

As this first phase progresses, the elements that bond the two people to each other unravel. Trust and support give way to isolation (physical and emotional) and blame. Each talks about the other with friends or family in ways not done before. Sexual infidelity is not uncommon, further adding to the distance between the couple. Each, intentionally or unintentionally, hurts the other, and each comes to feel bitter, isolated, inadequate, guilty, and betrayed all at the same time. The desire to want to be with the spouse, to identify with the spouse, to feel a complement to the spouse, gradually erodes.

Falling out of love with a spouse is not, however, the same as severing the bonds that tie spouses together. The attachment bond, which seems firmly in place by the second year of marriage (Kitson, 1982), erodes at a much slower rate than the bond of love. Separated spouses seem to experience the same sense of separation distress that Bowlby (1969) and Ainsworth (Ainsworth, Blehar, Waters, & Wall, 1978) have described in young children. There is a continual focus on the absent spouse. Little serves to compensate for the person's absence. Feelings of unhappiness, anxiety, and increased vulnerability are frequently reported (Golan, 1981). This emotional distress hinders the performance of everyday activities. Sleep becomes more erratic, and drinking may increase. Separated spouses are often preoccupied with learning about the other, in effect assuring themselves that the other is still available and therefore not completely separated from them.

The ambivalence the two people feel toward each other—wanting to be away from the other and yet find themselves uncomfortable when they are separated—prompts the attempts at reconciliation, the periods of sexual intimacy, and the confused and conflicting emotional states that are experienced by many separated couples. At the same time, some people also experience what Weiss (1975) refers to as a sense of euphoria. These individuals report that they feel a great excitement at the chance to be freed from an oppressive marriage and to be able to pursue their own goals. However, Weiss also notes that this euphoria is fragile and the slightest setback or disappointment can resurrect the separation anxieties.

As the separation continues and as it becomes increasingly apparent that the marriage should end and will end, separation anxieties decrease and feelings of loneliness ensue. The couple may be involved in negotiations over children and property. Most likely the adversarial nature of these negotiations hastens the dissolution of the attachment bonds. Gradually,feelings of loneliness (unlike separation distress) can be countered by being with other people.

During the initial phases of the divorce process, some people turn to alcohol in an effort to ease their anxiety and loneliness.

Phase II: Post-divorce adjustment. The second phase, the postdivorce adjustment period, is difficult for almost all adults (Golan, 1981; Hetherington, Cox, & Cox, 1976). Individuals find themselves confronted with an array of practical and personal issues that are either new to them or for which old coping strategies no longer seem effective. Moreover, they find themselves confronting these new challenges at a time when their sense of self-esteem is perhaps at its lowest. The most intensive investigation of this period is by Hetherington, Cox, and Cox (1976; 1977), and although it dealt only with couples having children, most of its findings are equally relevant to childless couples.

Hetherington and her colleagues studied children and their families from both intact and divorced homes. Their data included diary records and interviews, parent ratings of children's behavior, and direct observations of parent-child interactions. They found that, during the two-year postdivorce period, both men and women experienced a variety of distressful emotional upheavals. These reactions tended to be most severe one year after the divorce and then began to lessen in intensity. In general, these feelings were ones of depression, anxiety, and apathy. Fathers reported having little sense of structure in their lives. Mothers most frequently reported feelings of helplessness and of physical unattractiveness. Both fathers and mothers reported feelings of lost identity and of no longer being sure of their status within the adult world. The sense of lost identity was most acute in those who had been married the longest.

Feelings of loss of competence were also frequently reported. Most adults in the study felt they were not performing old tasks as well as before and that they were less comfortable in social situations. They also had serious doubts about their ability to successfully marry again. Perhaps as a way to cope with these feelings as well as to rebuild their lives, many of the adults in the study involved themselves in various recreational and educational programs, in self-help groups, and in social clubs. A number made serious attempts to present a more youthful image. The value of this flurry of social activity seemed short lived. Casual sexual encounters did increase during the first year, especially for men, and both men and women reported valuing the companionship these encounters provided. By the end of the first year following the divorce, however, it became apparent to both the men and the women that their casual sexual encounters were not an adequate substitute for the sense of intimacy they once had. These feelings were particularly strong among the women in Hetherington's sample. Many reported that these casual sexual encounters left them more depressed and further lowered their already floundering sense of self-esteem. Hetherington believes that this lack of a sense of intimacy serves as a powerful force toward the establishment of more permanent relationships.

Emotional and social issues were not the only ones that served to depress the newly divorced adults. Many found themselves confronted with tasks previously performed by their spouse, by a number of tasks that two had once managed, and by a more limited financial base than they had once enjoyed. Not surprisingly, men found themselves overwhelmed with cooking, laundry, and routine household tasks; women with home repairs, financial matters, such as insurance and taxes, and contacts with service providers, such as mechanics, electricians, and plumbers. Even those who had previously mastered these skills found that having to be responsible for all of them was a burden that left little time for personal pursuits. As a result, household schedules were sometimes chaotic. Bedtimes and meal-

Routine household tasks often prove distressful for recently divorced men.

times were less regular than before, menus less planned and more "creative."

Phase III: Reconstruction. Gradually, by the second year, circumstances began to improve for these divorced adults. The development of new social contacts, the acquisition of new skills, and the establishment of household routines signaled what Weiss (1975) sees as the beginning of the third, or reconstruction, phase of the divorce process. As these new tasks are encountered and mastered, the individual's sense of self-worth and self-identity begins to improve again.

A major element in this reconstruction phase is the development of an intimate relationship (Furstenberg, 1982; Weiss, 1975). Most divorced adults remarry. On the average, men do so within three years and women within five to six years. The time lag for women probably reflects the fact that they usually have custody of their children and therefore must find someone who is acceptable for their children as well as themselves.

Second marriages have a different quality from first marriages. They are seen by the participants as more practical, less romantic. Since many divorced adults come to see the failure of the first marriage as being due to youth or impressionability, the second one is perceived in more pragmatic terms. Issues about child care, division of labor, money, and work are discussed and often resolved before and not after the wedding. In some cases, especially when one person has considerable economic assets, the other might agree to relinquish inheritance claims in favor of children from the first marriage. Such an event reflects the fact that these marriages are viewed not only as more practical, but also as more risky. Adults in a second marriage report being less willing to remain in an unsatisfactory marriage than those who are in their first marriage (Furstenberg, 1982).

Most remarried adults report that the second marriage is quite different from the first. There is a greater overlap in roles, both in and out of the house. Communication is more open, supportive, and effective. Each person sees the other as being more tolerant. Both see the relationship as more egalitarian than the first one was (although the husband still thinks the relationship is more egalitarian than the wife thinks it is). Most people account for this improvement by saying that they found a better partner rather than by saying that they themselves became a changed person. Apparently, it is important to them that the spouse from the first marriage be pictured as the "heavy."

Ironically, the divorce rate for second marriages is actually higher than for first, particularly if the second marriage occurs soon after the divorce. It is still unclear whether the high divorce rate reflects a greater difficulty in making a second marriage work (Cherlin, 1978) or a greater willingness to end an unsatisfactory relationship (Furstenberg, 1982).

The Effects of Divorce on Children

Virtually all children are initially adversely affected by their parents' divorce (Hetherington, Cox, & Cox, 1979; Kelly & Wallerstein, 1976; Wallerstein, 1983; Weiss, 1979). Whether these initial reactions continue beyond the two-year postdivorce period in which they are typically found depends to a large extent on the ability of the parents to successfully maintain their parental responsibilities even though they are no longer married (Wallerstein, 1983). Even when divorced parents are able to communicate effectively concerning their children's welfare, however, the experience of divorce presents children with a set of developmental issues that continue to be confronted even into their early adult years.

The Age of the Child

How children react to the separation of their parents is largely a function of their developmental level. This is so because developmental level correlates with the nature of the parent-child relationship, the cognitive and emotional mechanisms through which children try to understand the changes that are occurring in their families, the behaviors children are likely to use to express their reactions to these events, and the settings in which these behaviors are most likely to be expressed.

Infants and toddlers. Infants and toddlers have little understanding of the meaning of separation and divorce. The reaction of these very young children is largely a function of changes in caretaking routines. Changes may occur both in the regularity of such routines as feeding, bathing, diapering, bedtimes, and playtimes and in the affective quality of these interactions. The decrease in the predictability of these routines and interactions often prompts heightened separation anxieties and increased dependence on the parent (for example, clinging, whining, and more frequent requests for help). These behavioral regressions may further emotionally overburden an already emotionally overburdened parent, again adding to the difficulty of the separation and divorce period.

Preschoolers. The egocentric nature of preschool-age children poses special problems in their initial reaction to parental separation and divorce. Unlike the infant who has little understanding of causal relations and unlike the older child whose reaction tends to take the forms of anger and denial, preschoolers often see themselves as being in some way responsible for their parents' problems. Because of their limited ability to appreciate the diversity of adult roles and relationships, preschoolers are prone to assume that the parents' problems must, in some way, reflect their behavior toward the parents. Changes in routine and periods of parental absence are often seen by preschoolers as in some way their fault.

Ironically, preschoolers' assumption that parental problems are, to some degree, their fault, doesn't translate into improved behavior. In fact, just the opposite is true, especially in the case of preschool-age boys and their mothers (Hetherington, Cox, & Cox, 1977; 1979). For the first year or two following the divorce, a coercive cycle is often noted in which the normally difficult task of parenting the preschooler (who shows high demands for attention, limited ability to understand parent requests, and limited reinforcement of the parent) is further exacerbated by the disruptions in family routines and the emotional distress experienced by all family members. Mothers report that their children, particularly their sons, are less able to play with peers, nag and whine more, make more dependency demands, and are less responsive to parental requests. This undesirable behavior, like their parents' adjustment pattern, also peaks at one year following divorce and then gradually decreases in intensity. However, Hetherington found that even at two years following divorce, the level of aggressive behavior for boys was still higher than that of same-age boys from intact families. Hetherington notes that the mothers' descriptions of their relationships with their sons included phrases such as "declared war, a struggle for survival, the old Chinese water torture, and like getting bitten to death by ducks" (pp. 32–33).

School-age children. School-age children don't see themselves as being the cause of their parents' marital problems. Rather, they express a great deal of anger toward their parents, often feeling victimized by the changes

Because they feel somehow responsible for their parents' problems, preschool children often have difficulty adjusting to divorce. These adjustment problems manifest themselves in negative behavior patterns.

the divorce is causing (such as having to move between two households), and sometimes demonstrate academic or interpersonal peer problems as a result of the emotional trauma caused by the separation and divorce. Kelly and Wallerstein (1976) note that these children often experience fears about an uncertain future now that the family is no longer together, continued hope for and even fantasies about their parents' reconciliations, and conflicts over loyalty to each parent (especially when custody is contested).

Adolescents. By adolescence, many children are able to understand the necessity of a parental divorce and accept it as preferable to continued family conflict (Reinhard, 1977). For some adolescents, the stress of separation and divorce results in a period of disruptive behavior that may include substance abuse, academic difficulties, sexual promiscuity, and delinquency (Kalter, 1977). On the other hand, many adolescents seem to fare quite well during this time. They show relatively little change in their behavior. These adolescents seem to have a variety of supports outside of the family (such as clubs, friends, and school groups) that they can turn to (Wallerstein & Kelly, 1974).

Learning New Family Patterns

Most studies show that most children suffer few prolonged negative consequences as a result of divorce. That is, on most developmental measures, they differ little from same-age children in intact homes. However, the divorce does present the child with a set of issues that the child from an intact family doesn't confront. These issues include learning to live in a single-parent household, adjusting to parents' remarrying, and ultimately deciding what meaning the parents' divorce will have in terms of their own future development.

Weiss (1979) finds that children in single-parent households experience a unique set of circumstances, both in terms of their responsibilities and in terms of their relationship to the single parent. These children tend to have more responsibilities, to be more self-reliant, and often to act more responsibly than children in two-parent families do. They assume a greater share of household chores and a more direct role in the decision-making process. In effect, children in single-parent families increasingly assume the role of partner rather than subordinate to the parent. They are not only responsible for doing more; they are also increasingly responsible for deciding what more needs to be done. According to Weiss (1979),

> the difference between the child's role in the one-parent and the two-parent household is made manifest when the child fails to perform an expected task. In the two-parent household, the parents may decide that the chore is beyond the child's capacity, or that getting the child to help is more trouble than it is worth, or that the child requires discipline or a lecture on household citizenship. In any event, a decision will usually be made on the parental level regarding how the issue is to be handled. In the one-parent household the child will be directly confronted by the parent. The child is failing to meet a partnership responsibility. . . . In the two-parent family a chore undone may mean more work for the parents. In the one-parent family it means this and more: the same chore represents a threat to the partnership understandings the parent has attempted to establish with the children in order to keep the household going. (Weiss, 1979, p. 101)

Children in single-parent families tend to have greater responsibility for the care of siblings and household chores than do children from two-parent families.

The consequence of this partnership role for the child seems to be a different perspective on the parent from the one that is typical in a two-parent family. Weiss finds that by adolescence, a much more realistic perspective is held of the parent. The adolescent has a fuller understanding of the weaknesses, as well as the strengths of the parent. One result of this more realistic perspective is that such adolescents are less likely to rely on parental advice, are somewhat more likely to feel the need to be self-reliant, and may even be envious of the "frivolous" life-styles of adolescents from two-parent families. In a sense, these adolescents are confronting a similar set of circumstances to that confronted by the early maturing males discussed in the chapter on adolescence. In both cases, adolescents find themselves in situations that catapult them into more adult roles.

Long-Term Postdivorce Adjustment

Wallerstein (1983) finds that children of divorce have eventually to deal with six issues. Only two, acknowledging the reality of the marital rupture and disengaging from parental conflict and distress and resuming customary pursuits, are resolved to a reasonable extent during the initial postdivorce adjustment period. The other four take a considerably longer time. These four seem to have as much to do with children's evolving concept of self as they do with their actual adjustment to the divorce.

Resolution of loss. The first of these four long-term issues is the ultimate resolution of loss. Children lose many things as a result of divorce. Their life-style invariably changes, often to one economically less advantaged. They may move, lose friends, and lose possessions. More important, they often lose extended family and, in many cases, a parent. Since most children reside most of the time with one parent, their relationship to the absent one is crucial to the resolution of this issue. Even in the best of circumstances where the noncustodial parent, usually the father, continues to maintain frequent contact with the children, and the custodial parent does not prejudice them against the absent parent, it appears to take a number of years for them to accept the loss of the parent and to understand its necessity. Children whose noncustodial parent is uninvolved in their lives find it extremely difficult to overcome the feeling that they are somehow to blame, at least partly, for the noncustodial parent's continued absence.

It is important to note that a noncustodial parent's absence does not necessarily reflect lack of interest in the children. Hetherington, Cox, and Cox (1977) found that for some fathers in their sample, especially those most involved with their children before the separation, the discontinuity of their parent-child relationship was very painful, and that as a defense, they increasingly withdrew from their children's lives.

Feelings of anger and blame. The second of children's four long-term tasks involves resolving feelings of anger and blame associated with the divorce—anger at their parents for having to divorce and blame of themselves for being partly responsible for the parents' inability to remain married. Wallerstein believes that children must gain a certain amount of cognitive and emotional maturity before they can understand the factors that actually contributed to their parents' divorce, the powerlessness of their parents to fully protect them from the trauma of the process, and their powerlessness to either cause or prevent the divorce. Wallerstein found that many of the children in her study, when they were finally able to resolve this issue in adolescence, reported developing much closer relationships with their parents.

Accepting the divorce as final. Wallerstein found that for many of the children in her sample, the fantasy of parental reconciliation, the third long-term issue, lasted for many years after the divorce was finalized, in many cases, even after both parents had remarried. This dilemma was especially common when one parent also held a similar fantasy, but it was not unique to this situation. This issue only began to resolve itself as adolescents gradually disengaged themselves from their parents and established their own independent lives. As the importance of their families of origin decreased, so too did their fantasies.

What about me? The fourth long-term issue focuses on the future. Children have to answer the question, What effect will my parents' divorce have on my ability to find and establish my own intimate relationships? Wallerstein offers no prediction as to how the children in her sample will ultimately answer this question. The only indication is that those who have successfully resolved the other issues will be more likely to successfully resolve this one (in this sense, Wallerstein's perspective is very similar to Erikson's conception of development). What is clear from the research of Wallerstein, as well as from the other research cited in this section, is that, like the effects of all other developmental events, the effects of divorce are measured not at one point but at many points across the life-span.

THE SIGNIFICANCE OF THE EARLY YEARS OF ADULTHOOD IN THE LIFE-SPAN

A great many things happen during the approximately fifteen-year period that comprises the early adult years. Perhaps the best way to fully appreciate the complexity and diversity of these events is to view them from several perspectives—from the perspective of the transition years, from a cohort perspective, and from a longitudinal perspective.

The Transition Years Perspective

How do the early adult years differ from the transition years of eighteen to twenty-five? First, and perhaps of greatest significance, the early adult years are marked by a greater diversity of roles than the transition years are. By the late twenties, most adults find themselves fully involved in the three careers of the adult years. They are parent, intimate, and worker. They find each career at the same time demanding and satisfying, and they sometimes find the three roles in conflict. This diversity of roles, as well as their potential for conflict, marks the greatest divergence between these two periods of the life-span.

The role of college student, in comparison, is not without its own stresses, but it is nevertheless one that makes fewer types of demands and one that is more integrated. Few college students are parents, and few find themselves as embedded in intimate relationships as married people do during the early adult years. This difference in embeddedness is reflected in the differing adjustments involved (both in terms of time and degree) when couples stop going together as opposed to when they divorce. Another way to sense the divergence between the two stages is to consider the degree to which individuals in each are free to schedule their own time. The college student may feel as many demands as the young adult but almost always is in a more flexible position for scheduling personal time to cope with these demands.

A second contrast between the two stages involves the degree of polarization between male and female roles. There is a greater parallel between the two roles during the college years than during the early adult years. The primary explanation for this variation in polarization is, of course, that during the early adult years people also assume the role of parent and that men assume this role in a very different way from the way that women in our culture assume it. Even though men's and women's roles are evolving, the difference is still great. A secondary explanation concerns the role of worker. Although it is likely that the major differences in the work patterns of men and women are a reflection of their distinct parenting roles, it is also true that through a variety of intentional and unintentional business practices, women have been less welcome and have received less support in many work settings. Again, this pattern is also changing, but the differences still clearly exist. The consequences of this sex-role divergence for relationships between men and women will be discussed later in this section. For now, it is enough to bear in mind that the degree of sex-role polarization is probably greater during the early adult years than it is during any other period of the life-span.

The differences between the early adult years and the transition years in terms of role diversity and polarization would leave one to think that the early adult years are not evaluated positively by the participants. As was noted at the beginning of the chapter, this doesn't seem to be the case. Complexity of life does not seem to be a correlate of life satisfaction. Rather, what adults value about this period, in comparison to the transition years, is the sense of full participation in the adult world. This sense of adult status comes more slowly than most people entering the stage expected it to, but as adults progress through this period, most increasingly feel society's recognition of them as full members of the adult community.

The surprising gradualness of this recognition may have as much to do with the individual's self-perception as it does with the perceptions of others. Levinson (1978) found that in order to perceive of themselves as adults, many of the men he studied first had to stop viewing others as teachers or mentors. This step often involved, sometimes painfully and sometimes with great ambivalence, the realization that they could no longer seek protection or support through the senior status of others.

The Cohort Perspective

A cohort perspective examines the early adult years from the perspective of successive generations. Is this period of the life-span today essentially the same as it was in the past and as it is likely to be in the future? The answer implied throughout the chapter is no. We seem to be in the midst of an evolution of early adult roles—these roles have already changed significantly with respect to what they were in the past and will continue to change in the future. This evolution has three principle components.

First, there seems to be a diminution of role standards as a force for defining adjustment (Veroff, Douvan, & Kulka, 1981) and an increasing tolerance for a diversity of adult life-styles. The range of life-styles that are considered normal or typical is broadening. Alternative life-styles are increasingly seen as just that, alternatives rather than failures of adjustment. This trend is evident in our changing views on homosexuality, voluntary childlessness, cohabitation without marriage, single parenthood, and males as primary care givers. Although alternative life-styles will continue to increase in number, they will probably never be adopted by more than a relatively small minority of the population. Their significance is not in terms of their numbers, however. Rather, for those who do not choose these alternatives, as well as for those who do, they are important because they make it possible to choose a life-style without fear of social censure or the label of pathology. They are also important because in making such conscious decisions about the course of their lives, people grow in maturity.

A second cohort pattern is an increasing emphasis on the importance of personal gratification. This change is most evident in the more economically advantaged segments of society and reflects the relative degree of affluence that the more educated segments of the population have acquired and are likely to continue to acquire. To identify this shift is not to

While advertisements depict career women as sharp, sophisticated, and well respected, many women find the realities of the workplace much less glamorous.

deny the vagaries of the economy or the fact that a college degree is no guarantee of full, adequate employment. Instead, it is to observe that as people become reasonably free from the worry of earning a decent living, they increasingly focus their attention on psychological rather than economic issues (Campbell, 1981). Opportunities for personal growth, for creativity, and for compatibility between work and other life roles are gradually becoming as much a consideration in the choice of employment as salary and fringe benefits are.

This shift in the importance of personal gratification is evident in the fact that, for those who have gained a measure of prosperity, this prosperity is remarkably unrelated to their personal sense of well-being. Rather, a sense that they are in control of their lives, not manipulated by outside forces, has become the major correlate of a sense of personal gratification (Campbell, 1981).

This shift is less evident among the less economically advantaged society. Material well-being is still the major correlate of life satisfaction among those for whom a decent standard of living is not taken for granted. For this reason, this cohort pattern may be increasing the distance between social classes in terms of their perceptions of the world (a gap not likely to close, given the fact that the high school dropout rate has actually increased over the past decade - from 23 to 27 percent).

Third, we appear to be in the midst of a shift in the importance of the traditional institutions in our lives (Veroff, Douvan, & Kulka, 1981). Surveys indicate that adults today are less likely than adults of previous generations were to view institutions such as the government, the schools, the courts, and organized religion as means through which to improve society. Just as they are attributing more and more importance to personal effort and personal control, adults are also increasingly likely to rely on their own efforts and the resources of support systems they create for themselves.

Where are these cohort changes likely to take future generations? First, the trend toward greater diversity of adult life-styles and greater symmetry in sex roles is likely to continue (Hoffman, 1977; Skolnick, 1978). Second, it is reasonable to expect that the workplace will continue to place greater importance on the psychological and interpersonal needs of employees. Third, the divorce rate probably will stabilize rather than decline to any significant degree. As is now true of those who end second marriages, people will be increasingly unwilling to maintain unsatisfying intimate relationships. Although it is likely that successive generations will be better able to develop the communication styles necessary to maintain intimate relationships, it is also true that the barriers to divorce (such as the employability of the wife, and social and legal mores) will also be lessened. The two changes are expected to balance each other, hence maintaining the present rate.

The Longitudinal Perspective

This view of the patterns of development during the early adult years focuses on the changes that occur to men and women across the age range of twenty-five to forty. The pattern that seems most significant is the U-shaped curve of marital satisfaction frequently reported by couples during this time. Although the slope of the curve correlates with the ages of children in the family, the deeper explanation for the pattern is the distinct developmental courses that men and women follow during this stage.

These patterns were described in some detail earlier in the chapter. At this point, however, it is worth repeating the fact that the role expansion process that many women are pursuing is placing increasingly greater demands on their time and energy. As demands become greater, the sense of personal control decreases and therefore so does satisfaction. This process is further complicated by the fact that the rate of sex-role change in men is considerably slower, thus creating a "sex-role gap." As this gap increases, the ability of each spouse to empathize with and therefore be appreciative of the efforts of the other decreases, and so too does the sense of marital satisfaction.

This polarization of sex roles begins to close toward the end of the early adult years. As children get older, enter school, and more generally become better able to care for themselves, many women find it possible to return to work or move from part-time to full-time work. Those who have continued to work full-time often find that the decreasing number of demands on their time associated with child care allows them to make a fuller personal commitment to work. Women who choose not to work are now in a position to pursue other, more personal and self-directed interests. At the same time as the wife's routines are changing, in ways that often involve a greater investment in work, husbands may be finding their commitment to work less than before and feeling a renewed interest in family. The two transitions decrease the distance between the roles of husband and wife, increasing the potential for shared activities and for the appreciation they have of each other's efforts. The result is an increase in marital satisfaction, usually continuing into the middle years. It seems reasonable to predict that, as the degree of sex-role reciprocity increases, this U-shaped pattern will become less common among married couples.

Before this section ends, mention should be made of the events of the early adult years that reinforce the notion that development across the life-span is best understood from a relative, constructivist, contextual perspective. The first bit of evidence comes from studies of personal happiness across the life-span (e.g., Campbell, 1981). Invariably, adults' perception of their personal well-being is relative rather than absolute in nature. How much they earn in relation to their peer group is more significant a variable than the actual amount itself. Other measures of life satisfaction, such as those pertaining to marriage or work, have more to do with the degree of role congruence than with actual status. The degree to which people are doing what they hoped to be doing and the degree to which they are doing what others important to them had expected of them is a more significant factor in determining life satisfaction during all of the adult years than the actual nature of their roles.

A second bit of evidence comes from the research on the importance of a sense of personal control during the adult years. Adults seem to be able to deal successfully with a great many tasks as long as they feel they have some degree of personal control over their situations. In this sense, it is not hard to understand why parenting a very young child is often a trying experience. Beyond the time and energy demands themselves is the fact that these demands are controlled more by the infant and young child than by the parent. As will be discussed in the concluding section, one of the most crucial elements in developing support programs for adults is to ensure that they foster a sense of personal control. The most effective parent education programs may be the ones that convince parents that what they do with their children does make a difference.

The final bit of evidence comes from research on the effects on children of divorce and of having working parents. Historically, the evidence on both issues was very negative. Over the last twenty years, these findings have changed. In the case of working parents (mothers really), the data now show little overall difference between the children of working mothers and the children of nonworking mothers. Rather, differences now seem to be determined by whether or not the mother finds her work or nonwork role satisfying. In the case of divorce, although there are clearly adverse consequences, children seem to cope with them quite well—well enough, in any case, to allow the argument that the effects of divorce on children are no worse and, in the long run, probably not as bad as the effects of continuing to endure an unhappy relationship between their parents.

Why have the findings in these two areas changed? One reason is certainly that we now have better supports systems for working families and families undergoing divorce. An even more basic reason, however, is that as these events have become more common, they have shifted from being atypical to being typical events. Because they have become typical, children need no longer feel that some sort of rare affliction has singled them out and made them different from their peers. In other words, the impact of these events, as well as the impact of others across the life-span, may have changed not because the events themselves have changed but because our perception of them has. The clear implication of all of these bits of evidence is that to fully understand the impact of life events, it is necessary to know not only the nature of the events themselves but also the context in which they occur.

HUMAN DEVELOPMENT AND HUMAN SERVICES

Making comparisons between developmental stages as to the degree of stress associated with each is always a difficult task. Not only are the developmental tasks different, but so too are the people. Are the developmental tasks of the preschooler more difficult to the preschooler than those of the adolescent are to the adolescent? Such comparisons are not easy to make, but it does seem that the variety, complexity, and discontinuity of the tasks of the early adult years must certainly make this period one of the potentially most difficult of the life-span.

Adults make use of a great variety of services during this period. Some find themselves in need of specific services to deal with specific problems such as divorce. Others find themselves in need of support services to maintain normal family functioning. Dual-earner families, which are increasing in number, find themselves in particular need of support services. Finally, all families find themselves influenced by the attitudes and values of the community, by social and legal policy, and by the workings of the economic system.

Special-Population Services

One example of a family support program designed for families dealing with specific issues is the divorce counseling program developed by Weiss (1975). As Weiss describes the program, it's aim is to "help recently separated individuals manage the emotional and social challenges of marital separation by providing them with information regarding the experiences they might encounter, by making available concepts and theoretical frameworks that might help them interpret these experiences, by describing others' experiences, and by making available a setting in which they could talk with others in their situation" (p. 311). The individual sessions deal with the emotional impact of divorce, the continuing relationship with the former spouse, the reaction of friends and family to the separation, changing parent-child relations, the emotional reaction of children, ways to put a new life structure together, and the establishment of new intimate relationships.

Weiss feels that programs such as these are valuable to participants for four reasons. First, by providing information, they reassure people that the experiences they are encountering, the feelings they have, and the difficulties they are confronting are not unique to them but rather are typical of such situations. This kind of information is extremely important, not only in the case of divorce, but in all life transitions. One of the most valuable aspects of the new parent support groups is the sharing of information and fears and, in the process, the realization that such feelings are normal and expected.

Second, these seminars provide participants a sense of community. The divorce rate may be very high, but not everyone divorces at the same time. Rather, what often happens is that a person feels isolated and depressed. Meetings with others in the same situation provide a measure of empathy not always available elsewhere. A related benefit is that since these sessions involve both men and women, each sex has the opportunity to understand "the other perspective" in a nonemotional setting. In the process all participants are given the opportunity to realize that their inability to deal effectively with their spouse does not necessarily imply an inability to deal with all members of the other sex.

Third, such programs offer participants the support they need to examine the particular circumstances of their lives, to understand why their marriage ended in divorce, and to recognize what this suggests about planning and preparing for the future. Finally, they help participants to actually begin the process of building a new life structure. This motivation to action is an especially important aspect of support groups. Lethargy is very common when people find themselves confronting more problems than they had before with fewer resources for dealing with them.

The format of the Weiss program is essentially similar to those used in dealing with other life transitions (Golan, 1981). The transition to parenting was already mentioned as a life event sometimes prompting the use of support groups. Support groups are also common among adults returning to school and among people confronting major job changes or unemployment or any one of the many events typical of this period.

General-Population Services

Most families don't divorce, and most weather transitions such as parenthood without undue stress or the need for supports other than family or friends. However, many families find themselves in need of more general supports as a result of their life-styles. The dual-earner families provide a good example. Their stresses are better seen as chronic than as acute, and as a result, the kinds of services they need are continuous in nature.

The most obvious of these are child-care services. It is not uncommon for working parents with children under the age of five to use a patchwork arrangement of child-care providers, often having to move the child from one setting to another during the day or use different settings on different days. The problems are not restricted to young children (Kamerman, 1980). Parents of school-age children often find that school hours aren't always compatible with work hours. What do you do if you have to leave for work by eight but your child doesn't have to be at school (a fifteen-minute walk) until nine? What about after-school care? Most schools do not provide after-school programs for children. Parents who can afford to send their child to an organized recreation program have a solution (even though it still requires the child to move from one setting to another), but in families less fortunate, the children join the growing ranks of "latchkey children"—children, who with their house key tied around their necks, are responsible for looking after themselves and sometimes a younger sibling until their parents come home from work. Other child-related problems concern the absence of in-school lunch programs at many elementary schools, and the problems related to care of children when school is not in session (Kamerman, 1980).

Working parents frequently report that personnel practices at their place of work are often in conflict with family needs. They feel greater effort should be made by employers to remove or at least lessen potential conflicts. In particular, working parents wish for greater flexibility in their work schedules, both in terms of starting and stopping times and in terms

Finding adequate and financially attainable childcare services is still one of the major problems faced by working parents.

of hours per day worked (for example, the choice of working four ten-hour days or five eight-hour days). Many also would like to see large employers set up in-house day-care programs or at least subsidize the cost of private programs. They would also like to see an expansion of sick leave time so that it includes the illness of children, more extensive maternity leave, and also paternity leave.

The fact that parents often have problems coordinating their schedules with those of the school system or the workplace indicates that parents are usually the ones who find themselves forced to make accommodations: the school system and the workplace rarely have to be accommodating. This lack of power will continue as long as adults fail to get the institutions they deal with to consider the impact of their policies on their families (Piotrkowski, 1979). One important advocacy role for all of us is to get institutions whose policies influence family function to consider the family impact in the same sense as our federal and state laws must now consider environmental impact.

As difficult as these coordination problems may be for dual-earner families, they are even more so for single-parent families (Ross & Sawhill, 1975). Single-parent families have increased largely as a result of the increasing divorce rate. Virtually all single-parent families are female headed, and although some are headed by unmarried mothers, most are headed by divorced women—this group now makes up as much as 15 to 20 percent of all families.

Money is the single greatest problem these families face. Since financial issues are the single greatest source of strain within a marriage and a major contributor to divorce, many of these families had relatively low incomes even when intact. Now divorced, the women find themselves poorly prepared to compete for jobs, many of which pay considerably less than comparable work for men. Ross and Sawhill (1975) suggest that two forms of intervention for single-parent families are first to increase their access to traditionally male occupations and lobby for more equitable pay systems and second to help adolescent girls understand the importance of having skills that make them employable. This second form of intervention may be particularly difficult given the reluctance of adolescents to consider the future except to imagine what they want it to be, but if more adolescents were made aware of the possibility that they might be the sole economic provider for their children, they might not find themselves in such difficult circumstances when the situation actually arises.

Social Policy Considerations

It should be clear by now that, given the developmental tasks of the early adult years, many people find themselves in need of support services. These services might come in the form of, for example, direct interventions or information dissemination or family advocacy. Irrespective of the form of the service, one characteristic they should all have is that they should leave a large degree of control over the situation in the hands of the individual or family receiving the service. Virtually all research on development across the life-span shows that people's behavior is to a large degree reflective of their actual or perceived control over the circumstances of their lives (Clarke-Stewart, 1977). The greater the sense of personal control is, the more deliberate and effective the behavior. With the exception of those circumstances in which such a policy would actually endanger someone (as in the case of child abuse), adults should always feel that they have a measure of control over their own development and their involvement in social support programs (Clarke-Stewart, 1977).

We turn next to the middle years of adulthood, roughly the age period of the early forties to the middle sixties. For some, middle age offers a satisfaction that can come only through maturity, like a fine red wine. But, for others, it marks the time that age is seen in terms of years to death rather than years since birth.

SUMMARY

THE COURSE AND CONTEXT OF DEVELOPMENT

1. Cohort changes over the last few decades have produced a greater diversity of family life-styles and greater variability in the timing of family transitions than were present in earlier generations.
2. Marital relations can be classified in terms of division of power, degree of task orientation, and division of labor. Marital relationships are gradually evolving from head/complement patterns toward equal partner patterns.
3. Marital satisfaction is largely a subjective estimate reflecting the degree to which

the relationship seems fair to each partner. In general, at all ages, husbands report higher degrees of marital satisfaction than do wives.

4. The transition to parenthood involves coping with new demands on time, changing divisions of labor, and redefining sources of satisfaction within the marital relationship.
5. The percentage of women in the work force, including those with young children, has increased dramatically over the past decade.
6. Women view their role as worker not as a substitute for their parent and intimate roles but as an additional component of their adult identity.
7. For men, work serves as the primary component of adult identity. It is often viewed as a means through which the responsibilities of the intimate and parent roles are partially fulfilled.
8. Children of working mothers differ little from those whose mothers do not work. A more significant difference exists between children of women who feel satisfied with their adult lives and children of women who do not.

DIVORCE

1. The dramatic increase in the divorce rate over the past few decades most likely reflects a shift in the expectations adults have about the benefits derived from marriage and a shift in the expectations society has about the place of marriage within the adult culture.
2. The process of divorce is complicated and traumatic; few adults who undergo this process experience little stress.
3. The major predictor of children's adjustment to parental divorce is the parents' ability to effectively maintain their parental relationship even after the spouse relationship has ended.

THE SIGNIFICANCE OF THE EARLY YEARS OF ADULTHOOD

1. The early years of adulthood are characterized by a greater degree of role conflict and a greater degree of sex-role differentiation than what occur during other stages of the life-span.
2. Future early adult cohorts are likely to demonstrate a greater diversity of lifestyles, place increasing emphasis on personal gratification, and place greater reliance on individually defined support systems.
3. The U-shaped pattern of marital satisfaction across the early adult years is largely a reflection of the pattern of sex-role differentiation commonly found during this stage.
4. Adult estimates of life satisfaction are more a reflection of degree of role congruence than they are a reflection of actual status.

HUMAN DEVELOPMENT AND HUMAN SERVICES

1. Because of the continuing evolution of adult roles, adults will require an ever greater diversity of support services and work patterns.
2. To ensure maximum benefit, social support services should be seen by adults as voluntary.

KEY TERMS AND CONCEPTS

INTIMACY

Scanzoni and Scanzoni's Patterns of Spouse Interaction
- Head/Complement Pattern
- Senior Partner/Junior Partner Pattern
- Equal Partner/Equal Partner Pattern

PARENTING

Engrossment

THE INTERPLAY OF CAREERS

Role Expansion Process

DIVORCE

Six Facets of the Divorce Process
- The Emotional Divorce
- The Legal Divorce
- The Economic Divorce
- The Coparental Divorce
- The Community Divorce
- The Psychic Divorce

SUGGESTED READINGS

Bernard highlights the distinction between "his marriage" and "her marriage." The research by Rhyne suggests some of the origins of men's and women's differing perceptions. The book by Rubin provides very vivid examples of this distinction for those least fortunate in our society.

Bernard, J. (1982). *The future of marriage* (1982 ed.). New Haven: Yale University Press.

Rhyne, D. (1981). Bases of marital satisfaction among men and women. *Journal of Marriage and the Family, 43*, 941–955.

Rubin, L. B. (1976). *Worlds of pain*. New York: Basic Books.

The transition to parenthood produces major changes in the lives of both parents. These four articles offer some insight into the nature of these changes.

Greenberg, M., & Morris, N. (1974). Engrossment: The newborn's impact upon the father. *American Journal of Orthopsychiatry, 44*, 520–531.

LaRossa, R., & LaRossa, M. M. (1981). *Transition to parenthood: How infants change families*. Beverly Hills: Sage Publications.

May, K. A. (1982). Factors contributing to first time fathers' readiness for fatherhood: An exploratory study. *Family Relations, 31*, 353–361.

Rossi, A. S. (1968). Transition to parenthood. *Journal of Marriage and the Family, 30*, 26–39.

As the following three reports demonstrate, juggling the sometimes conflicting demands of family and work can be a trying experience for both parents and children.

Hayes, C. D., & Kamerman, S. B. (Eds.). (1983). *Children of working parents: Experiences and outcomes*. Washington, DC: National Academy Press.

Hoffman, L. W., & Nye, F. I. (1974). *Working mothers*. San Francisco: Jossey-Bass.

Kamerman, S. B. (1980). *Parenting in an unresponsive society: Managing work and family time*. New York: The Free Press.

The reading of any of these four reports should provide you a better understanding of the reasons why the divorce experience is so traumatic for adults.

Cherlin, A. J. (1981). *Marriage, divorce, remarriage*. Cambridge: Harvard University Press.

Hetherington, E. M., Cox, M., & Cox, R. (1976). Divorced fathers. *The Family Coordinator, 25*, 417–428.

Kitson, G. C. (1982). Attachment to the spouse in divorce: A scale and its application. *Journal of Marriage and the Family, 44*, 379–393.

Weiss, R. S. (1975). *Marital separation*. New York: Basic Books.

The research of Hetherington and the research of Wallerstein provide two of the most complete explanations of the impact of divorce on children.

Hetherington, E. M., Cox, M., & Cox, R. (1979). Play and social interaction in children following divorce. *Journal of Social Issues, 35*, 26–50.

Wallerstein, J. S. (1983). Children of divorce: The psychological tasks of the child. *American Journal of Orthopsychiatry, 53*, 230–243.

9
THE MIDDLE YEARS OF ADULTHOOD

AGES FORTY TO SIXTY

CHAPTER OUTLINE

There is a riddle often told by school-age children. It asks how far a person can walk into a forest. The answer is only half way; after that the person starts walking out of the forest. For some adults, middle age is the first time the "humor" of the riddle is fully appreciated. For these adults, middle age comes to mean a time of decline, a time of aging rather than merely a time of growing older. Fortunately, however, most middle-aged adults do not see the middle years as a time of decline. Rather, in defiance of the logic of riddles, the middle years of adulthood serve as a period of reflection, as a time of appreciating what is gained by being older rather than what is lost, and as a time of confronting new developmental tasks.

As in other stages in the life-span, the quality and character of development during the middle years is not random. Since a major component of the middle years is reflection—on past and present, as well as future—the actual and anticipated nature of these periods is a significant factor. In this respect, both social class and sex role continue to exert a powerful influence on the course of development. There is also a cohort component to the developmental events of the middle years. In fact, those adults experiencing middle age today are experiencing a slice of life that literally did not exist fifty or sixty years ago. And given the significant increase in the percentage of young adult women entering the work force and the changing patterns of married life for both men and women, it is also likely that the current generation of young adults' experience of middle age will be different from that of the present cohort.

From a longitudinal perspective, the middle years can be distinguished from earlier stages in terms of the degree of variability between individuals. As people move farther away from the age-graded childhood years, the timing of life events becomes more individual. And as the effects of these individually timed events cumulate, so too does variability. As Atchley (1975) puts it, the life course is like a successively bigger equation made up of a series of contingent probabilities. As people get older the cumulative effects of decisions concerning such variables as age at marriage, years of education, number and spacing of children, career shifts, and change of residence all sum to make individual life histories increasingly unique.

The middle years of adulthood comprise two relatively distinct periods. The first, usually occurring during the forties, concerns the transition from the young adult years to the middle years. This is the time of the much heralded mid-life crisis. Actually relatively few adults experience a mid-life crisis, but most report this period of the life-span as a time of reflection. The increased reflection is prompted by the fact that some of the major components of young adulthood take on different values during the middle years. First, children have grown and parents find themselves with an "empty nest." Second, the biological event of menopause in women and the increasing awareness of the more gradual biological changes that have actually been occurring for many years in both men and women force adults to confront the fact that biologically, in some respects, they are less fit. Third, the meaning of work typically changes during the middle years. In essence, the dreams of youth are confronted by the realities of experience. The fact that most adults do not experience crisis suggests that these transitions are seen as resulting in a "net gain." Parents like having the house to themselves; workers find that although they haven't become president of the company, they haven't done badly either.

The second phase of the middle years occurs during the fifties. Most adults report this time as very positive (Campbell, 1981). Perhaps reflecting the decreasing sex-role polarization that accompanies the postparental years, many adults report marital satisfaction to be very high. Many take special joy in assuming the role of grandparent, which furthers their sense of generativity. At the same time, many adults find themselves caring for aged parents, a responsibility that can be both emotionally and financially draining. Planning for retirement is an increasingly important element of life during the fifties. But rather than seeing this time as "the beginning of the end," adults in their fifties are very much involved in the world. In fact, in comparison with other age groups, middle-aged adults have the highest rate of voting in all elections and show the most participation in the political process at all levels of government; they have the highest rate of involvement in social and recreational groups and of participation in civic and community organizations, and they are a larger source of philanthropy than any other age group (Smith & Macaulay, 1980).

The successful resolution of the developmental tasks of the middle years is sometimes made difficult by the fact that we live in a youth-oriented society, one that equates vigor and competence and attractiveness with characteristics most typical of the young adult. Although this age bias will probably lessen for future cohorts, it must seem ironic to many presently middle-aged adults that even though their age group holds the positions of power and responsibility in virtually all aspects of society, most advertisers still think that their product will sell better if graced by a thirty-year-old than if graced by a fifty-year-old.

This chapter is divided into four sections. Since variability is an important aspect of development during the middle years, the first section examines the factors that contribute to this variability. Social class and sex role are prominent aspects of this discussion. The second section focuses on the transition period of the middle adult years. Since the patterns are somewhat distinct for men and women, each will be discussed separately and then integrated in the discussion of the decade of the fifties. The third section will examine the significance of development during the middle years of adulthood for the course of development across the entire life-span. The chapter will close with a discussion of the provision of services to middle-aged adults.

THE CONTEXT OF DEVELOPMENT DURING THE MIDDLE YEARS OF ADULTHOOD

The chapter on adolescence presented the notion that our present conception of adolescence is, in a sense, a by-product of the historical and cultural changes that have occurred over the past hundred years. As a result of the decreasing age for onset of puberty and the increasing number of years of education required, a unique space has been created within the life-span. A parallel set of circumstances has contributed to our present conception of development during the middle years of adulthood.

The Cohort Context

Around the turn of the century, a woman could expect to marry in her early twenties. Her first child would be born within the next year or two and her last child in her mid to late thirties. Given the relatively high infant and childhood mortality rates, she might expect to lose one or two children and see three or four reach adulthood. Her last child would leave home to marry when she was in her late fifties. Her husband would probably not attend the marriage of their last child. The likelihood is that he would have been dead at least two years (Skolnick, 1977). If the core of the middle years is defined as a period beginning with the youngest child's leaving home and ending with the death of the spouse or retirement or a significant decline in health, then it is clear that no such period existed at the turn of the century (Treas & Bengston, 1982).

The space of the life-span that we think of as middle age has occurred as a result of a declining birthrate, which is causing the postparental period to begin sooner, and improvements in health care, which are resulting in people's living longer and doing so in good health. As in the case of adolescence, two events, each influencing normative age expectations but in opposite directions, have produced a unique set of circumstances that we think of as middle age.

The death of a spouse is not a component of our present conception of middle age. Middle-aged adults certainly die, husbands more often than wives, but with the average life expectancy now moving into the seventies, most adults enter middle age with the notion that they will experience the entire period with their spouse alive. Unfortunately one of the correlates of this expectation is that the death of a spouse during the middle years, because it is unexpected, is often more traumatic.

Just as this present generation of middle-aged adults is finding the experience different from that of their parents, so too will future generations find the experience somewhat different from that of the present generation. The increasing individuality in the timing of life events such as marriage, childbirth, and career, the increasing divorce and remarriage rate, and the movement toward second careers during the middle years all mean that the character of the developmental events of the middle years will become more individual and variable (Sarason, 1977). Some people may find themselves parents of young children again; others may start out in a new business (Neugarten, 1979).

The increasing importance of the interpersonal aspects of marriage now evident in the young adult population suggests that when this generation reaches middle age, the companionship aspects of the marital rela-

Improved health care and a growing emphasis on physical exercise have helped to change the nature of middle age.

tionship may assume greater importance than they have in the past. To the extent that the empty nest transition is difficult for middle-aged adults, the shift in marital roles should make it less so. The difficulty will be further lessened for women as more continue to enter the work force and choose to do so not only for financial reasons but also for reasons concerning personal growth and identity. In essence, future cohort changes may result in a greater ease of transition between the early and middle years of adulthood than is now the case.

The Longitudinal Context

If the transition to middle age is a period characterized by reflection, then the nature of earlier experiences, especially those of the early adult years, must be a significant factor in the transition. Pearlin (1980) finds that the early adult years are particularly stressful, often more so than middle age. Young adults have less control over their lives, are often less able to predict the timing of life events, and frequently find themselves with minimal resources for coping. In particular, Pearlin notes that in the workplace, young adult workers experience greater job strain and a greater sense of work overload. They are more likely to feel alienated from other workers and to perceive the workplace as depersonalized. There is a greater degree of conflict between the demands of home and the demands of work, especially for mothers. Young adult workers are more susceptible to the effects of shifting economic conditions on the workplace. They are the ones most likely to be fired or laid off or demoted. They are the ones most likely to have their work schedules shifted or to be transferred. They are also the ones most likely to be promoted, but even a positive step such as promotion entails a degree of stress as new responsibilities are mastered.

Marriages seem more vulnerable during the early adult years than they do later on. Young adults are less likely than middle-aged adults to feel that their spouse understands and appreciates them. There is usually a greater discrepancy between the actual circumstances of the marriage and the expectations of each spouse. Divorce and separation are more common during the early adult years, and the demands of child rearing are ever present. The nature and quality of the events of the early years set the tone of the transition to middle age as well as significantly influencing the quality of the middle years.

The Sex-Role Context

Since the developmental paths of men and women take different courses during the early adult years, it is not surprising that the core elements of the mid-life review emphasize different issues for men and women. Men are more likely to focus on occupation; women, on family life. Whether this differentiation will continue to hold for future cohorts is yet to be determined, but the fact that full-time working women still place a greater emphasis on family life than full-time working men do suggests that there will probably continue to be some degree of difference.

Although sex-role polarization is perhaps at its greatest during the early adult years it does not begin there (Eichorn, Clausen, Haan, Honzik, & Mussen, 1981; Gutmann, 1981). Rather, it reflects the continuation of a process of socialization that extends back through adolescence into childhood (Gilligan, 1982; Komarovsky, 1976). Specifically, it reflects a process that encourages males to emphasize the instrumental or autonomous aspects of their personalities and women to emphasize the nurturant or interpersonal aspects of their personalities. The transition to middle age marks a reversal in this socialization pattern. Gutmann (1977) refers to it as the androgyny of later life. Eichorn and her colleagues (1981), in a follow-up of the subjects in the original Berkeley longitudinal studies begun in the 1920s, found that by middle-age, men and women felt comfortable enough about their life circumstances "to permit themselves to reveal the less conventional aspects of themselves more fully and directly" (p. 422). Apparently, when the socialization pressures of the early adult years are finally lifted, men feel more comfortable expressing the nurturant side of their personalities and women feel more comfortable expressing their autonomous side. One consequence of this sex-role pattern reversal is that middle-aged couples place greater emphasis on the companionship aspects of relationships rather than on the sexualizing aspects more typically emphasized during the early adult years (Peck, 1968).

The increasing androgyny of the middle years is a major factor in the more positive perception both husbands and wives have of their marriages, but the fact that men still view their marriages as more satisfying than women do indicates that not all of the discrimination felt by many young adult women is removed during the middle years. In particular, Troll and Nowak (1976) find that too often middle-aged men were stereotyped as being at the peak of their potential while middle-aged women were thought to be past theirs. This misperception of the status of middle-aged women reflects the view that their primary, if not exclusive, adult role is that of parent. As this chapter will show, the parent role is only one facet of the identities of women during the middle years of adulthood.

The Social Class Context

The experience of middle age is perceived in different ways by individuals in different segments of society. The more economically advantaged the social context, the more likely it is for middle age to be viewed as a positive transition, one providing new opportunities for growth and one distinct from old age (Giele, 1980; Karp & Yoels, 1982; Myers, Lindenthal, & Pepper, 1974). The key element in these differing perceptions of middle age is the role of work in the life structure.

People who by virtue of education and occupation find themselves at middle age in positions of power and responsibility see middle age as the time of the emergence of executive processes (Neugarten, 1968). These advantaged middle-aged adults see middle age as a time of increased self-awareness, of maximum capacity for handling highly complex tasks, and of "driving rather than being driven." These are individuals who see themselves as distinct both from young adults and from old adults. They see their experiences as providing them a level of competence not possible during the young adult years, and they sense in themselves little evidence of the biological declines one would associate with advanced age.

For those less advantaged, work offers fewer opportunities for advancement and challenge. It involves a shortened career ladder, a greater emphasis on physical capabilities in job performance, a lower salary structure, and fewer opportunities for autonomy, creativity, or new learning (Karp & Yoels, 1982). Although most would choose to work even if they

BOX 9-1 SEX ROLE ACROSS THE ADULT YEARS

Our understanding of the concepts of masculinity and femininity and the roles of men and women develops gradually over the course of the entire life-span. Although research indicates that the concepts are well established during the childhood years, it also indicates that these notions are modified as adults confront the various developmental tasks of the adult years. One of the best insights into the nature of these continuing modifications during the adult years is provided through a series of studies by Feldman and her colleagues at Stanford University.

Feldman used a variety of measures of sex-role orientation with adults across the early and middle years range. One of the measures used, the Bem Sex Role Inventory, is especially useful in helping clarify the patterns of change in sex role across the adult years. The Bem Inventory consists of a list of sixty items (see Table A). People are asked to rate on a scale of one to seven the degree to which the item is descriptive of them. The sixty items represent three distinct sets of twenty items each (on the inventory all the sixty items are arranged randomly). One set of twenty consists of words that independent raters judged to be more appropriate in our society for men, another twenty were judged more appropriate for women, and the remaining twenty were judged to be equally appropriate for either sex. The sixty items are not unique to this inventory. They are words most of us might use to describe and measure masculine and feminine behavior. What is distinct about the inventory, and what makes it so useful as a tool for measuring changes in sex role across the adult years, is its scoring.

Traditionally, masculinity and femininity have been conceptualized as opposites. Thus people who are very masculine have to be low on measures of femininity. Bem argues, as do now many others, that in fact the two kinds of behavior are really independent. That is, there is no reason why an individual cannot value and demonstrate those traits that have traditionally been viewed as masculine (column I) and at the same time those traditionally viewed as feminine (column II). Therefore the inventory provides both a masculinity score and a femininity score, each reflecting the self-ratings given on the relevant items. The relationship between the masculine and feminine scores provides a third score. This third score is a measure of androgyny and reflects the degree of difference in the self-ratings on the masculine and feminine items. A highly sex-typed individual would score high on one and low on the other. An androgynous individual would show little difference on the self-ratings given for the masculine and feminine items.

Feldman's work using the Bem highlights a very important pattern in the course of sex-role development across the adult years. Specifically, the developmental tasks of the early adult years usually increase the degree of sex typing found in both young adult males and females. As the nature of these developmental tasks changes through the middle years, in particular as the parent role shifts and

Sources: Abrahams, Feldman, & Nash, 1978; Bem, 1974; Feldman & Aschenbrenner, 1983; Feldman, Biringer, & Nash, 1981.

TABLE A Selected Items from the Masculinity and Femininity Scales of the BSRI

MASCULINE ITEMS	FEMININE ITEMS
Aggressive	Affectionate
Analytical	Cheerful
Has leadership abilities	Loves children
Self-reliant	Sympathetic
Willing to take risks	Yielding

Note: Reproduced by special permission of the Publisher, Consulting Psychologists Press, Inc., Palo Alto, CA 94306, from The Bem Sex Role Inventory by Sandra Bem, PhD., © 1978. Further reproduction is prohibited without the Publisher's consent.

the prominence of occupational needs decreases, there is a decline in sex-typed behaviors. This does not mean that men become less masculine and women less feminine. Rather it means something much more significant from a developmental perspective. It means first that there is no change in the masculinity levels for males and femininity levels for females and second that there is an increase in the self-ratings given on those traits typically judged as more appropriate for the opposite sex. When people talk about the "unisex" of middle age, this is what they are talking about. It is not that the middle aged are somehow neutral; it is just the opposite—they are now best able to express most fully all aspects of the self.

This man's work requires a combination of masculine and feminine characteristics, such as self-reliance and gentleness.

Middle age often brings with it a sense of freedom to explore new interests and develop new personality traits.

were financially able to quit, they would do so because of the company of fellow workers and because work gives them something to do with their time. The character of their work is less often cited as an asset. For these adults, middle age is not a time of new opportunity but more typically a time associated with growing old.

The research on the relationship of work and development shows that these patterns involve more than an association between type of work and sense of aging. They show that the cumulative experience of work influences the likelihood of continued intellectual growth through the adult years (Giele, 1980; Kohn & Schooler, 1978). In particular, the cumulative effects of the complexity of work have been shown to influence intellectual flexibility in the middle years—a major correlate of attitudes and values, of self-perception, and of social orientation.

The cumulative impact of the workplace extends beyond one's perception of age and continued intellectual flexibility. It also influences the ability of families to cope with stress. The less advantaged continue to be more vulnerable to the vagaries of the economy and have fewer resources for coping with them when they do occur. In Skolnick's words, "the realistic fear of unexpected economic danger imposes a heavy psychological burden in the form of nagging anxiety and a loss of sense of freedom" (1977, p. 146).

Using these four contexts as a frame of reference, we are now in a position to examine more fully the developmental events that constitute the transition from the early adult years to the middle adult years for both men and women.

THE COURSE OF DEVELOPMENT DURING THE MIDDLE YEARS OF ADULTHOOD

Because the developmental experiences of men and women during the early adult years tend to be relatively distinct, the core components of their transition experiences are often different also. Since occupation is for most men the core component of their identity during the early adult years, it serves as the core component of their transition experience. The parallel statement for women would identify marriage and the family as core components. However, as was mentioned earlier, the increasing investment women are making in their role as worker is almost certainly changing the character of their transition experiences.

All transitions initially involve a sense of loss, which ideally is replaced by a sense of optimism about what is to be gained (Golan, 1981). The greater the investment is in what is about to be lost, the more difficult the process. For those who have put all their eggs in one basket, irrespective of whether the basket is at home or in the office, this transition period can be especially stressful (Giele, 1980; Levinson, 1978; Osherson, 1980). The stress of loss is heightened by the realization that often little can be done to counter it. For example, there is really no way to stop children from growing up and wanting to be on their own. Nor is there much that can be done to counter the promotion and personnel policies of business and industry that typically slow the rate of promotion with increasing age. The loss of some roles and the loss of control over other roles make this transition a potentially treacherous crossing.

Transition Patterns for Men

The research literature suggests that the mid-life transition for men can prove a difficult period. Changes in the degree of satisfaction derived from work, the increasing evidence of biological aging, and the launching of grown children into their own adult worlds usually prompt middle-aged men to reflect on their own values and priorities.

Changes in the Role of Work

Work plays such a central role in the transition process for men because it is the core component of what Levinson (1978) calls the dream. The dream exists in different forms, but it is as present in the life of the assembly worker (Chinoy, 1955) as it is in the life of the corporate executive (Sarason, 1977).

> The illusions, the sense of omnipotence and the excitement of heroic drama give the Dream its intensity and its inspirational qualities. But they contribute also to the tyranny of the Dream. Reducing this tyranny is a major task of the Mid-Life Transition, whenever the dream has had an important place in a man's life and he is in the grip of its myth. The task is not to get rid of the Dream altogether, but to reduce its excessive power: to make its demands less abso-

lute; to make success less essential and failure less disastrous; to diminish the magical-illusory qualities. Later, a man may continue to seek excellence, but he gains more intrinsic enjoyment from the process and the product of his efforts and he is less concerned with recognition and power. (Levinson, 1978, p. 248)

The tyranny Levinson talks about seems to arise for three reasons. First, success in the workplace serves as the ultimate justification for many of the socialization experiences of males. The value placed on success is reflected in the behaviors that are reinforced and modeled from childhood and in more direct efforts concerning education and entry into and advancement in the job market.

Second, the tyranny's power is enhanced by the types of defense mechanisms typical of the adolescent and young adult. Vaillant (1977) found in his longitudinal study of adult males a common tendency in young adults to use projection and rationalization as psychological defense mechanisms. **Projection** refers to the tendency to deny the existence of feelings or states within yourself and instead pass them off or project them onto another. **Rationalization** is a mechanism by which logical or rational but not necessarily accurate explanations are offered to explain life events. Arguing, for example, that a promotion was denied because the foreman was prejudiced against me is an example of projection. By projecting the failure to get the promotion onto the foreman, I am able to absolve myself of any responsibility for the decision. Since the lack of promotion has nothing to do with my competence as a worker, I have no reason to reflect on my work skills. The significance of both these defense mechanisms is that neither really requires the individual to seriously confront himself and, as a result, the dream is not easily threatened.

The third reason that the dream becomes so powerful is that much of what happens during the early adult years fosters the belief that the dream is in fact coming true. The rate of promotion and advancements is greatest early in the work history, across all levels of the work force (Clausen, 1981; Farrell & Rosenberg, 1981). Within highly technical industries for example, Graves, Dalton, and Thompson (1980) note a four-step ladder of job advancement. The lowest level is what they call the apprentice level. Such entry level positions usually involve the assignment of one piece of a larger project. The work tends to be routine and detailed, to provide the exact specifications for materials on a construction project, for example. The second level gives the worker a degree of responsibility over the development of a small project. It represents the first opportunity to show someone what he really can do. As such, it is an important opportunity to establish a degree of credibility. The third step provides an opportunity to serve in a mentoring or sponsoring capacity. Now the worker is not only responsible for demonstrating continuing competence in his own right but also for developing competence in others. Finally, if successful at these three rungs of the ladder, the individual is given the opportunity to exert a significant influence on the future direction of the entire company or, in very large corporations, a significant portion of the corporation. In this capacity, he might be given the opportunity to decide how company resources will be allocated in the future.

Advancement relative to others becomes the social clock men use to judge their progress. Graves, Dalton, and Thompson (1980) found that as long as a man's advancement was in step with that of others of the same ability and seniority, commitment to work remained high. When the rate

Research has indicated that men gauge their career advancement in relation to the advancement of their peers. Their perception of advancement, in turn, influences their level of work commitment. As advancement slows, commitment declines.

slowed compared to that of others, so too did commitment. However, since there is usually less room at the top of the ladder than at the bottom, the rate for most must slow because fewer and fewer are able to continue to advance.

It might seem reasonable to ask why, if most others' rate of advancement is also slowing, a person might not simply conclude that even though he is not making it to the top, he is not doing any worse than most others and therefore have little problem with his slowing rate. That is, why is the basis for comparison the few who continue to advance rather than the majority who don't? It's a good question to ask, but apparently this insight escapes many men during the middle adult years. Perhaps this is what Levinson means by the tyranny of the Dream.

In some ways, coming to terms with the realities of the workplace may actually be harder for those in the more advantaged segments of society (Tamir, 1982). It is not that those in less prestigious occupations are more likely to accomplish their dreams; rather it is that their personal identity may be less wrapped up in their dreams (Clausen, 1981). Factory workers or construction workers or department managers or delivery people may have less of a personal investment in work and therefore may be better able to define themselves independently of their work. In contrast, Sofer (1970) found that males in high-status positions in large corporations often had so much invested in their work that they sometimes even had trouble differentiating their personal goals from company role expectations. It is not that the blue collar worker is any less unhappy about his slowing advancement than the executive is; it is rather that the blue collar worker may be able to take it less personally.

Another way to appreciate this social class difference in terms of the meaning of work is to consider Sarason's (1977) findings on the transition to middle age for those men who occupy executive and professional positions and who have been successful—those who have come closest to fulfilling their dreams. Strangely enough, he found that many report an empty feeling. They had accomplished a lot, but it didn't seem to bring them what they thought they were looking for. Certainly, for some, the problem was compounded by a poorly articulated sense of what they were looking for. But even when their goals were well defined, these men were realizing that there is no necessary connection between material or professional success and one's personal sense of satisfaction in life. Sarason suggests that, for some, the consequences of success may be just as devastating as those of failure. As an example, he notes that the suicide and substance abuse rates among physicians, a group having one of the highest mean incomes of any in middle age and enjoying high status within society, is four or five times higher than the average rates for other adults of the same age.

Biological and Physical Changes

Although work occupies the pivotal position in the male mid-life transition, it is not the only component of the transition. In general, any event that forces the middle-aged male to consider the limits of his power, his degree of control, and his sense of omnipotence serves the same purpose.

Death must certainly be the ultimate ego deflater. It becomes considerably more common during middle age (five or six times more so than during young adulthood) and middle-aged men seem very aware of this fact (Jacques, 1981). An increased awareness of death comes about through having to deal with the death of parents, with the realization that after middle age is old age, a time associated with death, and with the fact that same-age friends, colleagues, or neighbors are dying. No wonder that Neugarten (1968) found in her studies of the middle-age transition that it was marked by a shift from viewing one's age as time from birth to viewing it as time until death.

The increasing preoccupation with death during the middle-age transition is probably a little melodramatic. Most middle-aged men do not die; quite the contrary, their average life-span continues to increase, and illnesses such as cancer and heart disease are considerably more treatable now then they were in the past. However, this preoccupation may reflect an increasing awareness of the variety of biological changes that have been occurring throughout the adult years. Men become more sensitive to these biological changes because the rate of change does begin to increase during the middle years and because of their increasingly self-reflective focus. Although the same pattern of biological change during the middle years is found in most men, both the rate and degree of change are strongly associated with the cumulative impact of life events during the early and middle adult years. It is perhaps not too much of an exaggeration to argue that the way we choose to lead our lives through the middle years determines, in middle age, the degree to which we biologically resemble young adults or old adults.

Changes in physical appearance. There are a variety of changes in physical appearance and biological functioning that occur during the middle years. These changes certainly don't begin during middle age, but they progress far enough by this time to make people increasingly aware of them. Some of the most noticeable include graying of hair color and balding, redistribution of weight (usually to the trunk), and a loss of elasticity of the skin, which produces a more furrowed appearance on the forehead and an increase in skin folds under the arms and chin and around the eye. The increasing looseness of the skin around the eyes sometimes gives the appearance that the eyes are set farther back into the skull.

Changes in sensory acuity. Both the eye and the ear decline in level of functioning during the middle years. Some degree of hearing loss, especially in men and especially in the upper registers is common during middle age (Bromley, 1974; Smith, Bierman, & Robinson, 1978; Stevens-Long, 1979). The greater incidence of hearing loss in men is generally assumed to reflect an increased exposure to noise over the adult years.

The changes in the eye tend to involve a decreasing ability to focus. Specifically, the lens becomes more opaque and less elastic. The increase in opacity decreases the clarity of the image that falls on the retina, and the loss of elasticity decreases the eye's ability to shift focus or accommodate to objects viewed at different distances. In addition, the muscles that control the iris and therefore the amount of light that enters the eye become less efficient. This results in a decrease in the size of the pupil and therefore the amount of light entering the eye, a change that probably has its most noticeable effect on the ability to drive at night.

Neither the changes in the eye nor the changes in the ear are of such magnitude as to seriously impede level of functioning. In most instances, a pair of bifocal glasses, in some cases a hearing aid, and perhaps less long-distance night driving are more than adequate compensations.

Changes in organ systems. There are also a variety of changes that occur in the structure and function of the organs of the body. In general, there is a thickening in the walls of structures such as blood vessels, the air sacs in the lungs, and muscle fibers. Further, there is a decrease in the degree of elasticity in the tissues that make up these structures. One result is that elevated blood pressure becomes a more common problem in middle age. The thickened walls of blood vessels decrease the diameter of the passageway resulting in an increased rate, or pressure. The decreasing ability of the lungs to expand and contract decreases the amount of air that can be taken in and exhaled. The thickening of the structures within the lungs reduces the efficiency of the oxygen/carbon dioxide exchange process. The consequence of these two changes in the lungs is that less oxygen-rich blood is available to the other parts of the body. The structural changes that reduce the exchange efficiency of the lungs are also evident, but to a lesser degree, in the kidney. Whereas when compared to a thirty-year-old, a fifty-year-old may have experienced a 10 percent loss in the level of functioning of the kidney, that same individual may have experienced a 15 or 20 percent loss in cardiac function and as much as a 25 percent loss in respiratory function (Smith, Bierman, & Robinson, 1978).

During the middle years, there is also a noticeable decrease in the capacity of the musculature to perform work. Speed and power are reduced, and the amount of time required for the muscle to recover from exertion is increased (Bromley, 1974). The effect of these changes is heightened by a greater stiffness in joints with increasing age. To a large extent, these changes can be offset by continued programs of exercise in the middle years and more attention's being given to loosening-up exercises prior to physical exertion.

While declines in overall biological status are part of middle age, the degree of decline is largely a reflection of diet, exercise, and life-style. The impact of these declines is most evident in levels of respiratory and cardiac functioning.

Cumulative impact of biological changes. The consequences of the changes in biological status that accompany middle age are relatively minimal in degree. Their cumulative effects are often of greater consequence. In particular, there is a shift in illness patterns from those described as acute to those that are chronic. **Acute illnesses** are those that are cured; after some period of time, they are no longer present. Infections are a good example. The appropriate use of an antibiotic eliminates the source of the infection, and as a result the body returns to its preinfectious level of function. **Chronic illnesses** are ones that are more likely managed than cured. They will not go away, but with the appropriate use of medication, diet, and exercise, their effects can be minimized. High blood pressure and joint-related ailments are examples of chronic conditions. The treatment strategies used to deal with each are not seen as removing the condition, rather they are seen as reducing the effects of the condition.

Paralleling the shift from acute to chronic conditions is a more general tendency of the vital capacity of the body to be reduced (Marshall, 1973). Men in their late fifties can do hard physical labor at about 60 percent of the rate achieved by men at forty. Since the impact of these cumulative factors becomes disproportionately greater as the degree of effort increases, a common coping strategy in middle age is to reduce the degree of physical exertion or to give more attention to the pacing of effort.

Effects of life-style on biological functioning. The patterns of change in physical appearance and biological status discussed in the preceding paragraphs are those typical of most middle-aged men and women. However, the degree to which any particular individual reflects these normative patterns is in middle age, probably more so than in other stages, a reflection of individual life-style. Indeed, the inclusion of this topic in the discussion on males reflects the fact that male life-styles often leave males more vulnerable to health-related problems than women are. As an example, one large-scale study conducted by Belloc and Breslow (1972) found that older adults who did not smoke, were not heavy drinkers, participated in a regular program of exercise and were not overweight, had regular sleep and breakfast habits, and did not snack frequently between meals had the average health status of individuals thirty years their junior who did not necessarily practice any of these health habits. They also reported that the degree of difference between those who practiced good health habits and those who did not increased with age—in part because the latter group tended to die sooner.

Studies such as that of Belloc and Breslow (1972) emphasize the fact that the cumulative influence of life-style factors on health during middle age—and therefore on life satisfaction since the two correlate highly during middle age—is essentially the cumulative influence of excess. Smoking, excessive use of alcohol, other forms of substance abuse, stressful life-styles, and overeating all correlate negatively with physical well-being and longevity. Smith, Bierman, and Robinson (1978) point out that obesity in middle age is the single most common preventable factor associated with excess mortality and morbidity.

The mechanisms through which life-style factors exert a cumulative influence on health and well-being have been most completely studied by Selye (1978). Selye conceptualized all life experiences, irrespective of their nature, as requiring some degree of adaptation on the part of the individual. The event may be perceptual or cognitive; it may be an illness or something as mundane as digesting a meal. It may be frequent in occurrence, it may require a great deal of adaptation to return the body to a homeostatic balance, or it may be of relatively trivial impact. But in all cases, the event or stressor requires some degree of adaptation.

Adaptation occurs at two levels. One, which Selye refers to as the **local adaptation syndrome,** involves those aspects of the body that are involved in the specific, immediate response to the stressor. For example, a cut in the skin produces some degree of inflammation in the surrounding tissues; the blood vessels and connective tissues secrete substances to kill invading microorganisms, and the temperature of the immediate area increases (Greenwood & Greenwood, 1979).

Of more relevance to the discussion of the cumulative effects of life events is the second level of adaptation—the **general adaptation syndrome.** In addition to responding to the needs of the local adaptation syndrome, the body also initiates a series of endocrine and nervous system changes that influence not only the local site but also the entire body. Depending on the specific form and degree of stress, the body may respond by elevating body temperature, raising blood pressure, increasing basal metabolic rate (resting rate of oxygen consumption), increasing the heart rate and rate of respiration, dilating the pupil of the eye so that more light can enter, and inhibiting the onset of muscle fatigue. In addition to the endocrine and nervous systems, the circulatory system, the liver, and the kidneys also play significant roles in the general adaptation system.

Selye (1978) believes that when stress is excessive and occurs over long periods of time, its cumulative impact on the body is to reduce its effectiveness in responding to events and restoring homeostasis. It is during middle age that this reduced effectiveness first becomes significant.

Many of the disorders of middle age (such as gastrointestinal problems, hypertension, and osteoarthritis) are characterized as nonspecific in origin (see Figure 9-1). They occur because of the way we choose to live our lives and because of the way the body reacts to this choice (Dohrenwend & Dohrenwend, 1974). Coronary heart disease (CHD), the most prominent of these nonspecific chronic conditions, is the single greatest killer of middle-aged adults (Friedman, 1978; Glass, 1978; Holmes & Masuda, 1974) (see Figure 9-2).

The research on the link between coronary heart disease and stress has been of two types. One has focused on the types of life events that produce the greatest amounts of stress (Dohrenwend, 1979; Hinkle, 1974; Holmes & Masuda, 1974; Holmes & Rahe, 1967) and the other, the influence of personality on CHD (Cohen, 1978; Friedman, 1978; Rosenman, Brand, Jenkins, Friedman, Straus, & Wurm, 1975).

Much of the life stress and CHD research is based on the Social Readjustment Rating Scale (see Table 9-1). The scale represents an attempt to indicate the relative degree of stress created by life events. The inclusion of both positive events such as marriage and negative events such as being fired from work emphasizes the fact that the degree of adaptation required by a change is often as meaningful an index of stress as the nature of the

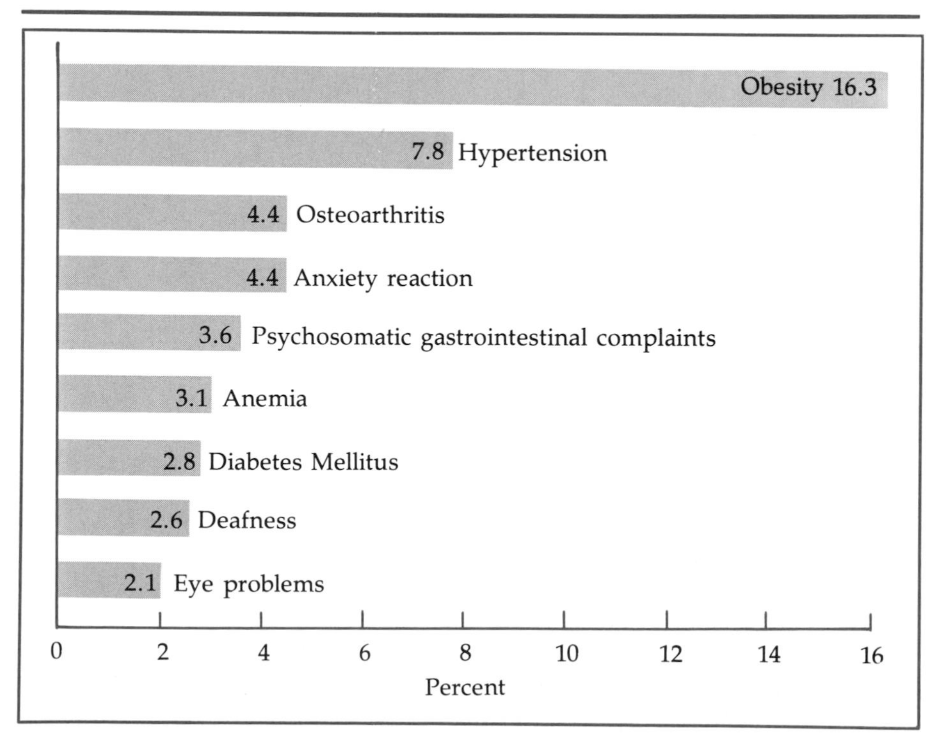

FIGURE 9-1
Disorders of middle age (adapted from Smith, Bierman, and Robinson, 1978).

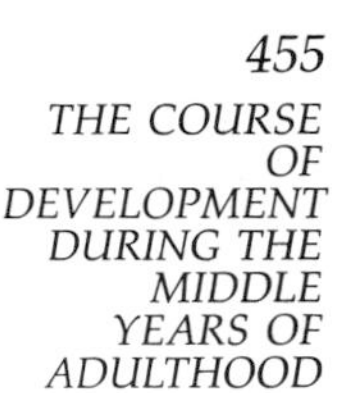

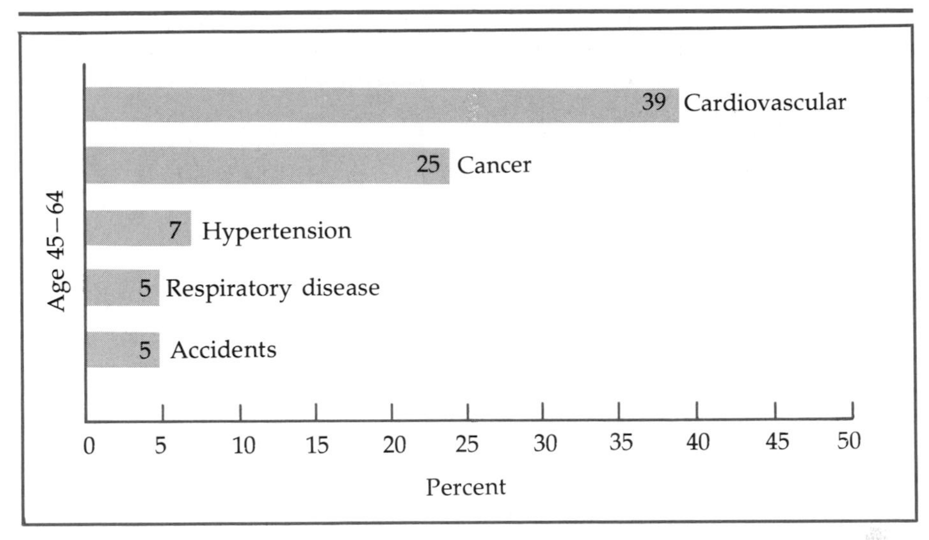

FIGURE 9-2
Common causes of death in middle age (adapted from Smith, Bierman, and Robinson, 1978).

event itself. The research indicates that as the total number of life crisis units increases, so too does the probability of illness. Although the mechanisms are still unclear, the evidence suggests that the greater the life change or adaptive requirement, the greater the vulnerability, or lowering of resistance, to disease, and the more serious the disease that can develop (Holmes & Masuda, 1974). From this perspective, it is not surprising that Hinkle (1974) found in a survey of life-style and health that healthy people are generally those who are doing what they want to be doing, like their work, like their family and associates, and are generally content and comfortable with their lives.

The second line of research on CHD, that concerned with personality, has found that some adults in middle age find it very difficult to enjoy what they have accomplished. Rosenman (1978) describes these adults as constantly striving to reach goals that in fact are usually poorly defined. This striving is as evident in leisure activities as it is in the workplace. It is as if every minute of the day must be spent in some sort of goal-directed and purposeful activity. The relentlessness of their pace often makes them easily frustrated by delays and intolerant of the efforts of others.

Such **Type A individuals,** as they have come to be known, are twice as likely to have a heart attack as **Type B individuals** are, and given a first attack, they are even more likely to have a second. Further, these ratios hold independent of other risk factors such as heredity, diet, stress, and excessive smoking and alcohol consumption (Rosenman, 1978). Type B individuals are not exactly the opposite of Type A's. In fact, they exhibit many of the same traits but not to the same degree. Rosenman (1978) describes them as not involved in a chronic struggle against time although they may at times feel time pressure; they may be ambitious but not overly

TABLE 9-1
Social Readjustment Rating Scale

RANK	LIFE EVENT	MEAN VALUE
1	Death of spouse	100
2	Divorce	73
3	Marital separation	65
4	Jail term	63
5	Death of close family member	63
6	Personal injury or illness	53
7	Marriage	50
8	Fired at work	47
9	Marital reconciliation	45
10	Retirement	45
11	Change in health of family member	44
12	Pregnancy	40
13	Sex difficulties	39
14	Gain of new family member	39
15	Business readjustment	39
16	Change in financial state	38
17	Death of close friend	37
18	Change to different line of work	36
19	Change in number of arguments with spouse	35
20	Mortgage over $10,000	31
21	Foreclosure of mortgage or loan	30
22	Change in responsibilities at work	29
23	Son or daughter leaving home	29
24	Trouble with in-laws	29
25	Outstanding personal achievement	28
26	Wife begin or stop work	26
27	Begin or end school	26
28	Change in living conditions	25
29	Revision of personal habits	24
30	Trouble with boss	23
31	Change in work hours or conditions	20
32	Change in residence	20
33	Change in schools	20
34	Change in recreation	19
35	Change in church activities	19

TABLE 9-1 (CONT'D)
Social Readjustment Rating Scale

RANK	LIFE EVENT	MEAN VALUE
36	Change in social activities	18
37	Mortgage or loan less than $10,000	17
38	Change in sleeping habits	16
39	Change in number of family get-togethers	15
40	Change in eating habits	15
41	Vacation	13
42	Christmas	12
43	Minor violations of the law	11

Source: Reprinted with permission from *Journal of Psychosomatic Research*, 11, T. H. Holmes & R. H. Rahe, The Social Readjustment Rating Scale, Copyright 1967, Pergamon Press, Ltd.

Obsessed with work and goal achievement, Type A personalities are prime candidates for heart attacks.

competitive; they are goal directed but not aggressively so. Although the impact of these two personality styles on CHD has been studied most with regard to the middle years, most evidence indicates that the two patterns are evident during the early adult years and even in childhood. To some degree, they reflect socialization patterns, an argument that is strengthened by the fact that Type A patterns are more evident in men than in women.

Changes in Family Roles

The changes that accompany the transition to middle age for men in their roles as workers and in their physical appearance and biological status are the two major defining issues of the transition period. There is also, however, a third issue—the postparental or **empty nest transition.** We actually know very little about this element of the transition in men's lives. Typically, the issue has been seen as more relevant to the lives of women, and as a result most of the research has focused on women. What little information is available (Lewis, Freneau, & Roberts, 1979; Lowenthal & Chiriboga, 1970) indicates that for most men seeing their children leave home is not a particularly difficult event. In particular, when father-adolescent conflict has been high, the transition may be quite welcome (Osherson, 1980).

The relative ease of the postparental transition for most men probably reflects the fact that they are typically less intertwined in the lives of their children than are their wives. As a result, the departure of children requires fewer changes for them than it might for their wives. Further, the departure of children is often seen by husbands as an opportunity to heighten the emphasis on the couple aspect of the family constellation (an issue that will be discussed more fully later in the chapter).

One exception to this pattern has been reported by Rubin (1979). Some of the middle-aged women she interviewed observed that the empty nest transition was actually harder for their husbands than it was for them. Rubin suggests that it is the abruptness of the transition for these men that makes the transition difficult. Unlike their wives who, in a sense, had been watching their children from birth on leave home for longer and longer periods, these men were so involved in their work that in Rubin's words, "One day . . . [the children] are gone—gone before . . . [their fathers] really had a chance to get to know them" (p 36).

Lewis, Freneau, and Roberts (1979) found that one factor influencing the father's adjustment to his children's leaving home was the quality of his relationship to his wife. Those who had the most difficulty adjusting to the youngest child's leaving home were the ones who reported their marriage as least satisfying. This group (approximately 25 percent of the total sample) also shared three other characteristics. They had the fewest children, they were the oldest members of the entire sample, and they perceived themselves as more nurturant or loving than did the total sample. In effect, they stood the most to lose by the departure of their children. Having fewer children, they suffered a greater loss at the departure of each; being older, they feared having fewer years to maintain contact with their grown children; seeing themselves as highly nurturant, they grieved about the loss of that aspect of their identity; and not seeing their marriage as particularly satisfying, they felt that they had little opportunity to compensate for the loss of their children.

It is useful to consider the findings of Lewis and his colleagues in relation to the transition-to-parenting studies discussed in the previous chapter. Both suggest that given the opportunity, fathers can become as involved in the lives of their children as mothers do. Both areas of research emphasize the dominant role that socialization patterns play in the evolution of male and female role identities across the life-span.

Transition Patterns for Women

When the two transition patterns are compared, one thing that stands out is the comparative ease of the transition for women. The notion of the mid-life transition as one of crisis is more common in the literature on men (e.g., Gould, 1978; Levinson, 1978; Vaillant, 1977) than in the literature on women (e.g., Neugarten, 1968; Rossi, 1980). This difference probably reflects the fact that the transition is more gradual and more expected in women than in men. The evidence suggests that many men, particularly those having the most difficult transitions, seem less willing to acknowledge that young adult life-styles will not necessarily continue through the middle years.

Two events mark the transition to middle age in women. The first involves the empty nest transition—grown children leaving home. The second is **menopause**—the termination of reproductive capability. The fact that these two events typically occur anywhere between the early forties and the early fifties again reinforces the importance of the social clock as well as the biological clock as a marker of stage transitions during the adult years (Neugarten, 1968).

This section will review first the changes in family role accompanying the middle-age transition and then the biological changes that are encountered. Following these discussions, the limited literature on the place of work in the lives of middle-aged women will be presented.

Changes in Family Roles

The most general conclusion that can be drawn from the research on the effects of grown children's leaving home is that most women find the transition easily made (Borland, 1982; Campbell, 1981; Harkins, 1978; Lowenthal & Weiss, 1976; Rubin, 1979). Researchers often report surprise at this finding because they feel the popular literature paints a more negative picture. The discrepancy probably reflects the fact that the popular media are more prone to report those studies involving clinical populations (rather than broad-based samples) that do find the transition to be difficult. Why then do most women find the transition relatively stress free, and what are the circumstances that cause a few to have difficulty?

The first reason that most women find the empty nest transition relatively stress free is that middle-aged women typically place greater emphasis on their role as spouse than on their role as parent. Therefore a major part of their family role identity continues into middle age (Frankel & Rathvon, 1980; Robertson, 1978; Saunders, 1975). In fact, the departure of children provides these women more opportunity to focus on their role as wife—an increasingly important emphasis as health-related issues in men become more common. As mentioned in the last chapter, the shift in emphasis toward the spouse is a recent pattern. It reflects the decrease in family size, the increasing length of the average life-span, and the increasing attention given to the interpersonal as well as the instrumental aspects of a marriage. Whereas at the turn of the century, it was not uncommon for the husband to die before the youngest child left home, the present cohort anticipates many years of postparental marriage.

The relative amounts of emphasis given to the roles of wife and mother are partly a function of the degree of satisfaction derived from each. As is true of men, those women who view their marriage as unsatisfying are probably more likely to have difficulty with children leaving home (Saunders, 1975). This problem may be compounded when the mother has become overly involved in or overly dependent on her role as mother. Such women might experience feelings of loss with the termination of their primary parenting role that would not be seen as compensated for by their relationship with their spouse (Borland, 1982; Robinson, 1977).

A second reason that the empty nest transition is not traumatic for most women is that they look forward to a time free of the worries of parenting. Again, as discussed in the last chapter, the parental role places a heavy burden on women. Even though there are wonderful compensations for parenting, it is nevertheless a time- and energy-demanding job that does not always get the recognition it deserves. Many women welcome the greater sense of personal control over their time and schedule that accompanies the transition. They look forward to pursuing new interests that were previously unavailable because of child-care responsibilities (Campbell, 1981; Robertson, 1978). This divestiture of the primary parenting role is partly a function of the mother's perception of her adequacy as a parent (Saunders, 1975). Those who think they have done a good job seem best able to launch their children and go on to other pursuits. It may also be a function of social class (Giele, 1980; Targ, 1979). The more economically advantaged the background is, the more opportunities may be present for growth in new directions.

The third and final reason why the transition is accomplished so readily is that it may not be that much of a transition after all. The empty nest is a gradual and continuous process that neither begins nor ends when the child leaves home. As Targ (1979) points out, children leave the nest as soon as they can crawl away from their mother's arms. Each developmental milestone that follows involves a progressively greater degree of separation. By the time children reach late adolescence, most function quite independently of their parents, and parents have developed routines that are independent of their children. After children do leave, they don't stop being children, and parents don't stop being parental (Lowenthal & Chiriboga, 1970). College students come back home during the summer; those who work may still live at home; and even when they are no longer physically present, most families maintain frequent contact. Moreover, as will be discussed later in the chapter, some parents are able to compensate for the loss of their parenting role by making a great investment in their role as grandparent.

Biological and Physical Changes

Although menopause is the most pronounced biological change in middle-aged women, it is not the only change. The variety of changes that were discussed in the section on biological changes in men are for the most part also evident in women. One common misperception is to consider all these changes as to one degree or another a correlate, or side effect, of menopause (Goodman, 1980; Nathanson & Lorenz, 1982; Neugarten, Wood, Kraines, & Loomis, 1968). This error probably explains why many premenopausal women envision the transition to be more stressful than it is actually reported to be by those experiencing it.

Changes in physical appearance. Women experience the same types of changes in physical appearance as do men. There is often an increase in weight and a shift in weight distribution to the torso, hair grays, and skin takes on a more wrinkled and somewhat transparent appearance (Stevens-Long, 1979). Since physical appearance is typically a more salient aspect of self-esteem in women than in men, the psychological reactions may be quite different even though the physical changes are similar (Turner, 1977). The changes that occur during middle age are likely to be seen as enhancing the appearance of men while detracting from the appearance of women (Nowak, 1977). Graying hair is said to lend men an air of distinction; facial lines convey experience and wisdom; even a paunch suggests a measure of prosperity. Although distinction, wisdom, and prosperity should be no less valued in women, our cultural values rarely support such equality. Nowak (1977) found this sense of inequality true of young adult men and women as well as of middle-aged men and women. Indeed, middle-aged women would seem to be their own worst critics with regard to physical attractiveness and youthful appearance. Women in their forties rate their own physical appearance lower than they do that of any other age group and, perhaps of greater importance, they rate it lower than other age groups rate the physical appearance of women in their forties (Turner, 1977). But in stark contrast middle-aged women in their fifties rate their sense of self-esteem as high as, if not higher than, same-age men rate theirs. It would seem that middle-aged women in their forties are responding more to the change in their physical appearance than to the new appearance itself.

Biological aspects of menopause. Menopause typically occurs during the late forties or early fifties. It involves a set of changes in hormone levels, which in turn results in the cessation of ovulation and menstruation. These changes are best understood by first reviewing the menstrual cycle in women. A typical cycle is initiated when the pituitary gland produces follicle-stimulating hormone. This hormone stimulates the maturation of an egg in the ovary and the production of estrogen in the ovary. The increase in estrogen production stimulates the hypothalamus to increase production of the luteinizing hormone (LH), which causes ovulation. Once ovulation occurs, the follicle in which the egg matured produces another hormone, progesterone, which inhibits further production of LH and estrogen.

At the same time as the egg is maturing, the increased production of estrogen also stimulates the lining of the uterus to grow and thicken in preparation for possible implantation of a fertilized egg. If the egg is not fertilized, the decreasing level of estrogen secretion causes the shedding of the lining of the uterus, which is experienced as menstruation.

Menopause occurs when the ovary is no longer able to respond to the follicle-stimulating hormone. Eggs no longer mature, and estrogen is no longer produced. Since estrogen production is halted, the uterus no longer undergoes change to prepare for possible implantation, and in turn, menstruation no longer occurs. The transition is gradual rather than abrupt, occurring over a period of a few years. During this interval, cycles become increasing irregular and infrequent, although pregnancy is still possible. Menopause is said to have occurred when the cycle stops completely (Perlmutter, 1978).

Historically, menopause has been associated with a wide variety of secondary effects including **vasomotor instability** (hot flashes), depres-

Long accosted by the under-thirty ideal of female beauty, middle-aged women now face a threat from such over-forty "glamour girls" as Joan Collins. Both views of womanhood tend to undervalue the other characteristics possessed by the vast majority of middle-aged women.

sion, headaches, insomnia, vertigo, weight loss, and palpitations (McKinlay & Jefferys, 1974). In fact, as McKinlay and Jefferys as well as a number of other researchers report (e.g., Nathanson & Lorenz, 1982; Notman, 1980; Perlmutter, 1978), most of these side effects are either unrelated to menopause or occur in only a small percentage of women. The women in the latter case are the ones most likely to seek medical intervention (and therefore are visible to researchers) and often demonstrate a history of having problems with all transitions during the adult years.

The only symptom that is reliably reported to occur in most women (approximately 75 percent) is hot flashes. Perlmutter (1978) describes the symptoms as a sudden sense of warmth that pervades the upper part of the body. The face may become flushed, areas of the arms or torso may become reddened, and perspiration may follow. As the body attempts to adjust to this sudden shift in temperature, the woman may then feel cold and clammy. The frequency, intensity, and duration of hot flash symptoms seem highly individual. They reach their peak as the actual time of menopause approaches. Following menopause, the symptoms gradually decrease. Although the actual causes of hot flash symptoms are unclear, they are related to the changing patterns of endocrine functioning.

The decrease in estrogen production following menopause is associated with a number of changes in women. These changes include a somewhat greater dryness of the skin, a decrease in the size of the uterus and vagina, a decrease in the ability of the vagina to lubricate itself during sexual stimulation, and a gradual increase in the brittleness of the bones. Some of these changes go undetected; others such as dry skin can be countered with lotions. Changes in the vagina are often countered with the use of vaginal creams and an increase in sexual foreplay. There is no evidence that menopause influences sexual responsiveness (Perlmutter, 1978). Those couples who have established a satisfying pattern of sexual activity previous to menopause continue to derive pleasure from their sexual relationship, some finding that the absence of fears about pregnancy heighten sexual enjoyment.

The most extreme intervention for the relief of postmenopausal symptoms is estrogen replacement therapy. Although the use of synthetic estrogen does alleviate a number of the postmenopausal symptoms, there is some evidence (Bart & Grossman, 1982; Perlmutter, 1978) that the therapy is associated with increased incidence of uterine cancer and phlebitis (inflammation of the blood vessels).

Psychological aspects of menopause. How do women react to the menopause experience? The general finding is that although women don't enjoy the side effects of menopause, especially the hot flashes, most don't regret their loss of reproductive capability and welcome being rid of the bother of menstruation (Neugarten, Wood, Kraines, & Loomis, 1968). Most report that their perceptions about the consequences of menopause are usually worse than the actual pattern of events. Neugarten and her colleagues believe this negative perception of premenopausal women reflects a less differentiated view of the second half of the life-span than is typical of postmenopausal women. Younger women are less likely to distinguish middle age from old age and therefore see menopause as the start of being old. In contrast, postmenopausal middle-aged women report feeling calmer, freer, and more confident (once the perimenopausal symptoms subside) then when they were younger.

Reviews of more recent research on women's adjustment to menopause have confirmed the findings that most women do not experience undue stress during menopause (on the average, only 25 percent of all women report experiencing physical or psychological symptoms sufficient to warrant medical intervention) and that most have little difficulty adjusting to the loss of reproductive capacity (e.g., Bart & Grossman, 1978; Severne, 1982). The reviews have also clarified some of the characteristics associated with ease of menopausal transition.

The highest rates of depression were most commonly found when the women were overly involved in the lives of their children. These women tended to be very protective of their children, often unduly limiting the range of their activities, and very intrusive in their lives, rarely providing an age-appropriate degree of autonomy. In effect, their depression at the time of menopause had more to do with the concurrent launching of children and the corresponding loss of a meaningful role than it did with the hormonal changes typical of menopause (Bart & Grossman, 1978).

The likelihood of such a pattern's occurring is correlated with level of education, social class, and work history (Notman, 1980). Well-educated women who are employed in satisfying occupations are least prone to depression and other severe reactions. Working women from less economically advantaged backgrounds are, in contrast, somewhat more prone to adverse reaction (Woods, 1982). These findings are reminiscent of the research on work and parenting presented in the last chapter. In both cases, life satisfaction was more associated with role congruence (doing what you want to do) than with the nature of the tasks themselves. At the same time, however, both sets of research also indicate that role congruence may be more easily achieved in the more economically advantaged segments of society. For those who value traditional feminine roles, the earning power of their husbands makes it possible for them not to need to work to supplement family income. For more nontraditional women, the nature of the jobs available offers the potential for greater intrinsic and extrinsic reward.

Changes in the Role of Work

The significance of work in the lives of middle-aged women today is different from what it is likely to be for the present generation of young adult women when they become middle-aged (Applebaum, 1981: Block, Davidson, & Grambs, 1981). For many women today, middle age represents an opportunity to enter the labor market for the first time or to return to school to prepare for an occupation. For this current cohort, middle age, particularly, the empty nest transition, is seen as a chance for development in new areas and an opportunity to acquire new sources of satisfaction. For the present generation of young adults, so many of whom are already in

Having interrupted their own educations to raise families and launch their husbands on careers, many women go back to school during middle age.

the labor market, the transition, in terms of work, will probably have a different meaning.

The issue of finding new sources of satisfaction in middle age is an important one, especially for a generation that placed great importance on the roles of spouse and parent. Pearlin (1975) argues that one of the major reasons that women have higher rates of depression than men is that the structure of their lives may offer fewer resources for gratification. This problem appears to be particularly acute among the present generation of college-educated middle-aged women who have not sought employment. Again, the issue is less one of working versus not working than of the potential gratification individuals are able to find in different social roles.

Work for middle-aged women may offer new avenues for intrinsic and extrinsic reward, but it also offers the same liabilities it does for middle-aged men. Type A behavior is more common in employed than in nonemployed women, especially among college-educated women (Waldron, 1978). And although Type A women are less prone to CHD than Type A men, they are four times more likely to develop coronary problems than Type B women (Haynes, Feinleib, Levine, Scotch, & Kannel, 1978).

The different impacts of work on the health of middle-aged men and women probably reflect the degree to which work is an integral part of different individuals' self-identity. In Rubin's book *Women of A Certain Age* (1979), she describes an interview with a middle-aged woman who worked as a senior executive in a large social service agency.

> She waved me to a chair, apologized for the mess, and asked me to be patient while she completed an expected telephone call that required her immediate attention. I sat down and waited. The phone rang, she identified the caller and began to speak rapidly, forcefully. "No, George, that won't do," she insisted. "It simply has to be *this* way." The man at the other end of the line demurred. They argued; she remained adamant. She won. I sat, admiring her performance, her forthrightness, when it occurred to me that George wasn't just some ordinary associate; he was the mayor of the city. I was delighted—no joyous, gleeful. Finally, I thought, a setting and a circumstance in which a woman can't possibly fail to acknowledge her work identity. As she replaced the telephone in its cradle, she turned to me smiling. "Okay, let's get started, I've told my secretary to hold all other calls for the next few hours." Given what I had just witnessed, I decided to start the interview with the question of identity. [Before we get into the rest of the interview, could you briefly describe yourself in some way that would give me a good sense of who and what you are?] She leaned back in her chair, looked thoughtfully out the window for a moment, then said: "Okay, I guess that's not too hard. I'm a very considerate, generally agreeable person. I'm a good wife, good mother, and a very good daughter. In general, I think it's accurate to say I'm kind, loving, warm, and very nurturant." (pp. 56–57)

Rubin says that she could only "gasp in disbelief."

Rubin's dramatic description highlights what may be the character of the major cohort difference between the current generation of middle-aged women and the current generation of young adult women. Of more significance than the differences in the actual structure of their adult life-styles, especially in terms of work, may be the different degrees of importance the careers of intimate, parent, and worker have in the adult identity of women in the two groups. A large number of women in both generations work, but it appears that women of the young adult generation are considerably more likely to incorporate their role as worker into their adult identities (Stroud, 1981).

Resolutions of the Mid-Life Transition

For most adults, the successful negotiation of the transition to middle age results in a significant shift in the way they see themselves in relation to other people, in relation to the value they place on their experiences, and in relation to the forces that influence their lives (Peck, 1968). Some make the shift more willingly than others and some make it sooner than others, but most eventually make it and by their fifties come to believe that what they have gained is a reasonable compensation for what they have lost. Those who are either unwilling or psychologically, at least, unable to make the shift, that is, those who cling to the value structure of young adulthood, find themselves having to become more and more "armored," or defensive, as their perception of the world becomes increasingly dissonant with its reality (Levinson, 1978).

Shifts in Relations to Other People

The resolution of the mid-life transition brings a reversal in the pattern of increasing sex-role differentiation that first becomes evident during middle childhood. Men become less aggressive, more affiliative and compassionate, and less dominant in the marital relationship; women become more instrumentally competent and more autonomous (Feldman, Biringen, & Nash, 1981; Targ, 1979). The net effect of this sex-role reversal is to make middle-aged men and women more androgynous, or like each other, in sex-role orientation and style. This in turn produces a shift in patterns of relationships, from one that places a value on the sexualizing aspects of human relationships to one that places a value on their socializing aspects (Peck, 1968).

Keep in mind that Peck is not suggesting that middle age is a time of celibacy. On the contrary, he is emphasizing the fact that the type of intimacy possible during the middle years, one that highlights the extent to which men and women can be alike rather than the extent to which they can be different, may not have been possible earlier.

Why does this type of intimacy seem to need to wait until the middle years to develop? Why isn't it possible sooner? The answer to the question has two parts. The first is that the sex-role socialization patterns evident during adolescence and the early adult years reinforce different values in men and women. These different socialization experiences foster the view that men and women are quite distinct from each other. By middle age, many of these distinctions have faded a bit; men aren't quite so masculine; women aren't quite so feminine.

The second part of the answer concerns the impact parenting has on young adult men and women. The task of caring for the young in a society such as ours, both in terms of earning enough money to care for them and in terms of ministering to their needs and wants, forces most young adults to assume a division of labor that again serves to differentiate the sexes. Once children leave home, the circumstances that prompted the division are less present, and behavior becomes more androgynous (Abrahams, Feldman, & Nash, 1978; Hyde & Phillis, 1979).

Shifts in the Value Placed on Experience

The formal operational logic that first becomes evident during adolescence makes possible the initial identity formation typical of that stage. It also makes possible the development of a life plan during the early years of adulthood. By providing the cognitive means necessary for conceptualizing a variety of "futures," formal operational logic makes it possible for young adults to conceptualize and make the series of tentative commitments that Levinson (1978) talks about.

But this logic also creates some problems for young adults (Labouvie-Vief, 1980). First, the sense of logical necessity that makes it possible for the concrete operational nine-year-old to realize that changing the shape of a real object such as a ball of clay will not change its weight (conservation of substance) doesn't necessarily have a parallel in the social world of the adult. The behavior of people is not nearly as lawful as that of a ball of clay, and as adults apply their logic to the problems of the everyday world, they soon confront the dialectic nature of most adult situations (Riegel, 1973). That is, they soon realize that there are contradictions inherent in situations that can't always be resolved through the power of formal operational logic. The two most notable examples are the intimacy/autonomy dialectic of marriage and the dependence/independence dialectic of parenthood.

The second problem the logic of youth creates for the adult is that it is potentially more useful in conceptualizing provisional life plans than in committing oneself to a real one. Whereas youth is is a period of flexibility and keeping options open, adulthood is a time of commitment to self and others. Labouvie-Vief (1980) believes that the transition from youth to adulthood prompts adults to shift their decision-making process from one that emphasizes the generation of multiple options and contingencies to one that emphasizes a focused management of resources given the nature of each adult's commitments to the careers of parent, worker, and intimate.

This transition occurs gradually over the early adult years. Contradictions come to be seen less often as problems to be solved and more often as circumstances requiring adaptation (Clayton, 1975). Experience becomes as valued a problem-solving tool as logic (Clayton & Birren, 1980). Middle-aged adults realize that the cumulation of their experiences, their wisdom if you will, gives them a perspective that is unavailable to younger people. They come to place increasing value and reliance on this new perspective. Since this more experience-based view places less emphasis on formal logic, middle-aged adults sometimes appear disadvantaged on ability tests that emphasize logic—a finding that is sometimes misinterpreted as reflecting a decline in ability during middle age (Horn, 1978; Schaie, 1979). But as Labouvie-Vief (1980) notes, tests that emphasize logic fail to appreciate and therefore fail to reflect the specialized gains that experience offers.

There are two important qualifiers in this shift toward the valuing of experience with age. It should come as no surprise by now that the qualifiers are sex and social class. If scholars such as Gilligan (1982) are correct, that is, if the socialization patterns of men and women create a fundamental difference in the way young men and women view themselves in relation to others, then it is quite possible that this shift may occur earlier in the lives of women than in the lives of men. Because women are more likely to confront the contradictions inherent in the parenting experience or in juggling the demands of career and family, they may sooner come to realize that logic is not necessarily sufficient to comprehend the nature of adult social roles.

Individuals of all social classes place increasing value on experience with age. What differs, however, is the way in which experience is conceptualized (Kohn, 1977; Miller, Schooler, Kohn, & Miller, 1979). The nature of the occupations associated with different degrees of education and therefore different social class levels influence the intellectual flexibility of adults in the middle years. The more economically advantaged the background is, the more intellectually flexible the individual. The more intellectually flexible the individual, the more likely experience is to be seen as relative rather than absolute. That is, the more likely it is that things experience has shown to be true will be seen as things that further experience will show to be in need of revision. The stereotype of middle-aged adults being rigid and fixed in their views, or of their being reactionary, if true at all, may be describing that segment of the population whose life circumstances limit the opportunities for self-direction and intellectual challenge (Farrell & Rosenberg, 1981; Gribbin, Schaie, & Parham, 1980).

Shifts in Influential Forces

The changes in sex-role orientation and the role of experience in decision making are part of a more general shift in self-concept. Put most simply, middle-aged adults are considerably more likely to feel that they "own themselves," that they "pull their own strings," that they "are in control of

Although this couple seems concerned about the openness of the young couple's display of affection, the research indicates that they are more likely the exception than the rule among the middle aged. Middle-aged adults are no more likely to be rigid or reactionary than any other age group.

their lives," and that they "are comfortable with themselves" than are young adults (Tamir, 1982). Levinson (1978) notes this comfortableness in his sample of middle-aged men. He describes them as having an optimal balance between being young and being old. He describes them as being in "the center of the lifecycle"—a phrase reminiscent of another period in the middle, middle childhood. Vaillant (1977) reports that the middle-aged men in his sample no longer blame others for their shortcomings and show greater consistency than younger people do between thought and action. He notes that at 25 they often forgot to vote, but by 47 they always remembered. Brim (1974) suggests that this comfortableness comes about partly because people realize that there are many things simply beyond their control, death being one of them. Fate becomes a more acceptable notion than it once was.

A particularly good example of this notion of comfortableness in middle age is provided by Livson (1981). Using data gathered over a period of many years through the longitudinal studies of the Institute for Human Development at the University of California at Berkeley, she was able to chart the relationship of life satisfaction to sex-role between adolescence and the completion of the mid-life transitions. Her sample consisted of both men and women. Approximately half of each group had a sex-role orientation described as conventional; the other half was nonconventional.

The traditional women in Livson's sample were feminine and conventional during adolescence. They were popular and sociable and developed personalities consistent with feminine role expectations. The nontraditional women were more intellectual and ambitious during high school. They placed a greater value on individuality. The traditional men were productive and ambitious, analytic and rational, and valued action over feeling. Finally the nontraditional men were emotionally expressive and undercontrolled. They tended to be more rebellious, attention seeking, and anxious.

Livson wanted to know how each of these four groups weathered the early adult years and the mid-life transition period. What she found was that those favoring a conventional sex-role orientation followed a smoother developmental course that those who did not. The traditionals, both men and women, seemed to fit better into society and therefore experienced less conflict between their personal value structure and that of the culture around them. This difference was particularly evident at age forty. The nontraditionals seemed more bitter and hostile; they seemed resentful that they had succumbed to the social forces of the time and had outwardly adopted the life-styles of their more traditional cohorts. Not surprisingly, on various measures of psychological health, the nontraditionals scored significantly lower than the traditionals.

Livson then asked if this marked difference at forty foretold later differences. Were the nontraditionals destined to be bitter, angry, and resentful for the balance of their middle years? The answer was no. By age fifty, all four groups scored equally well on the same measures that had distinguished them so clearly just ten years earlier. They still maintained their traditional and nontraditional values, only now they felt more comfortable with them. They felt less dissonance between what was expected of them and what they expected of themselves.

The changes in self-concept that result from the mid-life resolutions are evident in the way middle-aged adults pursue the three careers of the adult years. The decreasing emphasis on achievement and the increasing emphasis on relationships is reflected in changing attitudes toward work, family, and spouse. The sense of generativity that Erikson (1980) describes as the psychosocial task of the middle years is evident in the value placed on the mentoring aspects of relationships. A sense of generativity may involve helping new workers learn their jobs; it may involve helping an adult child adjust to parenthood; it may take the form of caring for aged parents; it may be expressed through greater participation in civic and community affairs. Whatever the context, all of these activities reflect a desire to influence events beyond one's own generation.

The Three Careers during the Later Part of the Middle Years

The resolution of the issues raised during the mid-life transition provides most middle-aged adults a new perspective on their adult roles. Marriage, family, and work continue to be important avenues of self-definition, but the sources of satisfaction derived through these roles take on a different character.

Work

The place of work in the lives of the middle aged differs in two ways from what it is in the lives of young adults. First, especially for men, work occupies a less central place in one's personal identity during middle age (Chinoy, 1955; Cohen, 1979). Second, although less central, it is at the same time viewed as more satisfying (Campbell, 1981; Clausen, 1981; Glenn, 1980). The two changes may appear contradictory, but they really aren't. They are a reflection of the shift in attitudes and values that were discussed in the preceding section.

By their fifties, most workers have reached their most advanced position (by the forties for many in low-income jobs). Few have any expectation of further promotion, and perhaps as a defense, few report any desire for further advancement (Bray & Howard, 1983; Clausen, 1981). Increasingly, degree of satisfaction is defined in terms of how far they have come rather than in terms of how much farther they would have to go to reach the top. Since after thirty years or so, most workers have come a fair way, it is reasonable that most now feel satisfied. Some also base their sense of satisfaction on a comparison between their present occupational status and that of their parents when they were middle-aged. Given the generally improved economic conditions over the past fifty years, most people also feel satisfied given this criterion.

The shift in reference point, from distance to go to distance traveled, provides an interesting counterbalance to the shift in age reference from years from birth to years till death noted by Neugarten (1968). Erikson (1980) suggests that a sense of ego integrity, the final psychosocial task, comes about through being satisfied with one's life work. Perhaps this counterbalance is a necessary step in that direction.

Social class differences in occupational level continue to be a significant influence on perception of work. The more economically advantaged the background, the less likely job security is to be a major criterion of satisfaction with work (Clausen, 1981). Job security is usually taken for granted as is a decent salary. For these individuals, security and decent income are concerns restricted to the young adult years.

For those with less advantaged backgrounds, concerns about adequate income and job security continue into the middle years. Although they are also less likely to be out of work as they acquire more seniority, the chance is still higher for them than for people in more prestigious occupations, and given the lower salary, they are less likely to have had the opportunity to create a financial cushion to fall back on. As will be discussed more fully in the section on the significance of development during middle age, the greater segmentation of the life-span for those in more economically advantaged circumstances comes about largely because their circumstances (specifically having a reasonable degree of job security) free them from old concerns and provide the opportunity for them to pursue new avenues of growth.

Family

By the time they reach their fifties, most middle-aged adults have seen their youngest child leave home. Their family roles do not cease, however. Most maintain frequent contact with their adult children, and many find themselves taking on two new family roles. The first is the role of grandparent, and the second is the role of caretaker for aged parents. These additional roles are at the same time very satisfying and also potentially very draining in terms of time, energy, and resources (Aldous, 1977). This problem can be especially acute for women since they are typically more involved in these intergenerational roles.

> It is one of those demographic ironies that whereas one set of trends promises increased freedom from family responsibilities and hence new opportunities for satisfying alternatives in later life, another set makes it difficult for many women to pursue such possibilities with total commitment. They are at risk of having their nest refilled. Whatever plans the middle-aged couple may have had for newfound decades of life together are increasingly likely to be upset by instabilities in the lives of either their children or their aged parents, which may require financial or personal attention and often an investment of emotional energy. The middle-aged woman seeking a new set of role obligations in the world of work or education must be prepared to reassume the maternal or daughter role when the once emptied nest begins to refill. Similarly, the middle-aged man may find decisions regarding retirement and residence affected by the need to provide temporary refuge for a grown child or elderly relative. (Hess & Waring, 1978, p. 250).

What are the trends that account for this ironic situation? First, the same set of circumstances that has improved the health of those in the middle years has also increased the life-span of those in old age. Although most older adults are able to maintain their own residence, the cumulative effect of the various chronic ailments associated with aging means that as these individuals continue to live longer lives, an increasing number of them will need some degree of physical and emotional care and financial support (Sprey & Matthews, 1982).

Second, the same set of social changes that has dramatically increased the number of women in the work force and has prompted couples to place greater emphasis on the interpersonal aspects of relationships has also prompted the significant increase in the divorce rate. Because of the financial burden associated with divorce, many young adults find them-

BOX 9-2
ROOM AT THE TOP?

You have probably seen pictures of giant snakes such as pythons swallowing large animals whole. Usually the picture shows the snake at full length with a great bulge in its middle. The impact of the post–World War II baby boom is often pictured in the same manner. The drastic increase in the birthrate following the war and lasting through the 1950s led to the unprecedented construction of new elementary and then secondary and finally college classrooms to accommodate the numbers of people and is the primary reason for so many dire predictions about the fate of the Social Security System when this "bulge" reaches retirement age in the early part of the next century.

For the moment, the leading edge of the baby-boomers is approaching middle age, and some experts wonder if many people will find themselves trapped in relatively low or middle level positions in the work force. What prompts this concern is the pyramidal structure of most businesses. As one goes up the corporate ladder, the competition for increasingly fewer and fewer positions becomes greater, especially if there are many people competing for these positions. A constantly expanding birthrate, the argument goes, would foster expansion of the economy, creating more senior positions. But with the exception of the baby-boomers, the birthrate has remained stable. Are they about to "hit the wall?"

"No," says an article in a recent edition of *Fortune* magazine. The article compares the career paths of a group of baby-boomers with those of their very successful fathers. Although acknowledging the potential bias of its sample (not everyone has a very successful father), the article contends that as these sons and daughters approach forty, their career development is

selves having little choice but to return home, at least temporarily. Such a renesting can rekindle old parent-child conflicts, and for adults who are no longer used to the pitter-patter of little feet, coping with young children can be a trying experience.

Third, the decreasing childbirth rate that gives middle-aged adults the opportunity to truly enjoy the prime of life also means that there are fewer people to share the burden of the generation in the middle. There are fewer siblings to share the care of aging parents, and there are fewer grown children who can help out if one separates and divorces (Hess & Waring, 1978).

Grandparenthood. The role of grandparent is an ambigious one in our society. Since three-generation households are rare, contact between grandparent and grandchild is often on an intermittent or scheduled basis. Often the degree of contact is a reflection of the quality of the parent–adult child relationship. In particular, it is a function of the character of the mother–adult daughter relationship since the mother–daughter link is usually the strongest (Fischer, 1981; Robertson, 1977).

equal to or better than that of their fathers when they were approaching forty. Further, the article claims that the experiences of this group are typical of their generation.

The article explains its conclusions by noting that predictions based solely on demographics fail to appreciate the changing nature of the business world. In particular, as more and more companies diversify into many products and services and as many decentralize their operations to create more "profit centers," the need for senior level managers is actually increasing. For this reason, the supply and demand ratio is staying about the same.

However, the article also notes that even though the number of people competing for each position may not be changing, their level of qualification is. There are more and more well-educated people competing for the senior positions. In 1960, for example, only 4,600 MBA degrees were awarded; last year the number was 65,000. One result of this increased competition is, the article notes, a more individualistic quality to career goals. The baby-boomers are less likely than their fathers were to wait their turn for promotions, more willing to make "lateral moves" or try making it on their own. The "organization man" may be a thing of the past.

Is there a price to pay for this entrepreneurial spirit? The article isn't sure. As long as the baby-boomers' level of confidence in themselves remains high, probably not, but as one soon-to-gray boomer put it, "Now I'm challenging people because they don't have the ability. What do I do when I reach that point and my ability's less and my drive's less—are they going to do the same thing to me?"

Source: Magnet, M. (1985). Baby-boom executives are making it. *Fortune Magazine,* September 2, 22–28.

Grandparents value their new role for a number of reasons (Neugarten, 1968). For many, it offers a sense of biological renewal, a chance to feel young again and to see the family line carried into a new generation. It is almost as if a degree of immortality is offered through the grandchild (Kivnick, 1982). Others see the role as a second chance to parent. In some cases, this second chance offers a form of compensation for feelings of not having been as good a parent to their children as they would have liked; for others, it represents a chance to use free time in an enjoyable fashion. Some grandparents talk about the chance to be a resource to their grandchildren in terms of time or experience or money. The notion of being able to establish a trust fund so that a grandchild will someday be able to attend college without creating a financial burden on the parents must certainly be a very pleasant thought. Not all grandparents look forward to frequent contact with their grandchildren, however. Some seem to prefer a more vicarious, or indirect, form of satisfaction. These grandparents bask in the glow of their grandchildren's accomplishments, not because the accomplishments are so noteworthy but simply because their grandchildren did them.

Removed from the day-to-day problems of parenting, many grandparents develop a special bond with their grandchildren.

The actual interaction patterns between grandparent and grandchild often depend on the age of the grandparent. Those that are still middle-aged seem to prefer either a lot of contact or very little (Neugarten, 1968). Those that maintain frequent contact typically establish a playmate type of relationship. Going to the zoo, going to the ball game, building model airplanes, going fishing, and watching the grandchildren play on the Little League team are the types of activities that are typical of this relationship. The emphasis is on leisure, nonessential activities. The contact is of choice rather than necessity. Other middle-aged adults, particularly those very much involved in their own lives, don't seem very interested in frequent contact. It is not that they don't care about their grandchildren; rather it is that they don't derive much pleasure from spending time with children. They may in fact be very involved in providing other forms of support (such as money) for the child.

Older grandparents, those over the age of sixty-five demonstrate different patterns of contact. Because both they and their grandchildren are older, trips to the zoo and other such outings are less common. Some see themselves as dispensers of information and experience; others, as family historians. Some take a more formal attitude and see themselves hosting periodic family get-togethers such as Thanksgiving, Christmas, or a Passover Seder. Such occasions are special occasions, requiring for children a bath, clean clothes, and often a lecture from parents about the kinds of words they can't use at their grandparent's house.

There is one final role that grandparents, both those of middle age and those who are older, are assuming in increasing number. That is the role of surrogate parent. As the number of dual-working families has increased and as the number of single parents has increased, more and more grandmothers are again finding themselves parenting. The role may be temporary, and it is not without its satisfactions, but it can add stress to the mother–adult daughter relationship. This role is more common in less advantaged families where money for day-care may be less available.

Generally, the birth of a grandchild improves the mother–adult daughter relationship (Fischer, 1981). Mothers feel that their daughters are assuming their appropriate responsibilities. Daughters come to see their mothers as potential sources of advice on child rearing. Most daughters report a new appreciation for their mothers now that they find the task of parenting not as simple as they once thought it to be. There is a dramatic shift in frequency of contact. Of those families that live within a local call of each other, few talk daily before the daughter becomes a mother; most talk daily after the birth of the grandchild (Fischer, 1981).

Care of aged parents. The demise of the multigeneration household has also affected the way middle-aged adults maintain contact with their aging parents. Because they live in their own residences, contacts are less likely to be casual and spontaneous and more likely to be planned and arranged (Hess & Waring, 1978). Depending on the distance between households, travel time can be an issue—sometimes involving more time than the length of the visit itself. When households are far apart and thus contact infrequent, responding to emergencies or locating necessary support services such as a visiting nurse or meals-on-wheels programs can be a difficult experience.

Most adult children who live within easy driving distance of aged parents do provide assistance. Cicirelli (1981) finds that the most common forms of assistance involve helping with home repairs, housework, and shopping, and serving as intermediaries with the various agencies that their parents receive services from. These forms of assistance are more likely to be provided by daughters than by sons. As parents continue to age, however, these nonspecialized services become increasingly less adequate, and the question of specialized care arises more frequently.

Because most of us no longer live in multigeneration-size houses, the question of what to do with aged parents who are no longer able to care for themselves can create severe emotional, physical, and financial trauma. In particular, Brody (1981) notes that although most adults over the age of sixty-five are able to completely care for themselves, approximately 10 percent require some form of supplemental service. Of those seventy-five to eighty-four years of age, 7 percent require full institutional care, and that figure increases to 20 percent in the post–eighty-five-year-old population. For middle-aged adults who are saving for their own retirement and who are fully aware of the medical costs they are likely to encounter as they grow old, wrestling with the emotional burden of deciding how to apportion money to meet the needs of parents and at the same time be prepared to meet their own eventual needs must be an extremely painful experience.

Cicirelli (1981) believes that one consequence of the various cohort changes over the past fifty years will be that same-age cohorts will increasingly assume responsibility for the care of each other. That is, as the birthrate continues to remain low, as families live apart from each other in increasing numbers, as the life-span continues to increase, and as many middle-aged adults find themselves confronting the prospect of providing for their own retirement years and also those of their aged parents, the aged themselves will assume a greater share of the care and support of each other. It is beginning to be common to find retirement communities or a series of efficiency apartments where the residents have grouped together in a cooperative fashion to help meet their everyday needs for food and household maintenance, social and emotional support, and nontechnical medical services. The effects such cooperative efforts have on intergenerational relations is an interesting question. Do, for example, such efforts affect the way middle-aged parents deal with their young adult children? Are they more likely to distance themselves in anticipation of shifting their focus away from family and toward cohort as a source of support in old age?

Marriage

For most married middle-aged adults, the later part of the middle years represents a time of high marital satisfaction for both husband and wife (Hayes & Stinnett, 1971; Petranek, 1975; Skolnick, 1981). And well it should. Most of the conditions that contribute to a high degree of marital satisfaction are usually present during this phase of the life-span. Degree of sex-role differentiation is relatively low, the competing demands of the parent role are no longer present, disposable income is relatively high, and socialization has replaced goal-direction as a primary value in relationships. Given such a set of circumstances, it is easy to understand that the feeling of a sense of equity and mutuality in the relationship and the feeling that each person understands and appreciates the efforts of the other (the major criteria of marital satisfaction) would be high (Lowenthal & Weiss, 1976).

The rather drastic improvement in degree of marital satisfaction lead Skolnick (1981) to argue that situational factors play as significant a role, if not a more significant one, in determining marital satisfaction as any enduring qualities of the couples themselves. Present level of job satisfaction, the relationship between current income and expenses, and the factors influencing both nuclear and extended family relations each serve to influence individual behavior and, in turn, couple relationships. As the specifics of these circumstances change, so too will perceived marital satisfaction. Skolnick believes that a major cohort factor distinguishing the couples followed in the Berkeley studies from the present generation of young married couples is the extent to which the situational nature of marital relationships is appreciated and at least implicitly understood.

> Perhaps the most striking impression from the case histories is the great potential for change in intimate relationships. The other impression concerns the impact of situational factors on marriage, especially work pressures on the husband and early child-rearing responsibilities of women. In reading some of the early adult interviews, it occurred to me that if the same marital problems arose today, a couple might conclude that they were incompatible and seek a divorce. As divorce becomes more frequent and more acceptable, commitment to marriage and to working through its problems inevitably lessens. Certainly the increased acceptability of divorce has released many people from deeply unhappy relationships. But it may also encourage too many people to discard relationships that have the potential for change and growth. (p. 296)

Although marital satisfaction improves for most couples, its meaning may not be the same for all of them. Swensen, Eskew, and Kohlhepp (1981) suggest that for some couples the increase in reported degree of marital satisfaction is primarily a reflection of a lessening in the number of potential sources of conflict. In other words, the marriage improves not because the two people spend more time with each other and enjoy each other's company more; it improves because there are fewer things for them to fight about. Because there are fewer things to fight about, less contact is necessary. Such marriages are seen as "devitalized." Swensen, Eskew, and Kohlhepp believe this type of marriage is most likely when partners main-

The postcard pose of these Florida vacationers presents the quintessential image of the happy middle-aged couple. While we might not all share their taste in dress, their obvious regard for each other is a goal sought by all of us in our relationships.

tain very stereotypic sex-role expectations of each other, primarily in terms of the roles of parent and worker. As the sources of conflict in these roles diminish (for example, when children leave home and income becomes adequate), the couple doesn't see the change as an opportunity to seek newer, more active forms of companionship. Rather, they just don't seem to see any need to have much contact with each other.

Nevertheless, the marriage partner, even in these seemingly devitalized marriages, is a very important source of security and support. One only need consider the severe trauma caused by the death or divorce of a spouse during middle age to appreciate the degree to which lives become intertwined over the years and to appreciate the fact that marriage, for both men and women, is the most influential element in their adult lives (Lowenthal & Weiss, 1976).

The discussion of the three careers during middle age as well as the earlier one on the transition to middle age emphasized the fact that the middle years of adulthood are a period of major developmental changes. These changes reflect the cumulative influence of life events as well as the context in which these events are experienced. As Vaillant (1977) noted in his study of Harvard graduates in their middle age, "what makes or breaks our luck seems to be the continued interaction between our choice of adaptive mechanisms and our sustained relationships with other people" (p. 368). The significance of the developmental events of middle age is the topic of the next section.

THE SIGNIFICANCE OF THE MIDDLE YEARS OF ADULTHOOD IN THE LIFE-SPAN

This is an interesting stage of the life-span. People enter it thinking they are still young and leave it wondering if they are yet old. It involves for most a period of reflection, which for some takes on rather traumatic proportions, and it involves for many a significant shift in the way they think of themselves. From a developmental perspective, the changes that occur during this approximately twenty-year period are perhaps more profound than those of young adulthood.

The Young Adulthood Perspective

When compared to the young adult years, those of middle age appear to be more integrated and less intense. The time-consuming and often energy-draining task of caring for young children is over. The sense of urgency to move ahead, to become established in the workplace is lessened. The dilemmas of scheduling, of finding personal time, and of finding time to enjoy marriage are less common. But there is an excitement that comes with the tasks of young adulthood, and that excitement also appears less common during the middle years. The action-oriented sense of industry is replaced by a more mellow, more reflective orientation. Clearly each has its advantages and disadvantages, but when older people talk about being younger, they no doubt are talking about what must seem to have been a more exciting time than the one they are now experiencing.

On the other hand, at least for the more economically advantaged segments of the population, the transition to middle age brings with it the opportunity for responsibility and authority, for greater self-direction, for a feeling of better self-understanding, and for more enjoyable interpersonal relationships. The male and female executives that Neugarten (1968) interviewed talked about middle age as a time of maximum capacity, of being able to function effectively in a highly complex environment, and of driving rather than being driven. The best adjusted among Vaillant's (1977) sample of middle-aged Harvard graduates were described as having more mature adaptive styles. These people felt that they were in control of their lives and as a result felt a stronger personal commitment to career, to the growth and development of others, and to their intimate relationships. Vaillant describes them as feeling increasingly comfortable with people and placing a greater value on human relationships.

The transition to middle age appears to offer, for some people at least, the potential for a fundamental shift in the way they see themselves. The potential for this type of shift seems less evident in the passage from the 18-to-25 period to the 25-to-40 period. It is not that people don't change from one of these periods to the other. Rather the change is more in knowing how to do something better (that is, in gaining greater experience) than in redefining what to do. The dreams of young adulthood are more realistically defined than those of the college years, but they are basically the same dreams. With the transition to middle age, the dream itself is questioned (Levinson, 1978).

We usually think of the mid-life transition as a twentieth century phenomenon, as something that plays such as *Death of a Salesman* or popular books such as *Passages* (Sheehy, 1976) make visible. Datan (1980) argues, however, that throughout history, there has been evidence of the series of events that we label the mid-life transition. It may be that the basic issues that adults confront as they enter their middle years have always existed and that only the context and the form they take have changed. Consider Datan's description of the classic tale of King Midas:

> In Midas we see a man who has gone beyond the struggle for subsistence to accumulate an effortless surplus. Midas was born of a union between a nameless satyr—the species which gave its name to the state of surplus sexual passion—and the Greek Goddess of Ida. While he was still an infant in the cradle, an omen foretold a fortune of wealth. The prophecy was fulfilled when Midas grew into adulthood: he extended hospitality to the debauched satyr Silenus, and thus found favor with Dionysus, whom the Greeks credit with discovering wine and divine madness. Dionysus asked Midas how he would like to be rewarded, and Midas requested that all he touch turn to gold. His wish was granted, and before long, he had touched his lunch and deprived himself of the sustenance of food and touched his beloved daughter and deprived himself of the sustenance of love. Like so many other folk heroes, Midas begged to be released from his wish; Dionysus, much amused, granted his wish.
>
> Midas, in going beyond subsistence, achieved wealth—and with it complete impoverishment, a perversion of excess. And yet the excess of Midas was simply an excess of mastery, a complete instead of a partial victory in the struggle to win subsistence. Nevertheless, the parallels between Midas and those whom we describe today as workaholics, whose dedication to work and its rewards at the expense of the inner self and its needs, are sufficiently close to persuade me that the mid-life crisis of affluence and accomplishment is not new.[1]

The story of Midas is useful in two ways. First, it again highlights the interplay of developmental events and the context in which they occur. Second, it provides some insight into the effect of social class on the character of life during middle age.

The mid-life transition is similar to puberty in that both mark the passage between developmental stages. As in the case of puberty, the context in which this transition occurs defines the nature of the stages involved. Once, puberty marked the passage to adulthood; today, it marks the passage to adolescence. Once, perhaps, mid-life marked the passage to the beginning of old age; today, it marks the start of a stage of "no longer being young but not yet being old." The changes in the expected life-span, in the size of families, and in the age at which childbearing occurs have all cumulated to forge this unique period. Both our present conception of adolescence and our present conception of middle age point out that a "side effect" of the technological evolution of a culture is the increasing segmentation of the life-span.

[1]Reprinted, by permission of the publisher, from N. Datan, "Midas and Other Mid-life Crises," in W.H. Norman & T.J. Scaramella (Eds.), *Mid-life Developmental and Clinical Issues* (New York: Brunner/Mazel, 1980), 12–13.

The Social Class Perspective

I'm not sure how social class would have been defined in the time of King Midas, but I'm fairly sure that he would have been considered privileged even before he got the "Midas touch." This advantage defined the character of his middle age, just as social and economic advantage continues to define the character of middle age today. In fact, social class may be the single most significant discriminator of developmental patterns during the middle years. In this sense, middle age is very different from young adulthood. In young adulthood, sex role is the most significant discriminator, a reflection of the parenting role. But once the parenting role ends, the polarized nature of sex-role orientation lessens, and it therefore becomes less of a discriminator between individuals. Social class background, on the other hand, has a continuous and cumulative impact on the life course.

Social class influences the time at which the mid-life transition is likely to occur (Giele, 1980), the way individuals are likely to adjust to it (Tamir, 1982), the way they are likely to view the opportunities of middle age (Giele, 1980), the perceptions of satisfaction and happiness in middle age (Campbell, 1981), and the degree of intellectual flexibility and cognitive competence evident (Gribbin, Schaie, & Parham, 1980; Kohn & Schooler, 1978; Schaie, 1983).

The more advantaged the background is, the more delayed are the factors that prompt the transition. Since education is extended, marriage and childbearing are delayed, and therefore so is the empty nest transition. The same extension of education defines the nature of the work pursued. The career ladder is longer, it is less dependent on physical skill, and it provides more opportunities for mentoring roles. Further, the greater degree of affluence makes possible a decreasing concern with financial security, a turning to new interests or pursuits, and the enjoyment of "the good life." In this sense, it is important to note that although there are no differences in degree of life satisfaction between social classes during middle age, there is a difference in reported degree of happiness (Campbell, 1981). Satisfaction is a measure of acceptance; happiness, more a measure of opportunity. The implication is that the more advantaged the background, the more likely the individual is to see middle age as a time for new opportunity and new growth, as a time when what is gained is more than adequate compensation for what is lost.

These differing perceptions of middle age exist because the cumulative impact of the structuring of roles within social classes influences the level of intellectual functioning through middle age (Schaie, 1983). As a result, the flexibility necessary for adjusting to changing life circumstances and for achieving the integrated abstract sense of self that is a major component of middle age is more likely to be found within the more economically advantaged segment of the population (Peck, 1968).

As important as social class is, it is equally important not to overstate its impact at the expense of other factors. Almost all middle-aged adults report that this is a more positive time of the life-span than earlier stages were—irrespective of social class background. This age-related shift is consistent with Skolnick's (1981) observation about the situational determinants of marital satisfaction. It also emphasizes the importance of distinguishing the degree to which a person's perception of and adjustment to a situation reflect some enduring quality of that person from the degree to which they reflect the specifics of the situation itself. The events typically encountered during middle age seem more enjoyable (perhaps because

For the successful executive, secure in his position and worth, the mid-life transition period may give rise to few problems

they are more easy to cope with) than many of those encountered during young adulthood.

It is also important not to understate the impact of sex role during middle age. It is certainly true that men and women are more alike at this time than during young adulthood, but they are not identical. Women still report their marriages in less satisfying terms than men do, they are still the ones more likely to be involved in family issues such as care of aged parents, and in the case of those with very traditional orientations, they are particularly vulnerable to depression (Pearlin, 1975). It is also women who are more likely to experience the death of a spouse during middle age (Lopata, 1980).

HUMAN DEVELOPMENT AND HUMAN SERVICES

The adjustments in life-style typically encountered during the middle-age transition are in part prompted by the life patterns of the young adult years and in part by the biological and social changes that correlate with chronological age. Making the transition less stressful requires a consideration of both sources of potential stress.

Fostering the Transition to Middle Age

Young adult life-style is a significant predictor of the transition experience. The more likely individuals are to invest all of their time and energy in a pattern that is stage specific, the more likely they are to encounter difficulty during the transition. For men this pattern is usually in terms of career; for women, it is in terms of family. When in either case one pattern becomes virtually the sole source of gratification, it becomes a potential source of trouble. The various studies reviewed in the chapter (Farrell & Rosenberg, 1981; Giele, 1980; Gribbin, Schaie, & Parham, 1980; Osherson, 1980; Pearlin, 1975; Tamir, 1982) all support this conclusion. The situation is, in a sense, more resolvable for men than for women. It is possible to change the promotion policies of companies so that employees don't necessarily peak while still young adults (Bailyn, 1980). There is no comparable way to slow the growth of children.

It is possible that future cohorts will be less affected by this issue. As the workplace becomes more skilled and technological in nature, employers will place emphasis on prolonging career growth, on providing opportunities for retraining within the company, and on maintaining employee morale. When it only takes a day to train someone to perform a task—as is typical on an assembly line—there is relatively little incentive for employer investment in employee satisfaction. But when the employee is highly skilled and performs a difficult task, the incentive is certainly greater (Shostak, 1980).

The role expansion process typical of young adult women should leave them less prone to extreme stress with the ending of their primary parenting role. The continuing role of worker as well as the renewed opportunities for spouse relations should serve to offset the loss of gratification from the parenting role.

The biological changes that accompany age are more or less inevitable. They are what happens to us as we age. The social meaning of these biological changes are not, however, inevitable, and it is here that a difference can be made. Specifically, we need to place greater emphasis on disentangling our notions of fitness and health from our notions of youth. When being fit and healthy and well and even normal means to be like someone in their twenties and early thirties, then it is not surprising that the middle aged sometimes feel "over the hill." When people are told that "they are in fine health—for a person of fifty-five," they are being told more than a simple statement about their biological status. They are also being told they aren't as fit as someone younger. That's probably accurate, but the issue is one of the standard. When the standard is absolute—the typical status of a young adult—those middle-aged and older must feel a degree of age discrimination. If the status were relative, that is if each age group were judged relative to its own norms, this discrimination might be lessened. The statement "you continue to be in excellent health" has a less prejudicial tone to it. It tells fifty-five-year-olds that they continue to be in better health than their age mates.

The issue of age discrimination involves more than simply the basis upon which people are judged. It also involves a prescription for intervention that is controversial. The best example during the middle years is estrogen replacement therapy for postmenopausal women. The issue is as basic as our definition of normalcy. If the loss of estrogen with menopause is seen as a condition requiring intervention, then menopause becomes a condition in need of treatment rather than a normal part of the aging process. Although it is true that estrogen replacement therapy reduces the likelihood of some age-related diseases in women, it has also been linked to the possibility of increased risk of uterine cancer, and as Notman (1980) argues, it leaves the impression that a high estrogen level is a normal condition across the adult years and therefore should be maintained. She believes that this impression in turn reflects the equating of femininity with fertility. Such an unfortunate equation leaves the impression that there is something abnormal about growing older, that it is a condition needing treatment or correction.

One of the most important values of cross-cultural research is that it helps us understand the meaning of the events in our culture. One thing such research has consistently demonstrated (Troll, 1982) is that in those cultures where older adults are valued, the transitions to middle age and old age are less stressful. One significant task for each of us is making sure that the assets and liabilities of each stage in the life-span are presented in their own right and not solely in reference to those of the young adult years. Perhaps now that advertisers are generally willing to include minority group members in their ads, we ought to lobby for a rule requiring that for every "seductive-looking svelte young adult" pictured, there also be "a mature and androgynous but perhaps somewhat rounded middle-aged adult" pictured.

Facilitating Development during the Middle Years

Two issues loom as particular impediments to the enjoyment of what middle age has to offer. One concerns the variety of social class-related issues and the other the increasingly common problem of providing adequate care for aged parents.

Again, social class defines the degree to which individuals in middle age continue to worry about the issues typically encountered in young adulthood. The more advantaged the background is, the more likely the individual is to have adequate income and job security. The more advantaged the background, the more likely it is that the personal and occupational environment has continued to foster a sense of cognitive competence and intellectual flexibility (Schaie, 1983). The combination of these two assets, economic security and intellectual flexibility, makes possible the reflection on the three careers of the adult years (intimate, parent, and worker) and the subsequent adjustment in attitudes, values, and behavior patterns that seem correlated with happiness during the middle years of adulthood (Campbell, 1981).

The impact of social class, as presented in this chapter, is largely defined in terms of working conditions. Therefore, attempts to improve the lives of middle-aged adults must focus on their conditions of employment. Specifically, employment policies must place greater emphasis on job security with increasing seniority and on policies that counter the intellectual numbing that comes from highly supervised and structured roles. Shostak (1980) notes that as the work force continues to become better educated, the threat of demoralized and bored workers becomes greater. However, he also notes that the potential for gains in productivity and indigenous worker reforms are also greater. Which direction the workplace takes will depend on the degree to which employers foster policies consistent with our understanding of development during the adult years. And the extent to which this happens in part will depend on how well those committed to bettering people's lives advocate and lobby and support such reforms.

The care of aged parents is an issue that is likely to become more common in subsequent years. As discussed earlier in this chapter, concerns exist at a variety of levels. People with aged parents often must assume the financial burden of expensive medical care and supplement an inadequate retirement income, find time to meet their parents' needs and at the same time their own and those of their children and perhaps even grandchildren, and in some cases have their parents live with them. Cicirelli (1981) finds that one factor complicating the provision of assistance and care to aged parents is that the two generations think that different types of assistance are needed. Middle-aged children tend to think in terms of issues of health; aged parents tend to be more concerned about day-to-day problems such as crime and dealing with bureaucratic agencies. In other words, the children see themselves as available to help when large issues emerge; the parents want help in dealing with the more banal issues. Clearly one intervention would be to help each generation better understand the needs, limitations, and perspectives of the others. Cicirelli suggests that the degree of contact between generations is a major determinant of the "generation gap" and that programs or services that increase this contact will result in better services to the aged, a feeling of efficacy for their children, and a closer bond between the two generations.

Programs that bring together different age groups—such as this home-sharing program sponsored by the Gray Panthers—help to lessen isolation among the aged and strengthen the bond between generations.

Cicirelli also suggests that agencies can ease the burden of middle-aged children by providing a variety of supplemental or backup services. These services might include day-care programs for the aged, respite services for those requiring intensive care, noninstitutional services such as visiting nurses, housekeepers, and transportation. These services are invariably cheaper than custodial care, and they are more likely to maintain the sense of dignity and independence that is so important in old age. Further, they relieve the middle-aged child from many time- and energy-draining activities. Cicirelli suggests that instead of providing these services themselves, middle-aged children should assume the role of "service managers." These are the people who can best advocate for the needs of their parents, who have the best relationship with them, and who are most committed to making sure that the services are provided in a reasonable fashion. Cicirelli finds that the aged welcome having their children serve in this capacity and that middle-aged adults also value the role. Those who find themselves dealing with this population should clearly support the development and implementation of such programs.

The next chapter concerns old age. It is a stage of the life-span filled with a variety of images, most of them negative. Since old age is the time of death, it is sometimes hard to see it as an opportunity for continued development.

Perhaps these words from Longfellow's "Age Is Opportunity" will help set the stage for a developmental view of old age.

> But why, you ask me should this tale be told
> Of men grown old, or who are growing old?
> Ah, nothing is too late
> Till the tired heart should cease to palpitate;
> Cato learned Greek at eighty; Sophocles
> Wrote his grand *Oedipus,* and Simonides
> Bore off the prize of verse from his compeers,
> When each had numbered four score years,
> And Theophrastus, at four score and ten,
> Had just begun his *Characters of Men.*
> Chaucer, at Woodstock with the nightingales,
> At sixty wrote the *Canterbury Tales;*
> Goethe at Weimar, toiling to the last,
> Completed *Faust* when eighty years were past.
> These are indeed exceptions, but they show
> How far the gulf-stream of our youth may flow
> Into the arctic regions of our lives
> When little else except life itself survives.
> Shall we then sit us idly down and say
> The night hath come; it is no longer day?
> The night has not yet come; we are not quite
> Cut off from labor by the failing light;
> Some work remains for us to do and dare;
> Even the oldest tree some fruit may bear;
> For age is opportunity no less
> Than youth itself, though in another dress.
> And as the evening twilight fades away
> The sky is filled with stars, invisible by day.

SUMMARY

THE CONTEXT OF DEVELOPMENT

1. Our modern conception of middle age is a recent phenomenon, reflecting a decrease in family size and an increase in average life expectancy.
2. When compared to younger adults, those middle aged report greater life satisfaction and demonstrate a decrease in sex-role polarization.
3. Social class is a significant factor in determining the quality of the middle years, largely by reflecting the cumulative impact of the work career.

THE COURSE OF DEVELOPMENT

1. For males, the core component of the mid-life transition is occupation; for women, it more typically involves family and intimate relationships.
2. The cumulative impact of adult life-style is a significant determinant of the degree of biological change during middle age.
3. In general, the declines in biological status through the middle years are best described as a decrease in the capacity for work and a decrease in the body's ability to recover from stress or insult.
4. Most parents report feeling little distress about their grown children's leaving home. This generalization is more true when the parent-adolescent relationship has been conflict prone, less so when the marital relationship is seen as unsatisfying.
5. For an increasing number of middle-aged adults, soon after the primary parenting role ends, the responsibility of caring for aged parents begins.
6. Resolutions of the mid-life transition usually involve a greater androgynous sex-role orientation, a greater emphasis on the value of experience, and a stronger feeling of being in control of one's own life.
7. Most middle-aged adults welcome the grandparent role, seeing it as a source both of personal gratification and of family continuity.

THE SIGNIFICANCE OF THE MIDDLE YEARS OF ADULTHOOD

1. When compared to young adults, middle-aged adults tend to place a greater emphasis on the interpersonal aspects of their activities.
2. Sociohistorical circumstances are providing the present generation of middle-aged adults a distinctly different set of life circumstances from that experienced by previous cohorts.
3. Social class exerts a powerful influence on the nature of the middle-age experience. It influences the age at transition, the nature and degree of resolution, the sources of satisfaction, future opportunities for growth, and the maintenance of cognitive competence and intellectual flexibility.

HUMAN DEVELOPMENT AND HUMAN SERVICES

1. Middle-aged adults will increasingly find themselves in need of support services to help care for aged and often dependent parents.

KEY TERMS AND CONCEPTS

Projection
Rationalization
Acute Illnesses
Chronic Illnesses
Local Adaptation Syndrome
General Adaptation Syndrome
Type A Individuals
Type B Individuals
Empty Nest Transition
Menopause
Vasomotor Instability

SUGGESTED READINGS

The transition to middle age has taken on a "rite of passage" quality in recent years—one suggesting that if the transition isn't traumatic, it isn't much of a transition. The article by Datan helps put this exaggerated notion in perspective.

Datan, N. (1980). Midas and other mid-life crises. In W. H. Norman & T. J. Scaramella (Eds.), *Mid life: Developmental and clinical issues.* New York: Brunner/Mazel.

The cumulative impact of life experiences on both biological and psychological aspects of development becomes increasingly evident during middle age. These five reports provide support for this belief.

Goodman, M. (1980). Toward a biology of menopause. *Signs, 5,* 739–753.

Gribbin, K., Schaie, K., & Parham, I. A. (1980). Complexity of lifestyle and maintenance of intellectual abilities. *Journal of Social Issues, 36,* 47–61.

Haynes, S. G., Feinleib, M., Levine, S., Scotch, N., & Kannel, W. B. (1978). The relationship of psychosocial factors to coronary heart disease in the Framingham study. *American Journal of Epidemiology, 107,* 384–402.

Lowenthal, M. F., & Weiss, L. (1976). Intimacy and crisis in adulthood. *Counseling Psychologist, 6,* 10–15.

Selye, H. (1978). *The stresses of life.* New York: McGraw-Hill.

The resolution of the mid-life transition involves new perspectives on old roles, especially those relating to sex and work. Giele argues that roles significant in early adulthood become less so in middle age; Sarason suggests that mid-life might be an ideal time to consider a second career.

Giele, J. Z. (1980). Adulthood as transcendence of age and sex. In N. J. Smelser & E. H. Erikson (Eds.), *Themes of work and love in adulthood.* Cambridge: Harvard University Press.

Sarason, S. B. (1977). *Work, aging, and social change.* New York: The Free Press.

Middle age brings not only a change in self but also a change in relations to others. The article by Lewis notes the effect of children's leaving home on fathers, Sprey talks about the grandparent role, and both Cicirelli and Brody consider the consequences of caring for aged parents.

Brody, E. M. (1981). "Women in the Middle" and family help to older people. *The Gerontologist, 21,* 471–481.

Cicirelli, V. G. (1981). *Helping elderly parents: The role of adult children.* Boston: Auburn House.

Lewis, R. A., Freneau, P. J., & Roberts, C. L. (1979). Fathers and the post-parental transition. *Family Coordinator, 28,* 514–520.

Sprey, J., & Matthews, S. H. (1982). Contemporary grandparent: A systematic transition. *Annals of the American Academy of Political and Social Sciences, 464,* 91–104.

10
THE LATER YEARS OF ADULTHOOD

CHAPTER OUTLINE

A quiz to start this penultimate chapter. Read each of the following statements and decide which ones are true of those aged sixty-five and older and which ones are false.

1. Older persons have more acute (short-term) illnesses than persons under sixty-five.
2. Older persons have more injuries in the home than persons under sixty-five.
3. The life expectancy of men at age sixty-five is about the same as women's.
4. The aged have higher rates of criminal victimization than persons under sixty-five.
5. The majority of the aged live alone.
6. Older persons who reduce their activity tend to be happier than those who remain active.

7. The majority of old people are senile (that is, defective in memory, disoriented, or demented).
8. Most old people have no interest in, or capacity for, sexual relations.
9. At least one-tenth of the aged are living in long-stay institutions (nursing homes, mental hospitals, homes for the aged, and so forth).
10. Most older workers cannot work as effectively as younger ones can.
11. Most old people are set in their ways and unable to change.
12. The majority of old people are socially isolated and lonely.
13. It is almost impossible for most old people to learn new things.
14. Older people tend to become more religious as they age.
15. The majority of old people feel miserable most of the time.

Quizzes such as this one are really "teasers." That is, their purpose is to make a point rather than actually to test your knowledge. In this instance, the point is that most people's perception of older adults (including older adults themselves) is often inaccurate. All of these statements are false (Palmore, 1977; 1981). Even though the evidence clearly shows them to be so, most adults mark many as true. In the case where older adults mark the statements as true and then note that they are not true in their particular instance, they conclude that they must be the exception to the rule (Harris, 1975). In fact, they are more likely the rule than the exception.

Why are our perceptions so inaccurate, so stereotyped? There are probably two reasons. One is that we tend to hear and read most about those older adults who are the least fortunate. We come across stories about lonely, isolated, depressed individuals living in single-room hotels in the decaying centers of cities. There are such terribly unfortunate people, but the fact is that the vast majority of older adults live within a short drive of their family and even those who don't live close maintain frequent contact (Shanas & Sussman, 1981).

We see pictures of barren corridors lined with withered aged adults in wheelchairs, attended by uncaring, unscrupulous profiteers. Such places exist. They are as inhumane as they look. But less than 5 percent of the aged live in institutions, and of this small number, many have appropriate, quality care provided by caring, competent individuals.

We hear about retirement communities with names such as "Happiness Acres," "Golden Years Retreat," and "Sunset Village." We read so many stories about retired adults moving to warmer climates that I sometimes wonder why Florida hasn't sunk. Retirement communities do exist; their quality and nature vary greatly. Some retired adults do move to more temperate climates. But again, most don't move, and the ones that do usually move to a smaller dwelling in the very same community. Disproportionate images such as the ones just described, particularly when they concern those most in need, do serve the useful purpose of mobilizing necessary services, but as will be discussed more fully later, they also distort our image of what it is like to be old.

The second reason our perceptions are so inaccurate is that the nature of the later years of adulthood has undergone more change, and will probably continue to undergo more change, than that of any other stage of the life-span. Two of the ideas most commonly associated with being old, that an old person has lived for many years and that this is the time when he or she will die, are, like so many other aspects of the life-span, twentieth century associations.

In the past, being old was defined in terms of health and physical ability, not chronological age. There were no unique roles ascribed to those who had lived a long time. Although it was true that increasing age was associated with an increasing likelihood of debilitating illness, life was so harsh for so many people that many were "old" while they were still "young." Being old meant being a burden, needing the care of others. The stereotype of the old as a burden still exists today, even though the association is no longer present (Matthews, 1979).

Before 1900, death was virtually a random event (Fries, 1980). The harshness of life, poor health and sanitation measures, and the inability to control infectious disease combined to make death a part of all stages of the life-span, not just the last. Visit an old cemetery and notice the ages at death. They reflect the full range of the life-span. Notice how many infants died within a few days of birth, and in the plot next to each, how many mothers also. Notice how many people of different ages all seemed to die around the same date. Childbirth was once a perilous experience; epidemics, a common event. The dramatic increases in the average life expectancy over the past hundred years have not come about because in the past so few lived a long life (even in the days of ancient Rome, 20 percent of the population lived to the age of seventy). They have come about because once so many people died so young. Although few developmentalists think that our life-spans will ever be extended beyond the range of 80 to 120 years, most think that there will continue to be significant increases in the percentage of adults who are able to live their full life-spans in reasonably good health.

The thrust of the last few paragraphs has been to emphasize the fact that the images of the elderly many people hold are often inaccurate or, at best, disproportional. No doubt the fears that many people have of growing old reflect these misperceptions (Kalish, 1974).

At the same time, it is equally important not to romanticize old age. Old age is a time of loss more than a time of transition. In a culture such as ours that places value on productivity, the loss of the opportunity, through retirement, to be productive is a difficult change to accept. Even the more basic fact that there are few well-defined social roles for old people leaves many with a feeling that their lives have less meaning than before (Rosow, 1973).

Most adults sixty-five and older suffer from some sort of chronic ailment. In many cases, the severity of the disability may be minimal or moderate, but it is nevertheless present. Tasks that were once trivial, such as opening a jar, can now be a problem due to arthritis. Favored pastimes, such as reading, can now be a frustration due to failing eyesight.

Decreased income is also a common problem faced by the elderly. Social Security benefits combined with private pension plans rarely provide the same level of income that the individual had when he or she was employed. For those who have no private pension plan, especially women whose sole income is Social Security survivor benefits, making ends meet can be very difficult. Declines in health and income force many of the elderly to accept a greater dependence on others, especially adult children. This loss of independence and the self-esteem that typically accompanies it is a bitter pill for many.

The ultimate loss in old age is death. Perhaps when death was a random event occurring at all stages of the life-span, people were less likely to reflect on its meaning or its imminence. But death is now a part of old age, and most older adults think about it (Kalish, 1976). They confront the loss of friends. They grieve for the loss of a spouse with whom they have shared a lifetime. They confront the reality of their own death.

Inadequate retirement income limits the options of many elderly people, forcing them into substandard housing and increasing their social isolation.

The bulk of this chapter is divided into five sections. The first concerns the context of development during the later years. It will discuss the influence that social class, sex, and generation (cohort) have on the course of development during the later years of adulthood. The second and third sections examine the two relatively distinct periods of this stage—what Neugarten (1975; 1979) has referred to as the **young-old** and the **old-old**. Although the two phases overlap, the first begins with the retirement transition and lasts usually until the mid-seventies or until health begins to decline significantly. The old-old period follows, lasting until death. The fourth section reviews the significance of old age within the life-span, while the last section deals with the provision of human services to the elderly.

THE CONTEXT OF DEVELOPMENT DURING THE LATER YEARS OF ADULTHOOD

The nature of the later years of adulthood is undergoing such a radical change that it is difficult to make generalizations that hold for more than a single cohort. What is safe to say is that for the current cohort of young adult men and women from both economically advantaged and disadvantaged segments of the society, the experience of old age will be significantly different from that of their parents and their grandparents.

The Cohort Perspective

The continuing conquest of disease and the improvements in health, sanitation, and nutrition, as well as in the more general conditions of life, have allowed each succeeding cohort, or generation, of older adults to be larger than the previous one. At the same time, the declining birthrate of the past few decades (and the prospect that it will continue to decline) means that not only will the absolute number of older adults continue to grow but so too will their percentage relative to other age groups. In absolute terms, each year marks an increase of over three million in the number of older adults, when compared to the previous year. In relative terms, the percentage of older adults will increase from the present 11 percent to approximately 16 percent by the time the present generation of college students reaches retirement age (Karp & Yoels, 1982). Of even greater significance than the changes in the absolute and relative numbers of older adults are the changes that are occurring in their composition and roles.

The very notion of old age as a distinct stage of the life-span is a relatively recent phenomenon. It reflects the increasingly distinct association of death with long life and the evolution of a distinct role for older adults—retirement. In 1890, approximately 70 percent of those age sixty-five and above were still actively involved in the labor force. Of the remaining 30 percent, most were unable to continue work because of poor health. There were very few who were in the position to retire and live on previously earned income and investments (Treas & Bengtson, 1982). The percentage of adults in the work force beyond age sixty-five declined slightly to 55 percent in 1920 and to 40 percent in 1950. Today the figure is approximately 20 percent, and it is likely to continue to decline slightly in coming years. The decline is not due to increasingly larger percentages of older adults being in poor health. It is due to the creation of Social Security beginning in the 1930s, the growth of individual and corporate pension plans, and more generally the acceptance of the notion that retirement is an acceptable role for older adults.

In the future, the only difference between middle-aged adults in their fifties and the young-old defined by Neugarten (1975) will be the fact that the latter are retired. There will be relatively few changes in biological status between those in their fifties and those in their sixties and even early seventies. The significant biological declines that characterize the old-old will not be typical until the seventies or even the eighties. For those young-old that choose not to retire, this phase of the life-span will be a continuation of the developmental patterns of their middle age.

Successive cohorts of older adults will continue to be better educated than the ones that preceded them. In 1975, 63 percent of those aged sixty-five and above had not graduated from high school (Palmore, 1981). Compare this figure to that for the current generation of college age adults. By the time they reach sixty-five, almost half their number will have graduated from college and more than 75 percent will have completed high school. These educational statistics indicate changes of greater significance than a simple shift in the number of years of schooling completed. The discussions in the previous chapters have frequently made reference to the relationship of schooling to income and occupation, to attitudes and values, and to the maintenance of cognitive skills. The cumulative impact of these educationally related factors will be increasingly evident in the future status of the elderly.

There are also important changes in family life that will influence the patterns of living for future cohorts of the elderly. First, there will continue to be a decline in the number of multigeneration households. More older adults will either maintain their present housing or move into smaller units, but few will move in with grown children. There will be a significant increase in the number of single, older adults, and most of them will be women. The increase in the divorce rate over the past few decades (and the expectation that the rate will remain high) and the increasing difference in the life-spans of men and women (in 1900, women outlived men an average of less than one year; by 1975, the difference had increased to over four years) will result in a larger number of women living alone (Gillaspy, 1979). In 1980, 70 percent of all widows maintained their own residence. In 1950, the number was only 28 percent of that population (Treas & Bengtson, 1982).

The significance of these various cohort changes will be discussed in the fourth section. For now, however, it is worth noting that the impact of these changes will be evident not only in terms of the way future generations of older adults live their lives but also in terms of the way future generations of those under age sixty-five will be affected. As future generations of older adults become a more distinct group, as their absolute and relative numbers increase, as they become better educated and maintain physical vigor longer, and as they live longer, their needs for and effectiveness in obtaining services will probably increase. In communities that have a large retirement population, it is not uncommon to find newspaper articles describing debates over the use of revenues to finance a new elementary school or a new senior citizens' center. Hopefully, as a nation, we will allocate our resources in such a way that both can be built. But, in any case, the structure of future generations of older adults will clearly bring about a shift in the distribution of social services and resources.

The Social Class Perspective

In the most direct sense, social class defines life's chances (Bengtson, Kasschau, & Ragan, 1977). It reflects the resources that are available to each person, the values, attitudes, and behaviors most likely to be acquired, and the risks of exposure to various life circumstances. From this perspective, it might seem reasonable to assume that since the transition to old age involves a decline in income for most, it must also involve a decline in socioeconomic status.

In an absolute or statistical sense this is true, but the issue is more complex. More older adults don't perceive the transition to old age as involving a decline in social status (Streib, 1976). In part this subjective perception of continuity in social class status reflects the economic patterns of the elderly. The urge to spend and buy that is a characteristic of youth and the earlier years of adulthood is less evident in old age. The amount of disposable income may decline, but the ratio of "what I want to what I have to get it with" may stay the same.

A second reason for a high degree of continuity in perceived social status is that older adults' view of themselves in the social class hierarchy is not based solely on their present circumstances but on the cumulation of their life experiences and their goals and aspirations. Just as the struggling young married couple living on a very minimal income might describe

An individual's perception of social status is not always based on current conditions. This gentleman's shirt, tie, and pen in the pocket would suggest an image of himself based on years of successful employment rather than attendance at a senior center.

their present circumstances as only "temporary," retired professionals will report that they "are no longer practicing." In other words, they are still the lawyer, doctor, or professor.

Notwithstanding elders' views of themselves in the social order, social class continues to exert a significant impact on them. It influences their very perception of when they become old—around age sixty for those less economically advantaged and around age seventy for those more so. These differing perceptions continue the pattern of life-events timing (that is, age at marriage, age at birth of first child, age at completion of education) associated with socioeconomic status. The social clocks of Neugarten and Datan (1973) turn more quickly for the less advantaged.

The more economically advantaged your background is, the more likely you are to live longer and do so in good health. You will be less likely to feel lonely, to fear being a victim of crime, and to live in minimal housing. You will be more likely to be involved in community associations, to take an interest in world events, and to have a broader friendship network. You will not, however, be more likely to maintain contact with siblings, other extended family members, and children and grandchildren (Bengtson, Kasschau, & Ragan, 1977; Blau, 1981; Palmore, 1981).

These various correlates of social class membership are not unrelated. In fact, quite the opposite is true. Together they sum to produce a higher level of life satisfaction during old age in the more advantaged segments of society (Campbell, 1981). The association of life satisfaction and social class in old age is strongly mediated by health (a social class correlate) and by the quality of peer relationships during old age.

Peer relations assume a more significant role in the lives of the elderly than in the lives of younger people (Blau, 1981). For most of the elderly, the parent career has significantly lessened and work occupies a less central place in the life structure. Peers serve to fill the time and energy voids that these changing role commitments involve. If an individual is also widowed, this statement gains even greater meaning.

The differences in peer relation patterns across social class may partially explain why those from economically advantaged groups report a higher degree of life satisfaction during old age than those from less advantaged groups do (Campbell, 1981). Middle-class adults are more likely to be involved in social groups not directly tied to their work and marital status than are working-class adults. For this reason, these social contacts are less affected by retirement or the death of the spouse. The voluntary and reciprocal nature of these peer relationships provides older adults a mechanism for maintaining a sense of usefulness and self-esteem now that other roles are less available.

The research reviewed in this section gives the clear impression that social class variables may assume greater importance in old age than in middle age. Perhaps, however, not as great a role as sex.

The Sex-Role Perspective

The pattern of sex-role androgyny typical of the middle years is also found during the "young-old" phase of old age. Couples in retirement continue to overlap in their distribution of roles and responsibilities as well as in their interests and activities. The overlap is never total but is considerably greater than it typically is in most couples during the early adult years. The real differential impact does not become evident until one spouse dies.

Consider a few statistics. The **average life expectancy** for women is approximately, today, six years more than for men. By the year 2000, this six-year gap is expected to increase to eight years—women will be expected to live to age eighty-one, men to age seventy-three (Brubaker, 1983). To look at the issue from a different perspective, among males between the ages of sixty-five and seventy-four, only 8 percent are widowers. Of women in the same age bracket, 40 percent are widows. Among men over age seventy-five, only 22 percent have managed to outlive their wives. The women seem to have an "easier" time; 68 percent manage the feat. As striking as these survival figures are, they take on added impact when the remarriage statistics are considered. Men sixty-five years of age and older are *eight* times more likely to remarry than are women sixty-five and older (Bengtson, Kasschau, & Ragan, 1977). The remarriage statistics should not be surprising given our cultural tendency for men to marry younger women. But in addition many widows do not wish to remarry. Given the fact that many women report their marriage to be less satisfying than men report theirs to be and the fact that many report not wanting to nurse and watch another husband die, many prefer not to remarry even if they have the opportunity.

Why do women increasingly outlive men? They do so because the major causes of death in women—childbirth and cancers of the uterus and breast—have declined dramatically over the past decades. On the other hand, the major causes of death in men—car accidents, lung cancer, and coronary heart disease—have actually increased. Notice that these are not necessarily causes of death in the elderly. Rather, they are factors that contribute to the differential survival rates of 160 women to every 100 men by age seventy-five (Matthews, 1979). However, even if the leading causes

Forty percent of all women between the ages of sixty-five and seventy-four are widowed. The statistics are against most of them remarrying.

of death in men and women were equalized and even if the conditions of their lives were more similar, most experts (Brubaker, 1983) still predict that the average life-span of women would continue to be greater than that of men; in other words, there seems to be a significant genetic base to life expectancy (Eisdorfer & Wilkie, 1977; Everitt, 1983; Timiras, 1978).

Women may outlive men, but they do so at two costs (Bengtson, Kasschau, & Ragan, 1977). One is that they are much more likely to live alone than are widowers (36 percent of widows compared to 15 percent of widowers), and they are more likely to have a very meager income (45 percent of widows compared to 33 percent of widowers live below the poverty line). If you are now debating the relative merits of outliving your spouse, you should also keep in mind that most research reports that men find it more difficult to adjust to the loss of a spouse than women do (Bengtson, Kasschau, & Ragan, 1977). This relative ease of adjustment to the loss of a spouse reflects the more active social network of most women (an advantage enjoyed throughout most of the life-span), the fact that intergenerational family ties are usually maintained by women, and the fact that some men have never really acquired basic competence in many domestic routines. It will be interesting to see if this last issue becomes of less consequence as subsequent cohorts of men become more competent in household routines.

The Longitudinal Perspective

What is the relationship of the later years of adulthood to the middle years? The transitions of mid-life involve a fundamental shift in or at least a reevaluation of the values that guide the early years of adulthood. Most adults, as they move farther into the middle years, make shifts in their priorities and values. Does a similar pattern mark the transition to old age?

Unfortunately there is no simple answer to the question. First, so much change occurs during the period of old age that it is more hazardous to make generalizations about typical patterns for this period than it is to make generalizations about middle-aged adults. For example, at the level of individual personality, the evidence supports a notion of continuity (Atchley, 1980; Costa & McCrae, 1980; Palmore, 1981). The words one might use to describe someone in middle age would be equally appropriate to describe the person in old age. People who enjoy being with other people continue to enjoy being with other people; people who like to sit and watch television or enjoy a book continue to watch and read; people who are hostile and combative continue to be a pain to deal with. However, even though the same words are appropriate, implying continuity, the intensity often changes, implying discontinuity. On the other hand, even though intensity changes, implying discontinuity, the person's relative position remains the same, implying continuity. People may mellow in old age, but the feisty ones mellow less.

This absolute/relative differentiation is an important one to appreciate, especially when considering the elderly. There is a tendency to assume that since decline is an element common to all in old age, that all old people must come to be much alike. In reality, this is not the case. Variability between people in old age on most measures is as great as or even greater than it was at earlier stages (Maddox, 1979). Social services that reflect this error, that fail to appreciate the fact that variability is present at all stages of the life-span, are rarely very successful in meeting the needs of the elderly. These services tend to be very homogeneous, and as a result many people find what they have to offer inappropriate.

A second problem encountered in determining continuity between middle age and old age is the relative lack of role definition for the elderly. Ask people to describe the responsibilities of those in their thirties. They would most probably answer in terms of the three careers of adulthood, intimacy, family, and work. They might ask if the person you have in mind is male or female. They might qualify their answer to reflect the sex of the individual. In any case, they would have an answer for you, and they probably wouldn't have much difficulty in forming it. Now ask them about someone who is 65 or 70 or 75 or even 80. My guess is that their answers would be more varied and less specific and that they would take longer to form. Continuity both within and between developmental stages is partly a function of the social context in which individuals find themselves. In old age, individuals find themselves in a social context with poorly defined expectations, obligations, responsibilities, and opportunities (Atchley, 1980). For those who have always been self-directed, who have an internal locus of control, this lack of clear social roles presents little problem; many even welcome being free of social expectations. But for others, especially those whose sense of self is defined in terms of social roles and social expectations, old age can present serious problems, often including depression (Atchley, 1980). For the self-directed elderly, the lack of socially defined roles may actually foster continuity. The self-directed are free to chart their own course and, like Livson's (1981) nontraditional middle-aged adults, seem to grow as social expectations become less rigid. In those whose sense of identity reflects the expectations of others, there may be significant changes in behavior, even a degree of social breakdown (Kuypers & Bengtson, 1973) as they no longer have clear standards to guide their behavior.

A third problem further clouds the issue of continuity. It is sometimes difficult to distinguish descriptions of behavior in old age from prescriptions of behavior in old age, prescriptions that offer advice on "successful

For many elderly, social isolation becomes a serious problem.

aging." These prescriptions differ largely in terms of the degree of continuity with middle age they recommend. The two best-known views are the activity and disengagement theories of aging. Advocates of the **activity theory of aging** encourage a maintenance of middle-age life patterns into old age, hence a high degree of continuity.

Advocates of the **disengagement theory of aging** (e.g., Cumming, 1975; Cumming & Henry, 1961; Williams & Wirths, 1965); believe that there should be a process of mutual severing of the ties between the aging individual and society. This mutual severing should reflect the decreasing expectations society has of people in old age and the reduced obligations people feel as they become old. The severing of bonds is seen as consistent with the increasing introspection typical of old age, with the sense of turning inward (Butler, 1974; Erikson, 1980), as well as with the decreasing capacities of the elderly. Cumming (1975) sees the disengagement process as a narrowing of the life space, an individualizing of the reward structure, a decrease in role obligations, and a further shift toward expressive rather than instrumental roles. The result, to Cumming, is "a mutual circular freeing process in which society and the individual arrive at an equilibrium different from that of middle age both in amount and kind of involvement" (1975, p. 189).

Although advocates of the two positions sometimes place themselves in adversarial roles, the two views are not really contradictory or extreme. Activity theorists are no more suggesting that the elderly should play shuffleboard until they drop than disengagement theorists are suggesting that they should lie in the street and wait for the first car to come along. Rather each is trying to understand how best to maintain a sense of equilibrium within a context of changing social roles and changing individual capacities. It is likely that as the later years of adulthood continue to differentiate into two relatively distinct periods, each view will find an appropriate niche.

THE YOUNG-OLD: THE COURSE OF DEVELOPMENT DURING THE FIRST PHASE OF OLD AGE

Neugarten's (1975) differentiation of old age into two relatively distinct periods is based on two considerations. The first is the biological status of the individual, and the second is the survival of both spouses. The criteria are not all-encompassing; for example, they don't consider those individuals who are either divorced, have never married, or have lost a spouse previous to old age. Nevertheless, the notion that there is a period in the life-span, beginning as early as the late fifties or early sixties and often lasting into the middle seventies, during which individuals maintain reasonably good health and vigor and continue to maintain their marital status, is an important one to recognize. As mentioned previously, this period is essentially a continuation of the developmental patterns found in middle age, and were it not for one important difference, there would perhaps be little reason to discuss the young-old separate from the middle aged. This one important difference is retirement.

Retirement

The idea that society is rich enough to support nonworkers is a recent phenomenon, unique to industrialized nations. The legislation creating the Social Security System was only enacted in 1935 (Torrey, 1982). Public pensions did exist in this country and in other countries prior to 1935, but they were based on disability rather than age. The idea that people who are still able to work need no longer do so is relatively new.

Retirement is a controversial issue. There is controversy over the cost of retirement. Since the money for retirement benefits comes not from workers putting money into a retirement fund over the course of their employment but rather from deductions taken from current workers' paychecks, the increasing percentage of the old in relation to the young has steadily resulted in larger and larger deductions. At the same time, as society continues to impress older adults with the notion that retirement is a reward earned for years of contributing to society, retired adults lobby for larger benefits.

There is conflict over the mandatory aspects of retirement. There appears to be little if any justification for **mandatory retirement** policies based on age (Atchley, 1976; 1980). There may be certain jobs that require a level of stamina or reaction time or physical fitness (for example, police officer or airline pilot) that is less likely in old age, but they are clearly the exceptions. If there is any justification at all for mandatory retirement, it rests more with the need to make room for new workers than with the declining abilities of older workers. The raising of the mandatory retirement age in

While the majority of older workers do not resent retirement, many do resent being forced to retire. However, as people live longer and the costs of providing retirement benefits increase, it is likely that mandatory retirement ages will be dropped or at least relaxed.

many sectors of the economy reflects the increasing recognition of the competence of older workers. It also reflects, however, the cost of providing retirement benefits to more and more people living longer and longer lives.

Even though most older workers resent the notion of mandatory retirement, they don't resent retirement itself. Of those who are able, as many as 20 percent choose early retirement (Foner & Schwab, 1983). Most who work in settings having mandatory retirement policies report that they want to retire anyway and therefore don't feel forced out. Such statements may simply be a way of putting the best foot forward, but they nevertheless represent an important step since the relative ease of the transition to retirement is partly a function of whether the move is perceived as voluntary (Kimmel, Price, & Walker, 1978).

The Retirement Transition Sequence

The retirement transition sequence follows a pattern similar to other transition sequences across the life-span (Golan, 1981). There is a preretirement phase, which in a sense begins with the first Social Security deduction from a person's paycheck but becomes more deliberate during middle age. Discussions may involve early retirement or a move. They may involve plans to travel or do things not possible up until retirement. Usually these plans are made informally: most employers offer little in the way of formal, systematic retirement planning for their employees (Atchley, 1980). As in all preparatory phases, there is somewhat of an idealized quality to the planning. Discussions about retirement are no more likely to involve contingencies for serious illness than those of an engaged couple are likely to involve the distribution of child care. But in both cases they begin to orient the individuals toward the new role.

The retirement itself may be marked with a celebration, a formal presentation, or a family event, or it may not. The fact that there is no clear marker for the transition (compared to a wedding ceremony) reflects the ambiguity that surrounds the role of retired adult in our society.

All transitions seem to have a honeymoon phase usually followed by a period of disenchantment. Retirement is no exception (Atchley, 1976).

The length of the honeymoon is variable, but it involves doing all those things that you have been wanting to do for years but couldn't. Unfortunately even the most rabid golfer gets tired of chasing a ball; the most ardent traveler, of the open road. Somehow, the job you couldn't wait to leave doesn't look quite so bad after you've left it. It did provide a sense of order in your life. It did offer a sense of camaraderie. Suddenly couples who haven't seen each other on a weekday at eleven in the morning for the past thirty years find themselves tripping over each other. What is happening is that routines that have been established over a period of many years—routines that defined schedules and patterns of relationships—are changing. New, more individualized routines must be established. For many, these new routines involve some sort of part-time work. Others become involved in civic or voluntary activities. Some establish comfortable social patterns; others come to enjoy a genuine sense of leisure. The specific pattern is less important than the fact that a sense of routine is once again established. This newly established period of stability continues until serious illness, the death of a spouse, or financial hardship forces the elderly couple or individual to assume a more dependent, less self-directed role. This increased sense of dependency marks the transition to Neugarten's (1975) old-old period.

Variables Influencing the Retirement Transition Process

The preceding description of the retirement sequence is somewhat idealized. It does not occur for all people in the same manner. Rather a variety of factors serve to influence the rate and ease of the transition as well as its very continuation.

Work history. Probably the most central variable is the work history of the individual. Those who identify closely with their work and whose work is highly individualized tend to retire later than those with routine, impersonal jobs (Sheppard, 1979). Other things being equal, the assembly-line worker retires before the plant manager. However, the blue-collar worker's enjoyment of retirement is often tempered by limited financial resources and fewer opportunities through work to acquire the self-directing skills useful in retirement (Atchley, 1976).

It is important to remember that people retire from particular jobs, not from work. Many retirees use their new freedom to pursue new lines of work, particularly if they were able to retire after a predefined number of years of service rather than at a particular age. For example, those in the military can retire after twenty-five or thirty years of service. If they entered the service at age twenty, they are only forty-five or fifty when they retire. Others look for work as a means of supplementing retirement benefits or simply as a means of filling idle hours. Those with unique skills or professional backgrounds may choose not to retire completely but to continue to work part-time. Meltzer (1981), for example, found that many lawyers continue to maintain their practice of the law but do so in such a way as to reduce the more stressful aspects of the profession. They typically reduce their caseloads, especially limiting high-stress criminal cases. Further, they tend only to provide legal services to former clients, therefore reducing the possibility of adversarial contacts within the lawyer-client relationship. No doubt many physicians, academics, accountants, and other professionals pursue a similar course.

The work history is closely tied to another factor mentioned previously. This factor is the individual's perception of retirement. First, does the person want to retire? Second, is retirement mandatory? The more likely the individual is to view the retirement decision as voluntary and desirable, the better the transition is likely to be. This seems particularly true for high-status individuals who have a great deal of personal investment in their work (Kimmel, Price, & Walker, 1978). Peers can be an important influence on the retiree's perception (Atchley, 1980). The comments of coworkers who have retired a few years earlier as well as the support of those still working can strongly influence the way the person feels about the decision. Distaste for mandatory retirement can be greatly eased by previous retirees' assurances that "it's the best thing that ever happened to you—you'll love it."

Health and financial resources. Health and financial resources play a major role in determining life satisfaction during retirement. Many people retire because of poor health, and for them some of the potential benefits of retirement are unobtainable. On the other hand, those who worked in hazardous settings often report an improvement in their health (Streib & Schneider, 1971). Those who are in poor health at retirement are more likely to come from less economically advantaged backgrounds and therefore suffer the added complication of limited income (Szinovacz, 1982). The combination of limited health and limited money seriously restricts the range of activities available to the retiree, making the transition adjustment more difficult.

Even for those in good health, money is a serious concern in old age. Only about one-third of all current retirees are able to supplement their Social Security benefits with other pensions. This percentage will certainly increase in coming years, but even so most people in old age find themselves learning to live on less. There are, however, a number of factors that help to offset the financial burden of old age.

The elderly are allowed to take double exemptions on their income tax. Their Social Security benefits are not taxed. Those who sold a house after one spouse turned fifty-five are not required to pay capital gains tax on the profit from the sale. Many state and local governments offer additional financial benefits, and many businesses offer "senior citizen discounts." As in earlier stages of the life-span, the issue of money often has as much to do with one's perception of adequate income as it does with the actual income itself (Campbell, 1981).

The impact of reduced income takes on added proportions when the retiree is female (O'Rand & Landerman, 1984; Szinovacz, 1980; 1982). Because the work histories of women involve more interruptions, because their final salary levels are less, because they often find themselves stuck in lower-track positions, and because they often enter the work force later, they are less likely than men are to qualify for pensions, and for those that do qualify, the benefits are lessened because of their work histories. Federal legislation is currently being enacted that will reduce the negative impact of women's work histories. However, it will be many years before the full impact of these changes will benefit retired women. In the meantime, many women, particularly those who are single, find themselves in retirement at a serious economic disadvantage.

With the exception of the economic issue, the retirement patterns of men and women are remarkably similar (Atchley, 1976; 1980; Szinovacz, 1982). Investment in work, health, and perception of retirement have the same influence on retirement adjustment for women as they do for men. The research of Atchley and the research of Szinovacz ably put to rest the outmoded notion that since women enjoy domestic tasks more than men do, they find little difficulty in making the transition. Women who are involved in their work, irrespective of whether they have worked all of their adult lives or returned to work when their children were grown, are as likely to complain of a lack of meaningful activities or productive use of time as are men.

Retirement creates a variety of new circumstances for individuals. A major one is the use of time. Those who are able to decide how to use their time, who have the resources to carry out their goals, and who derive pleasure from the activity, make good adjustments and derive great pleasure from their retirements. It seems relatively unimportant what the activities are. Sitting on the porch watching the grass grow can be as satisfying, if that's what a person wants to do, as spending many hours in volunteer work.

Couple and Family Relations during Retirement

Marital satisfaction. Retirement, by changing an individual's social roles, also changes couple relationships. In general, those couples who have well-established companionship patterns, patterns involving a history of shared activities, who do not have a highly sex-typed division of labor, and whose decision-making and conflict resolution patterns primarily involve negotiation rather than power welcome retirement as an opportunity to further their relationship (Ade-Ridder & Brubaker, 1983; Atchley, 1980; Atchley & Miller, 1983; Keith & Brubaker, 1979). Day-to-day activities such as shopping are seen as simply an additional opportunity to enjoy a shared activity. Since time is of little concern, these couples often make a full day out of what was once a hurried activity.

For those whose marriages do not involve any of the characteristics described in the preceding paragraph, retirement often brings about a decrease in marital satisfaction. Wives may expect husbands, now retired, to assume a greater share of the housework. Husbands, unaccustomed to such a role, are often unwilling. The situation may be further complicated by the fact that declining health may prevent some men from doing the more traditional, more physical tasks such as yard work or exterior maintenance (Dobson, 1983). Since marital patterns tend to correlate with social class, there is evidence that the more economically advantaged the background, the greater the likelihood that marital satisfaction will improve (Atchley & Miller, 1983). However, the correlations are far from perfect, and therefore one needs to be careful in making generalizations about social class and marital satisfaction in old age. For example, Dobson (1983) notes that professional men who were highly involved in their work rarely find involvement in household routines much compensation for their former roles.

Intergenerational family relations. Intergenerational family relations during the young-old period don't seem very different from what they were during the later years of middle age (Beckman & Houser, 1982; Walker & Thompson, 1983). The grandparent-grandchild patterns continue as they did, although as grandchildren grow older, they may be less interested in maintaining the tie. Older adults continue to value contact with grown children and grandchildren, but a frequently noted finding is

Generational differences and the growing independence of adolescents may weaken the bond between grandparents and older grandchildren.

that they don't always derive a great deal of emotional satisfaction from family contacts (Lee & Ellithorpe, 1982). Parents maintain frequent contact with their grown children, each expresses concern and interest in the activities of the other, each gives aid to and receives aid from the other, each enjoys spending time with the other, but the relationships are not often described as intimate. A distancing seems to occur. Perhaps parents, not wanting to be a burden, maintain an emotional distance. Perhaps parents are only able to communicate effectively to a peer the feeling of growing old. Perhaps parents wonder if children maintain contact out of a sense of obligation rather than a sense of desire.

One possible explanation for this emotional distancing is that as parents age, the declines in health and, for some, financial status heighten each generation's awareness of the complementary nature of the parent-child relationship. Complementary ties are unequal in that neither can assume the role of the other. In childhood, such ties define the rights and obligations of each. As children grow and establish independent life-styles, the obligatory aspects of the relationship decline. Each generation is reasonably able to care for its own needs, to meet its own obligations. To be sure parents continue to help children and children continue to help parents, but what may be different about these exchanges is that they become more voluntary. Parents are obligated to feed and clothe their children. They have no such obligation to babysit for their grandchildren. The parent-child bond becomes more reciprocal, more peerlike in character.

But the continuing aging of the parent (or other family changes such as the divorce of the adult child) shifts the relationship away from the reciprocal pattern back again toward a complementary pattern. As roles again become more distinct and unequal, the ease of emotional exchange across generations lessens. Parents don't want to burden their children with the problems of growing old. Children don't want to burden parents with the problems of being the generation in the middle. Parents and children do not stop loving or caring for each other; they don't stop taking pride in the accomplishments of the other or delighting in being with the other. But the actual or even anticipated increase in dependency needs, for many aging parents, may place barriers across the parent-child exchange.

Sense of Self during the Retirement Transition

About two-thirds of all adults, both men and women, report little difficulty in making the adjustment to retirement and in establishing satisfying activity patterns during the early phase of old age (Atchley, 1980; Szinovacz, 1982). This group has a number of things in common. First, they were not "obsessively" wedded to their work. They were able to let go when they chose to or had to. Second, they continue to enjoy relatively good health and financial status, which suggests that they probably represent the more economically advantaged segments of the social class spectrum (Campbell, 1981). Third, and perhaps most important, they are able to maintain a sense of purpose in their lives (Ward, 1979). These older adults report that they feel they continue to have a degree of personal control over the course of their lives (Ward, 1977), and that they feel they are doing something they enjoy doing (Markides & Martin, 1979). These enjoyable activities may be of the leisure/recreation genre or more work related; they may involve a very busy schedule or a very casual one. Again, the specifics appear less important than the elderly's perception of them.

What about the other third? What about those who do not make a successful transition or who have considerable difficulty in doing so? Ultimately, the question of adjustment concerns more than the one-third who have trouble with the retirement transition. Since life satisfaction is tied to a sense of control over one's life and, in turn, this sense of control is tied to health, the question eventually concerns all the aged. As a cohort confronts the declines in health that accompany the aging process, more and more people will potentially find it difficult to maintain an acceptable degree of autonomy and self-direction and may increasingly turn to others for direction and role definition (Rosow, 1973). In a society such as ours that continues to have difficulty defining a clear role for the aged, the potential for what Kuypers and Bengtson (1973) have labeled the **social breakdown syndrome** is great for all older adults. The circumstances for such a breakdown are well described by Rosow (1973).

> Even though people are classified as old, they have almost no duties. Consequently, they tend to live in an imperfect role vacuum with few standards by which to judge themselves and their behavior. Others have few expectations of them and provide no guides for appropriate activity. There are only platitudes: take care of yourself, stay out of drafts, keep active, hold onto the banister, find a hobby, don't overdo, take your medicine, eat. The very triviality of these bromides simply documents the rolelessness of the aged, the general irreversibility of the losses, and their ultimate solitude in meeting their existential declines.
>
> In this sense, it is virtually impossible for them to be literal role failures. This is not necessarily reassuring, however, for psychologists know that unstructured situations create anxiety. Certainly with a broad horizon of leisure and few opportunities, many old people feel oppressively useless and futile. They are simply bored—but not quite to death. (p. 121)

The specifics of the social breakdown syndrome are detailed by Kuypers and Bengtson (1973) in a seven-step sequence. Step one, the precondition of susceptibility, refers to the degree to which individuals are unable to maintain an independent locus of control—the degree to which

Programs like SCORE—the Service Corps of Retired Executives—provide opportunities for retired adults to use their knowledge and experience in helping others to develop business opportunities.

they are unable to maintain an independent sense of direction and purpose. Most of the young-old are able to maintain an independent locus of control, but as people continue to age, more and more of them find it difficult to do.

Step two begins when those unable to maintain a degree of self-direction become increasingly dependent on social cues given by others. As long as these cues provide a clear **role definition,** then these people are able to maintain a sense of identity and a sense of purpose. The only difference is the source of the direction.

The potential problem for the aged comes at step three. The external cues provided the aged tend either to be vague or to portray individuals as incompetent. In other words, they either provide little foundation on which the elderly person can build a sense of identity, or they provide an identity that is negative and self-deprecating.

> From the dominant functionalistic perspective of Western society, the elderly person is informed—directly and indirectly—of his uselessness, obsolescence, low value, inadequacy, and incompetence. To the degree that these specific messages are conveyed and to the degree that the elderly person—rendered susceptible—adopts them as true for the self, a cycle of events is established which leads to a generalized self-view of incompetence, uselessness, and worthlessness. (Kuypers & Bengtson, 1973, p. 189)

Steps four through seven of the sequence then complete what becomes a self-fulfilling prophecy (see Figure 10-1). Old people begin to feel that maybe they really aren't able to do things for themselves any longer

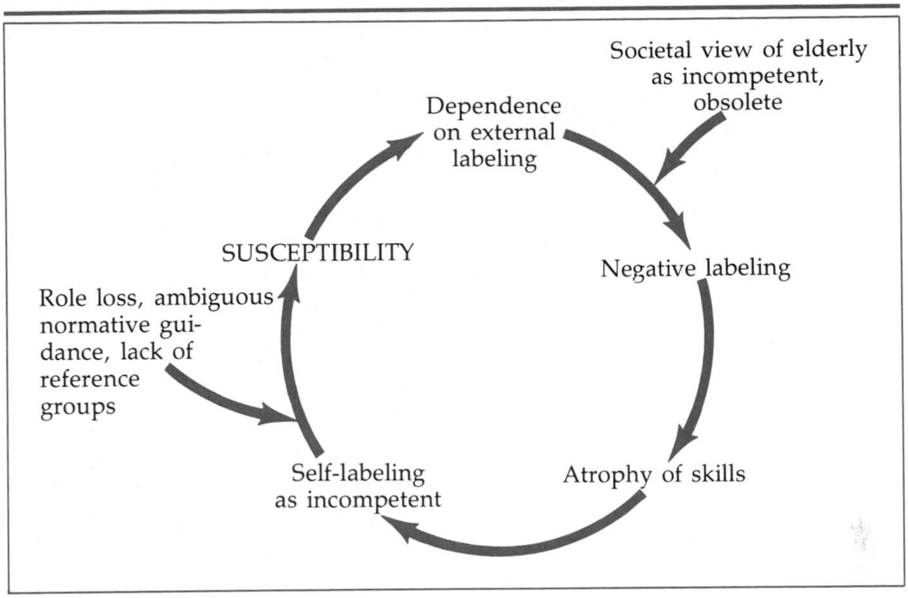

FIGURE 10-1
The social breakdown syndrome (from Kuypers & Bengtson, 1973).

(step four) and to acquire a set of behaviors akin to the dependent, help-seeking behavior of children (step five). As others take over tasks previously done by the elderly individual, individual competence does decline (step six) and finally the elderly person's self-concept is one of inadequacy, dependence, and incompetence (step seven).

The social breakdown syndrome is not the inevitable fate of the aged, but it is a sequence that describes the fate of a minority of the young-old and certainly a larger number of the old-old. Preventing or at least reducing the consequences of social breakdown requires a heightened awareness of the declining capacities of the very old and the necessity of modifying their environments in such a way that a positive self-concept can be maintained throughout the entire life-span. As the next section describes, however, this is not an easy task.

THE OLD-OLD: THE COURSE OF DEVELOPMENT DURING THE SECOND PHASE OF OLD AGE

This section on the old-old first reviews the variety of biological and cognitive changes that occur in the very old and that eventually result in death. Included here are explanations for these biological and cognitive changes and a discussion of how the aged confront death. It then reviews the grief, mourning and recovery process experienced by the surviving spouse, typically the wife. A last part explores how family, friends, and others provide support for the aged.

The Primary Aging Process

Biological decline and finally death are the most prominent aspects of this last period of the life-span. These inevitable events will probably occur later and later in the life-span as our knowledge about the biological, psychological, and social aspects of development continues to grow, but, again, they are inevitable, For this reason, it is sometimes difficult to think about the positive developments that seem unique to this time, that is, the life review process and the finding of new strength to cope with personal loss. Perhaps because it is difficult to think of new growth at the end of life, there is a tendency to simply patronize the very old, to make them comfortable. This is a pity because even in the process of dying, one's understanding of self and others continues to evolve.

Biological Aspects

The sensory and motor system declines that were discussed in the last chapter continue into old age. It is the rare individual past age sixty-five who does not suffer from some chronic ailment or decreased sensory or motor capacity. However, for most young-old, the effects of the **primary aging process** are not so severe that their ability to maintain everyday routines is seriously impaired (Eisdorfer & Wilkie, 1977). Those who are seriously impaired during this period more typically suffer from conditions associated with cumulative life experiences or secondary aging factors such as cancers that are not necessarily age related.

Sensory acuity. None of this, however, is meant to imply that the age-related deficits are of little consequence. The continuing decline in visual capacity, for example, renders the old particularly vulnerable to glare. This vulnerability occurs because the lens continues to become opaque and causes light entering the eye to scatter. Cataract surgery, the replacement of the lens, is quite common and usually alleviates the problem (Rockstein & Sussman, 1979). In addition to the problem of cataracts, the elderly are further bothered by the decreasing sensitivity of the retina, which causes decreased color discrimination and difficulties in seeing under low illumination.

Problems of hearing are common. In addition to the loss of sensitivity to high-pitch tones, the elderly have problems in discriminating between sounds, including speech, and also have difficulty in localizing the source of a sound (Rockstein & Sussman, 1979). Other sensory systems also show a gradual progressive decline. The ability to discriminate smells decreases as does the sense of taste. It is not uncommon for the old to complain about tasteless food and to make greater use of seasonings. The problem is particularly distasteful for those on restricted diets.

Motor skills. The skeletal system continues to decrease in mass, partially causing the stooped appearance of many old people. The aged are more susceptible to fractures, and when they occur, they take longer to mend than they do in younger people and often mend incompletely. A walk down a snow-covered sidewalk can be a perilous journey for some older adults.

The continuing decrease in muscle fiber also adds to the stooped appearance of the elderly as well as further reducing their strength. Movement is further impeded by the stiffening of joints and the loss of joint cartilage (Weg, 1975).

Organ systems. Basic biological systems also continue to decline during old age. The decreasing flexibility of muscle and cartilage reduces the ability of the rib cage to expand. As a result, the maximum amount of air that can be inhaled is reduced. Blood vessels continue to thicken and further lose elasticity. As a result, blood pressure increases, making a stroke a more common risk. The heart works less efficiently. Its walls become less elastic and more fibrous. The heartbeat slows, and each beat pumps less blood. To complicate matters further, the blood supply to the heart itself is often reduced, resulting in chest pains (angina), shortness of breath, and if severe enough, heart attack. Problems of the cardiovascular system are the single leading cause of death in old age (Rockstein & Sussman, 1979).

As marked as each of these biological changes may be, it is their cumulative effect that eventually produces serious incapacity and death. With increasing age, as individual organs lose their reserve capacity and become progressively less able to respond to changes in other organs, the body becomes progressively less able to maintain homeostasis and to respond adequately and quickly to stress, particularly when stresses are multiple (Palmore, Cleveland, Nowlin, Ramm, & Siegler, 1979; Shock, 1977). In effect, death occurs when the biological systems are no longer able to maintain an adequate level of functioning. Thus, although the "final" cause of death may be quite specific, the "actual" causes are more extensive.

Even in the absence of disease, biological functioning declines during old age.

Well before death, many of the very aged find themselves severely limited in what they are able to do. Simple tasks such as climbing stairs, getting about the house, and going outside become insurmountable barriers. Personal grooming tasks such as dressing, washing, bathing, and even cutting toenails become very difficult (Matthews, 1979). Most of the very aged who are in institutional care are there not because of serious or acute medical problems but because the cumulative effect of chronic conditions has left them unable to provide for their everyday needs.

Cognitive Aspects

The primary aging process is not restricted to physiological changes. There are also changes in cognitive functioning. These changes are of two varieties. The first involves specific decrements in ability level that have been associated with increasing age. The second involves changes in the way older adults confront tasks, that is, the way the elderly compensate for the declines that accompany old age.

Most of the declines are not large in magnitude. There is little evidence that any of the changes have a significant effect on the abilities of most individuals in their sixties and seventies. In fact, on the basis of longitudinal research, Schaie (1983) believes that it is not until age eighty that the average older adult will fall below the middle range of performance for younger adults.

Tasks demonstrating declines in performance. Where declines have been noted, they mostly involve speed of performance. Older adults show deficits on tasks that are timed; when they are allowed to work at their own pace, the deficit lessens (Arenberg & Robertson-Tchabo, 1977). For example, consider a task in which the subject is required to associate one of three numbers with one of three symbols. Each time one of the symbols is flashed on a screen, the subject has to press one of three buttons, each having one of the three numbers written on it. Older adults who are given a limited amount of time to make their choice make either more **errors of commission** (by pressing the wrong button) or more **errors of omission** (by failing to press any of the three buttons in the allotted time) than younger adults do. However, if both age groups were told to take all the time they needed before making their choice, there would be little if any difference in the performance of the two age groups.

Older adults show deficits in performance on tasks that require information processing with unfamiliar objects. Such a task might involve deciding which of four three-dimensional objects, when rotated, would look like the original (Siegler, 1983). Older adults show deficits on tests that involve making inferences, that is, drawing general conclusions from specific bits of information (Hartley, 1981). In such reasoning tasks, the elderly are less likely to consider all of the relevant information and are less likely to shift hypotheses when one is shown to be incorrect. This seeming "rigidity" is certainly detrimental to problem-solving skill, but as will be discussed shortly, it may be an adaptive component of the compensations the elderly make to the deficits encountered in old age.

The fact that the declines in ability noted among the elderly are more evident on laboratory tasks involving uncommon materials and procedures than they are in more natural settings involving everyday information has led to the speculation that the deficits are more apparent than real—that they are an artifact caused by the conditions of the experiments. As appealing as this hypothesis is, it is probably incorrect. The declines in information-processing ability are real but are of little consequence for well rehearsed, frequently used everyday knowledge and skills (Perlmutter, 1983). However, when the aged are in situations requiring the learning of new skills, they may be at a particular disadvantage. Research on memory change in old age has helped uncover the reasons for this decline in information-processing ability.

Changes in memory functions. The act of remembering involves three aspects. The data must first be organized or encoded in such a way that they can be placed in memory. Second, they must be stored, and third, there must be some process by which the data are brought out of memory, or retrieved (Smith, 1980). Most of the evidence indicates that the age-related deficit is in terms of the first aspect. With increasing age, there seems to be a decreasing ability to effectively organize information in such a way that it can be easily remembered and then retrieved when needed (Craik, 1977).

The encoding deficit is evident on learning tasks that require some form of information processing for efficient solution (such as, remembering diverse objects by grouping them into general categories such as color or shape) and on recall tasks that require efficient encoding for solution (the elderly have considerably more difficulty in recalling a correct answer than they do in recognizing the correct answer among a choice of possible answers). The question of whether the elderly are no longer able to encode efficiently or are simply less likely to do so is yet to be answered although the evidence tends to favor the second explanation for all but the very aged (Perlmutter, 1983). Since it appears that the elderly are less likely than younger people are to use efficient information-processing strategies to encode information, the next logical question must be why.

Changes in information-processing strategies. The answer provided by a participant in a study on conservation ability in old age conducted by Roberts, Papalia-Finlay, Davis, Blackburn, and Dellman (1982) offers some insight as to why the elderly are less likely to use efficient information-processing skills. Conservation problems were used by Piaget as a way of determining if children had acquired certain notions of permanence. For example, showing children two equal-numbered groups of checkers, each arranged in a different pattern, and asking if the number of checkers in one group is equal to that in another is a way to determine if children are aware that quantity is invariant irrespective of spatial arrangement, that is, of determining if they can conserve number. Preschoolers usually are unable to conserve number; school-age children rarely have a problem.

The reason these tasks were administered to the elderly has to do with a debate within the research community concerning the nature of cognitive skills in old age. There is a body of evidence indicating that the elderly are not able to solve conservation problems that present little challenge to children (see Denney, 1982). Three explanations are possible. One concerns a cohort effect. The current population of the elderly has never had the opportunity to acquire the categorical reasoning skills needed to solve conservation-type problems. If so, future cohorts, formally educated, should have little difficulty with the task. A second possible explanation concerns some sort of "undevelopment." This argument suggests that, with age, previously acquired skills become lost, that people become more childlike in their reasoning as they grow old. The third explanation gets us back to the participant in the study.

The particular problem under investigation was conservation of area. If an equal number of same-sized objects are arranged in different patterns on two different pieces of paper, is the amount of area covered on each piece of paper the same? For the actual experiment, the description was of same-sized barns on a pasture. In one case the barns were adjacent to make one large barn; in the other, they were spread apart. The actual question concerns whether cows grazing on the two fields would have the same amount of grass to eat. The "correct" answer is that since the same amount of area is covered, the same amount of grass must be available to the cows. The particular participant in question said, "no two fields ever grow grass the same way."

The point of all this is that the shift toward valuing experience over logic first evident in middle age may, by old age, become so dominant that the elderly find it difficult to deal with the hypothetical, logical problems typical of information-processing research. Perhaps as a compensation for declining sensory, motor, and physiological capabilities, the elderly favor a cognitive style that places emphasis on experience, on broad generalizations, and on well-rehearsed routines and deemphasizes those tasks that require speculation, hypothesis testing, an emphasis on detail, or shifts in orientation (Labouvie-Vief, 1982). As Hussian (1981) suggests, the aged may no longer seek or prefer newness but may instead prefer sameness as a way to conserve energy. Supposedly better ways are not genuinely better if they require an expenditure of energy, when energy is limited, to make the transition. What appears to some as a decline in cognitive skills may simply be a reasonable adaptation to the circumstances of old age.

Interplay of biological and cognitive changes in old age. These cognitive adaptations are not very evident among the young-old but become more pronounced among the old-old. The specific link between these cognitive changes and more basic physiological mechanisms is still unclear, but one likely link is through the reduced blood supply to the brain caused by declines in the cardiovascular system (Granick & Friedman, 1973; Woodruff, 1983).

One particularly intriguing line of research on the relationship between cognitive and physiological functioning concerns a phenomenon called **terminal drop**. When cognitive test scores collected longitudinally on an aged population are evaluated retrospectively (after most of the sample has died), a dramatic decline is evident one to two years preceding death, even before any physiological markers are present. In effect, it becomes almost possible to predict time of death from change in cognitive test scores (Steuer & Jarvik, 1981). The data are still very speculative and open to interpretation, but they reinforce the notion that cognitive and physiological functions remain relatively independent across the life-span as long as each is functioning within a fairly broad range of normality. However, when one moves out of this broad range, the other is affected (Woodruff, 1975).

Steuer and Jarvik (1981) believe that the terminal drop phenomenon may reflect the fact that level of cognitive functioning serves as a sensitive indicator of individuals' overall level of physiological functioning. The cumulative impact of changes in overall level of functioning may be evident before significant changes in individual organs or organ systems are detectable. The sources of these declines may include physiological factors such as decreased blood flow, metabolic changes, changes in activity patterns, or the interactive effects of the many medications the elderly commonly take. These changes may also reflect the variety of psychosocial changes that occur in the lives of the elderly. In this category, Steuer and Jarvik include such factors as feelings of helplessness and hopelessness, social isolation, and lack of intellectual stimulation.

Theoretical Aspects

The various advances that have been made in our understanding of the aging process have served to delay the onset of the primary aging process (Fries, 1980). None of them has prevented it from occurring and nothing short of a virtual "body transplant" will. All living organisms, plant or animal, have a finite life-span. In the case of humans, four factors have been viewed as accounting for our aging and ultimately our deaths.

The first is that individual cells seem to have a predefined life-span (Hayflick, 1977). Each is capable of a limited number of cell divisions, after which it stops dividing and dies. Just as there are factors that can hasten the death of cells, there are probably also those that can prolong the cell division process—but the process appears ultimately finite.

A second factor concerns the cumulative impact of simply being alive. This "wear and tear" notion (Shock, 1977) emphasizes the importance of how we lead our lives, what we eat, what pollutants we encounter, and the degree to which we maintain our bodies. Even the very process of metabolism produces by-products, called **lipofuscins,** that accumulate in the cells, possibly influencing cell efficiency. Other by-products of cell metabolism produce chemical compounds called **free radicals,** which increase the permeability of the cell membrane, making it more vulnerable (Rockstein & Sussman, 1979).

A third factor involves the immune system. For most of our lives, our immune systems serve to fight bacteria or viruses or other foreign agents. Individuals whose immune systems fail to function can die from the most trivial, the most common, the most generally harmless of bacteria. With advanced age, the immune system changes. It becomes less able to distinguish its own cells from those of foreign agents. It develops autoimmune properties—the body literally attacks itself (Shock, 1977). Three of the chronic conditions associated with aging—adult onset diabetes, anemia, and arthritis—are caused by **autoimmune reactions.**

The fourth factor is the genetic code itself. It is certainly a factor in our longevity. It is evident in the difference in the life-spans of men and women. It is reflected in the longevity correlations between parents and children and between siblings. It is possible that with age, the number of errors in the transmission of genetic information increases, resulting in cells that are less viable or even in mutant cells. It is possible that certain genes that have deleterious effects only become active late in life. The specific mechanisms are yet to be determined, but there is no question that there is a genetic component to the aging process.

These four factors do not function independently of each other. As indicated at other points in the chapter, primary aging is a cumulative, interactive process, each change influencing the functioning of other systems and processes. Timiras (1978) has proposed that the coordination of the primary aging process is determined by the balance of excitatory and inhibitory processes within the brain. These neurotransmitters regulate the functioning of neuroendocrine centers in the brain, which in turn regulate the endocrine secretions. Endocrine hormone secretions act on target organs in the body to regulate virtually all physiological functioning. With age, the ability of these neurological centers to maintain an appropriate

BOX 10-1
TERMINAL DECLINE

Terminal decline is a phenomenon frequently noted in the gerontology literature. It reflects the fact that measures of cognitive and emotional status are often earlier predictors of nearness to death than physical or physiological measures are. Nearness to death in this instance does not mean weeks or months but years. Among populations of the elderly, it is possible to predict likelihood of death as far from the actual date of death as five years. This finding has been demonstrated in both retrospective and prospective research designs.

Retrospective designs look backwards. For example, a group of elderly people would receive one or more tests of psychological functioning on one or perhaps two or three separate occasions. Given the age of the sample, it is likely that a high percentage would die during the next five to ten years. If people's test scores were examined when they died and if those who died within five years of the last testing had a significantly different test profile from that of the people who lived ten years after testing, then it would be established, retrospectively, that the test scores could serve as a predictor of nearness to death.

Prospective designs look forward. In the retrospective design, it is quite possible that the two groups differ in ways other than test scores. A way to control for this possibility is to match groups of people so that they are alike in as many ways as possible in terms of predictors of nearness to death. Age, sex, medical history, and social class background are the most common controls. Again the psychological tests are administered and age at death eventually noted. (I and the people who have done this work are very aware of the ethical implications of this research design. Now that the phenomenon has been established, efforts are no longer directed at documentation but at prevention and remediation.) Both the retrospective and the prospective designs demonstrate terminal decline, and they do it before measures of physical or physiological status do. In fact, those of the aged who are seriously ill at the time of these tests but subsequently recover do not demonstrate the pattern.

What kind of tests predict nearness to death? Surprisingly, they are the most mundane of measures. Items such as copying a design, drawing a person, learning to associate pairs

level of functioning declines. The impact of this decline becomes evident at all levels of physiological and psychological functioning. There is a decline in muscle tone, in level of cardiac and respiratory functioning, in endocrine secretion and immunological functioning, and in the variety of other systems that help maintain homeostasis. As these declines continue, the individual becomes less and less able to effectively respond to virtually any form of stress.

of words, or substituting numbers for symbols have all been shown to be predictors of terminal decline. What these tests have in common is that they all require the ability to organize and integrate information, which seems, in turn, to reflect a sense of purpose and meaning in life, or what others have simply called a zest for living. All of us differ on these measures, and these differences are present at all ages. For the aged, however, whose reserve capacity is diminished, whose life situation is often problem filled, and who occupy an ill-defined role in society, the ability to form a sense of purpose may be more difficult. The evidence suggests that as a sense of purpose becomes more difficult to establish and maintain, something as intangible but nevertheless real as the will to live fades.

Sources: Botwinick, West, & Storandt, 1978; Lieberman & Tobin, 1983; Riegel & Riegel, 1972.

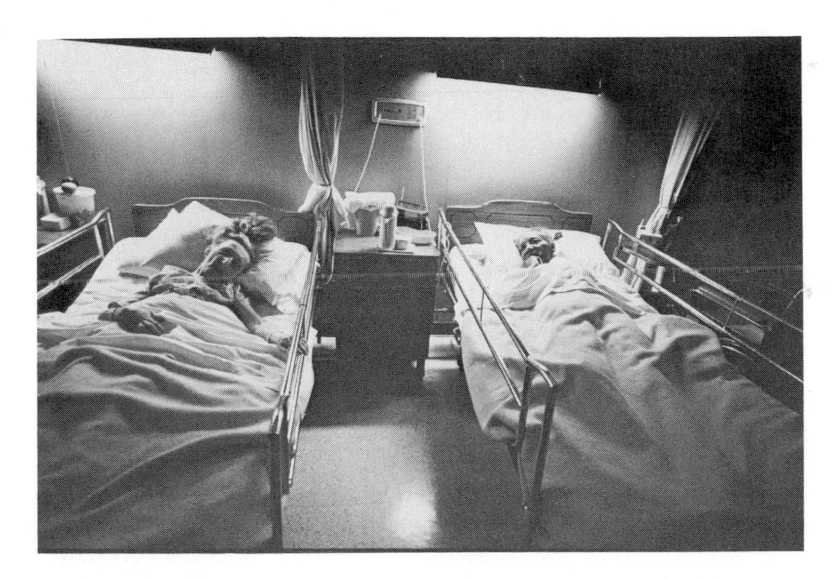

The loss of control over one's life that often accompanies confinement to nursing homes is among the factors that affect terminal decline.

Disengagement and Death

The adaptations that the old-old make in response to the various biological and cognitive declines serve to maintain a degree of identity, self-direction, and independence, hallmarks of adulthood in our culture. A frequent consequence, however, is a progressive disengagement from people and things and a turning inward. This increased introspectiveness is seen by some (Butler, 1974; Erikson, 1980) as a necessary step in coming to terms with one's own death.

The feelings of helplessness and isolation experienced by many nursing home residents may contribute to the phenomenon of terminal drop. Organized programs of contact between the young and the old help to combat these negative feelings.

Disengagement and the Life Review

The previous discussion on cognitive changes among the old-old indicated that cautiousness, the preference for well-rehearsed learning strategies, and the use of previous experience as the basis for current judgments are all means by which the very old attempt to adjust to the declines of advanced age. More generally, they are one component in the old-old's adjustment to the variety of losses that characterize advanced age—loss of spouse, loss of peers and kin, loss of biological and cognitive capacity, and loss of community. No longer having the capacity to actively confront these various losses, to establish new life patterns, and to redefine the self in response to these changes, the old-old seem to turn inward, making greater use of the past as a basis for maintaining a sense of continuity, integrity, and self-identity (Langer, 1981; Lieberman & Tobin, 1983). This disengagement, or turning inward, is one of the primary means that the old-old have of avoiding the sense of depression that can accompany feelings of loss and helplessness at any age (Breslau & Haug, 1983).

This turning inward is not unique to the old-old. The introspections of the young adolescent and the middle-aged adult are also attempts to remove themselves from the present. But there is an important difference. Disengagement from the present enables the adolescent to speculate on possible futures. It enables the middle-aged adult to reflect on the past in order to better cope with the future. For the old-old, there is no sense of future. For these adults, disengagement from the present and a turn to the past are the means through which a sense of self is maintained in the present. There may be an element of distortion in these reflections, especially when the past was not as they would have wanted it to be or when cognitive and biological capacity are seriously impaired. But, as Lieberman and Tobin (1983) note in their studies of relocation of the very old, this disengagement represents "an eloquent testimony to the coping ability of the elderly" (p. 260). Even at the very end of life, there is still the attempt to maintain a coherent sense of self.

Disengagement seems to be a universal feature of the old-old. Sex, social class, and ethnic and cultural background no longer serve to differ-

The tendency of many of the old-old to turn inward and live in the past is an attempt to deal with the physical and emotional losses that accompany old age.

entiate the patterns as they do throughout most of the life-span (Lieberman & Tobin, 1983). There is, however, some controversy over the degree of "premeditation" involved in the disengagement process. Lieberman and Tobin (1983) believe that it is primarily a process of simplification, one used by the old-old to cope with the assaults of time and loss.

Others see it as having a broader purpose. Erikson (1980) believes that a focus on the past is an attempt to come to terms with one's life, a prerequisite to accepting the inevitability of one's finiteness. Individuals accomplishing this task are said by Erikson to have achieved a sense of ego integrity. In Erikson's words,

> It is the acceptance of one's own and only life cycle and of the people who have become significant to it as something that had to be and that, by necessity, permitted of no substitutions. It thus means a new different love of one's parents, free of the wish that they should have been different, and an acceptance of the fact that one's life is one's own responsibility. (p. 104)

Butler's (1974) concept of the **life review process** is similar to Erikson's notion of integrity. Butler believes that the process is an active, deliberate attempt to adjust to the losses experienced by the very old. Rather than being prompted by the decreasing ability to deal with even day-to-day tasks, it is prompted by the realization of approaching dissolution and death.

> The life review is characterized by a progressive return to consciousness of past experience, in particular, the resurgence of unresolved conflicts which can now be surveyed and integrated. The old are not only taking stock of themselves as they review their lives; they are trying to think and feel through what they will do with the time that is left and with whatever material and emotional legacies they may have to give to others. They frequently experience grief. The death of others, often more than their own death, concerns them.

BOX 10-2
THE LIVING WILL

Our ability to develop new technologies often proceeds faster than our ability to successfully integrate them into the patterns of our daily lives. Each new technological advancement brings not only the promise of a new solution but also, in many cases, new questions about the moral, legal, and ethical uses of this new technology. This set of circumstances is no more clearly evident than in the case of the variety of life support mechanisms that are available for the critically ill person. When there is every hope for improvement, then there is little question as to their value. But what if there is little or no hope? Is what can be done what should be done? How does a family know if maintaining life support is what the person would want?

One solution to the dilemma is the "Living Will." It instructs that heroic measures are not to be taken when there is no reasonable expectation of recovery from extreme physical or mental disability.

The will is not a legal document but a moral and ethical one. Nevertheless, the pattern of legal precedents in this country is consistent with the idea that patients have the right to determine the course of their own treatment even if that course results in death. Concern for Dying, a nonprofit educational group, recommends that those using the Living Will make their wishes known to all in a position to act upon the request when the Will is signed. This group believes that this will further increase the likelihood that the individual's wishes will be honored. More information about the Living Will can be obtained by writing the Concern for Dying at 250 West 57th Street, New York, New York, 10019.

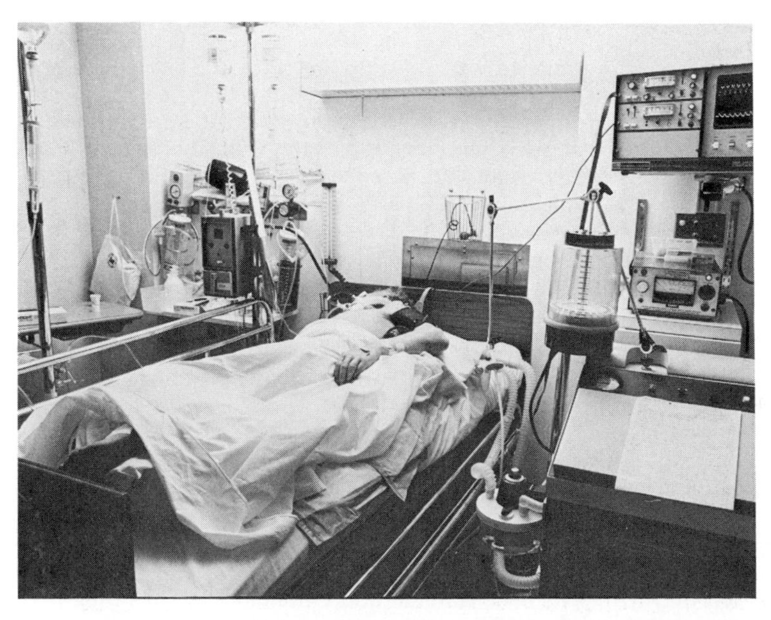

My Living Will
To My Family, My Physician, My Lawyer and All Others Whom It May Concern

Death is as much a reality as birth, growth, maturity and old age—it is the one certainty of life. If the time comes when I can no longer take part in decisions for my own future, let this statement stand as an expression of my wishes and directions, while I am still of sound mind.

If at such a time the situation should arise in which there is no reasonable expectation of my recovery from extreme physical or mental disability, I direct that I be allowed to die and not be kept alive by medications, artificial means or "heroic measures". I do, however, ask that medication be mercifully administered to me to alleviate suffering even though this may shorten my remaining life.

This statement is made after careful consideration and is in accordance with my strong convictions and beliefs. I want the wishes and directions here expressed carried out to the extent permitted by law. Insofar as they are not legally enforceable, I hope that those to whom this Will is addressed will regard themselves as morally bound by these provisions.

(Optional specific provisions to be made in this space — see other side)

DURABLE POWER OF ATTORNEY (optional)

I hereby designate ______________________ to serve as my attorney-in-fact for the purpose of making medical treatment decisions. This power of attorney shall remain effective in the event that I become incompetent or otherwise unable to make such decisions for myself.

Optional Notarization:

"Sworn and subscribed to

before me this ________ day

of _________, 19______."

____
Notary Public
(seal)

Signed _______________________

Date _______________________

Witness _______________________

____
Address

Witness _______________________

____
Address

Copies of this request have been given to _______________________

_______________________ _______________________

(Optional) My Living Will is registered with Concern for Dying (No. ________)

> Perplexed, frightened at being alone, and increasingly depressed, they at times become wary or cautious to the point of suspicion about the motivations of others. If unresolved conflicts and fears are successfully reintegrated they can give new significance and meaning to an individual's life, in preparing for death and mitigating fears. (p. 19)

Both Butler and Erikson give the personal awareness of death a more central role in explaining the behavior of the old-old than do Lieberman and Tobin. For Lieberman and Tobin, a preoccupation with personal death is not evident until one to two years preceding death. It becomes evident at this time on measures of personality and self-concept, probably reflecting the same terminal drop phenomenon seen on cognitive tasks. Lieberman and Tobin suggest that as the task of maintaining a sense of coherence becomes increasingly beyond the capacities of the individual even when that sense is rooted in the past, the self begins to disintegrate, and death soon follows. Whether death is a result of the disintegration or the disintegration is a reflection of the continuing loss of vital capacity and homeostasis is a question that as yet has no answer (Ziegler & Reid, 1983).

Death

There is an implicit assumption in all of this discussion of disengagement and death. Simply put, it is that the end of life comes at life's end. Although not correct for all, the assumption is correct for many today and will be even more so for future cohorts (Fries, 1980). More and more people will live into their seventies and eighties, and their perceptions of death in general, their feelings about their own mortality, and their actual dying experience will reflect the fact that they have lived long lives. The remainder of this section on disengagement and death will examine each of these aspects of death, both for those who do die at life's end and for those who don't. It is only when the two are contrasted that a full understanding of the disengagement process among the old-old can be achieved (Kalish, 1976).

When compared to all who are younger, the old-old have a relatively unique view of death and their own mortality. The different perception, in part, reflects the disengagement process. Having distanced themselves from present events, they may feel there is less to lose through death (Kalish, 1976), that death is the next expected event. In the sense that death comes to be seen as a normative, timely event (Neugarten & Datan, 1973), it generates less anxiety because there is less uncertainty involved. Although the old-old think more about death than other age groups, they report being less afraid (Kalish & Reynolds, 1981). Through the deaths of peers, they have seen a lot of people dying. They have some notion of what it appears to be like. They have fewer dependents than younger adults and therefore are less likely to be concerned about the consequences of their death for others. Most report having lived a full life, having done many of the things they had hoped to do. Many feel that the infirmities of age and the loss of social status make life feel less worth living. Few report a wish to fight death, to undergo painful treatment that might promise more years. They are in no hurry; they want to continue the things they are now doing. But few have the desire to start new things; when death comes, they will accept it—or are at least resigned to its inevitability. They are more concerned about the loss of dignity that can accompany a lingering terminal illness, about pain, and about fruitless heroic measures to prolong life (and in the process, dying) than they are about death itself. This perspective is evident in the retelling of an interview by Kalish and Reynolds (1981) with a ninety-two year-old man.

> After forty years as a teacher, he spent another quarter century developing new breeds of flowers. Now, he stated, he was bored with that; he also feared the changing political arena and the changing neighborhood. No one in his family needed him, he felt he was physically and financially unable to have new experiences (he had already done a good bit of living in his 92 years). He was able to express his total lack of interest in living to reach one hundred years. He did not even wish to see what he felt was the culmination of the present disruptions in the social order. With great calmness, he said he was totally ready to die. I believed him. Yet I also believe that he will do nothing to expedite death and that his readiness to die will only have one effect on him: he will not re-engage in new lines of activity but instead will pursue only those activities and relationships now on-going. (p. 72)

The work of Kubler-Ross. Probably the most well-known work on the way individuals cope with their own dying is that of Kubler-Ross (1969). Kubler-Ross, a psychiatrist, identified a five-step **sequence of response to death** to describe how the terminally ill confront their own deaths. Kubler-Ross labels the five steps as denial and isolation, anger, bargaining, depression, and, finally, acceptance.

The virtually universal first reaction of those given a diagnosis of a terminal illness is denial. The diagnosis must be wrong, my X ray or blood sample must have been mixed up with someone else's, it simply can't be true, I don't feel that sick, I don't want to die.

Denial serves two very important purposes. First, it provides a period of time for people to organize themselves, to prepare their defenses to cope with a terminal illness, and to let the diagnosis sink in slowly. Second, it usually prompts the getting of a second opinion. Blood samples don't often get switched, but some prognoses are less clear-cut than others. Perhaps another physician does not believe the situation is terminal. Perhaps there is a form of therapy not known or available to the first physician. Assuming the diagnosis is correct, Kubler-Ross finds that denial is relatively short-lived and quickly replaced by a strong sense of isolation. A sentence of death leaves people feeling cut off and alone, even from those closest to them. The other people are not going to die; they are therefore separate from those who are.

Anger comes to replace denial and isolation. Having begun to come to terms with the diagnosis, most think it unfair and feel bitter. Feeling discriminated against, they lash out at others, friends, relatives, intimates, hospital staff. Any minor accident—the television being too loud, the coffee being too hot—serves as a prompt for venting hostility. The anger only serves to make visits more painful. Friends and relatives, feeling guilty and grief stricken, visit less. Unfortunately this only serves to further heighten the patient's discomfort and anger.

The one thing that will stop the anger, a cure, is not in the power of others to provide. There is little they can do to help the person resolve the rage. But this too does seem to pass. Anger is the result of the first step in coming to accept death; bargaining is the result of the second.

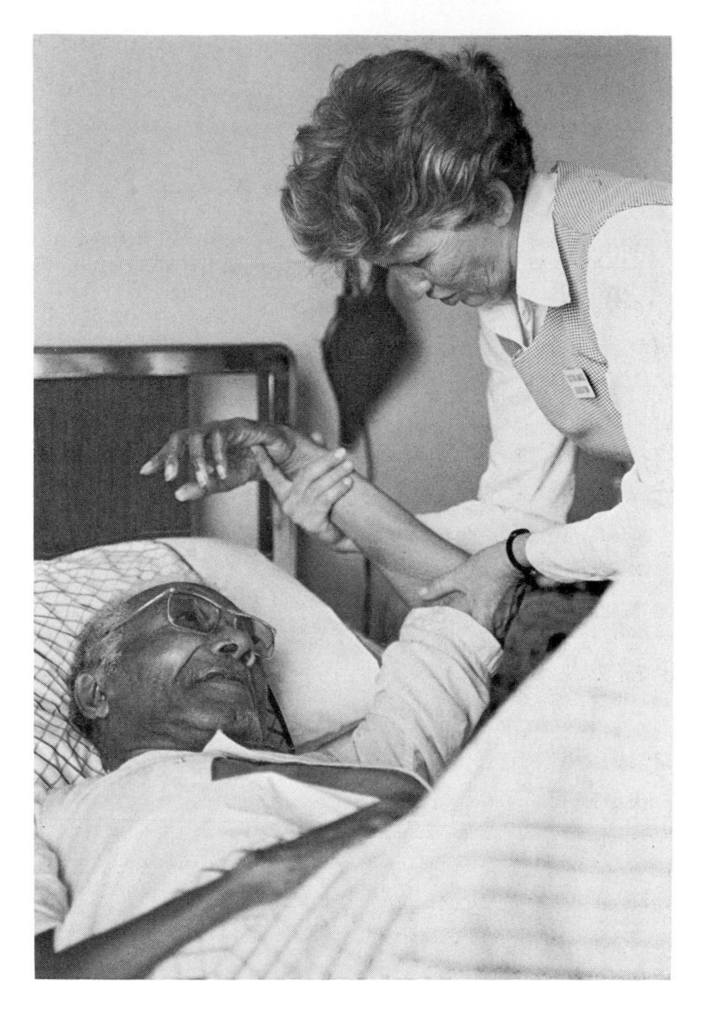

Kubler-Ross and others who work with the terminally ill believe that perhaps the most important aspect of their assistance is assuring the patient that there will always be someone available to share with them whatever time they have remaining.

If being angry at themselves and others doesn't get the terminally ill anywhere, then they think maybe there is another way. For example, they may reason like this: I'll make a deal with God; I'll promise to give up smoking, to go to church every Sunday, to live an honorable and decent life. I'll volunteer for some sort of exotic, experimental treatment. There must be something I can do to save my life or at least to live it a little longer. Bargaining serves a similar purpose as denial. It allows a degree of postponement; it further delays the necessity of having to come to terms with the reality of the situation. But it also may make coming to terms with reality a little easier in that there is just that much more certainty that everything that could be done has been done. Not surprisingly, the terminally ill are most vulnerable to medical quacks at this point in the sequence.

Depression follows bargaining. It is a kind of acceptance, albeit one of sadness and reluctance. There are two aspects to this depression.The first, what Kubler Ross calls **reactive depression,** reflects all the sadness and misery that has transpired. Painful treatments have been endured, sometimes great financial burdens have been accrued, the person's life as well as the lives of others has been drastically changed. Who wouldn't be depressed under these circumstances? The second aspect of depression is preparatory. The first concerns what has already transpired; the second, what is to occur. The first can be partially alleviated by assurances that others will be able to go on after the person's death. There is little that can alleviate the second. Attempts to lighten the mood may actually be counterproductive.

Preparatory grief is necessary if the dying person is to be able to achieve a sense of acceptance that death is near. There is no joy in this acceptance; it is rather one of inevitability. The person is tired and weak and withdrawn. There is little emotion. There is little desire to talk or even to have visitors. Kubler-Ross notes that the person appears to be in less pain. All that is wanted is to have those very closest present, to provide assurance that he or she will not die alone. As death approaches, the person may drift in and out of coma or sleep; there is little interest in food and little awareness of surroundings. It is a time that Kubler-Ross describes as the "silence that goes beyond words."

> Those who have the strength and the love to sit with a dying patient in the silence that goes beyond words will know that this moment is neither frightening nor painful, but a peaceful cessation of the functioning of the body. Watching a peaceful death of a human being reminds us of a falling star; one of the million lights in a vast sky that flares up for a brief moment only to disappear into the endless night forever. To be a therapist to a dying patient makes us aware of the uniqueness of each individual in this vast sea of humanity. It makes us aware of our finiteness, our limited lifespan. Few of us live beyond our three score and ten years and yet in that brief time most of us create and live a unique biography and weave ourselves into the fabric of human history. (Kubler-Ross, 1969, p. 276)

In presenting the sequence, Kubler-Ross stresses the fact that not all terminally ill patients achieve what she would define as a true sense of acceptance (perhaps no more than one-third do) and few proceed through the sequence in a linear fashion. The more typical pattern is to go back and forth between the middle three steps as each day brings new hopes, fears, and disappointments. In fact, Kubler-Ross notes that most of the patients she worked with kept some degree of hope up to the last few days of life.

The Kubler-Ross sequence is useful in understanding some but not all **dying trajectories** (Kalish, 1976; Kastenbaum & Aisenberg, 1972). Most of her work has been with cancer patients whose illness might run a fairly lengthy course but for whom the outcome of the illness, if only for those in the research, was known with a fair degree of certainty. These were people whose illness might be temporarily arrested or whose life might be prolonged but who were terminal. Her population was not primarily aged; many were young-old, middle-aged, and younger. Given what we know about the very aged, acceptance might be more common in them; denial, bargaining, and anger, less so. The therapeutic aspects of helping the dying person come to terms with the illness might be less feasible as reduced capacity makes resolution more difficult.

Other views of the process of dying. Glaser and Strauss (1968) believe that the Kubler-Ross sequence may only be useful when the illness has a long trajectory and certain outcome—as with some terminal cancers. Other circumstances may have different sequences or prompt different reactions to death. Consider the patient who knows that in two weeks he will be undergoing high-risk surgery. If the surgery is successful, he will be able to resume a normal life. If it is unsuccessful, he will die within a few months. Further, there is a chance that he might not even survive the surgery. What type of sequence might this man experience? What about a woman who has an inoperable condition that might at any time cause sudden death or might never affect her? What about a cancer patient whose condition is not terminal, who has had a malignant tumor successfully treated but lives with the threat of recurring episodes? How might this person react as each three- or six-month follow-up visit does or does not hold the promise of no new growths?

Pattison (1977) finds that it is the degree of uncertainty that often determines the nature of the reaction to the possibility of death. In this sense, the very old might have an "advantage" in that death becomes more certain with increasing age. Because less anxiety is generated by the threat of death, the very old might be better able to use more mature coping mechanisms to deal with their end. Denial, aggression, displacement, and emotional dissociation might be more common in the young or middle-aged adult; sublimation, anticipation, and even humor and altruism, in the very old.

In the most general sense, the question of how best to cope with death may have less to do with people's reaching a level of acceptance and more to do with their achieving what Weisman (1972; Kalish, 1976) terms an **appropriate death.** An appropriate death is described as one that is as free of pain and suffering as possible, that maintains emotional and social supports, and that helps the dying person resolve unfinished tasks, prepare others to continue in his or her absence, satisfy wishes, and gradually

In addition to offering support to the terminally ill, many hospices provide counseling and assistance to family members.

yield control of self to others. Within the limits of disability, the dying individual should be allowed to operate on as high and effective a level as possible even though only tokens of former fulfillment can be offered. **Hospices,** places where the terminally ill and their families can spend their time together, have as their goal ensuring that each person achieves as close to an appropriate death as possible. They have come into existence because meeting the needs of the terminally ill is often not possible within the home and because the primary concerns of hospitals are in terms of curing rather than comforting patients (Pattison, 1977).

Grief and Mourning

Death ends an individual's life-span. It does not end relationships, however. The others in those relationships must now come to terms with their loss and must go on with the process of living.

The loss, through death, of a spouse is one of the greatest sources of stress that adults encounter. It requires the survivor to undo all of the bonds of intimacy that have been created over a period of many years, to radically alter his or her self-image, to establish a new concept of self, to establish new social relationships, and to do all of this without the aid of the one person who has served as the primary source of support.

Anticipatory Grief

The process of bereavement begins before death. For all but those whose loss is sudden, there is a period of **anticipatory grief** (Kubler-Ross, 1969; Raphael, 1983). The experience of coping with a terminally ill spouse or one who suffers from a variety of serious, chronic ailments that have required periods of hospitalization and have seriously diminished individual capacity forces the other person to confront not only the realities of the present moment but also the likely future once the spouse dies. The surviving spouse shares the depression, pain, and anguish of the dying. Each confronts the losses and sorrows that precede death—the loss of body control, the loss of primary relations, the loss of identity, the pain and suffering, the fear of loneliness, and the fear of the unknown (Pattison, 1977).

The surviving spouse frequently rehearses the death, trying to imagine what it will be like to be alone, trying to anticipate the adjustments that will be required. Each issue that is confronted and each decision that is made further heighten the reality of the situation for the survivor as well as for the dying person (Kalish, 1976). Imagine yourself as the surviving spouse and consider how you would respond to each of these statements. "How shall the remainder of our time together be spent?" "How much medication for pain should I take? If I take large doses, I will be relatively free of pain but perhaps too incoherent to talk to you or even appreciate your presence." "If I lapse into coma, I want you to tell the doctors to do nothing to prolong my life." "We need to talk about the funeral."

The opportunity to experience anticipatory grief does seem to soften somewhat the actual impact of death. The most intense grief reactions are experienced by those whose loss is through sudden, unexpected death, as in the case of a middle-aged or even young adult (Ball, 1977). But the comparison is merely relative. The absolute intensity of the grief and mourning under all circumstances is great.

The Bereavement Process

Initial reactions. The first reaction to the loss of a loved one is shock, numbness, and disbelief (Raphael, 1983), not intense emotional reaction but rather an absence of feeling, as if the survivor was somehow removed from the situation. Many people report being unable even to remember much of those first few hours to weeks following the death. They participate in the rituals. They arrange the funeral. They receive calls of condolence. They go through the motions, but somehow the impact has yet to register. This initial reaction occurs even when death was expected and perhaps even welcomed as a way to free the loved one from pain and suffering. It may last as little as a few hours or as long as a few weeks. Like other transitions, disbelief and numbness protect the person from overwhelming stress; they provide an interval for mustering coping mechanisms.

Confronting the loss. Once the initial sense of disbelief passes, survivors begin to confront the full impact of their loss. They are prone to experience a variety of psychosomatic symptoms including insomnia, loss of appetite, fatigue, depressed mood, guilt, restlessness, and fits of uncontrolled crying. Of these symptoms, sleep problems, depression, and uncontrolled crying are the most commonly reported: they occur in around 80 to 90 percent of all cases (Clayton, Halikes, & Maurice, 1971). Anger and feelings of desertion are common, but so are feelings of guilt. Survivors feel confused, aimless. Their emotions are mixed and unpredictable. Everything seems to remind them of their loss, of the fact that the spouse is gone. Each thing they touch seems to evoke a long-forgotten memory. At the same time, they can feel guilty about still being alive and angry and resentful that they must go on alone.

This second reaction may last for a few months and seems to serve the purpose of helping the survivor psychologically bury the dead spouse. Just as the infant seems to need to drop everything before finally acquiring the insight that "things" drop when released, the survivor has to undo every bond, every link, every context for interaction before finally being able to accept the fact that the spouse is no longer present. It is a process through which one's internal reality aligns itself with one's external reality (Parkes, 1972). It is a process that can linger; not all bonds are immediately evident.

Being alone. Once this alignment occurs, most of the psychosomatic symptoms lessen and gradually disappear. The survivor has come to psychologically accept the death and now must confront a new life. The first reality of this new life is that it is a very lonely life. Loneliness is the single most frequent complaint of widows and widowers during the first year or two following the death of a spouse. Everything now seems to tell survivors how alone they are. The house seems empty. They go out for dinner with old friends but realize that they are the only one alone. The rest are all couples. The conversation is stilted. Somehow everything feels very different.

The loneliness does wane but only slowly. Day-to-day sadness may be gone. There may even be periods toward the end of the first year when the survivor is actually able to put the sadness out of mind, even periods of some happiness. But the first birthday, the first anniversary, the first birthday of each child, the first holidays alone each again bring back the sadness. Gradually the survivor adapts to being alone. Tasks that were done by the spouse are learned. New friendships are made. In many cases survivors feel very good about the fact that they not only have managed to survive the loss but have, in a way, actually grown as a result of it. They find themselves able to do things they never thought they could. They don't forget the death of the spouse, but they do learn to live with it.

Factors Influencing the Resolution of the Bereavement Process

A variety of factors influence the intensity, duration, and resolution of the bereavement process. These factors include the age of the survivor (Ball, 1977; Kalish, 1982; Raphael, 1983), the sex of the survivor (Barrett, 1978; Berardo, 1970; Bikson & Goodchilds, 1978; Gerber, Rusalem, Hannon, Battin, & Arkin, 1975; Kalish, 1976), the cause and duration of the death (Kalish, 1982), the character and quality of the relationship (Golan, 1981; Heyman & Gianturco, 1973), and the nature and availability of formal and informal support systems (Lopata, 1979; Lowenthal & Robinson, 1976; Silverman, 1972).

The age of the survivor. The younger the survivor is, the more difficult the transition is, but also the more likely the survivor is to remarry and establish a new life-style. The increased difficulty for younger survivors reflects both the "untimeliness" of the spouse's death and the fact that the young or middle-aged survivor may be saddled with a variety of family and financial burdens. For those survivors under the age of sixty five, there is a significant increase in the death rate, in the incidence of substance abuse, in the incidence of hospitalizations, in the frequency of visits to a physician, and in the number of prescriptions issued above what one would expect in a comparable same-age population (Clayton, Halikes, & Maurice, 1971; Jacobs & Ostfeld, 1977). The fact that most of the deaths are related to cardiovascular causes lends some credence to the notion of dying from a "broken heart" (Parkes, 1972). There is some evidence that the over sixty-five population also has an increased mortality rate, but since many older adults already suffer from a variety of potentially life-threatening conditions, etiology is harder to establish.

The causes of this increased risk of death or trauma following the death of a spouse are unclear (Jacobs & Ostfeld, 1977; Kalish, 1982). It is possible that changes in habits following the death increase the likelihood of death or trauma. It is also possible that in the process of caring for a terminally ill spouse, survivors neglect their own health needs. It is possible that the physiological consequences of the stress reactions are too much for the survivors' systems to adequately tolerate. Whatever the specifics, evidence such as this clearly indicates the centrality of intimate relationships during the adult years.

Very few of those over sixty-five remarry. Since most survivors are women, there are few potential mates. Since children are grown and self-sufficient, the need for additional income and additional aid in home and family tasks is lessened or removed. Since many people over the age of sixty-five are already widowed, there is an already established "community" for the newly widowed to enter. The thirty- or forty-year-old widow or widower, on the other hand, finds none of these conditions true for them. There are a variety of factors—children to raise, urging from friends and family, social norms, personal needs—that push these younger survivors toward remarriage (Golan, 1981).

For the old-old, bereavement often has a unique pattern. Having less of a sense of future, knowing that their own life-span is short, and having fewer capacities for adapting to change, the very old often choose to maintain a life-style as similar as possible to the one they had established before the death of the spouse. "The widow(er) holds on for the few years left, living on together with the lost person, maybe cherishing this image while awaiting reunion in some form of afterlife" (Raphael, 1983, p. 314). To the extent that the completion of the bereavement process involves the severing of former bonds so that new ones can be formed, it is probable that the old-old never actually complete it. Seeing an aged parent living as if the spouse were still figuratively if not literally present may be difficult, but such behavior may represent a very reasonable adaptation for many people at this time.

One of the more unique aspects of this partial resolution of the bereavement process among the very old is what Lopata (1979; 1980) refers to as the **sanctification process.** She finds that widows tend to idealize their late husbands, to cleanse them of all their shortcomings, and to remember only their good points. When shortcomings can't be ignored, they are often rationalized. A widow talking about a husband who was an alcoholic might explain that he really couldn't help it, it wasn't his fault, they put so much pressure on him at work. For the very old, sanctification appears to be an important component of their partial resolution of the bereavement process. Portraying the late husband to be almost saintly allows the widow to bask in his memory rather than to confront the realities of structuring a new identity and a new life-style.

The sex of the survivor. There is very consistent evidence that, at all ages, men have a more difficult time adjusting to the loss of a spouse than women do. Partly this evidence reflects the fact that if men think about the loss of a spouse at all, they probably assume they will go first. For this reason, the death of their wife must be unexpected and therefore additionally stressful.

Barrett (1978) found that widowers were more likely than widows to feel lonely and depressed, to evaluate community services as inadequate, to need help with household tasks, to eat poorly, and to not want to talk about the death of the spouse. Since men are less skilled at day-to-day domestic routines, the death of a spouse may create even more discontinuity in their lives than it does in the lives of women (Berardo, 1970).

Probably the greatest single reason for the sex difference in recovery patterns is the fact that women usually have more extensive friendship and kinship ties than do men (Lopata, 1980; Lowenthal & Robinson, 1976). Because these ties are more extensive and intimate, there is less of the feeling of isolation and loneliness and therefore greater support for getting on with life. Further since showing dependence and need are less consistent with a traditional masculine sex role, even for the aged, men are less likely to seek the help or establish the friendships that would help alleviate the feelings of loneliness and isolation.

As Longino and Lipman (1981) suggest in the following quote, these sex differences probably reflect lifelong patterns of interpersonal relations.

Because many men lack experience in handling routine household chores, the death of a spouse may prove particularly disruptive.

> In contemporary society, and perhaps historically as well, women, particularly married women, take responsibility for keeping up the family and friendship ties of their households with correspondence, with holiday and special occasional greetings, and with little gifts and thoughtful actions. Most adult men find this kind of attention to network maintenance tedious, if not downright distasteful. As a consequence they usually have invested less in maintaining the ties, the links in the support network. And in old age, it is the woman who reaps the rewards of her investment. So long as they are married, the man shares in the dividends. Spouseless men, however, have the fewest primary relations in their support networks, and as a corollary, the fewest family members giving them support of all kinds. Perhaps this helps to explain why older men suffer more upon the loss of spouses than do older women and why the man's emotional recovery from widowhood has been found to be slower than that of women. (p. 175)

Social Supports for the Old-Old

Both family and friends take on added importance for the widowed but in different ways, ways that again reflect the distinction between the voluntary, reciprocal nature of peer relationships and the complementary, obligatory nature of family ties.

Family support. The importance for older adults of not feeling themselves to be a burden on grown children and of maintaining comfortable routines serves to place certain limits on parent–grown child interaction patterns. Most older adults do not want to move in with their children (Lopata, 1980; Shanas & Sussman, 1981). Rather they prefer to maintain frequent contact with family but in such a way that they don't feel themselves to be under too many obligations or to be too much of a burden. A parent might go out of her way to stress the fact that she would appreciate a ride to a friend's house only if the daughter had plenty of free time. The daughter in turn might emphasize how little disruption in her schedule would be caused by the request. This "posturing" is more evident in more economically advantaged families (Streib & Beck, 1980) and certainly becomes harder to maintain at all levels as failing health renders the aged, widowed parent less able to maintain an independent life-style.

Posturing serves the valuable purpose of enabling the parent to maintain a sense of independence and in so doing protects her from having to adjust to just one more change in her self-concept (Lieberman & Tobin, 1983). The parent knows that she really has no choice but to ask the daughter. Neither she nor any of her friends still drive, and there is no public transportation. The daughter honors the request because it comes from her mother. She has a busy day, and the mother's old friend lives just far enough away that dropping the mother off, returning home, and then going back to get her is out of the question. She is stuck for the afternoon. Each is, in reality, obligated to help the other. But the situation isn't presented this way; it is presented as a voluntary arrangement, and for this reason, the parent is able to maintain the degree of dignity that comes through independence and is often lost through old age. As the surviving parent's health continues to decline, the illusion becomes harder to maintain and the obligatory nature of the parent-child bond again becomes dominant as grown children, out of a sense of love as well as obligation, assume an increasingly greater degree of care of the aged parent, ultimately bearing the responsibility of the funeral arrangements.

Peer supports. Peers serve a different role for the aged. Perhaps because peer associations are voluntary, a sense of dignity and independence is easier to maintain in these relationships. Perhaps because peers are better able to empathize, there is also more support derived from these relationships. Both Silverman (1978) and Vachon (Vachon, Lyall, Rogers, Freedman-Letofsky, & Freeman, 1980), for example, report that widow-to-widow self-help programs reduce the intensity and duration of the bereavement process. To a degree, peers are able to substitute for the dead spouse in terms of shopping, sharing meals, conversation, and other age-related activities (Cantor, 1979). Here, too, social class is a factor. The more economically advantaged the background is, the more likely the widow is to have social contacts independent of the husband and the more likely she is to be comfortable traveling alone, or even simply going to restaurants or theaters (Lopata, 1979). Heyman and Gianturco (1973) have found that coping with the loss of a spouse is easier for more economically advantaged widows, in part, because they establish more independent social contacts and therefore the death of the husband causes less disruption in their lives. These social class distinctions are probably less relevant in the lives of the old-old, but they do reemphasize the long-term consequence of marriage and friendship patterns across the adult years.

For the old-old, as health continues to decline and mobility is further restricted, the distinction between friend and neighbor tends to blur, especially when neighbors are also aged (Cantor, 1979). Neighbors become friends largely because they come to rely on each other for mutual support. When this is no longer possible, the aged adult either lives with grown children or other family members or is placed by the family in an institutional setting. At the end of life, we ultimately come to depend on the obligations and emotional bonds that extend across generations.

The previous chapter on middle age ended with a poem by Longfellow. It spoke of the ability to be productive and creative in the "arctic regions of our lives" and of seeing things "invisible by day." The material reviewed in this chapter certainly supports Longfellow's view, especially for future cohorts. But the material also speaks of sadness, loss, depression, and death, events far from "the gulf-streams of our youth." The interplay of these two sets of characteristics gives the later years of adulthood their unique character—older adults must adapt to loss in order to maintain continuity while at the same time coming to accept the discontinuity that comes with death.

THE SIGNIFICANCE OF THE LATER YEARS OF ADULTHOOD IN THE LIFE-SPAN

There is a great variety of changes that occur during the later years of adulthood. Each is of interest in its own right, but when considered collectively, these changes further add to our understanding of the patterns of development that occur across the life-span. Three kinds of changes are particularly evident during this stage. The first, the cognitive changes, reinforces the notion of development as a series of successive adaptations involving the evolving capabilities of the individual and the changing demands of the social and physical environment. The second, the role of family and peers as social supports, again emphasizes the distinction between reciprocal and complementary relationships. The third, the self-concept in old age, stresses the pivotal role that people's concept of self plays in their attempt to maintain a sense of integrity and coherence across the life-span.

Cognitive Changes

It is possible to view the process of development from either a quantitative or a qualitative perspective. As was discussed in the beginning of the text, a quantitative perspective holds that the same laws or factors that regulate development at any one stage also do so at all others. The advantage of this approach lies in its ability to make comparisons across stages. We can talk about the developmental status of the school-age child compared to that of the adolescent, the physical strength of the preschooler compared to that of the infant, and the vocabulary of the toddler compared to that of the preschooler. In essence, the study of development becomes the study of the additions and subtractions occurring across the life-span.

There is nothing incorrect about such an approach. Many of the changes that occur across the life-span are readily understood as something more or less than what was present before. There is a drawback to this quantitative view, however. It inevitably leaves the impression that development consists of a period of growth, followed by a period of stability, followed by a period of decline. But isn't this in fact what happens? Wouldn't people have to be very myopic not to acknowledge that in many ways, by the time they become very old, their best years are behind them? This is certainly true, and the very old would be the first to acknowledge it. The drawback, however, does not lie in the answer but in the question.

A qualitative view depicts development not as a progression of stages but as a succession of stages (Baltes, Dittmann-Kohli, & Dixon, 1984; Labouvie-Vief, 1980). Progression implies improvement; succession implies sequence. Certainly the developmental accomplishments of one stage are incorporated into those that follow, but a qualitative perspective argues that development across the life-span is better understood in terms of the contemporary interactions of the individual within the environment than in terms of comparisons with future or former levels of competence. Children are best understood not as incomplete or imperfect adults but rather as individuals using their unique abilities to cope with the demands of their environment. The aged are not best understood as merely a shadow of their former selves but as individuals using their unique abilities to cope with the demands of their environment. A reference to the cognitive changes that accompany old age should help clarify this point.

The research presented in this chapter shows that the old do seem to have difficulty with learning new skills, especially when the material is unfamiliar, when they are unable to set their own pace, and when efficient learning requires some sort of information-processing strategy. The research also shows that the old seem to be hesitant about even trying to learn new things, that they are more likely to make errors of omission, and that they are less likely to switch learning strategies even when one is found to be ineffective. The old sometimes have problems remembering details; sometimes what is remembered appears jumbled with other bits and pieces of the past. In what sense can these deficits be considered adaptive?

Go back to the chapter on the early years of adulthood and find the daily schedules of the two working, married mothers (pp. 405–406). To make such a schedule work or, more generally, to accomplish the variety of sometimes conflicting tasks that they encountered required a set of cognitive skills that emphasized flexibility, attention to detail, an orientation in time and space shared with others, and able encoding, retention, and recall of factually correct information. Clearly, this set of competencies is not typical of the old. But neither is it as necessary to the old, especially the very old, as it is to the two working mothers. The gradual disengagement, the increasing interiorization, the attempts at simplification, the declining sense of future, and the focus on reflection and life review require a different cognitive focus. In particular they require a focus emphasizing broad themes rather than specific detail. In fact, organization in terms of time and place (spatial-temporal orientation), so important to the two young mothers, may actually now be a hindrance as more basic themes and patterns are sought. Even the cognitive boundaries that foster a sense of self as distinct from other (an egocentric view) may prove maladaptive as the aged consider their place in the broader scheme of life, and ultimately the loss of sensory and psychomotor abilities may help prepare them to accept the inevitability of their own death (Labouvie-Vief, 1980).

Social Relationships

The distinction between reciprocal and complementary social relations has been used throughout the text as a way to explain the impact different social relationships have on people across the life-span. Complementary relationships are essentially unequal. The participants each perform a unique role. The parent-child relationship is the classic example. Reciprocal relationships exist between peers. Each is able to take the role of the other. Close friendships are the best examples of reciprocal relationships. Marriage relationships can take either form, although the historical trend is from complementary to reciprocal patterns. Each form of relationship entails a unique set of rights, obligations, rewards, and disappointments. Each can be thought of as being in or out of balance. Complementary relationships are in balance as long as a degree of equity is maintained. Reciprocal relationships are in balance as long as a degree of equality is maintained. Both are fundamental contexts of development across the life-span; each fosters a distinct perception of self; each serves as the foundation for a broader understanding of cultural institutions.

The research reviewed in this chapter indicates that maintaining a sense of intimacy with others is one of the most effective means of coping with the transitions of aging (Blau, 1981). The difficulties encountered in adjusting to the loss of a spouse, especially for men, clearly reflect this fact (Traupmann & Hatfield, 1981). It is also evident in the role that peer-based self-help groups such as widow-to-widow programs play in helping the bereaved cope with loss (Silverman, 1978).

Older people who have the companionship of their spouse and other family members generally have an easier time adjusting to the aging process.

The research also indicates that even though intergenerational contact, aid, and support are common between the aged and their grown children, the elderly are less likely than their adult children are to report these contacts as intimate (Walker & Thompson, 1983). In fact, many aged parents actually choose or at least feel the necessity to maintain a degree of distance between themselves and their children (Lowenthal & Robinson, 1976).

What do these sets of research findings indicate about the nature of interpersonal relationships in old age? The value of reciprocal peer relations is not hard to understand. It is a pattern that has been evident throughout virtually the entire life-span. Peers share a sense of time and place (Rowles, 1980), and they provide the comfort of empathy. But what about the parent-child bond? There actually seems to be a distancing taking place, when the old-old and their children are compared to the middle-aged and their children or even to the young-old and their children. What accounts for this distancing?

Emotionally satisfying relationships have two things in common. First, they exist by choice rather than by obligation. Second, they are able to maintain a mutually acceptable degree of balance. As parents age, each prerequisite becomes less attainable. Declines in finance, health, and mobility force the aged parent into a more dependent role. Whereas once the parent and the children took turns driving to the ball game or picking up the check at dinner, now the grown child always drives and always pays. Whereas once the parent would baby-sit for the new parents and in return would get some help redecorating the living room, now there is little the parent can offer to maintain a sense of equity in the relationship. The imbalance becomes even more glaring as grown children find themselves putting even more time and effort into their relationship with their parents. The complementary roles of parent and child have reversed themselves and each feels uncomfortable assuming the mantle of the other.

Parent as child and child as parent are not roles that our culture prepares us for or easily allows us to accept (Jarvik, 1983). Young parents long for the day when they can fully enjoy sharing the fun of a shopping trip with their children rather than constantly worrying about them getting lost or hurt crossing the street. Now, many years later, they find themselves unable to cross the street without the help of the children they once helped do the very same thing. Neither seems comfortable with the new roles. The parents feel themselves to be a burden on the children; the children are distraught over seeing the parents so dependent. Talking it over, a common solution in reciprocal relationships and when parents and children were each much younger, now simply serves to further emphasize the role reversal. If the relationship were voluntary, it would end. But the parent-child bond involves obligation as well as choice. The relationship continues, but as the research indicates, the gratification derived from it often declines.

Sense of Self

Throughout the text, the sense of self has been portrayed as one of the most important, if not the most important, organizing principles of individual development across the life-span. But it is perhaps not until we examine the course of development during old age that its centrality can be truly appreciated.

We live in a culture that equates maturity and adulthood with independence, self-direction, purpose, and social utility. It does not matter that at times the institutions of our culture don't always reinforce or foster this notion of maturity in some or all. The fact remains that if people of any age are asked to define what it means to be an adult in our culture, these words or ones very similar would be given in answer. A strong sense of personal control correlates with high life satisfaction and happiness across the entire range of adulthood (Campbell, 1981). Conversely, a low sense of personal control or efficacy prompts feelings of despondence, futility, and anxiety during the adult years (Bandura, 1982) and in old age also correlates with poor health and even likelihood of death (Lieberman & Tobin, 1983; Ostfeld, 1983; Ziegler & Reid, 1983).

The sense of self is a reflection of the individual's cumulative interactions with the environment. It is not a mirror image of those interactions, but certainly those whose skills match well the demands and expectations of the environment come to feel better about themselves than those whose don't. For the elderly, the decline in skill level and the heightened ambiguity of social demands and expectations together directly assault the sense of self. The lengths to which the aged will go, intentionally or unintentionally, to maintain a sense of self attest to the importance of maintaining a sense of self in old age.

The increasing value placed on peer contacts (Satariano & Syme, 1981) and the emotional distancing between parents and grown children (Jarvik, 1983) are two responses to this assault on the sense of self. Contact with family prompts an increased awareness of dependency whereas the voluntary, reciprocal nature of peer relations helps maintain a sense of value and purpose (Langer, 1981).

The importance of continuing to feel that one's actions remain voluntary is also evident in the research on relocation during old age. Relocations that are perceived as voluntary maintain or even enhance a sense of self-esteem (Kahana & Kahana, 1983; Kimmel, Price, & Walker, 1978; Williams & Wirths, 1965). Involuntary relocations produce the opposite effect. Ostfeld (1983) reports that when the aged are relocated involuntarily, even when they report that the new location is actually an improvement over the old, they are more likely to experience nursing home admission, hospitalization, surgery, stroke, angina attacks, increased visits to physicians, a general decline in health, and feelings of sadness and depression. Again, these consequences, that is, feelings of loss of control, are more likely to occur even though the relocated agree that the new residence is nicer than the old. The point is that they didn't want to leave the old residence, and because they were made to do so, their adult perception of self as one who is self-directed was compromised.

The old don't accept their loss of status willingly. They fight back as best they can with the resources that are available to them. First, they are more likely to see their loss of status as reflecting an unresponsive or even hostile environment than they are to see it as reflecting some change in self (Gurin & Brim, 1984). Second, they may develop what Gutmann (1977) calls the "adaptive paranoia of later life." By projecting the decline in status onto the actions of others, the aged adult is better able to maintain a stable sense of self and avoid the depression that can accompany the feeling of loss of control. Gutmann even suggests that this active vigilance may actually improve cardiovascular function and thus avoid "the vegetative position that is perhaps prodromal to death" (p. 309).

Lieberman and Tobin (1983) note a similar phenomenon in their studies of aged adults coping with relocation and institutionalization. They describe those who dealt best with the transitions as being "cantankerous." These troublemakers, by being quarrelsome and aggressive, managed to exercise a greater degree of control over their situations; that is, others had to contend more with them than with those who passively accepted their situation, resigned themselves to their circumstances, and in many cases declined further in psychological and physical vigor. Lieberman and Tobin are the first to acknowledge that aggressiveness is not an ideal solution or even a preferred one. But its presence in old age again stresses the importance of the sense of self and the lengths that individuals will go to to maintain it.

When all interactive strategies for maintaining a coherent sense of self fail, Lieberman and Tobin (1983) find that the very old withdraw into themselves and their memories of the past—even to the extreme of distorting these memories—as a coping strategy.

It would be wrong to leave this section giving the impression that illusion is a coping strategy unique to the aged. It is not. It is one used at all stages of the life-span. In a sense, illusion is merely the negative side of the image of individual development as an active, constructive process. What is unique about the process in old age is that many of the checks and balances have been removed. Having fewer social expectations to respond to, having fewer interpersonal contacts to maintain, having fewer responsibilities, and having fewer means to exercise control over the environment, the old person is "freed" from many of the constraints that our realities place on us. The fact that such a strategy comes as a last resort indicates how valuable maintaining interaction with the social and physical environment is to the aged. It also indicates the importance of finding ways for the aged to maintain interactions with the social and physical environment without the loss of a sense of self-esteem.

HUMAN DEVELOPMENT AND HUMAN SERVICES

An earlier section presented Weisman's (1972) notion of an "appropriate death." An appropriate death is one that allows the terminally ill the maintenance of a sense of personal dignity and self-worth, one that frees them from as much pain as possible and, to the degree possible, helps them cope with their losses. Although much that happens during the later years of adulthood has relatively little to do with death, this definition of an appropriate death could just as easily be used to describe an "appropriate old age." For each person to experience an appropriate old age, the social and physical environments that the elderly find themselves in must continue to remain responsive (that is, change) to their needs. For this to happen, social policy and action at all levels of government and society must reflect the value of maintaining a sense of dignity, independence, and self-worth. And for this, in turn, to happen, many of the stereotypes of the aged will need to be replaced by a more realistic image. Like Russian dolls, each of these goals is nested within the other, from the level of the microsystem to the level of the macrosystem (Bronfenbrenner, 1979).

Lawton (1980) believes that the actual and perceived changes that most of the elderly find in their dealings with their physical environment often lead to an increased sense of vulnerability, which in turn fosters a sense of dependency. An appropriate antidote is to modify the physical environment so that it continues to be as responsive to the needs of the old as it is to the needs of younger people. Many of the architectural barriers that impede the handicapped also affect the elderly—doors too heavy to open, steps too steep to climb, streets too wide to cross before the lights change, inaccessible public transportation, poorly lit corridors, barely legible signs (Parr, 1980; Windley & Scheidt, 1980). Other concerns, such as fear of crime and decaying housing, are not unique to the elderly, but because the elderly have fewer resources with which to counter these factors, they tend to be especially vulnerable. Each type of physical environment seems to offer its own unique challenges to the old. The urban environment provides many essential services in reasonably close proximity, often accessible by public transportation, but as Gutmann (1977) argues, the modern urban environment is often more in tune with the needs of the young than with the needs of the old.

The rural or small town environment certainly offers a greater sense of community (Rowles, 1980) but often at the expense of easily accessible services, and for the isolated rural elderly person, this sense of community may be limited to a sense of place rather than including a sense of social contacts (Shanas & Sussman, 1981).

Relocation is one way to solve the dilemma of person-environment fit. For people who are in a position to make a voluntary move to a community offering a more responsive physical and social environment, this may be a reasonable solution. But as Lawton (1980) notes, these people represent a small minority of the total over–sixty-five population. Even more specifically, they are most likely to be the recently retired who are renters rather than home owners, who have high incomes, and who have a history of moves during their adult years. This group should become larger in subsequent cohorts, but will probably continue to be a minority. And, even under these most favorable circumstances, many find the adjustment to a

Because their populations are homogeneous, planned retirement communities can gear activities and services to meet the needs of the elderly. Sun City, Arizona, with its population of 48,000 and average age of sixty-five, is a prime example of such a community. However, not all adults over the age of sixty-five want to become "Sun City Pom Pom Girls," and of those that do, many find the cost of such communities beyond their reach.

new location sufficiently stressful that health and morale are at least temporarily affected (Lawton, 1980).

Involuntary relocation, whether the actual forced relocation as in an urban renewal situation or the "we all think you would really be happier in a smaller apartment" variety, is frequently associated with an increase in serious physical and psychological problems and even with an increase in likelihood of death (Lieberman & Tobin, 1983; Ostfeld, 1983). There are clearly situations that warrant involuntary relocation, including situations requiring various types of institutional care, but such moves should be seen as a last resort.

What then are we left with? Each circumstance seems associated with its own unique set of barriers for the aged. An appropriate aging implies the maintenance of as high a degree of continuity across time as possible (Covey, 1981). Continuity is not the same as identity; the goal is not necessarily the maintenance of middle-age life patterns into old age. Aging does necessitate compensations, both in terms of modified environments and in terms of altered life-styles. But where such compensations are required, they should be treated in such a way as to maintain as high a degree of autonomy and independence as the situation allows and as the individual requires.

It is unlikely that aging individuals, alone, can effectively make public and private agencies be continually responsive to the changing needs of old age. One solution is the variety of special interest groups that lobby for the needs of the elderly. These groups, composed mostly of elderly people, have been particularly effective in legislation dealing with retirement age discrimination, Social Security, and medical benefits. Their effectiveness will probably grow as the aged continue to become a larger percentage of the population.

But aging seems to be an individual experience. Each aging individual encounters his or her own unique set of problems and limitations. Large special interest groups may not always be able to meet these individual needs. Peers seem best equipped to meet some of these individual needs. The research presented in this chapter indicates their importance in old age, both for day-to-day practical support and for coping with some of the losses of old age (Silverman, 1978). But the value of peers may be inversely related to their age. As more and more people live to see eighty and beyond, the cumulative effects of the circumstances of old age may leave many only minimally able to meet their own needs, much less those of others. The alternative would seem to be family.

True, the research indicates a distancing between aged parents and grown children. Parents do not wish to be a burden on children. Grown children find themselves caught between meeting the emotional and financial needs of their children and meeting the emotional and financial needs of their own parents. Being cared for by those you once cared for is a measure of dependency that many are unwilling or unable to take. But at the same time, Cicirelli (1981) finds that given a choice, aged parents would want to turn to grown children for assistance. In other words, the distancing, when it occurs, is not by choice. The solution must lie in findings ways for grown children to provide support to parents without fostering, unnecessarily, a sense of dependency for parents and a sense of burden for children. Finding such ways leads to the practices of human service agencies.

Agencies that provide services to the aged need to be aware of the way their services affect the autonomy and dignity of the recipient. They need to consider whether the manner in which the service is provided actually makes the recipient more needy or less needy (Kuypers & Bengtson, 1973; Langer, 1981). Programs that provide total care when such care is unnecessary may actually hasten the aging process. If grown children are to assume the responsibilities of advocate and representative, then agencies must be responsive to their recommendations concerning service delivery. It will not always be possible to be responsive to such recommendations, but if an appropriate aging involves a high degree of continuity, then every effort should be made to provide flexible services.

The practices of human service agencies reflect prevailing social policies. A change in practice first requires a change in policy. With regard to the elderly, many practices reflect policies concerning social status and chronological age. Whenever services are targeted for a particular age group, there is a tendency to assume that all members of that age group are appropriate recipients of that service or benefit. In terms of the elderly, this assumption may play a role in the feelings of vulnerability and dependency found in old age. As the evidence reviewed in this chapter demonstrates however, there is as much, if not more, variability between people over the age of sixty-five as there is between younger people. In fact, as Neugarten (1979) argues, age is probably becoming a less significant correlate of behavior and development with each successive cohort. Increasingly, social policies need to be defined in terms of social status rather than age status.

Just as practice reflects policy, policy reflects the attitudes and values of a society, and these attitudes and values sometimes reflect stereotypes. This dilemma seems especially true for the elderly. They are seen as a homogeneous group when in fact they are a highly diverse group. Kalish (1979) calls this stereotype the "new ageism."

> The message of the new ageism seems to be that "we" understand how badly you are being treated, that "we" have the tools to improve your treatment, and that if you adhere to our program, "we" will make your life considerably better. You are poor, lonely, weak, incompetent, ineffectual, and no longer terribly bright. You are sick, in need of better housing and transportation and nutrition, and we—the nonelderly and those elderly who align themselves with us and work with us—are finally going to turn our attention to you, the deserving elderly, and relieve your suffering from ageism. (p. 398).

Some of the elderly are poor, lonely, weak, and all the rest, but most aren't. The young-old continue to become more distinct from the old-old. The old-old continue to be a diverse population. Only when we are able to fully appreciate the diversity and growth as well as the decline and loss that characterize the later years of adulthood will we be able to ensure an appropriate old age. And perhaps only when we are able to do that will we be in a position to ensure an appropriate life-span.

This chapter completes the review of the stages of development across the life-span. One task still remains, however. This task requires an understanding of life-span development from a perspective broader than that of individual stages, from a perspective that cuts across stages. In particular, it requires an understanding of the patterns of development that characterize the entire life-span. This is the topic of the next, and last, chapter.

SUMMARY

THE CONTEXT OF DEVELOPMENT

1. Most people, including the elderly, have a very distorted image of this age group. Most old people are seen as withered and in need of constant care. Although some are this way, most continue to live independent lives.
2. The character of the over–sixty-five population is probably changing faster than that of any other age group, evolving into two relatively distinct groupings—the young-old and the old-old.
3. Social class and sex continue to be significant correlates of life circumstance during old age.

THE COURSE OF DEVELOPMENT: THE YOUNG-OLD

1. Most older adults welcome retirement but resent the fact that it is often mandatory. There is little evidence to support the notion of age-based mandatory retirement policies.
2. Adjustment to retirement depends on previous work history, health and financial resources, and degree of marital satisfaction.
3. Intergenerational aid is frequent between the aged and their families. There does, however, tend to be a greater degree of emotional distancing reported during old age than there is earlier in the life-span.
4. Maintaining a sense of purpose in old age is a significant correlate of physical and psychological well-being.

THE COURSE OF DEVELOPMENT: THE OLD-OLD

1. The declines in biological status first evident during middle age become more pronounced during old age.
2. Most of the aged have to cope with one or more chronic debilitating conditions.
3. Declines in cognitive functioning are not evident until the late seventies or eighties. Those evident earlier are more a reflection of changes in rate of information processing than they are of changes in its level.
4. The primary aging process reflects changes in cell viability, the cumulative impact of life-style, changes in the functioning of the immune system, and the regulatory ability of the genetic code.

DISENGAGEMENT AND DEATH

1. The life review is a common preoccupation among the very aged.
2. The association of death with old age is a fairly recent event, largely reflecting changes in health care and life-style for those younger.
3. An individual's personal coping with death reflects age, the nature and course of the terminal condition, and the availability of appropriate support systems.
4. The loss of a loved one is one of the most difficult of life's transitions. It often takes one to two years for the survivor to begin to reestablish a stable and satisfying life-style.
5. The resolution of the bereavement process reflects the age and sex of the survivor and the availability of peer and family support systems.

THE SIGNIFICANCE OF THE LATER YEARS OF ADULTHOOD

1. The nature of the changes typical during old age helps reinforce the value of a qualitative view of life-span human development and the significance of the distinction between reciprocal and complementary relationships.
2. The relationship between self-concept and well-being in old age highlights the central role of the self in the developmental process.

HUMAN DEVELOPMENT AND HUMAN SERVICES

1. The increasing longevity of the population will require significant changes in family roles, provision of services, and allocation of resources to all age groups.

KEY TERMS AND CONCEPTS

AGING

Neugarten's Young-Old and Old-Old
Average Life Expectancy
Activity Theory of Aging
Disengagement Theory of Aging
Mandatory Retirement
Kuyper and Bengtson's Social Breakdown Syndrome
Role Definition
Primary Aging Process
Errors of Commission
Errors of Omission
Terminal Drop
Lipofuscins
Free Radicals
Autoimmune Reactions
Butler's Life Review Process

DEATH AND GRIEF

Kubler-Ross Sequence of Response to Death
Reactive Depression
Preparatory Grief
Dying Trajectories
Weisman's Appropriate Death
Hospices
Anticipatory Grief
Lopata's Sanctification Process

SUGGESTED READINGS

There are three books that provide a clear appreciation for what in Harris's words are the myth and reality of aging in America.

Atchley, R. C. (1976). *The sociology of retirement.* New York: Schenkman.

Blau, Z. M. (1981). *Aging in a changing society* (2nd ed.). New York: Franklin Watts.

Harris, L. (1975). *The myth and reality of aging in America.* Washington: National Council on the Aging.

An article by Timiras provides an excellent summary of the biological factors that account for the primary aging process.

Timiras, P. S. (1978). Biological perspectives on aging. *American Scientist, 66,* 605–613.

Old age involves more than biological change. Change in family relations and change in self are also integral components. The following two sources deal with these issues.

Brubaker, T. H. (Ed.). (1983). *Family relations in later life.* Beverly Hills: Sage Publications.

Lieberman, M. A., & Tobin, S. S. (1983). *The experience of old age.* New York: Basic Books.

Death is as much a part of life as is birth. These four sources provide insight into the ways people deal with their own impending deaths as well as with the ways those close to them cope with the loss.

Glaser, B. G., & Strauss, A. L. (1968). *Time for dying.* Chicago: Aldine.

Kubler-Ross, E. (1969). *On death and dying.* New York: Macmillan.

Lopata, H. Z. (1979). *Women as widows: Support systems.* New York: Elsevier.

Raphael, B. (1983). *The anatomy of bereavement.* New York: Basic Books.

11
PATTERNS OF DEVELOPMENT

CHAPTER OUTLINE

In the course of writing the first ten chapters of this book, I was struck by the fact that certain topics, findings, variables, and qualifiers seem to appear and reappear at almost every stage of the life-span. In a sense this fact shouldn't be very startling at all. The process of development is cumulative and interactive. Events that happen at one point in the life-span should influence events at other points in the life-span. True enough, but the tendency, either in terms of research or in terms of service, is more typically to focus on a limited segment of the life-span—to understand all the factors that influence the neonate's first few days of life or to provide comprehensive services to the aged widow. Questions such as whether the neonate and the widow have much in common or even if what happens to the neonate will someday influence his or her bereavement are usually not asked. To attempt to answer such questions or even to attempt to frame the questions in such a form that they can be addressed requires a step back away from the specific circumstances of the neonate or the widow. It requires a look at the "forest" rather than the "trees." Or, more to the topic, it requires a look at the patterns of development that give shape and meaning to the developmental events of the life-span.

POWERS OF MAGNIFICATION: A LIFE-SPAN VIEW OF DEVELOPMENT

In one of the early chapters, I drew a distinction between considering the events of a particular stage from the perspective of that stage alone and considering them from the more general perspective of the entire life-span. I now want to return to that distinction because a complete understanding of development involves understanding the relationship of the whole to its parts. In the earlier discussion, I made the analogy between on the one hand a life-span view and a stage-specific view and on the other viewing a microscope slide under two powers of magnification. The greater the degree of magnification is, the more precise the detail. Structures that are not evident under low powers of magnification are clear under higher powers. If the magnification is switched to low power, the precise details are gone. But the view is not a void. New detail appears, new structures are evident, and new relationships emerge. It is not a question of one power of magnification being more correct or truer than the other. Both are correct and both are useful, but each in its own way.

The Psychological Organism

What then are we studying when we examine the course of human development across the entire life-span? In effect, what is it that is developing? What is developing is a **psychological organism.** This organism is the product of the continuing interaction of a **biological organism** with a **sociohistorical context.** The fact that the organism changes across its life-span is a reflection of the continuing evolution of its two sources. If by some circumstance, the two sources remained static, so too would the organism.

What is the nature of this organism and how is it distinct from its two sources? These questions are best answered through analogy. Consider first the process involved in the formation of water or ordinary table salt. Each is formed through the interaction of more basic elements: hydrogen and oxygen in the case of water and sodium and chloride in the case of table salt. Neither the salt nor the water resides in one or the other of its constituents. None of the basic elements tastes salty or feels wet. Rather, the salt and the water are created when their respective elements are combined. They come into existence as a result of an interaction, and each continues to exist as long as the interaction continues. If the elements are dissociated, the product of their interaction ceases to exist. The psychological organism comes into existence in the same manner. It is neither inherent in nor reducible to either its biological host or its social context.

There is a limit to this analogy, however. Salt and water are material. They can be seen, weighed, touched, and tasted. The psychological organism, however, isn't material in this sense. It can't be weighed or photographed. Appreciating the nature of its existence requires a retreat to a second analogy—to the physicist's concept of torque.

Torque is a force. Technically, it is defined as "a twisting or wrenching effect or moment exerted by a force acting at a distance on a body, equal to the force multiplied by the perpendicular distance between the line of action of the force and the center of rotation at which it is exerted" (*Webster's New World Dictionary*, 1976). From a more practical perspective, the concept of torque is evident in the tightening of a screw with a screwdriver or a nut with a wrench. The failure to appreciate its presence usually results in the stripping of the threads of the screw or bolt or the mangling of the slot on the head of a screw. Like water and salt, torque is the result of an interactive process. Unlike water or salt, torque is not material. Its presence is inferred rather than observed. The very unhappy nine-year-old who has just stripped the threads on the hub of his or her tire rim doesn't see torque but the effect of torque.

In a sense, the psychological organism is like torque. It exists as the result of an interaction, and its presence is inferred rather than observed. In the case of the psychological organism, the inferences are based on behaviors and changes in patterns of behaviors, that is, development. One of the reasons that the study of human development is such a difficult endeavor is that we are studying something that can only be known indirectly, through inference. When we say that the two-year-old is experiencing separation anxiety during the absence of the parent, we are making an inference based on the child's crying and reluctance to be held by others. When we say that adolescents are searching for an identity, we are making an inference based on their frequent changes of wardrobe, manner of behaving, and stated plans for the future. When we say that the aged disengage as a means of maintaining a coherent sense of self, we are making an inference based on a decrease in their frequency of being with people or going to once favorite places. There are limits to this analogy too, however. Torque is a passive notion. It is the inevitable result of a confluence of specific actions and materials. The psychological organism across the lifespan is anything but passive and inevitable. It seems to have a life of its own, one that sometimes proves very hard to predict.

These analogies are useful but limited. They help to define some of the characteristics of the psychological organism, but they don't define all of them. The relationship between the psychological organism and its components is two-way, not one-way. In this sense we describe the psychological organism as active. The **active organism** influences and is influenced by its component parts. It does so in ways that are at times deliberate and at times unintentional. It does so through actual behavior as well as through a continuing construction and reconstruction of the nature and meaning of the events it encounters. It is not inorganic and inert and stable; it is organic and at times unstable and even volatile. But it is not random. There is some rhyme to its behavior, some pattern to its development, and the remainder of this chapter examines this rhyme and pattern of the psychological organism.

The Components: The Biological Organism

The psychological organism may not be reducible to either the biological organism or the sociohistorical context, but it nevertheless is a reflection of these two components. An understanding of the nature of these components is prerequisite to a full understanding of the product of their interaction.

Species-Specific Characteristics

I can stand at the window of my office and flap my arms as hard as I can for as long as I can and I will not lift myself up and soar like a bird. It simply isn't in the cards for humans (not to mention the arms). Like any other species, our unique set of **species-specific characteristics** contributes to our unique definition as a species and to our unique developmental course. A species-specific characteristic is one universal to the species. Characteristics such as walking on two feet or using language are ones that readily come to mind. Other species may share some of our characteristics (or we some of theirs), but the sum total of these characteristics distinguishes us from other species and makes each species unique. For most of our lives we are bipedal; we walk upright without the aid of our hands. We have an opposable thumb which allows us to grasp and hold firmly. We have the ability to use language as a means of communication and as a means of symbolically representing experience. We appear to be a social species, few of us preferring total isolation for any period of time. We are a highly

Among our species-specific characteristics is our ability to communicate through the use of language.

adaptable species, able to survive over a range of environments greater than that over which virtually any other species can survive. We are a species capable of self-awareness: we can see ourselves as objects and reflect on our own nature (Nash, 1970). This last characteristic, which seems unique to our species, no doubt provides us the advantage of the ability to adapt over a wide range of circumstances but also the anxieties and fears that can accompany self-awareness. It makes us the only species whose members are able to contemplate their own eventual deaths and to feel the need to search for some sense of purpose or meaning in their lives.

Compared to other species, we have a very slow rate of maturation. That is, the amount of time it takes to achieve adult physical and physiological maturity is longer for us than it is for other species. Montagu (1981) argues that this characteristic makes possible the long period of socialization needed for effective functioning in our cultures. From an evolutionary perspective, this prolonging of the juvenile phases of development allows more time for the childlike qualities of curiosity, playfulness, imagination, joyfulness, and a sense of wonderment to become a permanent part of our psychological selves. Montagu even goes so far as to suggest that the impact of these youthful virtues could be further enhanced if the juvenile period could be extended still further by increasing the period of gestation—a suggestion many women might well take exception to.

As a species we come equipped with a variety of "built-in" behavioral patterns. Many of the ones present at birth such as the tonic neck reflex or the sucking reflex or the rooting reflex gradually disappear as the nervous system develops, but other patterns such as the emotional expressions of smiling, fear, startle, surprise, and crying remain. These patterns don't define our behavior, but they certainly influence its definition.

The Role of Biological Structure and Function

Even our physical structure influences our behavior. Our height and weight and body proportions influence our movement and coordination, our strength and stamina, and our means of responding to situations. Our sensory and nervous systems define the range of stimulation we are capable of dealing with and therefore some of the limits on the development of the psychological organism across the life-span. The changes in the rate of biological change across the life-span influence our psychological growth. During periods of rapid biological change, such as puberty, more of our psychological self seems to be tied up responding to our biological changes than during periods of comparatively slow biological changes more typical of the adult years. We possess a variety of qualities that have been collectively referred to as temperament traits (Thomas & Chess, 1980). These qualities such as intensity of reaction, adaptation to new situations, and regularity were discussed in the chapter on infancy. Although the origin of these temperament traits is still unclear, they are nevertheless an aspect of our biological makeup and as such influence our mode of functioning as well as the behaviors of others toward us.

There is, however, no one-to-one correspondence between the biological and psychological organisms. In fact there are remarkably few correlations between biological and psychological functioning across the life-span that are not a reflection of culturally imposed values. Height and weight don't correlate with cognitive abilities; blood pressure doesn't correlate with personality, and so on. When correlations are found, for example between physical appearance and self-esteem, they are always culture specific. What passes for beauty is a very fleeting quality. When cross-cultural correlations are evident, they are most likely to occur at the beginning or the end of the life-span where level of biological functioning does seem to impose specific limits (Birren, Kinney, Schaie, & Woodruff, 1981; Bower, 1979) or during periods of extreme stress to or abuse of biological structures (Dohrenwend, 1973; Selye, 1978).

The fact that there are few correlations does not mean that level of biological functioning is unimportant. Rather, it means that most people most of the time maintain an adequate level of biological functioning to adequately support the functioning of the psychological organism. Further, it should not be assumed that the relationship is unidirectional, that is, that the psychological reflects the biological and not vice versa. The psychological organism is an active organism and as such can influence as well as be influenced by the biological. Consider the value of prepared childbirth techniques as a means of countering the stressful effects of uterine contraction or Lieberman and Tobin's (1983) observation that the more "cantankerous" nursing home residents maintained better health in part because their psychological state actually resulted in an increased blood flow and therefore an increased oxygen supply to the brain or Campbell's (1981) finding that life satisfaction and activity patterns correlate more highly with perception of health in old age than with any actual measure of health, and you can gain some appreciation for the bidirectionality of effects.

The Role of the Gene

Two final points need to be made in this discussion of the biological organism. The first concerns the role of genetics. In a sense, the functioning of the biological organism can be thought of also as the product of an interaction process. Specifically, it is the product of the interaction of one's genetic endowment with a particular environment, initially intrauterine and subsequently extrauterine. The relationship between genotype and phenotype is a reflection of the nature of this interaction process since any genotype is capable, in interaction, of a variety of phenotypes. The genetic code is the primary regulator of the functioning of the biological organism across the

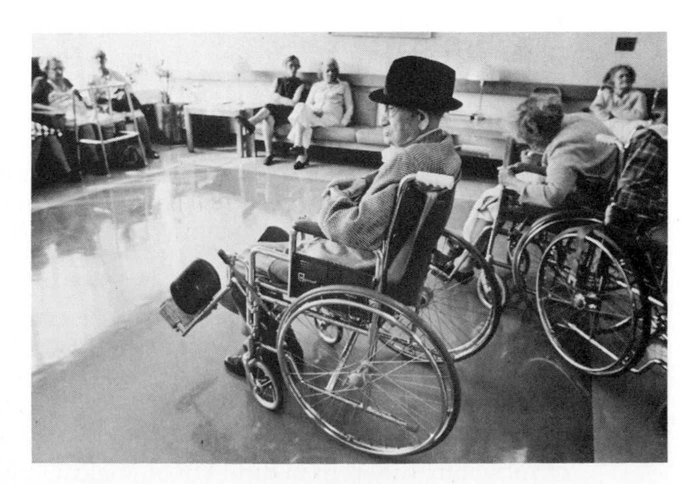

Psychological states can have an effect on biological functioning. Cantankerous nursing home residents may be in better health than their submissive and inactive counterparts partially because of the increased oxygen flow to the brain that is a byproduct of their agitated psychological state.

entire life-span. It is equally evident in the differentiation and growth of structures, in the maintenance of level of functioning, and in the biological declines that are evident in later life.

Since genetics influences the functioning of the biological organism, it also plays a role in influencing the functioning of the psychological organism. However, it should be clear that its role is indirect and mediated by a variety of other factors. The influence is direct only when a particular gene exerts such an overwhelming direct effect on the biological organism that its range of expression (that is, its potential for unique interactions with diverse environmental factors) is so severely limited that it severely limits the range of expression of the psychological organism. The variety of inherited genetic diseases and the variety of errors of replication during cell mitosis discussed in Chapter 2 are good examples of such direct effects. In this regard, it is important to note that such direct influences are almost always negative in nature.

Individual Differences

The second point that still needs to be made concerns the relationship between species-specific characteristics and individual differences in the level of functioning of the biological organism. With the exception of those whose development is severely abnormal, species-specific characteristics are present in all individuals, irrespective of culture or historical time. But they are not present to the same degree or in the same form in all. All people talk, but we talk in a variety of languages and dialects within languages. Further, some people are more skilled at the use of language than others. In other words, people show a range of individual differences on all characteristics, both species specific and context specific. A full understanding of development across the life-span requires an appreciation both of those developmental patterns that are characteristic of the species and of those that account for the differences between members within a species (McCall, 1981).

The Components: The Sociohistorical Context

Context defines time and place. It emphasizes the importance that the particular slice of history our individual life-spans traverse has on the course of our development and further the importance of the place where that particular traverse occurs. Living in New York City between the years 1800 and 1875 defines a very different life from the one defined by living there between the years 1900 and 1975. Similarly, living in New York City during the period of 1900 to 1975 defines a very different life from the one defined by living in the bush country of Africa during the same interval. Knowing the characteristics of the sociohistorical context is parallel to knowing the range of wavelengths of light that the eye can detect. That is, both forms of knowledge specify the nature of the information that is available to the psychological organism.

Cohorts

If the impact of the biological organism on development is best appreciated through an understanding of species-specific characteristics and the range of variability inherent in different biological systems, then the impact of the sociohistorical context is best appreciated through a focus on cohorts (Kuhlen, 1963; Schaie, 1965; 1983).

Probably the classic cohort example of the impact of the sociohistorical context is the data on changes in intelligence test scores across the adult years. The specifics of the issue have already been discussed in earlier chapters, but what is relevant in the present context is that successive cohorts of adults show less and less decline in intellectual competence across the middle and later years of adulthood. The changing nature and impact of their life experiences—of their educational experiences, their health, medical, and nutritional history, and the nature of their work—have summed to allow the maintenance and, in an increasing number of cases, the further growth of the intellectual skills first acquired during the childhood years.

The Human Ecosystem

At present, the writing of Bronfenbrenner (1979) on the ecology of human development best conceptualizes the sociohistorical context. The structure was originally presented in the first chapter and has served as a framework for integrating the various experiences encountered at each stage of the life-span. It is probably useful to again review the basics of the human ecology model.

The microsystem. The human ecology model defines four system levels that influence the course of our development. The first, the microsystem, concerns the structures of our immediate environments. It is defined

The microsystems that comprise a child's immediate environment have a profound effect on development. Imagine how different these children's world would be if they were attending a small rural school rather than this fortress-like inner-city school.

both in terms of its material and physical characteristics such as number of rooms in the house and in terms of the social roles and relationships that exist between those who share our microsystems. Most of us encounter across our life-spans the microsystems of our family of origin and later of our family of procreation, our educational setting, our neighborhood and workplace, and our settings for leisure and recreation.

The mesosystem. The level of the mesosystem represents the relationships that exist between specific microsystems. Many of the dilemmas encountered during the early years of adulthood can be thought of as dilemmas at the level of the mesosystem. That is, many of these problems involve the coordination, or more specifically the lack of coordination (Kamerman, 1980), between the home and the workplace or the home and the school, between the roles of mother, wife, and worker, between the roles of husband and wife, and so forth.

The exosystem. The level of the exosystem represents the place where events occur that influence our lives but with which we have no direct contact. The events that occur within broad economic, legal, or governmental settings exemplify such exosystem events. The stock market crash of 1929 was such an exosystem event; so too were the Japanese attack on Pearl Harbor in 1941, the 1954 Supreme Court desegregation decision declaring the "separate but equal" doctrine unconstitutional, and the Vietnam War. Exosystem events need not be on such a grand scale as these. A local shopkeeper who discovers that a huge shopping mall is to be built in a neighboring community is soon to feel the impact of events created by such a decision in which he did not participate.

The macrosystem. The broadest level is that of the macrosystem. The macrosystem is a level of values and ideals, goals and assumptions, blueprints and ideologies. It is the level of interpretation of the Constitution and other similar documents. It is the level that guarantees individual freedoms and family rights, even though at times such guarantees can have a negative impact (for example, the difficulty in severing parental ties in chronically abusive families). The structure of the macrosystem changes, the changing views of the appropriate roles of men and women being a current example of a macrosystem change, and it exists in different forms in different places. Countries favoring a socialist orientation see the state as appropriately being more intrusive in the lives of citizens than capitalist systems see it as being. This intrusiveness might be seen as beneficial (as in the case of state-funded total medical care) or deleterious (for example, the limiting of freedom of choice as to work, marriage partner, or family size), but in either case the assumption is that the proper role of the state is to play a more direct role in the lives of its citizens than it does in capitalist societies.

The Influence of the Ecosystem on the Psychological Organism

The individual's perception of these four levels is as much a subjective as an objective matter. We choose what elements constitute our microsystems. The choice is not arbitrary, but neither is it simply a carbon copy of the elements that objectively constitute our microsystems. For some, casting a vote in an election is sufficient to make the government part of their microsystem; for others no such sense of efficacy exists. They may simply feel that their life is influenced by events that will always remain out of their control. The shopkeeper may decide that the new mall will hurt business and see no choice but to relocate. However, the shopkeeper may also fight the construction by participating in the zoning board hearings on the proposed mall, that is, by redefining a part of the exosystem as mesosystem, a developmental process aptly described by Bronfenbrenner (1979) as "where exo was there shall meso be."

The relationship between the psychological organism and the sociohistorical context is like the relationship between the psychological organism and the biological organism. That is, there is no one-to-one correspondence, and the influence is bidirectional. The shopkeeper who becomes active in the zoning board hearings is acting on his social ecology. The outcome of his actions, in turn, will have bearing on his future endeavors. It will not only determine if he will continue to operate his business in its present location but in so doing will also influence many other aspects of his life and the lives of those he touches. It will almost certainly influence his pattern of involvement in future political activities. Perhaps he will even run for a seat on the zoning board.

But it is more than simply the outcome of the zoning board deliberations that will influence his life. It is also his perception of the proceedings. Does he think the proceedings were fair? Does he think the decision to grant the mall a building permit was already made in the minds of the zoning board members and therefore that the hearings were merely a charade? As discussed throughout the text, one of the major factors distinguishing people from economically advantaged backgrounds from those whose backgrounds are less advantaged is their perception of the amount of power they have within their environments (Kohn, 1977), and again, although people's subjective perceptions are not formed independently of what others might call the objective reality of a situation, neither are they carbon copy.

The Components: Their Interactions

The psychological organism is a **self-regulating system.** It tries to maintain a steady state or degree of balance among and between itself and its components. This notion is reflected in Piaget's process of assimilation and accommodation, in Schaie's dialectical balance between thesis and antithesis, and in Levinson's notions of being young and being old, or the adult's dilemma of maintaining a balance between affiliation and individuation. In each case, the psychological organism behaves in order to maintain or to restore and maintain a degree of balance, or equilibrium.

The process of maintaining a degree of equilibrium is not, however, a static one. Instead, it is a process of **dynamic equilibrium.** In other words, it is not one akin to placing weights on a balance beam. The psychological organism, in the course of maintaining a sense of equilibrium, does not return to its original state; it does not come full circle. The helix rather than the circle is the appropriate metaphor. The psychological organism can maintain a steady state (equilibrium) only by changing, by constantly adjusting to changes within the biological organism and the sociohistorical context. The sequence of these changes thus becomes the course of development, and the study of this course becomes the study of the cumulative, bidirectional interactions of the biological organism and the sociohistorical context with the psychological organism.

THE COURSE OF DEVELOPMENT

The course of development of the psychological organism has been presented as a succession of stages. Each stage represents a relatively unique pattern of organization among the various elements that constitute the psychological organism. Further, the rate of succession across stages bears some relationship to time. Individuals of a particular age are more likely to be at one stage of development than at another. This section reviews the structure of these developmental stages, the factors that influence their succession, and the relationships that exist across them.

Patterns of Organization

The notion of stage as a structure, or pattern, of organization among diverse elements (Noam, Kohlberg, & Snarey, 1983; Wohlwill, 1973) emphasizes the importance of coherence and integration in understanding the concept of development. It highlights the fact that the child who knows that pouring water from one glass into another has no effect on the amount of water in the glass is also more likely to know, than the child who cannot conserve continuous quantity, that changing the spatial arrangement of a group of objects has no bearing on the number of objects in the arrangement, that a group of objects differing in terms of shape and color can be sorted on either dimension and even sorted on both dimensions simultaneously, that fairness has more to do with equality than with personal desire, and so on. It also highlights the fact that parents who value highly the virtues of respect for authority, hard work, and the maintenance of traditional values are also those who tend to value obedience in the behavior of their children, who are not likely to allow their children a significant voice in child-rearing decisions, who are most likely to use strong measures to curb children's behavior that is inconsistent with their values, and so on. The correlations are not perfect, but neither are they random. The patterns do appear, their sequence does bear some relationship to time, and they therefore are useful in helping to sift through and order the vast variety of behaviors that constitute a person's life-span. Two examples of this usefulness seem appropriate. One concerns parallels in stage structure across the life-span and the other the role the components play in defining stage structure.

The Helix

If you compare behaviors typical at different ages, such as those of the two-year-old, the fourteen-year-old, and the forty-five-year-old, they seem to have remarkably little in common. If, however, you go a little deeper and examine the pattern that exists among each of the three sets of behaviors and if you go even deeper yet and consider how these three superficially distinct patterns of behavior may in fact be related, you are asking a question about parallels between developmental stages. These three particular ages are all often described as stages of adolescence, that is, as times when the individual seems to place great emphasis on the importance of autonomy and independence. The metaphor of development as a **helix** also reflects this notion of stage parallels. Figure 11-1 presents a helix used by Kegan (1982) to depict the course of development of self across the life-span. Through the path of the helix weaving back and forth, each time at a new level (that is, stage), Kegan depicts the notion of themes recurring across the life-span.

Kegan's sequence of stages, the incorporative, the impulsive, the imperial, and so forth, corresponds to the age-related stages discussed throughout the text. The impulsive infant and preschooler are reflective of

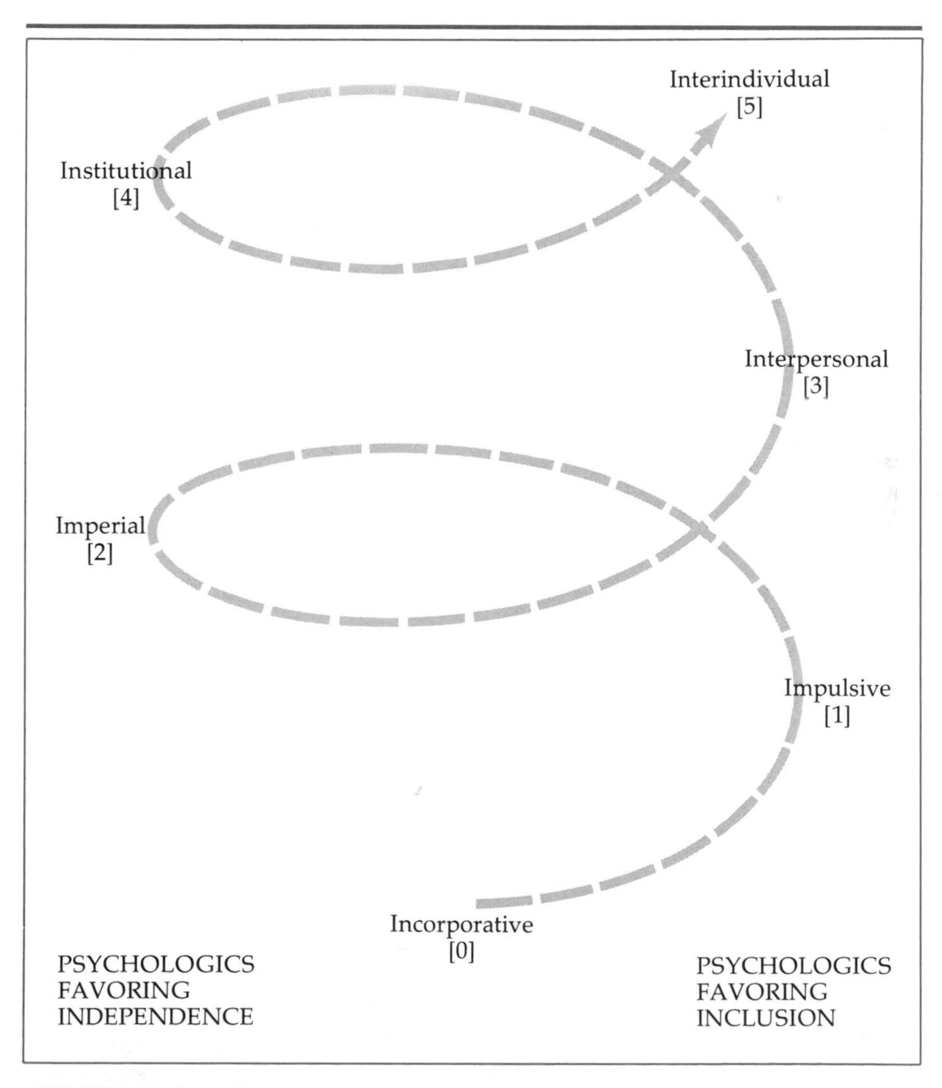

FIGURE 11-1
The helix of developmental truces (from Kegan, 1982).

Erikson's focus on autonomy and initiative and on Piaget's discussion of the way in which the preoperational child is so easily overwhelmed by the power of the present. The imperial child is Erikson's industrious child and Piaget's concrete operational child. Interpersonal concerns reflect the dilemmas of adolescence and the transition years; institutional concerns reflect the dilemmas of young adulthood.

Like the theorists discussed in the chapters on middle age and old age, Kegan identifies an interindividual level that reflects a coming to terms with oneself. He sees each stage as a "developmental truce" between two often conflicting desires: the desire to feel that one is part of a social group and the desire to remain distinct and autonomous. Each stage resolves the dilemma in a different way, sometimes giving greater weight to one desire, sometimes to the other. In Kegan's words,

> we moved from the overincluded, fantasy embedded impulsive balance to the sealed up self-sufficiency of the imperial balance; from the overdifferentiated imperial balance to the overincluded impersonalism; from interpersonalism to the autonomous, self-regulating institutional balance; from the institutional to a new form of openness in the interindividual. (p. 108)

The notion of parallels in stage development is an important one. It reminds us that certain basic issues, such as identity formation, are repeatedly confronted, at successive levels and in different contexts, over the course of a life-span. It may even be that the individual's own awareness of these parallels is itself a significant factor in determining the mode and ease of resolution of recurring developmental tasks such as identity formation.

The Changing Role of the Components

The second example of the usefulness of stage structure concerns the role that the biological organism and the sociohistorical context play in defining the successive stage structures of the psychological organism across the life-span. Although both components are reflected in stage structure at all stages, their roles seem to change. In particular, from infancy through childhood and across the adult years, the sociohistorical context assumes an increasingly greater role in differentiating people. It is in this sense that Neugarten and Datan (1973) talk about the shift from childhood to adulthood as involving a shift from the biological clock to the social clock as a major correlate of development. It is in this sense that normative life events become associated with cultural norms rather than with biological timetables. It is in this sense that people are seen as increasingly becoming less like each other, or more variable, with increasing age.

None of this implies that the functioning of the biological organism becomes irrelevant to adult development or even that the adult biological organism does not change. Neither statement is true. What is does mean, however, is that the events that are seen as significant developmental milestones during the adult years (age at marriage, age at parenthood, age considered middle age, and so forth) are more a reflection of context than of biology. What it also implies is that if variability increases with age, then the timetable of developmental events based on the sociohistorical context is more variable than the one based on the biological organism. That is, there is more potential flexibility in the sociohistorical factors that define, for example, age at marriage, than in the biological factors that define the timing of the sequence of motor development skills.

Sequence of Stages

The preceding discussion on the structure of developmental stages indicates how best to envision the organization of the psychological organism, but it says nothing as to why we change. Even the discussion on parallels across stages provides little insight as to the motivating force for change. This section will provide some answers to the question of why the psychological organism changes over time.

Simply put, the psychological organism develops because its component parts—the biological organism and the sociohistorical context—develop. Development is an integral part of all life. All forms of living organisms have a finite life-span, and during that life-span, they develop. They are each in some way different at death from what they were at birth. Different theorists may characterize the changes in different ways, but all acknowledge that change is an inherent part of all biological organisms.

At the same time that biological change occurs across a life-span, sociohistorical change is occurring across successive life-spans. We may differ in our opinions as to whether our ability to annihilate ourselves with nuclear weapons and our ability to control population through safe, effective contraception represent an advance over a time when annihilation was attempted through the use of sticks and stones and population control through the tossing of infants over the sides of cliffs, but we almost certainly must agree that change has occurred. In fact, the human biological organism is not seen as having changed in virtually any manner over the past 35,000 to 50,000 years (Fishbein, 1976). If we are different from our prehistoric "forepersons," it is because of the changes in our sociohistorical context.

We don't need to go back 35,000 years to appreciate the significance of these sociohistorical changes. The changes in the timing and structuring of developmental stages over the past 150 years, largely reflecting the increasing impact of urbanization, technology, and industrialization on our lives, are evidence enough (Hagestad, 1981). The developmental stages of adolescence and, increasingly, of youth, the modern structure of middle age, the association between old age and time of death, even the dramatic increase in average life expectancy itself are all testament to the changing sociohistorical context. Again simply, the psychological organism develops because its component parts develop.

If development is as inherent a characteristic of the psychological organism as it is of the biological organism, how can the developmental sequence be described? The orientation of the text has been to describe the stage sequence as a succession of qualitatively distinct patterns of organization. Each pattern reflects the best attempt that can be made at a particular time to establish and maintain an optimal degree of equilibrium between the psychological organism and its environments. At the same time, I have mentioned that development can also be viewed from a quantitative perspective. From the quantitative perspective, change is seen in terms of degree rather than kind. The difference between the two perspectives is not one of right and wrong but rather one of emphasis. There are certainly changes over the life-span that are better described as quantitative than qualitative. Many changes in the biological organism (such as height and weight) can be so characterized. Many of the skills measured on standardized aptitude and achievement tests (such as vocabulary and information-processing rate) also fit a quantitative model. Why then the preference for a qualitative model?

The designation of the years between age twelve and age nineteen as adolescence is an outgrowth of modern society. Prior to the Industrial Revolution, children passed from a short childhood directly into adulthood.

For me, the answer to the question has as much to do with aesthetics as it does with science. From a quantitative perspective, as I mentioned previously, it is hard to see the study of development as anything other than the study of improvement followed by decline. It is certainly true that on many developmental measures, such as speed of performance, the course of improvement is followed by gradual decline. But if that is all that is seen, the remarkable coping skills of both the young and the old are left unappreciated, and services and interventions are seen largely as helping the young grow up and keeping the old from growing older. There is beauty at all stages of the life-span, and a qualitative perspective is better able to capture it.

Interdependence of Stages

If stages are a sequence of successive adaptations, what of the third of the three questions asked at the beginning of this section? What is the relationship between events at different stages? How does one stage go about succeeding another? How do the events of one stage influence the events of another? The question of the relationship between events at different stages is one that generates great debates among developmentalists (see Clarke & Clarke, 1976, and Sroufe, 1977, for contrasting perspectives). The conventional wisdom has been that earlier experiences influence later ones and therefore exert a disproportionate impact on the course of development. However, as was discussed in Chapter 1 and again in Chapter 7, the stark contrast in the nature of events encountered across an entire life-span makes it hard to see how earlier experiences would necessarily determine or even influence later ones. In other words, alternative perspectives see the course of development as more discontinuous than it was once thought to be.

If development is seen as a succession of reconstructions of previous experiences interacting with present encounters as well as evolving conceptions of future status, then what is carried from the past is not absolute but relative. Preschool children respond to the actual behaviors of the parent, but adolescents are more likely to respond to their perceptions of the meaning behind their parent's behavior. Further, that meaning continually undergoes reconstruction as adolescents move into adulthood. When adults become parents, the experience of coping with an infant invariably results in reflections on their memories of themselves being parented. Typically, these reflections result in a more favorable evaluation of their own parents. This reevaluation, in turn, may well result in an improved present or future relationship between adults and their parents as well as perhaps an improvement in the adult-infant relationship. From such a conceptualization, it is evident that new experiences influence our remembrance of old ones at the same time that old experiences influence current behavior. The effect is truly transactional (Sameroff, 1975).

Even if development across the life-span is a transactional experience and even if new experiences change our remembrance of old ones, the past itself cannot be changed. Earlier experiences may not influence all new experiences at all stages of development and those earlier experiences may be subject to reinterpretation, but they nevertheless do exert a significant influence on the course of development. How does this happen?

Using a scheme originally suggested by Anastasi (1958) to describe the relationship between heredity and environment, we can depict four modes in which earlier developmental experiences can influence subsequent ones.

Mode I

The most direct mode is one in which an earlier event or sequence of events leaves such an indelible mark on a person that no subsequent experience or stage reorganization has any appreciable offsetting influence. The impact of this earlier event is so strong that virtually all areas of a person's future function, at all future points, are affected by the earlier event. Fortunately, such happenings are relatively uncommon (Goldhaber, 1979). The mechanisms of continuity in the psychological organism appear to be less direct than those of the biological organism, no doubt reflecting the more qualitative nature of the psychological organism. This is not to suggest that such devastating early experiences cannot and do not happen. Children who carry the psychological scars of years of physical and mental abuse will always bear some burden. But fortunately there are few psychological events that leave as sudden and profound an impact on development as assaults on the biological organism (for example, paralysis, amputation, congenital defects).

Mode II

At the other extreme of the continuum are situations in which what happens at one stage of development has no bearing on future functioning. Age-specific fashions represent such a case. Adolescents dress in ways that are consistent with the dress of their peer groups. Often these dress patterns are distinct and different from adult fashions. And they are not predictive of adult dress patterns. They are merely a reflection of influences that are specific to time and place. They remain specific to time and place because there is no vehicle of continuity available. The biological organism is not altered by the clothing. The sociohistorical context actually supports discontinuity in dress between the adolescent and adult stages, and the psychological organism, although not forgetting clothing once worn, probably has little reason to incorporate these memories into current decision making.

Mode III

So much for the extremes. What about the two modes in the middle? If one extreme represents unlikely cases and the other inconsequential cases, then the two modes in the middle must account for most of the continuity that exists across the life-span. The first of these two middle modes represents situations in which early experiences do carry over into later ones but their breadth and intensity of impact depends on two additional factors. The first is the current situation and the second the nature of the events during the interval between the original event and the current situation.

The research on the effects of divorce on children provides a particularly good example of Mode III situations. The findings, discussed in Chapter 8, indicate that parental divorce has both short- and long-term influences on children's behavior and development. The findings also point out that the age of the child at the time of the divorce is a significant factor as is the quality of the parents' relationship to each other following the divorce. Further, the findings indicate that the impact differs at different developmental stages because at each stage new tasks are encountered and the legacy of the divorce experience must be incorporated into each. For the adolescent, this legacy might influence identity formation. For the young adult, it might influence decisions surrounding marriage and later decisions concerning parenting strategies. For the middle-aged adult, the divorce experience might influence involvement in the care of now aged parents. In all these instances, the continuing impact of the original event (divorce) depends on the circumstances surrounding the original event (such as the age of the child), the intervening circumstances (such as the continuing relationship of the parents), and the current circumstances (such as the decision to marry).

Consider a second example. A couple is very involved in preparing for the birth of their first child. They attend prepared childbirth classes, carefully consider the options that are available in terms of actual labor and delivery techniques, and believe that the shared experience of their child's birth will always be one of the most significant events in their married life. Although shared childbirth is a valuable experience, the couple is probably overestimating its long-term impact. There is actually little direct, long-term relationship between involvement in shared childbirth and other aspects of the parent-child relationship. But this doesn't mean that there isn't any continuity (nor does it mean that adults not involved in prepared

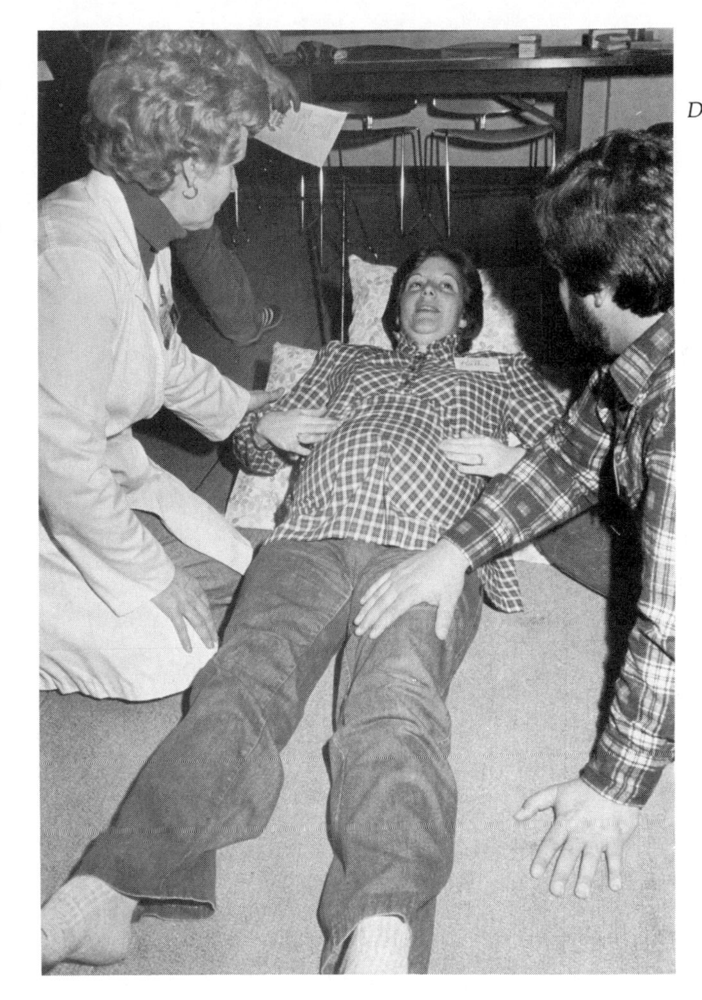

The influence of a particular experience on future development is sometimes overestimated. While participation in prepared childbirth classes has positive benefits, research has shown that participation has few direct, long-term effects on the parent-child relationship.

childbirth programs place their child at a disadvantage). Unlike the example of divorce in which the event itself continues to influence the course of development, in this example the shared childbirth experience may in turn lead to a second experience (perhaps greater involvement on the part of the father), which may lead to a third experience (increased marital satisfaction), and so on. In other words, continuity is maintained through setting in motion a series of events that are each distinct from the others. There is continuity between these events because the presence of one influences the likelihood that the next will occur.

Intervening experiences can serve to heighten the impact of the original event or to lessen it. They can influence the pervasiveness of the impact. Put another way, the probability that an event will have long-term impact on the course of development depends as much on the continuing circumstances of an individual's development as it does on the nature of the original event (Werner & Smith, 1977). Introduce a significant change in the individual's environment and the impact of earlier events is likely to be reduced. Maintain the same set of forces, as in the case of the culture of poverty discussed in Chapter 2, and the earlier effects such as low birth weight, poor diet, and greater susceptibility to illness will combine with poor educational opportunities and limited parental coping skills to lead to low achievement, limited job skills, early pregnancy, and still another generation's being trapped in the cycle described in Rubin's (1976) *Worlds of Pain*.

Mode IV

The last mode differs from the third in terms of degree of continuity. In the example of divorce provided in the discussion of Mode III, continuity of influence across stages was present, but the nature and degree of that influence changed. In this instance, continuity depends on the match between the earlier experience and the present circumstance. Maccoby (1980) recognizes this pattern when she notes that the parent's impact as a model is often not evident until children, many years later, find themselves in the same circumstances in which they once observed their same-sex parent. Kagan and Moss (1962) observed a similar pattern, which they dubbed **the sleeper effect**. Their longitudinal research design investigated patterns of development from infancy through the early adult years. They were particularly interested in the long-term, cumulative impact of developmental experiences. They found that in some cases parent behavior toward the infant was often a better predictor of preadolescent and young adult behavior than was parent behavior at the two later times of measurement.

In a sense the original experience lies dormant until such time as a future event makes the original experience relevant. If no such future event occurs, if for example the adult never decides to parent, then the original experience of observing the parenting behavior or the mother or father remains dormant and is not predictive of current or future behavior, and little if any continuity is demonstrated. The fact that predictability across the life-span decreases the farther apart the two points are is no doubt a reflection of this generalization. The more distant the two periods are, the less relevant the experiences of one are to the experiences of the other.

This discussion on the structure, sequence, and interplay of developmental stages across the life-span is intended to leave you with four lasting impressions.

1. Life-span human development is best understood as a qualitative process involving a succession of relatively unique attempts at adaptation between the psychological organism and its environments.
2. The process of adaptation is bidirectional—it is accomplished both through the psychological organism's responding to the demands of its environments and through the organism's trying to modify its environments to be more consistent with its structure.
3. The process of adaptation involves both actual changes in behavior and reconstructions of the meaning, importance, and interrelationships of events anticipated and experienced.
4. The interplay of developmental events across the life-span is a complex process, involving different degrees of impact at different points in the life-span, involving different areas of impact at different points in the life-span, and reflecting not only the nature of the original event but also the nature of the current situation and the sequence of events occurring during the interim.

THE CONTEXT OF DEVELOPMENT

The discussion of patterns of development, up until now, has focused on the general characteristics of the psychological organism. The descriptions of stage structure, stage sequence, and components are ones that appropriately apply to all people. However, a large part of this text has been devoted to demonstrating that not all people are alike. In particular, discussions have shown that two factors—social class and sex role—are particularly significant correlates of the course of individual development across the life-span. It is to these two factors that the discussion now turns.

Sex as a Correlate of Development

One of the most consistent patterns of development across the entire life-span concerns the different developmental patterns of men and women. At virtually any stage of the life-span, it is hard to find research that does not show males and females to be different in some way. The difference may be in terms of ability, it may be in terms of motivation or orientation, it may be in terms of health or life expectancy, or it may be in terms of interpersonal relationships. Whatever the focus, whatever the age, the differences seem to exist (see Block, 1984, for a complete review). Further, even though the ranges of variability for men and women on any particular measure may overlap considerably, the total aggregate, or pattern, of these specific aspects of development shows a greater degree of differentiation between men and women than is present for any individual aspect. In the most general sense the differences imply that the world as constructed by women and the world as constructed by men are probably not the same. The two worlds certainly overlap, but like two artists each painting a scene from a slightly different perspective, the final products reflect two unique orientations.

That differences exist in the behavior of men and women across the life-span is an empirical statement, a statement of what exists in a particular place and at a particular time. In and of itself it says little about the origin of these differences, the relative advantages and disadvantages these differences confer on men and women, or their impact on the ability of men and women to deal with other members of their own sex or with the opposite sex. Further, this empirical statement says little about the ease or desirability, from a developmental perspective, of attempting to change these patterns of development.

The behavioral differences that exist between men and women reflect the cumulative interplay of biological factors with the sociohistorical context. These differences tend to maintain themselves across the life-span largely because the culture continues to value distinct roles for men and women. As summarized by Block (1984), parents continue to place greater emphasis on their sons' achieving impulse control, independence, and personal responsibility than on their daughters' achieving these goals. Parents, particularly fathers, are more likely to use physical punishment with their sons. On the other hand, most parents place greater emphasis on maintaining close relationships with their daughters. They expect greater truthfulness from them than they do from their sons. Although parents are less likely to use physical punishment with their daughters, they are nevertheless more restrictive of their daughters' activities than of their sons'. This restrictiveness is especially true of mothers.

Even if the culture no longer valued distinct roles, even if, for example, the culture were as supportive of men assuming the primary parenting role as it now is of women, the differences, to the degree that they reflect species-specific biological factors, would still be present. Of much greater importance, however, if such a circumstance were to occur, is the fact that these continuing differences would best be appreciated as we now view temperament differences. That is, they would say more about the way one does a task than about how well one does it.

The interplay of these early biological and social differences continues to cumulate across the life-span. Differences are evident in educational and career goals and in the relative value of the three adult roles of intimate, parent, and worker. They are reflected in the form each of the three roles takes for men and women across the life-span, they are reflected in the satisfactions derived from each across the life-span, and they are reflected in the value and nature of the interpersonal bonds that men and women experience across the life-span. If the observations of researchers such as

The emphasis on achievement and individual effort that is characteristic of many men may increase their chances for advancement and reward but perhaps at the cost of peer emotional support.

Levinson (1978) and Gilligan (1982) are correct, then the orientation of men at least into middle age is one that continues to place primary emphasis on achievement and individual effort, and the orientation of women is one that values affiliation, compromise, and intimacy.

Not only do the developmental patterns of men and women differ across the life-span but so too do their sources of satisfaction and dissatisfaction and the relative values that the culture places on these developmental patterns. Even in early childhood, girls seem to be aware of the fact that society values males more than females, and this awareness continues throughout the adult years.

This discrimination is unfortunate for two reasons: first because it is discriminatory and second because it only tends to emphasize the advantages of the male role and the disadvantages of the female role rather than considering the advantages and disadvantages inherent in both roles. For example, the same emphasis on affiliation and nurturance that tends to interrupt the career development of women (via their greater role in parenting) also provides them a larger support system for dealing with stress. In other words, any particular aspect of male or female sex role has both inherent advantages and disadvantages. The achievement orientation of men may well lead to advancement and its attendant rewards, but the research indicates that it does so at the price of friendships and social supports. One only needs to review the discussion of male and female reactions to divorce or the death of a spouse to appreciate the importance of support. The very fact that women appear to be more in a quandary about how best to integrate their adult roles than men (who, the evidence suggests, are likely to argue that advancement in work also benefits the intimate and parent roles) suggests that the same set of social forces that makes interruption a common aspect of their adult lives may also make it more likely that they will understand the dialectic nature of adult issues sooner than men.

What then of change? If there are advantages as well as disadvantages to traditional male and female roles, might not people lose as much as they would gain through making sex roles less traditional? The answer is yes and no. Yes, if those valuing less traditional sex roles fail to appreciate the fact that behaviors are embedded in patterns and that a change in any one behavior invariably produces a change in others as well. (As mentioned in the previous paragraph, the price one pays for the rewards of greater achievement may well be a decrease in the social support of others. The adult who doesn't understand the fact that some behaviors appear "yoked" to others may be surprised to find success not very satisfying.) No, if a movement toward less traditional sex roles involves a shift from complementary to reciprocal relationships between couples. The more reciprocal the relationship is, the greater the potential for empathy. The greater the potential for empathy, the more likely each partner is to be supportive and appreciative of the efforts of the other. And the more likely this is to happen, the more likely it is that individuals will find relationships such as marriage more satisfying than they now do.

Over the past twenty to thirty years, there has been a consistent effort to make sex roles less traditional and stereotypic and more androgynous. Not surprisingly, the major supporters of these changes have been women. Nor is it very surprising that the greatest degree of change has occurred in the lives of women. Corresponding change in the lives of men is only now becoming evident (Pleck, 1983). These changes are not occurring in the lives of all women, however. They are much more evident in the lives of women from economically advantaged backgrounds than in the lives of those from less advantaged backgrounds. An appreciation of why this is so requires an understanding of the role social class plays in defining the patterns of development across the life-span.

Social Class as a Correlate of Development

Social class is essentially a correlate of economic advantage. Social class membership influences life's opportunities, rewards, and risks. It influences the way one parents, it influences one's perception of self in relationship to others, and most generally, it influences one's view of the world. In this sense social class membership exerts an influence similar to sex-role orientation.

Throughout the text, I have made extensive use of the work of Kohn and his colleagues as a basis for defining the impact of social class membership. According to Kohn,

> the essence of higher class position is the expectation that one's decisions and actions can be consequential; the essence of lower class position is the belief that one is at the mercy of forces and people beyond one's control, often, beyond one's understanding. Self direction—acting on the basis of one's own judgment, attending to internal dynamics as well as external consequences, being open minded, being trustful of others, holding personally responsible moral standards—this is possible only if the actual conditions of life allow some freedom of action, some reason to feel in control of fate. Conformity—following the dictates of authority, focusing on external consequences to the exclusion of internal processes, being intolerant of nonconformity and dissent, being distrustful of others, having moral standards that strongly emphasize obedience to the letter of the law—this is the inevitable result of conditions of life that allow little freedom of action, little reason to feel in control of fate. . . . Self-direction, in short, requires opportunities and experiences that are much more available to people who are more favorably situated in the hierarchical order of society; conformity is the natural consequence of inadequate opportunity to be self-directed. (Kohn, 1977, p. 189)

The research reviewed in the text has documented the impact of work in particular and social class–related life circumstances in general on measures of life satisfaction, on measures of self-esteem and locus of control, on parenting strategy, and even on the maintenance of cognitive skills through the middle years of adulthood.

Social class differences are present from birth, actually even from before birth. They cumulate across infancy, childhood, and adolescence. The completion of high school and the continuation of education through college are increasingly serving as crystallizers of these early social class differences. The futures of those who go on to college are strikingly different from the futures of those who do not. Four factors contribute to the increasingly pivotal role of a college education.

First, the research reviewed in Chapter 7 shows that one of the consequences of a college education is a shift in value orientation and information-processing strategies that parallels those defined by Kohn in the preceding quote.

Second, since a college degree is increasingly becoming the "union card" for many desirable jobs in our society, jobs that provide opportunities for self-direction as well as career growth and adequate income, it is coming to serve a function that once the high school diploma provided. In particular, it serves to solidify the positions of adults standing in the more economically advantaged and less economically advantaged tracks in society.

Third, college influences the timing of a number of life events that have consequences long after "Pomp and Circumstance" has ended. In particular, it influences the age at marriage, the age at first parenting, the number of children one is likely to have, and the spacing between children.

Fourth, as the numbers of men and women attending college have become approximately equal, college increasingly influences choice of spouse (or at least the educational level of the spouse) and probably the degree to which the couple adopts a complementary or reciprocal marriage pattern. In fact the greater change in sex-role orientation for economically advantaged women is largely a reflection of the immediate and prolonged impact of the college experience. And as college-educated couples continue to delay both marriage and parenting, the gap between the timing of these events for the college and noncollege populations is increasing. To the extent that these timetables correlate with income, job advancement, employment opportunities, and couple relations, we may be approaching a time in our history when the gap between the advantaged and disadvantaged will actually be increasing.

Before we get too heady about the virtues of college, a few words of caution are in order. First and foremost, I am talking about circumstance not ability. There is simply no reason to believe that given the opportunity, those from less advantaged backgrounds could not acquire the orientation and skills of those from more advantaged backgrounds. Second, as you probably well know by now, there is nothing magical about a college education. It is quite possible that other situations could lead to the same pattern of events. Third, as you also probably know by now, not all economically advantaged adults are open-minded, self-directed, morally autonomous individuals, nor are all less advantaged adults the opposite. The argument is rather that if such a pattern is to occur, its likelihood is greater in some life circumstances than in others and the probability of these circumstances' occurring correlates with college attendance. Fourth, as more and more adults enter degree programs, it is becoming increasingly clear that the benefits of a college education are not limited to those of "college age." Finally, even if a college education leaves one more open-minded and self-directed and even if the timing of major life events is consistent with the pattern just defined, there is no guarantee that the adult world will continue to maintain and nurture these values. If, for example, the economy tumbles as it did during the Great Depression of the 1930s, we could all be on the corner selling apples and pencils.

THE DEVELOPING SELF

All of the topics discussed in this chapter concern factors that influence the growth and development of the psychological organism. The focus on the two components—the biological organism and the sociohistorical context—clarified the relationship between the two components and its product. The discussion of developmental stages presented the notions of stages as a sequence of successive structures of adaptation. The preceding discussion of social class and sex role highlighted the two primary dimensions that people in our culture use to sort or classify individual patterns of development.

All of these factors influence the development of the psychological organism across the life-span. But, in the final analysis, what is it that is being influenced? I believe the answer is one's concept of self as a unique individual. Each chapter in the text has discussed the self-concept as it exists at that particular point in development, and it seems appropriate that the last discussion of this text should deal with the core element of the psychological organism—**the developing self.**

From the time that the infant, around eighteen months of age, looks in the mirror and shows surprise upon seeing the dot of rouge that the experimenter has unobtrusively placed on his nose (Lewis & Brooks-Gunn, 1979) to the time in very old age when the ability to maintain a sense of self seems to be a determinant of nearness to death (Lieberman & Tobin, 1983), the sense of self serves as the integrating component of the psychological organism. It is the fundamental frame of reference, the foundation upon which almost all actions are predicated. Few activities engage our lives so profoundly as the definition and enhancement of the self (Rosenberg, 1979).

Like all aspects of the psychological organism, the self develops. Over time the self changes from one defined primarily in terms of physical qualities to one defined in terms of subjective qualities (Rosenberg, 1979). Ask young children to tell you something about themselves, and they will tell you about the prizes they have won, the television shows they like to watch, what kind of house they live in, whether they can ride a bike, and so on. Older children, asked the same question, are likely to respond more as we would. They will talk about feelings and emotions, wishes and dreams, attitudes and secrets. Ask young children to tell you the best or the worst things about themselves, and they may say that they can run fast and that they once took some cookies that their mother told them not to touch. The older children will tell you, for instance, that they are kind or easygoing, and they will tell you that they sometimes lie or sometimes wish that people they don't like will fall and hurt themselves. For the young child, the self is external and objective; for the older child and adult, it becomes internal and subjective. Increasingly, judgments reflecting the self are based less on one's life circumstances and more on one's perception of those circumstances (Crosby, 1982).

Over time the self attains a degree of integration, or wholeness, not typical of children or, in some respects, of adolescents and youth. The behavior of preschool children has a splintered quality to it. There is comparatively less continuity across situations than is found in adults. Preschool children seem to be creatures of the moment, ones who can make demons appear and disappear, ones who can have more to drink simply by finding another glass. Even the behavior of school-age children has a strong situational quality to it. For them, the rightness or wrongness of an act has more to do with the likelihood of detection than with some inner sense of right and wrong.

By adolescence, we are aware that our public selves and our private selves need not be the same, that we can present ourselves as we want to be seen rather than as we think we really are. This awareness of the public and the private is also an integral part of the adult self-concept but what join it, also beginning in adolescence, are the self-reflecting capabilities that seem to lead to a search for a feeling of wholeness (Breger, 1974). We

When young children are asked to describe themselves, they are apt to respond in terms of physical qualities, such as telling you that they have a Mickey Mouse cup and towel or a two-story house.

become increasingly aware of our inconsistencies, we talk about being phony, we no longer are comfortable feeling scattered. We find it increasingly difficult to deal with the fact that our ideal self and our real self are so disparate. By middle age, we manage to make them closer—either by changing one or the other but in either case by resolving the discrepancy.

As the concept of self grows, so too does the concept of other (Maccoby, 1980). At first there seems to be little awareness of this notion. It is not until toward the end of the first year that the infant even seems aware that there is an other. Even having made the discovery, preschoolers still have difficulty seeing others as anything but an extension of themselves. If they want something, they simply assume others will want them to have it. How perplexed they appear when this proves not to be the case. They have no problem in seeing others in relationship to themselves. They can easily tell you how many brothers and sisters they have. But ask them about their sibling's siblings, and they will probably come up one short. They can't yet see themselves as others see them.

Through the school years and into adolescence, this awareness becomes more evident. School-age children and adolescents compare themselves to others, they try and best others, and they vie for others' friendships. As they come to appreciate that others don't necessarily see the world as they do, they become better able to appreciate the need for explanation and justification, increasingly designed with the other in mind. Into the adult years, this awareness of the interplay of self and other is best reflected in the development of the sense of intimacy. As Youniss puts it, "The product of this process is mutual understanding in which the self and the other become progressively articulated, not as 'I' or 'you,' but as 'we'" (1980, p. 3).

The process of self-definition, as it shifts from objective to subjective, as it becomes better integrated within the self and between self and other, also increases in scope and complexity. The self is defined over larger and larger domains with greater attention paid to consistency across domains. It is hard not to think of the willingness of adolescents to profess strong opinions on virtually any topic, almost irrespective of their knowledge of or interest in the topic, as anything else but an early attempt to extend the definition of the self into increasingly broader areas. Each of Erikson's psychosocial stages, each an aspect of self-definition, involves an increasingly broad "radius of significant relations." For the infant seeking a sense of trust, the radius is limited to the primary care giver. By middle childhood, the boundaries of the self are seen as extending to the school and the neighborhood; by the eighth stage, the stage of ego integrity, they have reached out to, in Erikson's terms, "my kind."

Early understandings of self and other are very simple. It is sometimes hard for the child to understand how someone can be both a parent and a lawyer or an American and a Catholic. With age, notions of self become more differentiated; more aspects of the self are made apparent. More ways of relating to another person become necessary. As the push for integration becomes greater, these distinct aspects of self are gradually integrated, each coming to be seen both in its own right and in its relationship to other aspects of the self. Increasingly the task of self/other integration is no longer one of coordinating one aspect of self (that is, child) with one aspect of other (that is, parent) but of coordinating one pattern of self (that is, child, adult, married, father, professor) with one pattern of other (parent, aged, widowed, mother, housewife).

Although the self follows a developmental path, two functional invariants are evident at each stage. One is the strong motivation to see the self in a favorable light, and the other, primarily evident from adolescence onward, is the even stronger motivation to see the self as a stable, continuous entity. We are reluctant to accept ourselves as changing, as no longer being what we have always thought we were (Rosenberg, 1979). The middle-aged man who looks in the mirror one day and says, "this can't be me, I can't look this old," has been looking at himself in the mirror daily for the past ten years. No one has suddenly put a spot of rouge on his nose. But his need to see himself as he has always thought of himself, as someone looking much younger, has somehow clouded his vision more than the steam from the hot water in the sink.

An unfavorable self-image is a correlate of depression. Intentionally or unintentionally, we try to avoid an unfavorable self-image. We selectively attend to some information and not to other information, we interpret information in the best light possible, and we are more likely to associate with those who like us and are like us. In all of these ways, we act to ensure the maintenance of a favorable image of self (Epstein & Erskine, 1983; Langer, 1983). Even our remembrance of the past evolves in the service of the self—the fish we caught keeps on getting a little bigger and the good old days keep on getting a little better (Cohler, 1982). The process is not without its limits. The more fanciful our image, the harder it is to find the people to endorse it. The aged adults studied by Lieberman and Tobin (1983) provide particularly poignant examples (see Chapter 10). Declines in health and loss of personal control seem to be the greatest assaults on our stable, favorable concepts of self. Lieberman and Tobin found that the very aged will do almost anything to maintain their image of themselves.

> Our premise then, is that psychological survival is equivalent to maintaining a sense of self-continuity, integrity, and identity, and it is toward this conservation of self that psychological work among the elderly is focused. Moreover, this work occurs within the unique psychological context of the aged—of a life lived and of personal finitude coupled with the fact of personal and structural losses. Roles do change, the body does fail, and important people die. The self is challenged to its very core because the opportunities for maintaining a coherent and consistent self, which are ultimately dependent on input from the external world, are radically altered. The sense of self does not change; rather, what we see is the utilization of strategies by the elderly to maintain this sense of selfhood. At its most general level, the strategies represent myths—the myth of control, the myth of self-constancy—and the blurring of the boundaries between the past and the present. (Lieberman & Tobin, 1983, p. 348)

What Lieberman and Tobin's finding highlights is that, at least within our culture, the maintenance of a favorable self-concept is largely dependent on the ability to maintain the perception of a sense of control over one's

Attempting to maintain a sense of control over one's life is a very personal effort, sometimes involving decisions to pursue nontraditional life-styles, as in the case of this man trying to make his point at the "Speaker's Corner" in Hyde Park, London.

life. The control may be expressed through action or merely through some sort of conceptual reorganization. It may be as much a matter of feeling able to understand the nature of the events that encompass our lives as it is an actual attempt to influence these events. The centrality of control is reflected in Gurin and Brim's (1984) finding that adults are more likely to interpret personal failure as being due to an unresponsive system than as being due to individual limitations. It is one of the most significant aspects of Kohn's (1977) explanation of the distinct developmental patterns of those from different social class backgrounds. The more economically advantaged their background is, the greater is their belief that they have control over their own lives. And in the final analysis, the greater the feeling of control people have, the greater the satisfaction they derive from their lives.

Almost from the first day that I began this text, I have been wondering how I was going to end it—what clever or wise or poetic phrase I would use to leave with you some lasting impression. Well, here I am, and I'm afraid I haven't been able to come up with anything original. So instead I will leave you with the words of a life-span developmentalist far wiser than I. In light of this last discussion, the words seem appropriate.

> This above all: To thine own self be true,
> And it must follow, as the night the day,
> Thou canst not then be false to any man.
> William Shakespeare, *Hamlet*, Act I, Scene III

SUMMARY

POWERS OF MAGNIFICATION

1. The development of the psychological organism is the product of the continuing interaction of a biological organism with a sociohistorical context.
2. Although the psychological organism reflects the structure and function of the biological organism, they are not identical.
3. The sociohistorical context is best appreciated through an understanding of cohort differences.
4. The equilibration process of the psychological organism is dynamic rather than static.

THE COURSE OF DEVELOPMENT

1. Each stage of development reflects a relatively unique pattern of organization among the various components of the psychological organism.
2. The roles that the biological organism and the sociohistorical context play in influencing the functioning of the psychological organism change over the course of the life-span.
3. The stage sequence is best appreciated as a succession of relatively unique patterns of organization, each pattern reflecting the best attempt that can be made at a particular time to maintain a degree of equilibrium between the psychological organism and its environments.
4. The process of adaptation is bidirectional, involving both changes in behavior and constructions and reconstructions of meaning.

THE CONTEXT OF DEVELOPMENT

1. At the most general level, social class and sex are the two most significant correlates, or predictors, of the course of development across the life-span.

THE DEVELOPING SELF

1. The concept of self is the core component of the developing psychological organism.
2. The maintenance of a favorable concept of self is largely dependent on the ability to maintain the perception of control over one's life.

KEY TERMS AND CONCEPTS

The Psychological Organism
The Biological Organism
The Sociohistorical Context
The Active Organism
Species-Specific Characteristics
Self-Regulating System
Dynamic Equilibrium
Helix
Kagan and Moss's Sleeper Effect
The Developing Self

SUGGESTED READINGS

A book by Montagu provides some interesting insights as to how our biological makeup influences our psychological functioning.

Montagu, A. (1981). *Growing young*. New York: McGraw-Hill.

One of the most complete discussions on the influence of sex-role socialization on development across the life-span is provided by Block.

Block, J. H. (1984). *Sex role identity and ego development*. San Francisco: Jossey-Bass.

The finding of one of the most comprehensive long-term longitudinal studies ever conducted provides a better understanding of the interplay of developmental events across the life-span.

Werner, E. E., & Smith, R. S. (1977). *Kauai's children come of age*. Honolulu: University of Hawaii Press.

The "developing self" is well described in each of these three books.

Kegan, R. (1982). *The evolving self*. Cambridge: Harvard University Press.
Langer, E. J. (1983). *The psychology of control*. Beverly Hills: Sage Publications.
Rosenberg, M. (1979). *Conceiving the self*. New York: Basic Books.

GLOSSARY

Italicized terms within the body of a definition are also found as separate entries in the glossary.

accepting parents Parents who are sensitive to their infants' needs; one of Ainsworth's three types of patterns of maternal behavior.

accommodation The process of modifying cognitive structures to adapt to new situations that cannot be assimilated into the existing cognitive structures.

active organism An organism that influences and is influenced by its component parts.

activity theory of aging A view whose advocates encourage maintenance of middle-age life patterns into old age, hence a high degree of continuity.

acute illnesses Illnesses that are cured, that go away after some period of time.

adolescence A period of the life-span during which physically mature individuals retain an essentially childlike role in society.

Ainsworth's patterns of maternal behavior See *patterns of maternal behavior.*

ambivalent parents Parents who are less consistently responsive to their infants than *accepting parents* are but who do not reject or resent their infants; one of Ainsworth's three types of patterns of maternal behavior.

analysis of variance A statistical procedure that makes it possible to determine the degree to which many different variables individually and jointly cause some behavior or behaviors to occur.

anticipatory grief A period in which a person whose spouse is dying or has diminished capacity must confront not only the realities of the present moment but also the likely future once the spouse dies.

Apgar score A simple index of the status of the major biological systems of the newborn. The Apgar score is the most common evaluation of newborn status.

appropriate death Weisman's term for a death that is as free of pain and suffering as possible, that maintains emotional and social supports, and that helps the dying person resolve unfinished tasks, prepare others to continue in his or her absence, satisfy wishes, and gradually yield control of self to others.

assimilation The process of integrating new experiences into existing cognitive structures.

attachment The process by which the parent–child relationship is formed, beginning with a kind of synchrony that develops between parent and child as a result of interaction during the first few weeks of life.

authoritarian parenting One of *Baumrind's parenting patterns.* The authoritarian parent emphasizes parental control, demands that behavior conform to a set standard, does not encourage verbal give-and-take, and believes in inculcating such instrumental values as respect for authority, respect for work, and respect for the preservation of order and traditional structure.

authoritative parenting The most desirable of *Baumrind's parenting patterns.* The authoritative parent exerts firm control at points of parent–child divergence, using reason as well as power, but does not hem in the child with restrictions. The authoritative parent makes high maturity demands, encourages verbal give-and-take, and, valuing both expressive and instrumental attributes, affirms the child's present qualities but also sets standards for future conduct.

autoimmune reactions Conditions in which the body's immune system attacks the body's own cells.

average life expectancy The average number of years of life expected for the members of a population, based on statistical probability.

awareness of transformations An understanding of what takes place in changing something from one state to another. This awareness, characteristic of *concrete operational thought,* is often lacking in *preoperational thought.*

Baumrind's parenting patterns Three relatively distinct patterns identified by rating four dimensions of parental behavior: parental control, maturity demands, the nature of the parent's communcation with the child, and nurturance. These patterns, each of which is in turn associated with a particular pattern of preschooler behavior, are *permissive parenting, authoritarian parenting,* and *authoritative parenting.*

biological clock A genetically regulated schedule by which development proceeds for the members of a species.

biological organism The component of the *psychological organism* that consists of the physical structure, body functions, and genetic makeup of the individual. See also *sociohistorical context.*

blastocyst The hollow, differentiated form the *morula* assumes after it has entered the uterus.

Brazelton neonatal behavioral assessment scale The scale that is probably used most frequently when the *Apgar score* indicates that a more complete neonatal evaluation should be made. It attempts to assess the full range of the infant's response to the environment.

Bronfenbrenner's human ecology model See *human ecology model.*

Butler's life review process See *life review process.*

canalization The mechanism by which the *genotypes* for certain characteristics and developmental patterns common to the species resist the environmental influences that could alter their expression.

central tendency See *measures of central tendency.*

centration The tendency of a preschooler, in solving a problem, to focus on only a limited part of the available relevant information.

cephalocaudal progression The sequence by which the development of voluntary motor control progresses from the head and upper torso to the lower trunk and legs.

chromosomes The twenty-three pairs of carriers of hereditary information. One member of each pair is inherited from the mother, the other member from the father. Each pair regulates a unique set of biochemical reactions.

chronic illnesses Illnesses that are more likely to be managed than cured. They will not go away, but with the appropriate use of medication, diet, and exercise, their effects can be minimized.

classical conditioning Learning that involves an association between a stimulus and a response; also known as Pavlovian conditioning.

classification The process of dividing a set of objects into categories; one of two skills tested by Piaget to study preschoolers' developing ability to accurately and effectively process information during the *preoperational stage.* See also *seriation.*

clinical strategy A research method frequently used by Piagetian researchers in which a researcher asks questions designed to gain insight into the motives or reasons for an individual's behavior or to determine the extent to which individuals understand various aspects of their lives. Also called the interview strategy.

cognitive development A pattern of changes in the types of mental operations available to an individual. The mental operations in use at any given stage form a unique construction of reality. See also *theory of cognitive development.*

cohort design A research strategy that uses longitudinal studies of several groups born at different times. Each longitudinal sample is a cohort.

collective monologues Parallel conversations, a typical pattern among preschoolers, in which each child of a pair talks about something, but the two things are different.

community divorce The aspect of divorce that involves announcing to family, friends, and the community in general that the marriage is over.

compensation A form of the mental operation called *reversibility* in which a change in one dimension is seen to counterbalance a change in another dimension and thus to neutralize its effect.

complementary reciprocity See *complementary relationship.*

complementary relationship A pattern of interaction between two or more people (e.g., parent/child) in which the behaviors of the participants are not interchangeable but nevertheless reflect the needs and rights of the other(s). The term *complementary reciprocity* is sometimes used to define this pattern.

concrete operational stage See *stage of concrete operational thought.*

concrete operational thought The type of thinking that characterizes the middle childhood years. Aspects of concrete operational thought include *awareness of transformations,* use of *inference, reversibility,* and *decentration.* See also *stage of concrete operational thought.*

conservation The concept that certain aspects of objects (e.g., weight, number, length) remain invariant irrespective of certain transformations (e.g., changes in shape, container, or position) performed upon them. Different levels of conservation are evident at different levels of cognitive development.

conservation experiment The classic demonstration of the transition from *preoperational* to *concrete operational thought,* in which the basic question asked is whether a child's reasoning is based primarily on the appearance of things or on the ability to use mental operations. In the conservation of continuous quantity task, liquid is poured from one glass to a glass of a different shape, and a child is asked whether the new glass contains the same quantity as the original glass. In the conservation of substance task, a ball of clay is molded into a different shape, and the child is asked whether the amount of clay in the new shape is the same as the amount in the original ball.

conservation of social identity The concept that even though a person changes at one level (dressing up like another person, pretending to be an animal, etc.), at another level the person remains the same.

controlled research experiment A method by which the influence of one variable on a particular behavior is examined independently of all other possible influences; the most rigorous of all research strategies.

coparental divorce The aspect of divorce that changes the way parental roles are maintained. This aspect involves decisions about where the children will live, how each parent will spend time with them, how each will provide for child support, and how future decisions about the children's welfare will be made.

correlation coefficient A measure of the degree to which uncertainty about one variable is reduced by knowledge about another.

cross-sectional design A research strategy that tests different groups of people at the same time.

cycle of poverty The process by which poverty is passed from one generation to the next, beginning with prenatal development and continuing through the child-rearing experience.

decentration A tendency away from focusing on a specific state or a particular aspect of a problem. Decentration in middle childhood marks a shift from early childhood in the relationship between perception and cognition: specific perceptions are seen within the context of a mental operation.

deductive reasoning The process in which a conclusion is reached by a logical progression of *inference* from the general to the particular.

deferred imitation Play that reflects the infant's ability to create a *mental representation* of some action, person, or object, store it in memory for some period of time, and then retrieve it from memory and reproduce it.

deoxyribonucleic acid (DNA) The genetic substance on each *chromosome,* composed of four molecules (adenine, guanine, thymine, and cytosine) arranged in a double helix, or twisting ladder, in such a way that adenine is always paired with thymine and cytosine is always paired with guanine.

descriptive statistics Methods of data analysis that give information about central tendency (the most common responses) and variability (the range of responses).

developing self The process of change over the life-span in the concept of oneself as a unique individual.

developmental norms Standards used to evaluate the normality of a child's rate of development.

developmental tasks theory A perspective, developed by Havighurst, in which developmental expectations are viewed as reflecting a culture's expectations for age-appropriate behaviors.

differentiation The process of specialization, first evident during the *period of the embryo* as the different organs are formed and subsequently evident postnatally as individuals make finer and finer distinctions between the various aspects of their experiences.

dilatation of the cervix See *effacement and dilatation of the cervix.*

disengagement theory of aging A view whose advocates believe there should be a process of mutual severing of the ties between the aging individual and society.

DNA See *deoxyribonucleic acid (DNA).*

dominant gene The expressed gene of a *heterozygous* pair.

dualistic world view The first of Perry's three *forms of experience construction.* In this phase students, most typically first year students, believe that, in essence, there are right and wrong answers to all questions and that these answers are knowable.

dying trajectories The different physical and emotional paths followed by patients with different courses of illness and different likelihoods of death.

dynamic equilibrium The process through which a system maintains a steady state by constantly adjusting to changes within the system.

ecological approach A method of study that focuses on ways in which the various institutions and people that an individual encounters both directly and indirectly in a lifetime influence the course of that person's development.

economic divorce The aspect of divorce that concerns the division of the property of the couple.

ectoderm See *embryonic ectoderm.*

effacement and dilatation of the cervix The first and longest phase of labor, in which changes in the cervix are brought about by a series of muscular contractions. By the completion of this phase, the opening of the cervix has reached a diameter of approximately four inches (10 cm).

egocentric thought A type of thought, characteristic of the *preoperational stage,* that implies a limited ability to take another's perspective.

ego integrity See *psychosocial stage of ego integrity.*

embryo See *period of the embryo.*

embryonic cell layers The three layers into which the cells separate early in the *period of the embryo* and from which the various organ systems develop.

embryonic ectoderm One of the three *embryonic cell layers* from which the various organ systems develop. From the ectoderm develop the hair, nails, mammary glands, eyes, ears, nervous system, and pituitary gland.

embryonic endoderm One of the three *embryonic cell layers* from which the various organ systems develop. From the endoderm develop the liver, pancreas, urinary bladder, and parts of the trachea, lungs, gastrointestinal tract, tonsils, and thyroid gland.

embryonic mesoderm One of the three *embryonic cell layers* from which the various organ systems develop. From the mesoderm develop the muscles, skeleton, urogenital system, spleen, blood and lymph cells, and the cardiovascular and lymphatic systems.

emotional divorce The aspect of divorce that concerns the failing relationship of the couple.

empty nest transition An element of the transition to middle age that involves the departure of grown children.

endoderm See *embryonic endoderm.*

engrossment The term used by Greenberg and Morris to describe the father's fascination with his newborn.

equal partner/equal partner pattern A spouse interaction pattern, still rare in our society, in which the husband and wife are equal in power, status, rights, and responsibilities; roles are no longer linked to the sex of the spouse; and (as in the other two patterns) both the instrumental and expressive aspects of the relationship are considered important. See also *patterns of spouse interaction.*

Erikson's psychosocial stage of autonomy See *psychosocial stages of trust and autonomy.*

Erikson's psychosocial stage of ego integrity See *psychosocial stage of ego integrity.*

Erikson's psychosocial stage of generativity See *psychosocial stage of generativity.*

Erikson's psychosocial stage of identity formation See *psychosocial stage of identity formation.*

Erikson's psychosocial stage of industry See *psychosocial stage of industry.*

Erikson's psychosocial stage of initiative See *psychosocial stage of initiative.*

Erikson's psychosocial stage of intimacy See *psychosocial stage of intimacy.*

Erikson's psychosocial stage of trust See *psychosocial stages of trust and autonomy.*

Erikson's theory of psychosocial stages See *theory of psychosocial stages.*

errors of commission Errors that involve choosing the wrong answer in a cognitive test.

errors of omission Errors that involve failing to choose any answer in the allotted time in a cognitive test.

establishment phase of marriage The period from the wedding to the birth of the first child, in which a couple must make the transition from thinking of themselves as relatively autonomous individuals to thinking of themselves as members of an interdependent pair.

event sampling A *naturalistic research strategy* in which the frequency of a particular type or types of behavior is observed and recorded.

evolving of commitments The third of Perry's three *forms of experience construction.* In this phase students no longer judge ideas and opinions according to their relation to absolute knowledge or according to an individual's right to have them, but rather according to the soundness of the foundation upon which they are based.

exosystem In Bronfenbrenner's *human ecology model,* the third level of the environment, which is outside the *mesosystem* and beyond the control of the individual but in which events occur that affect the *microsystem* and the mesosystem.

experience construction See *forms of experience construction.*

expulsion of the afterbirth The third phase of

labor, in which the *placenta* and fetal membranes are pushed through the birth canal.

expulsion of the fetus The second phase of labor, in which the fetus is pushed through the birth canal.

facets of the divorce process See *six facets of the divorce process.*

fetal alcohol syndrome (FAS) A set of conditions any one of which is more likely to occur in the offspring of a woman who uses alcohol excessively during her pregnancy. These conditions include low birth weight, a slower rate of growth through childhood irrespective of diet, a smaller head circumference, below-average scores on intelligence tests, and a variety of congenital anomalies, especially of the cardiovascular and urogenital systems.

fetus See *period of the fetus.*

fine motor skills Movements that require good prehensile skills and good eye–hand coordination.

formal operational stage See *stage of formal operational thought.*

forms of experience construction Three stages of intellectual development defined by Perry among college students. The stages, in sequence, are a *dualistic world view,* a *relativistic world view,* and an *evolving of commitments.*

free radicals By-products of cell metabolism that increase the permeability of the cell membrane, making it more vulnerable.

Frieze's stage theory of adult female development See *stage theory of adult female development.*

general adaptation syndrome The level of adaptation that involves a series of endocrine and nervous system changes that influence not only the local site but also the entire body.

generalized other A generally accepted standard or perspective against which individual views and actions can be weighed.

generativity See *psychosocial stage of generativity.*

genetic code The sequence of paired molecules on each *chromosome.* Each set of three pairs, called a codon, specifies the formation of a particular amino acid.

genotype The genetic makeup of an individual.

Gesell's maturational theory of child development See *maturational theory of child development.*

gross motor skills Movements that involve relatively large muscle masses.

Havighurst's developmental tasks theory See *developmental tasks theory.*

head/complement pattern The traditional marriage pattern, in which the division of labor is rather complete and decision-making power remains vested primarily in the husband. The difference between this pattern and the owner/property pattern common in earlier periods of history is a greater emphasis in the head/complement pattern on the expressive component of the relationship. See also *patterns of spouse interaction.*

helix A spiral shape used by Kegan and others to depict the course of development of self across the life-span. The path of the helix weaving back and forth, each time at a new level (that is, stage), is used to depict the notion of themes recurring across the life-span.

heterozygous Having the genes for a particular trait different on the two *chromosomes* of a pair.

homeostasis The maintenance of regular biological function.

homozygous Having the genes for a particular trait identical on the two *chromosomes* of a pair.

hospices Places where the terminally ill and their families can spend their time together. Hospices have as their goal ensuring that each person achieves as close to an *appropriate death* as possible.

human ecology model A model, developed by Bronfenbrenner, in which the environment is divided into four interconnected levels: the *microsystem,* the *mesosystem,* the *exosystem,* and the *macrosystem.*

hypothetical-deductive reasoning Reasoning in which a conclusion is drawn from premises that include an assumption independent of empirical observations.

identity A form of the mental operation called *reversibility* in which two equal opposite actions on an object are seen to cancel each other out.

identity achievement The highest level of identity formation among Marcia's four *identity statuses.* Identity achievers are the most self-accepting, show the least discrepancy between self-assessments and assessments made of them by others, tend to hold the strongest values, and show the most willingness to act on the basis of those values. According to Marcia, they are the only group capable of real intimacy with others.

identity diffusion The result of the failure of an adolescent to find a comfortable role. This failure constrains the adolescent's ability to plan for the future, and the balance between satisfying present needs and satisfying future needs is tipped in favor of the former. Among Marcia's four *identity statuses,* identity diffusion is the lowest level of identity formation. Persons in identity diffusion status, having no commitments to any value structure, will readily change to meet the expectations of those they are currently impressed by. They demonstrate the least amount of basic trust and are the least able to form close relationships.

identity foreclosure A pattern in which an adolescent assumes an identity without reflecting on the appropriateness of that identity. As one of Marcia's four *identity statuses,* identity foreclosure is a state in which one has resolved identity questions by adopting the value structure of another person rather than by choosing a value structure of one's own. Persons in identity foreclosure place a high value on maintaining traditions, avoid conflict, tend to be fairly rigid in their judgments, and seem more influenced by the judgments of others than by their own evaluations.

identity statuses Four unique patterns identified by Marcia in the responses of college males to interview questions on topics of identity formation. These statuses—*identity achievement, psychosocial moratorium, identity foreclosure,* and *identity diffusion*—correspond to patterns identified by Erikson among adolescents.

imitation See *social learning.*

inductive reasoning The process in which a conclusion is reached by a logical progression of *inference* from the particular to the general.

inductive techniques *Parenting techniques* characterized by (1) a set of standards that children can use to judge their future actions and intentions, (2) emphasis of the consequences of children's actions not only in terms of the children but also in terms of other people, (3) suggestion of alternative strategies that the children can use in dealing with similar problems in the future, and (4) ideally, a nonhostile, supportive environment for discussions of the children's behavior. These techniques appear to have a positive influence on the development of moral reasoning and behavior in middle childhood.

industry See *psychosocial stage of industry.*

infant temperament characteristics A measure of the way an infant goes about a task, compared with the ways other infants go about the same task.

inference The drawing of a conclusion based on all information available.

inferential statistics Methods of data analysis that can document the presence of causal relationships.

initiative See *psychosocial stage of initiative.*

instrumental conditioning A means of changing the patterns of behavior through the association of a response with its consequences.

interview strategy See *clinical strategy.*

intimacy See *psychosocial stage of intimacy.*

introspection A process in which adolescents think about themselves not only as they see themselves, but also as they would like to see themselves.

Jacob–Monad model A model of the mechanism of interaction between the gene and the environment. In this model the structural genes, which regulate particular biochemical reactions, are regulated by operator genes and regulator genes.

Kagan and Moss's sleeper effect See *sleeper effect.*

Kendler and Kendler's reversal-nonreversal shift task See *reversal-nonreversal shift task.*

Kubler-Ross sequence of response to death See *sequence of response to death.*

Kuyper and Bengtson's social breakdown syndrome See *social breakdown syndrome.*

legal divorce The aspect of divorce that involves the actual termination of the marriage contract.

Levinson's stage theory of adult male development See *stage theory of adult male development.*

life review process A process Butler describes as an active, deliberate attempt, prompted by the realization of approaching dissolution and death, to adjust to the losses experienced by the very old.

lipofuscins By-products of metabolism that accumulate in the cells, possibly influencing cell efficiency.

local adaptation syndrome The level of adaptation that involves those aspects of the body that participate in the specific, immediate response to the stressor.

longitudinal design A research strategy that identifies a group of people and at various points in their lives obtains relevant information about them.

Lopata's sanctification process See *sanctification process.*

love withdrawal techniques *Parenting techniques* in middle childhood that regulate children's behavior by appealing to the bond that exists between parent and child. These techniques may be effective in dealing with the immediate situation, and they can foster a sense of conscience, but in the extreme or as a steady diet they provide little guidance as to alternative ways of behaving, and they can produce considerable anxiety in adolescence and adulthood about self-worth.

macrosystem The fourth level of the environment in Bronfenbrenner's *human ecology model.* Events at this level, which define direction for a society, reflect people's shared assumptions about the way things should be done.

mandatory retirement A policy of requiring employees to retire by a certain age.

Marcia' sequence of identity statuses See *identity statuses.*

mass-to-specific progression The sequence by which the development of voluntary motor control progresses from the large motor groups to muscles involved in increasingly finer movement.

maternal behavior See *patterns of maternal behavior.*

maturation The process through which biological structures achieve their mature state.

maturational theory of child development A theory, developed by Gesell, that a child's development is governed to a large extent by heredity, which regulates the process of *maturation.*

mean The arithmetic average, obtained by summing all the scores and then dividing by the number of scores summed.

means-end separation The ability of an infant to sequence *schemes* to obtain some desired goal.

measures of central tendency Information that tells what is most typical, average, or common about the population being studied.

measures of variability Information about the *range* of scores or how frequently each possible score in a particular distribution is represented.

median The middle point of a distribution, separating the upper half of the cases from the lower half.

menarche The beginning of menstruation.

menopause The termination of female reproductive capability.

mental representations Images in the mind that are copies of real objects and actions.

mesoderm See *embryonic mesoderm.*

mesosystem The second level of the environment in Bronfenbrenner's *human ecology model.* It is defined by the interplay among the various *microsystems* in which an individual lives.

metacognitive skills An individual's awareness of various techniques of deliberately coordinating and controlling one's own attempts to learn and solve problems, including predicting the consequences of an action or event, checking the results of one's own action, monitoring one's ongoing activity, and testing reality.

metamemory skills An individual's awareness of the various techniques deliberately employed to retain information.

microsystem In Bronfenbrenner's *human ecology model,* one of the immediate environments, comprising both settings and experiences, that constitute an individual's day-to-day reality.

mode The single most frequent score or category in a distribution.

morula A structure of approximately sixteen cells that by the third day after conception has evolved from the fertilized ovum, or *zygote,* by a series of cell divisions.

naturalistic research strategy A strategy in which researchers assume the role of observers rather than experimenters.

negation A form of the mental operation called *reversibility* in which an object is seen to be the same as if an action performed on it had never happened.

neonate A newborn child.

Neugarten's young-old and old-old See *young-old* and *old-old.*

object permanence The concept that an object hidden from view continues to exist. The development of this concept demonstrates the infant's ability to form a *mental representation* of the object.

old-old Neugarten's term for the second phase of old age, a period that begins usually in the mid-seventies or when health begins to decline significantly and that lasts until death.

one-to-one correspondence A relationship between two sets that enables each member of one set to be matched with one member of the other set. Children of about six or seven years of age are able to show a one-to-one correspondence between two sets of objects.

operational definition The notion that an object, event, or relationship should be defined as the actual procedure used to collect the information about them.

orthogenetic development A process, following

the same general path in every person, in which people and events become distinct from each other and then gradually come to be understood in relationship to other people and events.

parenting patterns See *Baumrind's parenting patterns.*

parenting techniques Three types of methods used by parents to aid development of moral reasoning and behavior during middle childhood: *inductive techniques, power assertion techniques,* and *love withdrawal techniques.*

patterns of maternal behavior Three relatively distinct groups of attitudes and responses identified by Ainsworth as the ways in which different mothers relate to their infants. See also *accepting parents, ambivalent parents, rejecting parents.*

patterns of spouse interaction Three categories of marriage relationships defined by Scanzoni and Scanzoni by evaluating task orientation, division of labor, and distribution of power. The three patterns—the *head/complement pattern,* the *senior partner/junior partner pattern,* and the *equal partner/equal partner pattern*—differ in the degree to which they favor equality in the relationship.

Pavlovian conditioning See *classical conditioning.*

perinatal anoxia A birth-related problem that is a potential contributor to mental retardation and cerebral palsy and that may result from insufficient oxygenation of the mother's blood, problems in the transfer of oxygenated blood from the mother to the placenta or from the placenta to the fetus, a knot in the cord, or the cord's wrapping itself around the neck of the fetus.

period of the embryo The second phase of prenatal development, following the *period of the ovum* and lasting approximately seven weeks, during which all of the major organ systems are formed, limbs become evident, and facial features begin to emerge.

period of the fetus The third phase of prenatal development, beginning in the eighth week after conception, during which change takes the form of growth in size and weight and further maturation of the various systems and parts of the body.

period of the ovum The first phase of prenatal development, which begins at conception and lasts approximately ten days.

permissive parenting One of *Baumrind's parenting patterns.* The permissive parent avoids the exercise of control, makes few demands, explains rules and attempts to use reason to regulate the child's behavior, and provides the child with a great deal of support, presenting himself or herself to the child as a resource to be used as the child wishes rather than as an active agent responsible for shaping or altering the child's ongoing or future behavior.

Perry's three forms of experience construction See *forms of experience construction.*

phenotype The ultimate expression of the individual's unique *genotype.*

phenylketonuria (PKU) A major recessive gene effect that prevents the formation of an enzyme needed to metabolize, or break down, one of the amino acids formed by *DNA.*

Piaget's stage of concrete operational thought See *stage of concrete operational thought.*

Piaget's stage of formal operational thought See *stage of formal operational thought.*

Piaget's stage of preoperational thought See *stage of preoperational thought.*

Piaget's stage of sensorimotor development See *stage of sensorimotor development.*

Piaget's theory of cognitive development See *theory of cognitive development.*

PKU See *phenylketonuria (PKU).*

placenta The organ through which the developing fetus receives oxygen and nourishment and excretes waste products.

power assertion techniques *Parenting techniques* used during childhood in which parents impose punishment or withhold resources. These techniques may be effective in dealing with the immediate situation, but they provide children little information about the right ways to deal with situations, and the anger and resentment the techniques often generate tend to separate children from their parents.

preoperational stage See *stage of preoperational thought.*

preoperational thought See *stage of preoperational thought.*

preparatory grief The aspect of a terminally ill person's depression—the fourth stage in the Kubler-Ross *sequence of response to death*—that concerns what is to occur.

prepared childbirth training classes Classes, attended by both the father and the mother, whose primary purpose is to provide the mother with a variety of psychological and physical techniques to help cope with the emotional and physical stresses that are part of labor and delivery.

pretend play Play that reflects the infant's ability to mentally transform actions or objects, to have one thing stand for, or represent, another.

primary aging process The continuing biological and cognitive changes that are directly attributable to advancing age.

primary circular reaction A behavior pattern that involves an infant's attempt to replicate a particular event involving the infant's own body.

projection The tendency to deny the existence of feelings or states within oneself and instead pass them off or project them onto another.

proximal-distal progression The sequence by which the development of voluntary motor control progresses from the muscle groups of the midline of the body to those of the extremities.

psychic divorce The process through which people again come to think of themselves as single, independent adults.

psychological organism The product of the continuing interaction of a *biochemical organism* with a *sociohistorical context.*

psychosocial moratorium As used by Erikson, a period of suspension of certain mental constraints, allowing for a time of *role experimentation* without the limiting consequences of commitments. As used by Marcia, one of four *identity statuses.* Psychosocial moratorium is a state of unresolved crisis and conflict. Persons in psychosocial moratorium status are often uncertain and overly critical, are unwilling to accept advice or support from others, tend to have a low sense of confidence in their ability to make choices, and demonstrate high scores on tests of anxiety level.

psychosocial stage of autonomy See *psychosocial stages of trust and autonomy.*

psychosocial stage of ego integrity The period of development associated with advanced age, in which, according to Erikson, the primary task is for individuals to accept their lives as they have lived them, that is, to accept the past as the past.

psychosocial stage of generativity The period of development of middle adulthood in which, according to Erikson, the primary task is development of concern for the next generation.

psychosocial stage of identity formation The period of development in which, according to Erikson, circumstances—the development of sexual maturity, the cognitive capacity of *introspection* and *hypothetical-deductive reasoning,* and the increasing number of social demands for adolescents to choose a vocation and adult life-style—prompt adolescents to reflect on their own development.

psychosocial stage of industry The period of development of middle childhood, approximately between the ages of six and twelve, in which, according to Erikson, children come to see themselves as producers, doers, and makers of things.

psychosocial stage of initiative The period of development of early childhood, in which, according to Erikson, children add a goal or purpose to the desire to do.

psychosocial stage of intimacy The period of development of early adulthood in which, according to Erikson, the primary task is the development of a sense of closeness with others, a concern for others that equals the concern for self. Intimacy involves a fusing of identities, both sexual and nonsexual.

psychosocial stage of trust See *psychosocial stages of trust and autonomy.*

psychosocial stages of trust and autonomy The two periods of socialization identified by Erikson in the first three years of life. The task of the first stage is the development of a basic sense of trust; the task of the second stage is the development of a sense of autonomy.

puberty The physical maturing of the individual that results in adult physical and physiological stature and reproductive capability.

range The difference between the highest and the lowest score in a distribution.

rationalization A mechanism by which logical or rational but not necessarily accurate explanations are offered for life events.

reactive depression The aspect of a terminally ill person's depression—the fourth step in the Kubler-Ross *sequence of response to death*—that reflects all the sadness and misery that has already transpired as a result of the illness.

recessive gene The nonexpressed gene of a *heterozygous* pair.

reciprocal obligation The belief that a person who benefits from an act of kindness must repay it.

reciprocal relationships Relationships involving mutual exchange. Individuals in reciprocal relationships are each able to assume the right and responsibilities of the other. Peer relations serve as the classic example.

rejecting parents Parents who are relatively insensitive to their infants' needs, who are likely to reject close physical contact, and who seem to accept with resentment the responsibility of being parents; one of Ainsworth's three types of patterns of maternal behavior.

relativistic world view The second of Perry's three *forms of experience construction.* In this phase students begin to believe that their view is as valid as that of the professors, that all viewpoints are equally correct, and that even if there is a true answer to a question, it is probably not knowable.

response to death See *sequence of response to death.*

reversal-nonreversal shift task A discrimination learning task, devised by Kendler and Kendler, that requires children to learn to identify the correct object in each of two separate but related problems. The degree to which children's solution of the first problem facilitates or hinders their solution of the second problem is used as a measure of the degree of shift in information processing strategies common in preschoolers to those more typical of school-age children.

reversibility The ability to mentally represent some action on an object and then to reverse the action. This quality of thought, missing in the *preoperational stage,* implies the ability to make reciprocal mental operations.

role definition The establishment of a set of activities, behaviors, and duties as appropriate for a person in a particular life situation.

role expansion process A process in which women are attempting to broaden their range of competence and sources of satisfaction without abandoning traditional functions, obligations, and sources of satisfaction.

role experimentation The trying out by adolescents of new roles or ways of behaving that is part of their search for comfortable identities.

sample A selection of individuals from whom information is gathered for a research project. The sample should reflect the characteristics of the population from which it is chosen and should be large enough to permit generalizations about the entire population from the data gathered.

sanctification process A process identified by Lopata in which widows idealize their late husbands, cleanse them of all their shortcomings, and remember only their good points. Sanctification appears to be an important component of the partial resolution of the bereavement process among the very old.

Scanzoni and Scanzoni's patterns of spouse interaction See *patterns of spouse interaction.*

scheme According to Piaget, the set of behaviors together with the cognitive capacity necessary to both recognize the appropriateness of the behaviors being used in a particular setting and to generalize the behaviors to new settings.

secondary circular reaction A behavior pattern that involves an infant's attempt to make an interesting event reoccur in the environment.

self-regulating system A system, such as the *psychological organism,* that tries to maintain a steady state or degree of equilibrium among and between itself and its components.

senior partner/junior partner pattern A spouse interaction pattern, increasingly common, characterized by a greater degree of overlap in division of labor and a more equitable sharing of power than the *head/complement pattern.* This pattern is like the head/complement pattern, however, in the balance it maintains between instrumental and expressive aspects of the relationship. See also *patterns of spouse interaction.*

sensorimotor stage See *stage of sensorimotor development.*

separation anxiety A feeling that develops at about the same time as *object permanence* and is demonstrated whenever the parent leaves the infant with an unfamiliar person or in an unfamiliar setting.

separation protest The distress the infant begins to show at about six months of age when separated from the parent.

sequence of identity statuses See *identity statuses.*

sequence of response to death A five-step process identified by Kubler-Ross to describe how the terminally ill confront their own deaths. The five steps are denial and isolation, anger, bargaining, depression, and acceptance.

seriation The process of placing in order a set of objects, such as a series of sticks of differing lengths. Seriation is one of the skills examined by Piaget to study preschoolers' developing ability to accurately and effectively process information during the *preoperational stage.* See also *classification.*

sickle-cell anemia A major recessive gene effect resulting from the mutation, or distortion, of one of the codons, or sets of *DNA* pairs, that regulate the formation of red blood cells.

six facets of the divorce process Six interdependent but identifiable issues of divorce noted by Bohannan, which highlight the multiple bonds involved in a marital relationship. The six facets are *emotional divorce, legal divorce, economic divorce, coparental divorce, community divorce,* and *psychic divorce.*

sleeper effect A pattern observed by Kagan and Moss in which, in some cases, parent behavior toward the infant was a better predictor of preadolescent and young adult behavior than was parent behavior at the two later times of measurement.

social breakdown syndrome A breakdown to which many aged persons are susceptible, in which, according to Kuypers and Bengtson, a person who is unable to maintain an independent sense of direction and control becomes increasingly dependent on social cues for *role definition,* receives vague or negative cues, begins to feel incompetent and therefore to depend more on others, experiences a resulting real decline in competence, and finally is left with a self-concept of inadequacy, dependence, and incompetence.

social clock A culturally determined schedule of the time intervals in which the various life events are considered appropriate.

social learning Indirect learning in which behavioral change occurs through imitation.

sociohistorical context The economic, social, and political conditions of the world in which an individual lives; the component of the *psychological organism* that consists of the particular slice of history an individual life-span traverses and the place where that particular traverse occurs. See also *biological organism.*

spatial relationships The relative characteristics of space (up, down, left, right, in front of, behind, etc.). Initial understanding of spatial relationships is partly a reflection of the preschooler's continuing physical-motor development.

species-specific characteristics Characteristics that are universal to a given species.

spouse interaction See *patterns of spouse interaction.*

stage-based approach A method in which the process of development is studied by observing the changes that occur in individuals as they age.

stage of autonomy See *psychosocial stages of trust and autonomy.*

stage of concrete operational thought Piaget's third stage of *cognitive development,* which occurs approximately between the ages of seven and twelve and which begins with the acquisition of the ability to separate an operation, or mental action, from the content of that action. See also *concrete operational thought.*

stage of development A range of ages grouped together as a basis for analyzing development.

stage of ego integrity See *psychosocial stage of ego integrity.*

stage of formal operational thought Piaget's fourth stage of *cognitive development,* which begins in *adolescence* and in which an individual is able to use cognitive skills not simply on real or concrete experiences, but on any and all experiences, real or imagined.

stage of generativity See *psychosocial stage of generativity.*

stage of identity formation See *psychosocial stage of identity formation.*

stage of industry See *psychosocial stage of industry.*

stage of initiative See *psychosocial stage of initiative.*

stage of intimacy See *psychosocial stage of intimacy.*

stage of preoperational thought Piaget's second stage of *cognitive development,* which occurs approximately between the ages of two and seven, and in which a child's thoughts are to a large extent based on the appearance of things and tend to focus on one aspect of a situation to the exclusion of others.

stage of sensorimotor development Piaget's first stage of *cognitive development,* lasting from birth to approximately twenty-four months.

stage of trust See *psychosocial stages of trust and autonomy.*

stages of labor and delivery The three phases of the birth process.

stage theory of adult female development A theory, developed by Frieze, that the lives of adult women follow similar patterns and are divided into stages by varying cultural and social demands, usually reflecting the unique demands of the role of parent.

stage theory of adult male development A theory, developed by Levinson, that the life patterns of adult men are very similar and consist of three stages: early adulthood, middle adulthood, and late adulthood.

standard deviation A measure of how much, on the average, each individual score differs, or deviates, from the sample *mean.*

statistical significance The probability that two events are not related merely by chance. Within the social sciences, a statistically significant association is said to exist if the probability of a particular association occurring by chance is less than five in a hundred.

telegraphic speech The grammatically incorrect two-word sentences that are the toddler's first word combinations.

temporal relationships The characteristics of time, first understood by the preschooler in terms of sequence rather than duration. This understanding is developed partly through the preschooler's awareness of the series of actions or steps involved in various motor activities.

teratogens Conditions in the environment that can have a devastating influence on prenatal development.

terminal drop A dramatic decline that occurs in cognitive test scores one to two years preceding death.

tertiary circular reaction A behavior pattern in which an infant repeats an action with several objects or performs several actions with the same object, in an intentional effort to discover new means and new ends.

theory of cognitive development A theory, developed by Piaget, in which development is divided into four stages—the *sensorimotor stage,* the *preoperational stage,* the *concrete operational stage,* and the *formal operational stage*—each defined by the types of cognitive processes, or operations, that an individual has available at that stage, rather than by the behaviors exhibited. See also *cognitive development.*

theory of psychosocial stages A theory, developed by Erikson, that each stage of development presents a psychosocial task that must be adequately resolved if development is to continue.

three forms of experience construction See *forms of experience construction.*

time sampling A *naturalistic research strategy* in which, at predetermined intervals, whatever behaviors are taking place are recorded.

transductive reasoning The process through which a conclusion is reached by a logical progression of *inference* from the particular to the particular. This type of logic is typical of the *stage of preoperational thought.*

trophoblast The outer wall of the *blastocyst.* The trophoblast implants itself in the uterine wall and evolves into the *placenta.*

Type A individuals Persons who constantly strive to reach goals that in fact are usually poorly defined, who must always be engaged in some sort of goal-directed and purposeful activity, who find it very difficult to enjoy what they have accomplished, and who are twice as likely to have a heart attack as *Type B individuals* are.

Type B individuals Persons who may at times feel time pressure but are not involved in a chronic struggle against time, who may be ambitious but are not overly competitive, who are goal directed but not aggressively so, and who are half as likely to have a heart attack as *Type A individuals* are.

variability See *measures of variability.*

vasomotor instability A secondary effect of *menopause* commonly known as hot flashes.

Weisman's appropriate death See *appropriate death.*

young-old Neugarten's term for the first phase of old age, a period that begins with the retirement transition and lasts usually until the mid-seventies or until health begins to decline significantly.

zygote A fertilized ovum.

REFERENCES

Abel, E. L. (1984). *Fetal alcohol syndrome and fetal alcohol effects.* New York: Plenum.

Abrahams, B., Feldman, S. S., & Nash, S. C. (1978). Sex role self-concept and sex role attitudes: Enduring personality characteristics or adaptations to changing life situations? *Developmental Psychology, 14,* 393–400.

Adams, B. N. (1979). Mate selection in the United States: A theoretical summarization. In W. R. Burr (Ed.), *Contemporary theories about the family* (Vol. 1). New York: The Free Press.

Ade-Ridder, L., & Brubaker, T. H. (1983). The quality of long-term marriages. In T. H. Brubaker (Ed.), *Family relations in later life.* Beverly Hills, CA: Sage Publications.

Ahammer, I. M. (1973). Social learning theory as a framework for the study of adult personality development. In P. B. Baltes & K. W. Schaie (Eds.), *Lifespan developmental psychology: Personality and socialization.* New York: Academic Press.

Ahrons, C. R. (1980). Divorce: A crisis of family transition and change. *Family Relations, 29,* 533–540.

Ainsworth, M. D. S. (1969). Object relations, dependency and attachment: A theoretical review of the infant-mother relationship. *Child Development, 40,* 969–1025.

Ainsworth, M. D. S. (1973). The development of mother-infant attachment. In B. M. Caldwell & H. N. Ricciuti (Eds.), *Review of child development research* (Vol. 3). Chicago: University of Chicago Press.

Ainsworth, M. D. S., & Bell, S. M. (1974). Mother-infant interaction and the development of competence. In K. J. Connolly & J. S. Bruner (Eds.), *The growth of competence.* New York: Academic Press.

Ainsworth, M. D. S., Blehar, M. C., Waters, E., & Wall, S. (1978). *Patterns of attachment.* Hillsdale, NJ: Lawrence Erlbaum Associates.

Albrecht, G. L., & Gift, H. C. (1975). Adult socialization: Ambiguity and adult life crises. In N. Datan & L. H. Ginsberg (Eds.), *Life span developmental psychology: Normative life crisis.* New York: Academic Press.

Aldous, J. (1978). *Family careers: Developmental change in families.* New York: Wiley.

Almquist, E. M. (1974). Attitudes of college men toward working wives. *Vocational Guidance Quarterly, 23,* 115–121.

Anastasi, A. (1958). Heredity, environment and the question "How?" *Psychological Review, 65,* 197–208.

Andrews, S. R., Blumenthal, J. B., Johnson, D. L., Kahn, A. J., Ferguson, C. J., Lasater, T. M., Malone, P. E., & Wallace, D. B. (1982). The skills of mothering: A study of parent-child development centers. *Monographs of the Society for Research in Child Development, 47* (Serial No. 198).

Angrist, S. S., & Almquist, E. M. (1975). *Careers and contingencies.* New York: Dunellen.

Apgar, V. (1953). A proposal for a new method of evaluation of the newborn infant. *Anesthesia and Analgesia, 32,* 260–267.

Applebaum, E. (1981). *Back to work: Determinants of women's successful re-entry.* Boston: Auburn House.

Arenberg, D., & Robertson-Tchabo, E. A. (1977). Learning and aging. In J. E. Birren & K.W. Schaie (Eds.), *Handbook of the psychology of aging.* New York: D. Van Nostrand Co.

Astin, A. W. (1977). *Four critical years.* San Francisco: Jossey-Bass.

Astin, A. W., Green, K. C., Korn, W. S., & Maier, M. J. (1984). *The American freshman: National norms for 1984.* Los Angeles: American Council on Education.

Astin, A. W., King, M. R., Light, J. M., & Richardson, G. T. (1974). *The American freshman: National norms for 1974.* Los Angeles: American Council on Education.

Atchley, R. C. (1975). The life course, age grading, and age-linked demands for decision making. In N. Datan & L. H. Ginsberg (Eds.), *Life span developmental psychology: Normative life crises.* New York: Academic Press.

Atchley, R. C. (1976). *The sociology of retirement.* New York: Schenkman.

Atchley, R. C. (1980). *The social forces in later life* (3rd ed.). Belmont, CA: Wadsworth.

Atchley, R. C., & Miller, S. J. (1983). Types of elderly couples. In T. H. Brubaker (Ed.), *Family relations in later life.* Beverly Hills, CA: Sage Publications.

Ausubel, D. P., Sullivan, E. V., & Ives, S. W. (1980). *Theory and problems of child development* (3rd ed.). New York: Grune and Stratton.

Bailyn, L. (1980). The slow burn way to the top: Some thoughts on the early years of organizational careers. In C. B. Derr (Ed.), *Work, family and the career.* New York: Praeger.

Ball, J. F. (1977). Widow's grief: The impact of age and mode of death. *Omega, 7,* 307–333.

Ball, S., & Bogatz, G. (1972). Research on Sesame St.: Some implications for compensatory education. In J. Stanley (Ed.), *Compensatory education for children: Ages 2 to 8.* Baltimore: Johns Hopkins University Press.

Baltes, P. B., Dittmann-Kohli, F., & Dixon, R. A. (1984). New perspectives on the development of intelligence in adulthood: Toward a dual process conception and a model of selective optimization with compensation. In P. B. Baltes & O. G. Brim (Eds.), *Lifespan development and behavior* (Vol. 6). Orlando: Academic Press.

Bandura, A. (1971). *Social learning theory.* Englewood Cliffs, NJ: Prentice-Hall.

Bandura, A. (1982). Self-efficacy mechanisms in human agency. *American Psychologist, 37,* 122–148.

Barrett, C. J. (1978). Sex differences in the experience of widowhood. Paper presented at the annual meeting of the American Psychological Association, Toronto.

Barry, W. A. (1970). Marriage research and conflict: An integrative review. *Psychological Bulletin, 73,* 41–54.

Bart, P. B., & Grossman, M. (1978). Menopause. In M. T. Notman & C. C. Nadelson (Eds.), *The woman patient* (Vol. 1). New York: Plenum.

Baumrind, D. (1968). Authoritarian versus authoritative parental control. *Adolescence, 3,* 255–272.

Baumrind, D. (1971). Current patterns of parental authority. *Developmental Psychology Monographs,* 4 (No. 1, Pt. 2).

Baumrind, D. (1973). The development of instrumental competence through socialization. In A. D. Pick (Ed.), *Minnesota symposium on child psychology,* (Vol. 7). Minneapolis: University of Minnesota Press.

Baumrind, D. (1975). Early socialization and adolescent competence. In S. E. Dragastin & G. H. Elder (Eds.), *Adolescence in the life cycle.* Washington, DC: Hemisphere Publishing Co.

Baumrind, D. (1978). Reciprocal rights and responsibilities in parent-child relations. *Journal of Social Issues, 34,* 179–196.

Bayne, J. R. (1970). Environmental modification for the older person. *The Gerontologist, 10,* 1.

Beckman. L. J., & Houser, B. B. (1982). The consequences of childlessness on the social-psychological well-being of older women. *Journal of Gerontology, 37,* 243–250.

Bell, S. M., & Ainsworth, M. D. S. (1972). Infant crying and maternal responsiveness. *Child Development, 43,* 1171–1190.

Beller, E. K. (1979). Early intervention programs. In J. D. Osofsky (Ed.), *Handbook of infant development.* New York: Wiley.

Belloc, N. B., & Breslow, L. (1972). Relationship of physical health status and health practice. *Preventative Medicine, 1,* 409–421.

Belsky, J. (1980). Child maltreatment: An ecological integration. *American Psychologist, 35,* 320–336.

Belsky, J. (1981). Early human experience: A family perspective. *Developmental Psychology, 17,* 3–23.

Belsky, J., Spanier, G. B., & Rovine, M. (1983). Stability and change in marriage across the transition to parenthood. *Journal of Marriage and the Family, 45,* 567–577.

Bem, S. (1974). The measurement of psychological androgyny. *Journal of Consulting and Clinical Psychology, 42,* 155–162.

Bengston, V. L., Kasschau, P. L., & Ragan, P. K. (1977). The impact of social structure on aging individuals. In J. E. Birren & K. W. Schaie (Eds.), *Handbook of the psychology of aging.* New York: Van Nostrand Reinhold Co.

Benson, R. C. (1974). *Handbook of obstetrics and gynecology.* Los Angeles: Langley Medical Publications.

Benson, R. C. (1978). *Current obstetric and gynecological diagnosis and treatment.* Los Altos, CA: Lange Medical Publications.

Benson, R. C. (1983). *Handbook of obstetrics and gynecology* (8th ed.). Los Altos, CA: Lange Medical Publications.

Berardo, F. M. (1970). Survivorship and social isolation: The case of the aged widower. *The Family Coordinator, 19,* 11–25.

Bernard, J. (1982). *The future of marriage* (1982 ed.). New Haven: Yale University Press.

Bikson, T. K., & Goodchilds, J. (1978). Old and alone. Paper presented at the annual meeting of the American Psychological Association, Toronto.

Birch H. G., & Gussow, J. D. (1970). *Disadvantaged children: Health, nutrition and school failure.* New York: Harcourt, Brace and World.

Birren, J. E., Kinney D. K., Schaie, K. W., & Woodruff, D. S. (1981). *Developmental psychology: A lifespan approach.* Boston: Houghton Mifflin.

Blank, M. (1974). Cognitive functions of language in the preschool years. *Developmental Psychology, 10,* 229–245.

Blau, Z. M. (1981). *Aging in a changing society* (2nd ed.). New York: Franklin Watts.

Block, J. H. (1984). *Sex role identity and ego development.* San Francisco: Jossey-Bass.

Block, M. R., Davidson, J. L., & Grambs, J. D. (1981). *Women over forty: Visions and reality.* New York: Springer Publishing Co.

Bloom, B. S. (1964). *Stability and change in human characteristics.* New York: Wiley.

Bloom, L., & Lahey, M. (1978). *Language development and language disorders.* New York: Wiley.

Bocknek, G. (1980). *The young adult: Development after adolescence.* Monterey, CA: Brooks/Cole.

Bohannan, P. (1971). *Divorce and after.* New York: Anchor Books.

Borland, D. C. (1982). A cohort analysis approach to the empty nest syndrome among three ethnic groups of women: A theoretical position. *Journal of Marriage and the Family, 44,* 117–129.

Borow, H. (1976). Career development. In J. F. Adams (Ed.), *Understanding adolescence* (3rd ed.). Boston: Allyn and Bacon.

Botwinick, J., West, R., & Storandt, M. (1978). Predicting death from behavioral test performance. *Journal of Gerontology, 33,* 755–763.

Bower, T. G. R. (1979). *Human development.* San Francisco: Freeman.

Bowerman, C. E., & Kinch, J. W. (1959). Changes in family and peer orientation of children between the fourth and tenth grade. *Social Forces, 37,* 206–211.

Brackbill, Y. (1979). Obstetrical medication and infant behavior in J. D. Osofsky (Ed.), *Handbook of infant development.* New York: Wiley.

Bray, D. W., & Howard, A. (1983). The A.T. & T. longitudinal studies of managers. In K. W. Schaie (Ed.), *Longitudinal studies of adult psychological development.* New York: Guilford Press.

Brazelton, T. B. (1973). *Neonatal behavioral assessment scale.* London: Spastics International Medical Publications.

Brazelton, T. B. (1980). Behavioral competence of the newborn infant. In P. M. Taylor (Ed.), *Parent-infant relationships.* New York: Grune & Stratton.

Breger, L. (1974). *From instinct to identity: The development of personality.* Englewood Cliffs, NJ: Prentice-Hall.

Breslau, L., & Haug, M. (1983). Some elements in an integrative model of depression in the aged. In L. D. Breslau & M. R. Haug (Eds.), *Depression and aging.* New York: Springer Publishing Co.

Brim, O. G. (1974). The sense of personal control over one's life. Invited address to Divisions 7 and 8 at the 82nd Annual Convention of the American Psychological Association, New Orleans.

Brim, O. G., & Kagan, J. (Eds.). (1980). *Constancy and change in human development.* Cambridge, MA: Harvard University Press.

Brim, O. G., & Ryff, C. D. (1980). On the properties of life events. In P. B. Baltes & O. G. Brim (Eds.), *Life span development and behavior* (Vol. 3). New York: Academic Press.

Brody, E. M. (1981). "Women in the middle" and family help to older people. *The Gerontologist, 21,* 471–481.

Brody, J. E. (1975, December 17). T.V. violence cited as bad influence. *The New York Times,* p. 20.

Bromley, D. B. (1974). *The psychology of human aging.* London: Penguin.

Bronfenbrenner, U. (1970). *Two worlds of childhood.* New York: Russell Sage Foundation.

Bronfenbrenner, U. (1975). Is early intervention effective? In J. Hellmuth (Ed.), *Exceptional infant* (Vol. 3). New York: Brunner/Mazel.

Bronfenbrenner, U. (1979). *The ecology of human development.* Cambridge, MA: Harvard University Press.

Brophy, J. E., & Evertson, C. M. (1981). *Student characteristics and teaching.* New York: Longman.

Brown, A. L., & DeLoache, J. S. (1978). Skills, plans, and self-regulation. In R. S. Siegler (Ed.), *Children's thinking: What develops?* Hillsdale, NJ: Lawrence Erlbaum Associates.

Brown, B. (Ed.). (1978). *Found: Long term gains from early intervention.* Boulder, CO: Westview Press.

Brubaker, T. H. (Ed.). (1983). *Family relations in later life.* Beverly Hills, CA: Sage Publications.

Burke, R. J., & Weir, T. (1976). Relationship of wives' employment status to husband, wife, and pair satisfaction and performance. *Journal of Marriage and the Family, 38,* 279–287.

Butler, R. N. (1974, December). Successful aging and the role of the life review. *Journal of the American Geriatric Society.*

Caldwell, B. M. (1967). What is the optimal learning environment for the young child? *American Journal of Orthopsychiatry, 37,* 8–20.

Campbell, A. (1981). *The sense of well-being in America.* New York: McGraw-Hill.

Cantor, M. H. (1979). Neighbors and friends: An overlooked resource in the informal support system. *Research on Aging, 1,* 434–464.

Chandler, M. J. (1977). Social cognition. In W. E. Overton & J. M. Gallagher (Eds.), *Knowledge and development* (Vol. 1). New York: Plenum.

Cherlin, A. (1978). Remarriage as an incomplete institution. *American Journal of Sociology, 84,* 634–650.

Cherlin, A. (1980). Postponing marriage: The influence of young women's work expectations. *Journal of Marriage and the Family, 42,* 355–365.

Cherlin, A. J. (1981). *Marriage, divorce, remarriage.* Cambridge, MA: Harvard University Press.

Chilman, C. S. (1979). *Adolescent sexuality in a changing American society.* Department of Health Education and Welfare Publication, No. (NIH) 79-1426.

Chinoy, E. (1955). *Automobile workers and the American dream.* Garden City, NY: Doubleday.

Cicirelli, V. G. (1969). Project Head Start, a national evaluation: Summary of the study. In David G. Hays (Ed.), *Britannica review of American education* (Vol. I).

Cicirelli, V. G. (1981). *Helping elderly parents: The role of adult children.* Boston, MA: Auburn House.

Clarke, A. M., & Clarke, A. D. B. (1976). *Early expe-*

rience: Myth and evidence. New York: The Free Press.

Clarke-Stewart, A. (1977). *Child care in the family.* New York: Academic Press.

Clarke-Stewart, K. A. (1978). And daddy makes three: The father's impact on mother and young child. *Child Development, 49,* 466–478.

Clarke-Stewart, K. A. (1978). Popular primers for parents. *American Psychologist, 33,* 359–370.

Clausen, J. A. (1981). Men's occupational careers in the middle years. In D. E. Eichorn, J. A. Clausen, N. Haan, M. P. Honzik, & P. H. Mussen (Eds.), *Present and past in middle life.* New York: Academic Press.

Clayton, P. J., Halikes, J. A., & Maurice, W. L. (1971). The bereavement of the widow. *Diseases of the nervous system, 32,* 597–604.

Clayton, V. (1975). Erikson's theory of human development as it applies to the aged: Wisdom as contradictive cognition. *Human Development, 18,* 119–128.

Clayton, V. P., & Birren, J. E. (1980). The development of wisdom across the life span: A re-examination of an ancient topic. In P. B. Baltes & O. G. Brim (Eds.), *Life span development and behavior* (Vol. 3). New York: Academic Press.

Cohen, J. (1979). High school subcultures and the adult world. *Adolescence, 14,* 491–502.

Cohen, J. B. (1978). The influence of culture on coronary prone behavior. In T. M. Dembroshe (Ed.), *Coronary prone behavior,* New York: Springer-Verlag.

Cohler, B. J. (1982). Personal narrative and the life course. In P. B. Baltes & O. G. Brim (Eds.), *Lifespan development and behavior* (Vol. 4). New York: Academic Press.

Colby, A., Kohlberg, L., Gibbs, J., & Lieberman, M. (1983). A longitudinal study of moral development. *Monographs of the Society for Research in Child Development, 48* (Serial No. 200).

Cole, L., & Hall, I. M. (1970). *Psychology of adolescence* (7th ed.). New York: Holt, Rinehart & Winston.

Coleman, J. C. (Ed.). (1975). *Youth: Transition to adulthood.* Chicago: University of Chicago Press.

Coleman, J. S. (1966). *Equality of educational opportunity.* (U.S. Dept. of Health, Education and Welfare Publication. No. OE-38001).

Collins, W. A. (1979). Children's comprehension of television content. In E. Wartella (Ed.), *Children communicating.* Beverly Hills, CA: Sage Publications.

Condry, J., & Siman, M. L. (1974). Characteristics of peer and adult-oriented children. *Journal of Marriage and the Family, 36,* 543–554.

Conger, J. J. (1972). A world they never knew: The family and social change. In J. Kagan & R. Coles (Eds.), *Twelve to sixteen: Early adolescence.* New York: Norton.

Coopersmith, S. (1967). *The antecedents of self-esteem.* San Francisco: Freeman.

Corah, N. L., Anthony, E. J., Painter, P., Stern, J. A., & Thurston, D. L. (1965). Effects of perinatal anoxia after seven years. *Psychological Monographs, 79* (3, Whole No. 596).

Corsaro, W. A. (1981). Friendship in the nursery school: Social organization in a peer environment. In S. R. Asher & J. M. Gottman (Eds.), *The development of children's friendships.* Cambridge, England: Cambridge University Press.

Costa, P. T., & McCrae, R. R. (1980). Still stable after all these years: Personality as a key to some issues in adulthood and old age. In P. B. Baltes & O. G. Brim (Eds.), *Lifespan development and behavior* (Vol. 3). New York: Academic Press.

Covey, H. C. (1981). A reconceptualization of continuity theory: Some preliminary thoughts. *The Gerontologist, 21,* 628–633.

Cowan, P. A. (1978). *Piaget with feeling.* New York: Holt, Rinehart & Winston.

Craik, F. I. M. (1977). Age differences in human memory. In J. E. Birren & K. W. Schaie (Eds.), *Handbook of the psychology of aging.* New York: D. Van Nostrand Co.

Cratty, B. J. (1979). *Perceptual and motor development in infants and children* (2nd ed.). Englewood Cliffs, NJ: Prentice-Hall.

Crosby, F. J. (1982). *Relative deprivation and working women.* New York: Oxford University Press.

Cumming, E. (1975). Engagement with an old theory. *International Journal of Aging and Human Development, 6,* 187–191.

Cumming, E., & Henry, W. (1961). *Growing old.* New York: Basic Books.

Cvetkovich, G., Grote, B., Lieberman, E. J., & Miller, W. (1978). Sex role development and teenage fertility related behavior. *Adolescence, 13,* 231–236.

Dacey, J. S. (1982). *Adult development.* Glenview, IL: Scott, Foresman.

D'Amico, R. J., Haurin, R. J., & Mott, F. L. (1983). The effects of mothers' employment on adolescent and early adult outcomes of young men and women. In C. D. Hayes & S. B. Kamerman (Eds.), *Children of working parents.* Washington, DC: National Academy Press.

Damon, W. (1977). *The social world of the child.* San Francisco: Jossey-Bass.

Datan, N. (1980). Midas and other mid-life crises. In W. H. Norman & T. J. Scaramella (Eds.), *Mid-life developmental and clinical issues.* New York: Brunner/Mazel.

Davidoff, I. F. (1977). "Living Together" as a developmental phase: A holistic view. *Journal of Marriage & Family Counseling, 3,* 67–76.

Davies, M., & Kandel, D. B. (1981). Parental and peer influences on adolescents' educational plans: Some further evidence. *American Journal of Sociology, 87,* 363–387.

Day, D. E. (1983). *Early childhood education.* Glenview, IL: Scott, Foresman.

Denney, N. W. (1982). Aging and cognitive changes. In B. B. Wolman (Ed.), *Handbook of developmental psychology.* Englewood Cliffs, NJ: Prentice-Hall.

de Villiers, P. A., & de Villiers, J. G. (1979). *Early language.* Cambridge, MA: Harvard University Press.

Dick-Read, G. (1959). *Childbirth without fear: The principles and practices of natural childbirth.* New York: Harper & Row.

Diepold, J., & Young, R. O. (1979). Empirical studies of adolescent sexual behavior. *Adolescence, 14,* 45–64.

Dimsdale, J. E. (1974). The coping behavior of Nazi concentration camp survivors. *The American Journal of Psychiatry, 131,* 792–797.

Dobson, C. (1983). Sex-role and marital-role expectations. In T. H. Brubaker (Ed.), *Family relations in later life.* Beverly Hills, CA: Sage Publications.

Dohrenwend, B. S. (1973). Life events as stressors: A methodological inquiry. *Journal of Health and Social Behavior, 14,* 167–175.

Dohrenwend, B. S., & Dohrenwend, B. P. (Eds.). (1974). *Stressful life events: Their nature and effects.* New York: Wiley.

Dorr, A., & Kovaric, P. (1980). Some of the people some of the time—but which people? Television violence and its effects. In E. L. Palmer & A. Dorr (Eds.), *Children and the faces of television: Teaching, violence and selling.* New York: Academic Press.

Douvan, E., & Adelson, J. (1966). *The adolescent experience.* New York: Wiley.

Dunn, J., & Kendrick, C. (1979). Interaction between young siblings in the context of family relationships. In M. Lewis & L. A. Rosenblum (Eds.), *The child and its family.* New York: Plenum.

Duvall, E. M. (1977). *Marriage and family development* (5th ed.). Philadelphia: Lippincott.

Dweck, C. S. (1981). Social-cognitive processes in children's friendships. In S. R. Asher & J. M. Gottman (Eds.), *The development of children's friendships.* Cambridge, England: Cambridge University Press.

Dweck, C. S., Davidson, W., Nelson, S., and Enna, B. (1978). Sex differences in learned helplessness. *Developmental Psychology, 14,* 268–277.

Eichorn, D. H., Clausen, J. A., Haan, N., Honzik, M. P., & Mussen, P. H. (Eds.). (1981). *Present and past in middle life.* New York: Academic Press.

Eimas, P. D., Siqueland, E., Jusczyk, P., & Vigorito, J. (1971). Speech perception in infants. *Science,* 303–306.

Eisdorfer, C., & Wilkie, F. (1977). Stress, disease, aging and behavior. In J. E. Birren & K. W. Schaie (Eds.), *Handbook of the psychology of aging.* New York: Van Nostrand Reinhold Co.

Elder, G. H. (1974). *Children of the Great Depression.* Chicago: University of Chicago Press.

Elder, G. H. (1975). Adolescence in the life cycle. In S. E. Dragastin & G. H. Elder (Eds.), *Adolescence in the life cycle.* Washington, DC: Hemisphere Publishing Co.

Elder, G. H., & Rockwell, R. C. (1976). Marital timing in women's life patterns. *Journal of Family History, 1,* 34–53.

Elder, G. H., & Rockwell, R. C. (1979). The life course and human development: An ecological perspective. *International Journal of Behavioral Development, 2,* 1–21.

Elkin, F., & Handel, G. (1972). *The child and society* (2nd ed.). New York: Random House.

Elkind, D. (1976). *Child development and education.* New York: Oxford University Press.

Elkind, D. (1978). *The child's reality.* Hillsdale, NJ: Lawrence Erlbaum Associates.

Elkind, D. (1982). *The hurried child.* Reading, MA: Addison-Wesley.

Elkind, D., & Lyke, N. (1975). Early education and kindergarten: Competition and cooperation. *Young Children, 30,* 393–401.

Emde, R. N. (1980). Emotional availability: A reciprocal reward system for infants and parents with implication for prevention of psychological disorder. In P. M. Taylor (Ed.), *Parent-infant relationships.* New York: Grune & Stratton.

Emde, R. N., Gaensbauer, T. J., & Harmon, R. J. (1976). Emotional expression in infancy. *Psychological Issues, 10* (Monograph 37, No. 1).

Emmrich, W. (1977). Structure and development of personal-social behaviors in economically disadvantaged preschool children. *Genetic Psychology Monograph, 95,* 191–245.

Entwisle, D. R. (1966). *Word associations of young children.* Baltimore: Johns Hopkins University Press.

Entwisle, D. R., & Doering, S. G. (1981). *The first birth: A family turning point.* Baltimore: Johns Hopkins University Press.

Epstein, S., & Erskine, N. (1983). The development of personal theories of reality from an interactional perspective. In D. Magnusson & V. L. Allen (Eds.), *Human development.* New York: Academic Press.

Erikson, E. (1950). *Childhood and society.* New York: Norton.

Erikson, E. (1968). *Identity: Youth and crisis.* New York: Norton.

Erikson, E. (1980). *Identity and the life cycle.* New York: Norton.

Eron, L. D. (1982). Parent-child interaction, televi-

sion violence, and aggression in children. *American Psychologist, 37,* 197–211.
Etaugh, C. (1974). Working mothers. *Merrill-Palmer Quarterly, 20,* 71–98.
Evans, E. D. (1975). *Contemporary influences in early childhood education* (2nd ed.). New York: Holt, Rinehart & Winston.
Everitt, A. V. (1983). Pacemaker mechanisms in aging and the diseases of aging. In H. Blumenthal (Ed.), *Handbook of diseases of aging.* New York: Van Nostrand Reinhold Co.
Fantz, R. L., & Nevis, S. (1967). Pattern preference and perceptual cognitive development in early infancy. *Merrill-Palmer Quarterly, 13,* 77–108.
Farrell, M. P., & Rosenberg, S. D. (1981). *Men at midlife.* Boston: Auburn House.
Fein, G. G. (1978). *Child development.* Englewood Cliffs, NJ: Prentice-Hall.
Fein, G. G., & Clarke-Stewart, A. (1973). *Day care in context.* New York: Wiley.
Feldman, H. (1971). The effects of children on the family. In A. Michel (Ed.), *Family issues of employed women in Europe and America.* Leiden: E. J. Brill.
Feldman, K. A., & Newcomb, T. M. (1969). *The impact of college on students.* San Francisco: Jossey-Bass.
Feldman, S. S., & Aschenbrenner, B. (1983). Impact of parenthood on various aspects of masculinity and femininity. *Developmental Psychology, 19,* 278–289.
Feldman, S. S., Biringen, Z. C., Nash, S. C. (1981). Fluctuations of sex-related self-attributions as a function of stage of the family life cycle. *Developmental Psychology, 17,* 24–35.
Fine, G. A. (1981). Friends, impression management, and preadolescent behavior. In S. R. Asher & J. M. Gottman (Eds.). *The development of children's friendships.* Cambridge, England: Cambridge University Press.
Fischer, K. W. (1980). A theory of cognitive development. *Psychological Review, 87,* 477–531.
Fischer, L. R. (1981). Transitions in the mother-daughter relationship. *Journal of Marriage and the Family, 43,* 613–622.
Fishbein, H. D. (1976). *Evolution, development and children's learning.* Pacific Palisades, CA: Goodyear.
Flanagan, J. C. (1980). Quality of life. In L. A. Bond & J. C. Rosen (Eds.), *Competence and coping during adulthood.* Hanover, NH: University Press of New England.
Flavell, J. H. (1977). *Cognitive development.* Englewood Cliffs, NJ: Prentice-Hall.
Fogel, A. (1984). *Infancy.* St. Paul: West Publishing Co.
Foner, A., & Schwab, K. (1983). Work and retirement in a changing society. In M. W. Riley, B. B. Hess, & K. Bond (Eds.), *Aging and society: Selected reviews of recent research.* Hillsdale, NJ: Lawrence Erlbaum Associates.
Forman, G. E., & Hill, F. (1984). *Constructive play: Applying Piaget in the classroom* (rev. ed.). Menlo Park, CA: Addison-Wesley.
Forman, G. E., & Kuschner, D. S. (1977). *The child's construction of knowledge: Piaget for teaching children.* Monterey, CA: Brooks/Cole.
Fox, G. L. (1980). The mother–adolescent daughter relationship as a sexual socialization structure. *Family Relations, 29,* 21–28.
Fox, M. F., & Hesse-Biber, S. (1984). *Women at work.* Palo Alto, CA: Mayfield Publishing Co.
Fraiberg, S. (1975). The development of human attachments in infants blind from birth. *Merrill-Palmer Quarterly, 21,* 315–335.
Fraiberg, S. H. (1959). *The magic years.* New York: Scribner's.
Francoeur, R. T. (1982). *Becoming a sexual person.* New York: Wiley.
Frankel, F., & Rathvon, S. (1980). *Whatever happened to Cinderella?* New York: St. Martin's Press.
Freedman, M. B. (1967). *The college experience.* San Francisco: Jossey-Bass.
Freeman, R. B., & Wise, D. A. (1982). *The youth labor market problem: Its natural causes and consequences.* Chicago: University of Chicago Press.
Friedman, M. (1978). Type A behavior: Its possible relationship to pathogenic processes responsible for coronary heart disease (a preliminary enquiry). In T. M. Dembroski (Ed.), *Coronary-prone behavior.* New York: Springer Publishing Co.
Friedrich, L. K., & Stein, A. H. (1973). Aggressive and prosocial television programs and the natural behavior of preschool children. *Monographs of the Society for Research in Child Development, 38,* (4, Serial No. 151).
Fries, J. F. (1980). Aging, natural death, and the compression of morbidity. *The New England Journal of Medicine, 303,* 130–136.
Frieze, I. H. (1978). *Women and sex roles.* New York: Norton.
Furstenberg, F. F. (1982). Conjugal succession: Reentering marriage after divorce. In P. B. Baltes & O. G. Brim, Jr. (Eds.), *Life span development and behavior* (Vol. 4). New York: Academic Press.
Furth, H. G. (1980). *The world of grown-ups: Children's conceptions of society.* New York: Elsevier.
Gagnon, J. (1972). The creation of the sexual in early adolescence. In J. Kagan & R. Coles (Eds.), *Twelve to sixteen: Early adolescence.* New York: Norton.
Gagnon, J. H., & Greenblat, C. S. (1978). *Life designs.* Glenview, IL: Scott, Foresman.
Gallatin, J. E. (1975). *Adolescence and individuality.* New York: Harper & Row.
Garbarino, J. (1976). A preliminary study of some

ecological correlates of child abuse: The impact of socioeconomic stress on mothers. *Child Development, 47,* 178–185.

Garbarino, J. (1980). Changing hospital childbirth practices: A developmental perspective on prevention of child maltreatment. *American Journal of Orthopsychiatry, 50,* 588–597.

Garbarino, J. (1980). Some thoughts on school size and its effect on adolescent development. *Journal of Youth and Adolescence, 9,* 19–33.

Garbarino, J. (1982). *Children and families in the social environment.* New York: Aldine.

Garbarino, J., & Crouter, A. (1978). Defining the community context for the parent-child relations: The correlates of child maltreatment. *Child Development, 49,* 604–616.

Garbarino, J., & Gilliam, G. (1980). *Understanding abusive families.* Lexington, MA: Heath.

Garn, S. M. (1980). Continuities and change in maturational timing. In O. G. Brim, Jr., and J. Kagan (Eds.), *Constancy and change in human development.* Cambridge, MA: Harvard University Press.

Garvey, C. (1974). Some properties of social play. *Merrill-Palmer Quarterly, 20,* 163–181.

Geerken, M., & Gove, W. R. (1983). *At home and at work: The families allocation of labor.* Beverly Hills, CA: Sage Publications.

Gelles, R. J. (1978). Violence towards children in the United States. *American Journal of Orthopsychiatry, 48,* 580–592.

Gerber, I., Rusalem, R., Hannon, N., Battin, D., & Arkin, A. (1975). Anticipatory grief and aged widows and widowers. *Journal of Gerontology, 30,* 225–229.

Gerbner, G., Gross, L., Signorielli, N., Morgan, M., & Jackson-Beech, M. (1979). The demonstration of power: Violence profile no. 10. *Journal of Communication, 23,* 177–195.

Gerhardt, L. A. (1972). *Moving and knowing.* Englewood Cliffs, NJ: Prentice-Hall.

Gesell, A., Halverson, H. M., Thompson, H., Ilg, F. L., Castner, B. M., & Ames, L. B. (1940). *The first five years of life.* New York: Harper & Row.

Gesell, A., & Ilg, F. L. (1949). *Child development: An introduction to the study of human growth.* New York: Harper & Row.

Gibson, E. J., & Walk, R. D. (1960). The visual cliff. *Scientific American, 202,* 80–92.

Giele, J. Z. (1980). Adulthood as transcendence of age and sex. In N. J. Smelser & E. H. Erikson (Eds.), *Themes of work and love in adulthood.* Cambridge, MA: Harvard University Press.

Gillaspy, R. T. (1979). The older population: Considerations for family ties. In P. K. Ragan (Ed.), *Aging parents.* Los Angeles: University of Southern California Press.

Gilligan, C. (1977). In a different voice: Women's conception of self and morality. *Harvard Educational Review, 47,* 481–517.

Gilligan, C. (1979). Woman's place in man's life cycle. *Harvard Educational Review, 49,* 431–446.

Gilligan, C. (1982). *In a different voice.* Cambridge, MA: Harvard University Press.

Ginsberg, S. D., & Orlofsky, J. L. (1981). Ego identity status, ego development and locus of control in college women. *Journal of Youth and Adolescence, 10,* 297–309.

Ginsburg, H. (1972). *The myth of the deprived child.* Englewood Cliffs, NJ: Prentice-Hall.

Ginsburg, H., & Opper, S. (1969). *Piaget's theory of intellectual development.* Englewood Cliffs, NJ: Prentice-Hall.

Ginzberg, E. (1972). Toward a theory of occupational choice: A restatement. *Vocational Guidance Quarterly, 20,* 169–176.

Glaser, B. G., & Strauss, A. L. (1968). *Time for dying.* Chicago: Aldine.

Glass, D. C. (1978). Patterns of behavior and uncontrollable stress. In T. M. Dembroski (Ed.), *Coronary prone behavior.* New York: Springer-Verlag.

Glenn, N. D. (1980). Values, attitudes and beliefs. In O. G. Brim, Jr., & J. Kagan (Eds.), *Constancy and change in human development.* Cambridge, MA: Harvard University Press.

Glick, P. C. (1979). Children of divorced parents in demographic perspective. *Journal of Social Issues, 35,* 170–183.

Golan, N. (1981). *Passing through transitions.* New York: The Free Press.

Goldberg, S. (1977). Social competence in infancy: A model of parent-infant interaction. *Merrill-Palmer Quarterly, 23,* 163–179.

Goldberg, S. (1979). Premature birth: Consequences for the parent-infant relationship. *American Scientist, 67,* 217–220.

Goldberg, S., & DiVitto, B. A. (1983). *Born too soon.* San Francisco: Freeman.

Goldhaber, D. E. (1979). Does the changing view of early experience imply a changing view of early development? In L. G. Katz (Ed.), *Current topics in early childhood education* (Vol. 2). Norwood, NJ: Ablex Publishing Co.

Goldhaber, D. E. (1982). The breadth of development: An alternative perspective on facilitating early development. In L. A. Bond & J. M. Joffe (Eds.), *Facilitating infant and early childhood development.* Hanover, NH: University Press of New England.

Goldhaber, D. E., Goldhaber, J., Ishee, N., & Thousand, J. (1980). Working with developmentally diverse preschoolers. Paper presented at the World Assembly for Preschool Educators, Quebec City, Canada.

Good, T. L. (1980). Classroom expectations: Teacher-pupil interactions. In J. H. McMillan

(Ed.), *The social psychology of school learning.* New York: Academic Press.
Goodman, M. (1980). Toward a biology of menopause. *Signs, 5,* 739–753.
Goth-Owens, T. L., Stollak, G. E., Messe, L. A., Peshkess, I., & Watts, C. (1982). Marital satisfaction, parenting satisfaction and parenting behavior in early infancy. *Infant Mental Health Journal, 3,* 187–198.
Gottlieb, G. (1976). Conceptions of prenatal development: Behavioral embryology. *Psychological Review, 35,* 215–237.
Gottman, J., Gonso, J., & Rasmussen, B. (1975). Social interaction, social competence and friendship in children. *Child Development, 46,* 709–718.
Gould, R. L. (1972). The phases of adult life: A study in developmental psychology. *The American Journal of Psychiatry, 129,* 521–531.
Gould, R. L. (1978). *Transformations.* New York: Simon & Schuster.
Gould, S. J. (1980). *The panda's thumb.* New York: Norton.
Grambs, J. D. (1978). Schools, scholars and society (rev. ed.). Englewood Cliffs, NJ: Prentice-Hall.
Granick, S., & Friedman, A. S. (1973). Educational experience and the maintenance of intellectual functioning by the aged: An overview. In L. F. Jarvik, C. Eisdorfer, & J. E. Blum (Eds.), *Intellectual functioning in adults.* New York: Springer Publishing Co.
Graves, J. P., Dalton, G. W., & Thompson, P. H. (1980). Career stages in organizations. In C. B. Derr (Ed.). *Work, family and career.* New York: Praeger.
Greenberg, M., & Morris, N. (1974). Engrossment: The newborn's impact upon the father. *American Journal of Orthopsychiatry, 44,* 520–531.
Greenstein, F. I. (1965). *Children and politics* (rev. ed.). New Haven, CT: Yale University Press.
Greenwood, J. W., & Greenwood, J. W. (1979). *Managing executive stress.* New York: Wiley.
Gribbin, K., Schaie, K., & Parham, I. A. (1980). Complexity of lifestyle and maintenance of intellectual abilities. *Journal of Social Issues, 36,* 47–61.
Guarlnick, D. B. (Ed.). (1976). *Webster's new world dictionary of the American language.* New York: William Collins & World Publishing Co.
Gurin, P., & Brim, O. G. (1984). Changes in self in adulthood: The example of sense of control. In P. B. Baltes & O. G. Brim (Eds.), *Lifespan development and behavior* (Vol. 6). Orlando: Academic Press.
Gutmann, D. (1975). Parenthood: A key to the comparative study of the life cycle. In N. Datan & L. H. Ginsberg (Eds.), *Life span developmental psychology: Normative life crises.* New York: Academic Press.
Gutmann, D. (1977). The cross-cultural perspective: Notes toward a comparative psychology of aging. In J. E. Birren & K. W. Schaie (Eds.), *Handbook of the psychology of aging.* New York: Van Nostrand Reinhold Co.
Hagestad, G. O. (1981). Problems and promises in the social psychology of intergenerational relations. In R. W. Fogel, E. Hatfield, S. B. Kiesler, & E. Shanas (Eds.), *Aging: Stability and change in the family.* New York: Academic Press.
Harkins, E. B. (1978). Effects of empty nest transition on self report of psychological and physical well being. *Journal of Marriage and the Family, 40,* 549–559.
Harris, L. (1975). *The myth and reality of aging in America.* Washington: National Council on the Aging.
Hartley, A. A. (1981). Adult age differences in deductive reasoning processes. *Journal of Gerontology, 36,* 700–706.
Hartup, W. W. (1970). Peer interaction and social organization. In P. H. Mussen (Ed.), *Carmichaels manual of child psychology* (3rd ed.). New York: Wiley.
Havighurst, R. J. (1972). *Developmental tasks and education.* New York: David McKay Co., Inc.
Hayes, C. D., & Kamerman, S. B. (Eds.). (1983). *Children of working parents: Experience and outcomes.* Washington, DC: National Academy Press.
Hayes, M. P., & Stinnett, N. (1971). Life satisfactions of middle-aged husbands and wives. *Journal of Home Economics, 63,* 669–674.
Hayflick, L. (1977). The cellular basis for biological aging. In C. E. Finch & L. Hayflick (Eds.), *Handbook of the biology of aging.* New York: Van Nostrand Reinhold Co.
Haynes, S. G., Feinleib, M., Levine, S., Scotch, N., & Kannel, W. B. (1978). The relationship of psychosocial factors to coronary heart disease in the Framingham study. *American Journal of Epidemiology, 107,* 384–402.
Healy, A. (1972). The sleep patterns of preschool children. *Clinical Pediatrics, 11,* 174–177.
Henggeler, S. W., & Borduin, C. M. (1981). Satisfied working mothers and their preschool children. *Journal of Family Issues, 2,* 322–336.
Hess, B. B., & Waring, J. M. (1978). Parent and child in later life. In R. M. Lerner & G. B. Spanier (Eds.), *Child influences on marital and family interaction.* New York: Academic Press.
Hess, R. D., & Torney, J. V. (1967). *The development of political attitudes in children.* Chicago: Aldine.
Hetherington, E. M., Cox, M., & Cox, R. (1976). Divorced fathers. *The Family Coordinator, 25,* 417–428.
Hetherington, E. M., Cox, M., & Cox, R. (1977). Stress and coping in divorce: A focus on women. Unpublished manuscript.

Hetherington, E. M., Cox, M., & Cox, R. (1979). Play and social interaction in children following divorce. *Journal of Social Issues, 35,* 26–50.

Heyman, D. K., & Gianturco, D. T. (1973). Long-term adaptation by the elderly to bereavement. *Journal of Gerontology, 28,* 359–362.

Hill, C. R., & Stafford, F. P. (1980). Parental care of children: Time diary estimates of quantity, predictability and variety. *Journal of Human Resources, 15,* 221-239.

Hill, J. P. (1980). The family. In M. Johnson (Ed.), *Toward adolescence: The middle school years, 79th yearbook nsst,* Chicago: University of Chicago Press.

Hill, R., Foote, N., Aldous, J., Carlson, R., & McDonald, R. (1970). *Family development in three generations.* Cambridge, MA: Schenkman.

Hinkle, L. E. (1974). The effect of exposure to cultural change, social change and change in interpersonal relationships on health. In B. S. Dohrenwend & B. P. Dohrenwend (Eds.), *Stressful life events: Their nature and effects.* New York: Wiley.

Hirsch, J. (1963). Behavior genetics and individuality understood. *Science, 142,* 1436–1442.

Hodgson, J. W., & Fischer, J. L. (1979). Sex differences in identity and intimacy development in college youth. *Journal of Youth and Adolescence, 8,* 37–51.

Hoffman, L. W. (1977). Changes in family roles, socialization and sex differences. *American Psychologist, 32,* 644–657.

Hoffman, L. W., & Manis, J. D. (1978). Influences of children on marital interaction and parental satisfactions and dissatisfactions. In R. M. Lerner & G. B. Spanier (Eds.), *Child influences on marital and family interaction.* New York: Academic Press.

Hoffman, L. W., & Nye, F. T. (1974). *Working mothers.* San Francisco: Jossey-Bass.

Hoffman, M. L. (1975). Developmental synthesis of affect and cognition and its implications for altruistic motivation. *Developmental Psychology, 11,* 607–622.

Hoffman, M. L. (1980). Fostering moral development. In M. Johnson (Ed.), *Toward adolescence: The middle school years, 79th yearbook nsst.* Chicago: University of Chicago Press.

Hogan, D. P. (1981). *Transitions and social change: The early lives of American men.* New York: Academic Press.

Holmes, T. H., & Masuda, M. (1974). Life changes and illness susceptibility. In B. S. Dohrenwend & B. P. Dohrenwend (Eds.), Stressful life events: *Their nature and effects.* New York: Wiley.

Holmes, T. H., & Rahe, R. H. (1967). The social readjustment rating scale. *Journal of Psychosomatic Research, 11,* 213–218.

Horn, J. L. (1978). Human ability systems. In P. B. Baltes (Eds.), *Life span development and behavior* (Vol. 1). New York: Academic Press.

Horner, M. (1972). Toward an understanding of achievement related conflict in women. *Journal of Social Issues, 28,* 157–176.

Hornick, J. P., Doran, L., & Crawford, S. H. (1979). Premarital contraceptive usage among male and female adolescents. *Family Coordinator, 28,* 181–190.

Houseknecht, S. K. (1979). Childlessness and marital adjustment. *Journal of Marriage and the Family, 41,* 259–265.

Houseknecht, S. K., & Macke, A. S. (1981). Combining marriage and career: The marital adjustment of professional women. *Journal of Marriage and the Family, 43,* 651–661.

Houts, P. S., & Entwisle, D. R. (1968). Academic achievement efforts among females. *Journal of Counseling Psychology, 15,* 284–286.

Hunt, J. McV. (1964). The psychological basis for using preschool enrichment as an antidote for cultural deprivation. *Merrill-Palmer Quarterly, 10,* 209–248.

Hunt, J. McV. (1969). *The challenge of incompetence and poverty.* Urbana, IL: University of Illinois Press.

Huntington, D. S. (1979). Supportive programs for infants and parents. In J. D. Osofsky (Ed.), *Handbook of infant development.* New York: Wiley.

Hussian, R. A. (1981). *Geriatric psychology.* New York: Van Nostrand Reinhold Co.

Huston-Stein, A., & Higgins-Trenk, A. (1978). Development of females from childhood through adulthood: Career and feminine role orientation. In P. B. Baltes (Ed.), *Life span development and behavior* (Vol. 1). New York: Academic Press.

Hutt, C. (1976). Exploration and play in children. In J. S. Bruner, A. Jolly, & K. Sylva (Eds.), *Play—Its role in development and evolution.* New York: Basic Books.

Hyde, J. S., & Phillis, D. E. (1979). Androgyny across the life span. *Developmental Psychology, 15,* 334–336.

Iglehart, A. P. (1979). *Married women and work.* Lexington, MA: Heath.

Inhelder, B., & Piaget, J. (1958). *The growth of logical thinking from childhood to adolescence.* New York: Basic Books.

Inhelder, B., & Piaget, J. (1964). *The early growth of logic in the child.* New York: Norton.

Jackson, P. W. (1968). *Life in classrooms.* New York: Holt, Rinehart & Winston.

Jacobs, S., & Ostfeld, A. (1977). An epidemiological review of mortality of bereavement. *Psychosomatic Medicine, 39,* 344–357.

Jacobson, G. F. (1983). *The multiple crises of marital separation and divorce.* New York: Grune & Stratton.

Jacques, E. (1981). The midlife crisis. In S. I. Greenspan & G. H. Pollock (Eds.), *The course of life.* (DHHS Pub No. (ADM) 81-1000).

Jarvik, L. F. (1983). The impact of immediate life situation on depression, illness and loss. In L. D. Breslau & M. R. Haug (Eds.), *Depression and aging.* New York: Springer Publishing Co.

Jencks, C. (1972). *Inequality.* New York: Basic Books.

Jensen, A. R. (1969). How much can we boost IQ and scholastic achievement? *Harvard Educational Review, 39,* 1–23.

Johnson, D. W. (1980). Group processes: The influence of student-student interactions on school outcomes. In J. H. McMillan (Ed.), *The social psychology of school learning.* New York: Academic Press.

Johnson, S. J., & Jaccard, J. (1980). Career-marriage orientations in college youth: An analysis of perceived personal consequences and normative processes. *Journal of Youth and Adolescence, 9,* 419–439.

Jones, M. C. (1957). The later careers of boys who were early or late maturing. *Child Development, 28,* 113–128.

Jones, M. C. (1965). Psychological correlates of somatic development. *Child Development, 36,* 899–911.

Jones, M. C., & Mussen, P. H. (1958). Self concepts, motivations and interpersonal attitudes of early and late maturing girls. *Child Development, 29,* 491–501.

Josselson, R. L. (1973). Psychodynamic aspects of identity formation in college women. *Journal of Youth and Adolescence, 2,* 3–53.

Kadushin, A. (1970). *Adopting older children.* New York: Columbia University Press.

Kagan, J. (1971). *Change and continuity in infancy.* New York: Wiley.

Kagan, J. (1972). A conception of early adolescence. In J. Kagan & R. Coles (Eds.), *Twelve to sixteen: Early adolescence.* New York: Norton.

Kagan, J. (1978). The baby's elastic mind. *Human Nature,* 66–93.

Kagan, J. (1981). *The second year: The emergence of self.* Cambridge: Harvard University Press.

Kagan, J. (1984). *The nature of the child.* New York: Basic Books.

Kagan, J., Kearsley, R. B., & Zelazo, P. R. (1975). Apprehension to unfamiliar peers. In M. Lewis & L. A. Rosenblum (Eds.), *Friendship and peer relations.* New York: Wiley.

Kagan, J., Kearsley, R. B., & Zelazo, P. R. (1978). *Infancy: Its place in human development.* Cambridge: Harvard University Press.

Kagan, J., & Klein, R. E. (1973). Cross-cultural perspectives on early development. *American Psychologist, 28,* 947–962.

Kagan, J., & Moss, H. A. (1962). *Birth to maturity.* New York: Wiley.

Kahana, E., & Kahana, B. (1983). Environmental continuity, futurity and adaptation of the aged. In G. D. Rowles & R. J. Ohta (Eds.), *Aging and milieu.* New York: Academic Press.

Kalish, R. A. (1976). Death and dying in a social context. In R. H. Binstock & E. Shanas (Eds.), *Handbook of aging and the social sciences.* New York: Van Nostrand Reinhold Co.

Kalish, R. A. (1979). The new ageism and the failure models: A polemic. *The Gerontologist, 19,* 398–402.

Kalish, R. A. (1982). Death and survivorship: The final transition. *The Annals of the American Academy of Political and Social Science, 464,* 163–174.

Kalish, R. A., & Reynolds, D. K. (1981). *Death and ethnicity: A psychocultural study.* Farmingdale, NY: Baywood.

Kalter, N. (1977). Children of divorce in an outpatient psychiatric population. *American Journal of Orthopsychiatry, 47,* 40–51.

Kaluger, G., & Kaluger, M. F. (1974). *Human development: The span of life.* St. Louis: Mosby.

Kamerman, S. B. (1980). *Parenting in an unresponsive society: Managing work and family time.* New York: The Free Press.

Kamii, C., & Derman, L. (1971). The Engleman approach to teaching logical thinking: Findings from the administration of some Piagetian tasks. In D. R. Green (Ed.), *Measurement and Piaget.* New York: McGraw-Hill.

Kanin, E. J., Davidson, K. R., & Scheck, S. R. (1970). A research note on male-female differentials in the experience of heterosexual love. *Journal of Sex Research, 6,* 64–72.

Kantor, D., & Lehr, W. (1975). *Inside the family.* San Francisco: Jossey-Bass.

Karmiloff-Smith, A., & Inhelder, B. (1974–75). If you want to get ahead, get a theory. *Cognition, 3,* 195–212.

Karp, D. A., & Yoels, W. C. (1982). *Experiencing the life cycle.* Springfield, IL: Thomas.

Kastenbaum, R., & Aisenberg, R. B. (1972). *The psychology of death.* New York: Springer Publishing Co.

Katchadourian, H. (1977). *The biology of adolescence.* San Francisco: Freeman.

Katchadourian, H. A., & Lunde, D. T. (1975). *Fundamentals of human sexuality* (2nd ed.). New York: Holt, Rinehart & Winston.

Kaye, K. (1982). *The mental and social life of babies: How parents create persons.* Chicago: University of Chicago Press.

Kegan, R. (1982). *The evolving self.* Cambridge, MA: Harvard University Press.

Keith, P. M., & Brubaker, T. H. (1979). Male household roles in later life: A look at masculinity and marital relations. *The Family Coordinator, 28,* 497–501.

Kelly, J. B., & Wallerstein, J. S. (1976). The effects of parental divorce: Experiences of the child in early latency. *American Journal of Orthopsychiatry, 46,* 20–32.

Kempe, R. C., & Kempe, C. H. (1978). *Child abuse.* Cambridge, MA: Harvard University Press.

Kendler, H. H., & Kendler, T. S. (1962). Vertical and horizontal processes in problem solving. *Psychological Review, 69,* 1–14.

Kendler, T. S. (1979). Toward a theory of mediational development. In H. W. Resse & L. P. Lipsitt (Eds.), *Advances in child development and behavior* (Vol. 13). New York: Academic Press.

Keniston, K. (1975). Youth as a stage of life. In R. J. Havighurst & P. H. Dreyer (Eds.), *Youth.* (74th Yearbook of NSSE). Chicago: University of Chicago Press.

Keniston, K. (1977). *All our children: The American family under pressure.* New York: Harcourt Brace Jovanovich.

Kett, J. F. (1977). *Rites of passage.* New York: Basic Books.

Kimmel, D. C., Price, K. F., Walker, J. W. (1978). Retirement choice and retirement satisfaction. *Journal of Gerontology, 37,* 575–585.

Kitchener, K. S., & King, P. M. (1981). Reflective judgement: Concepts of justification and their relationship to age and education. *Journal of Applied Developmental Psychology, 2,* 89 116.

Kitson, G. C. (1982). Attachment to the spouse in divorce: A scale and its application. *Journal of Marriage and the Family, 44,* 379–393.

Kivnick, H. O. (1982). *The meaning of grandparenthood.* Ann Arbor, MI: U.M.I. Research Press.

Klaus, M. H. (Ed.). (1982). *Birth, interaction and attachment.* Skilman, NJ: Johnson and Johnson.

Klos, D. S., & Paddock, J. R. (1978). Relationship status: Scales for assessing the vitality of late adolescents' relationships with their parents. *Journal of Youth and Adolescence,* 353–371.

Kohlberg, L. (1964). Development of moral character and moral ideology. In M. L. Hoffman & L. W. Hoffman (Eds.), *Review of child development research* (Vol. 1). New York: Russell Sage Foundation.

Kohlberg, L. (1976). Moral stages and moralization: The cognitive developmental approach. In T. Lickona (Ed.), *Moral development and behavior.* New York: Holt, Rinehart and Winston.

Kohlberg, L., & Gilligan, C. (1972). The adolescent as a philosopher: The discovery of the self in a post conventional world. In J. Kagan & R. Coles (Eds.), *Twelve to sixteen: Early adolescence.* New York: Norton.

Kohn, M., & Schooler, C. (1978). The reciprocal effects of the substantive complexity of work and intellectual flexibility: A longitudinal assessment. *American Journal of Sociology, 84,* 24–52.

Kohn, M. L. (1977). *Class and conformity: A study of values* (2nd ed.). Chicago: University of Chicago Press.

Komarovsky, M. (1973). Cultural contradictions and sex roles: The masculine case. *American Journal of Sociology, 78,* 873–884.

Komarovsky, M. (1976). *Dilemmas of masculinity: A study of college youth.* New York: Norton.

Konopka, G. (1976). *Young girls.* Englewood Cliffs, NJ: Prentice-Hall.

Kubler-Ross, E. (1969). *On death and dying.* New York: MacMillan.

Kuhlen, R. G. (1963). Age and intelligence: The significance of cultural change in longitudinal and cross-sectional findings. *Vita Humana, 6,* 113–124.

Kuypers, J. A., & Bengtson, V. I. (1973). Social breakdown and competence. *Human Development, 16,* 181–201.

Labinowicz, E. (1980). *The Piaget primer.* Menlo Park, CA: Addison-Wesley.

Labouvie-Vief, G. (1980). Adaptive dimensions of adult cognition. In N. Datan & N. Lohmann (Eds.), *Transitions of aging.* New York: Academic Press.

Labouvie-Vief, G. (1980). Beyond formal operations: Uses and limits of pure logic in life-span development. *Human Development, 23,* 141–161.

Labouvie-Vief, G. (1982). Dynamic development and mature autonomy. *Human Development, 25,* 161–191.

Lamaze, F. (1958). *Painless childbirth: Psychoprophylactic method.* London: Burke.

Lamb, M. E. (1978). Influence of the child on marital quality and family interaction during the prenatal, perinatal and infancy periods. In R. M. Lerner & G. B. Spanier (Eds.), *Child influences on marital and family interaction.* New York: Academic Press.

Lamb, M. E., Chase-Lansdale, L., & Owen, M. T. (1979). The changing American family and its implications for infant social development: The sample case of maternal employment. In M. Lewis & L. A. Rosenblum (Eds.), *The child and its family.* New York: Plenum.

Lamb, M. E., & Easterbrooks, M. A. (1981). Individual differences in parental sensitivity: Origins, components and consequences. In M. E. Lamb & L. R. Sherrod (Eds.), *Infant social cognition.* Hillsdale, NJ: Lawrence Erlbaum Associates.

Lamb, M. E., Frodi, A. M., Chase-Lansdale, L. & Owen, M. T. (1978). The father's role in non-traditional family contexts: Direct and indirect effects. Paper presented at the 1978 meeting of the American Psychological Association, Toronto.

Langer, E. J. (1981). Old age: An artifact. In J. L.

McGaugh & S. B. Keisler (Eds.), *Aging.* New York: Academic Press.

Langer, E. J. (1983). *The psychology of control.* Beverly Hills, CA: Sage Publications.

Langer, J. (1969). *Theories of development.* New York: Holt, Rinehart & Winston.

LaRossa, R. (1983). The transition to parenthood and the social reality of time. *Journal of Marriage and the Family, 45,* 579–589.

LaRossa, R., & LaRossa, M. M. (1981). *Transition to parenthood: How infants change families.* Beverly Hills, CA: Sage Publications.

Larson, R., & Csikszentmihalyi, M. (1978). Experiential correlates of time alone in adolescence. *Journal of Personality, 46,* 677–693.

Larson, R., Csikszentmihalyi, M., & Graef, R. (1980). Mood variability and the psychological adjustments of adolescents. *Journal of Youth and Adolescence, 9,* 469–491.

Lawton, M. P. (1980). Environmental change: The older person as initiator and responder. In N. Datan & N. Lohmann (Eds.), *Transitions of Aging.* New York: Academic Press.

Lazar, I., & Darlington, R. B. (1982). Lasting effects of early education. *Monographs of the Society for Research in Child Development, 47* (3, Serial No. 195).

Leboyer, F. (1975). *Birth without violence.* New York: Knopf.

Lederer, W. J., & Jackson, D. D. (1968). *The mirages of marriage.* New York: Norton.

Lee, G. R., & Ellithorpe, E. (1982). Intergenerational exchange and subjective well-being among the elderly. *Journal of Marriage and the Family, 44,* 217–224.

Lefkowitz, M. M., Eron, L. D., Walder, L. O., & Huesmann, L. R. (1977). *Growing up to be violent.* New York: Pergamon Press.

Lein, L. (1979). Male participation in home life: Impacts of social supports and bread winners responsibility on the allocation of tasks. *Family Coordinator, 28,* 489–495.

Lemon, B. W., Bengtson, V. L., & Peterson, J. A. (1972). An exploration of the activity theory of aging: Activity types and life satisfaction among in-movers to a retirement community. *Journal of Gerontology, 27,* 511–523.

Lenneberg, E. (1967). *Biological foundations of language.* New York: Wiley.

Lerner, I. M., & Libby, W. J. (1976). *Heredity, environment and society.* San Francisco: Freeman.

Lesser, G. (1974). *Children and television: Lessons from "Sesame Street."* New York: Basic Books.

Lester, B. M. (1979). A synergistic process approach to the study of prenatal malnutrition. *International Journal of Behavioral Development, 2,* 377–395.

Lever, J. (1976). Sex differences in the games children play. *Social Problems, 23,* 478–487.

Lever, J. (1978). Sex differences in the complexity of children's play and games. *American Sociological Review, 43,* 471–483.

Levine, L. (1973). *Biology of the gene.* St. Louis: Mosby.

Levinson, D. J. (1978). *The seasons of a man's life.* New York: Knopf.

Levinson, D. J., Darrow, C. M., Klein, E. B., Levinson, M. H., & McKee, B. (1976). Periods in the adult development of men: Ages 18–45. *The Counseling Psychologist, 6,* 21–25.

Levinson, H. (1980). An overview of stress and satisfaction: The contract with self. In L. A. Bond & J. C. Rosen (Eds.), *Competence and coping during adulthood.* Hanover, NH: University Press of New England.

Lewis, M., & Brooks-Gunn, J. (1979). *Social cognition and the acquisition of self.* New York: Plenum.

Lewis, R. A., Freneau, P. J., & Roberts, C. L. (1979). Fathers and the postparental transition. *Family Coordinator, 28,* 514–520.

Liben, L. S. Memory in the context of cognitive development: The Piagetian approach. In R. V. Kail & J. W. Hagen (Eds.), *Perspectives on development and memory.* Hillsdale, NJ: Lawrence Erlbaum Associates.

Lieberman, M. A., & Tobin, S. S. (1983). *The experience of old age.* New York: Basic Books.

Liebert, R. M., Neale, J. M., & Davidson, E. S. (1973). *The early window: Effects of television on children and youth.* New York: Pergamon Press.

Liebert, R. M., Sprafkin, J. N., & Davidson, E. S. (1982). *The early window: Effects of television on children and youth* (2nd ed.). New York: Pergamon Press.

Lipsitt, L. P., Kaye, H., & Bosack, T. N. (1966). Enhancement of neonatal sucking through reinforcement. *Journal of Experimental Child Psychology, 4,* 163–168.

Livson, F. B. (1981). Paths to psychological health in the middle years: Sex differences. In D. H. Eichorn, J. A. Clausen, N. Haan, M. P. Honzik, & P. H. Mussen (Eds.), *Past and present in middle life.* New York: Academic Press.

Livson, N., & Peshkin, H. (1980). Perspectives on adolescence from longitudinal research. In J. Adelson (Ed.), *Handbook of adolescent psychology.* New York: Wiley.

Lohmann, N. (1980). Life satisfaction research in aging: Implications for public policy. In N. Datan & N. Lohmann (Eds.), *Transitions of aging.* New York: Academic Press.

Lomax, E. M., Kagan, J., & Rosenkrantz, R. (1978). *Science and patterns of childcare.* San Francisco: Freeman.

Longino, C. F., & Lipman, A. (1981). Married and

spouseless men and women in planned retirement communities: Support network differentials. *Journal of Marriage and the Family, 43,* 169–177.

Lopata, H. Z. (1979). *Women as widows: Support systems.* New York: Elsevier.

Lopata, H. Z. (1980). The widowed family member. In N. Datan & N. Lohmann (Eds.), *Transitions of aging.* New York: Academic Press.

Lorenz, K. (1971). Part and parcel in animal and human societies. In *Studies in animal and human behavior* (Vol. II). (R. Martin, Trans.). Cambridge: Harvard University Press.

Lowenthal, M. F. & Chiriboga, D. (1970). Transition to the empty nest. Paper presented at the annual meeting of the American Psychological Association, Miami Beach.

Lowenthal, M. F., & Robinson, B. (1976). Social networks and isolation. In R. H. Binstock & E. Shanas (Eds.). *Handbook of aging and the social sciences.* New York: Van Nostrand Reinhold Co.

Lowenthal, M. F., & Weiss, L. (1976). Intimacy and crisis in adulthood. *Counseling Psychologist, 6,* 10–15.

Lyle, J. G. (1970). Certain antenatal, perinatal and developmental variables and reading retardation in middle-class boys. *Child Development, 41,* 481–491.

Maccoby, E. E. (1980). *Social development.* New York: Harcourt Brace Jovanovich.

Maccoby, E. E., & Jacklin, C. N. (1974). *The psychology of sex differences.* Stanford, CA: Stanford University Press.

MacFarlane, A. (1977). *The psychology of childbirth.* Cambridge, MA: Harvard University Press.

Macklin, E. J. (1972). Heterosexual cohabitation among unmarried college students. *The Family Coordinator, 21,* 463–472.

Maddox, G. L. (1974). Aging and individual differences: A longitudinal analysis of social, psychological and physiological indications. *Journal of Gerontology, 29,* 555–563.

Marcia, J. E. (1966). Development and validation of ego identity status. *Journal of Personality and Social Psychology, 3,* 551–558.

Marcia, J. E. (1980). Identity in adolescence. In J. E. Adelson (Ed.), *Handbook of adolescent psychology.* New York: Wiley.

Marcia, J. E., & Friedman, M. L. (1970). Ego identity status in college women. *Journal of Personality, 38,* 249–263.

Markides, K. S., & Martin, H. W. (1979). A causal model of life satisfaction among the elderly. *Journal of Gerontology, 34,* 86–93.

Markman, E. M. (1977). Realizing that you don't understand: A preliminary investigation. *Child Development, 48,* 986–992.

Marshall, W. A. (1973). The body. In R. R. Sears & S. S. Feldman (Eds.), *The seven ages of man.* Los Altos, CA: Kaufmann.

Marshall, W. A., & Tanner, J. M. (1970). Variations in the pattern of pubertal change in boys. *Archives of the Diseases of Childhood, 45,* 13.

Maslow, A. H. (1968). *Toward a psychology of being.* Princeton, NJ: Van Nostrand Reinhold Co.

Masnick, G., & Bane, M. (1980). *The nation's families: 1960–1990.* Boston: Auburn House.

Masters, W. H., Johnson, V. E., & Kolodny, R. C. (1985). *Human sexuality* (2nd ed.). Boston: Little, Brown.

Matteson, D. R. (1975). *Adolescence today: Sex role and the search for identity.* Homewood, IL: Dorsey Press.

Matthews, S. H. (1979). *The social world of old women.* Beverly Hills, CA: Sage Publications.

May, K. A. (1982). Factors contributing to first time fathers' readiness for fatherhood: An exploratory study. *Family Relations, 31,* 353–361.

McCabe, M. P. (1984). Toward a theory of adolescent dating. *Adolescence, 19,* 159–170.

McCall, R. B. (1976). Toward an epigenetic conception of mental development. In M. Lewis (Ed.), *Origins of intelligence: Infancy and early childhood.* New York: Plenum.

McCall, R. B. (1977). Childhood IQ as predictors of adult education and occupational status. *Science, 197,* 482–483.

McCall, R. B. (1981). Nature-nurture and the two realms of development: A proposed integration with respect to mental development. *Child Development, 52,* 1–13.

McCall, R. B., Eichorn, D. H., & Hogarty, P. S. (1977). Transitions in early mental development. *Monographs of the Society for Research in Child Development, 42* (3, Serial No. 171).

McClelland, D. C. (1973). Testing for competence rather than for "intelligence." *American Psychologist, 28,* 1–14.

McDill, E. L., & Rigsby, L. C. (1973). *Structure and process in secondary schools.* Baltimore: Johns Hopkins University Press.

McKinlay, S. M., & Jefferys, M. (1974). The menopausal syndrome. *British Journal of Preventive and Social Medicine, 28,* 108–115.

Mead, G. H. (1934). *Mind, self, and society: From the standpoint of a social behaviorist* (C. W. Morris, Ed.). Chicago: University of Chicago Press.

Meier, H. C. (1972). Mother centeredness and college youths' attitudes toward social equality for women: Some empirical findings. *Journal of Marriage and the Family, 34,* 115–121.

Melson, G. F. (1980). *Family and environment: An ecosystem perspective.* Minneapolis, MN: Burgess Publishing Co.

Meltzer, M. W. (1981). The reduction of occupational stress among elderly lawyers: The creation

of a functional niche. *International Journal of Aging and Human Development, 13,* 209–219.
Miller, B. C., & Sollie, D. L. (1980). Normal stress during the transition to parenthood. *Family Relations, 29,* 459–465.
Miller, J., Schooler, C., Kohn, M. L., & Miller, K. A. (1979). Women and work: The psychological effects of occupational conditions. *American Journal of Sociology, 85,* 66–94.
Mills, C. J. (1981). Sex role, personality, and intellectual abilities in adolescents. *Journal of Youth and Adolescence, 10,* 85–113.
Minuchin, P., Biber, B., Shapiro, E., & Zimiles, H. (1969). *The psychological impact of school experience.* New York: Basic Books.
Montagu, A. (1981). *Growing young.* New York: McGraw-Hill.
Montemayor, R., & Eisen, M. (1977). The development of self-conceptions from childhood to adolescence. *Developmental Psychology, 13,* 314–319.
Moore, D., & Hotch, D. F. (1981). Late adolescents' conceptualization of home leaving. *Journal of Youth and Adolescence, 10,* 1–11.
Moore, K. L. (1974). *Before we are born.* Philadelphia: W. B. Saunders.
Morrison, G. S. (1980). *Early childhood education today* (2nd ed.). Columbus, OH: Merrill.
Mosher, F. A., & Hornsby, J. R. (1966). On asking questions. In J. S. Bruner, R. R. Olver, & P. M. Greenfield (Eds.), *Studies in cognitive growth.* New York: Wiley.
Mueller, E., & Lucas, T. (1975). A developmental analysis of peer interaction among toddlers. In M. Lewis & L. A. Rosenblum (Eds.), *Friendship and peer relations.* New York: Wiley.
Murray, F. B. (1979). The conservation paradigm: The conservation of conservation research. In I. Sigel, R. Golinkoff, & D. Brodzinsky (Eds.), *New directions and applications of Piaget's theory.* Hillsdale, NJ: Lawrence Erlbaum Associates.
Murstein, B. I. (1976). *Who will marry whom?* New York: Springer Publishing Co.
Murstein, B. I. (1980). Mate selection in the seventies. *Journal of Marriage and the Family, 42,* 777–792.
Mussen, P., & Eisenberg-Berg, N. (1977). *Roots of caring, sharing and helping.* San Francisco: Freeman.
Mussen, P. H., & Jones, M. C. (1957). Self conceptions, motivation and interpersonal attitudes of late and early maturing boys. *Child Development, 28,* 243–256.
Myers, J. K., Lindenthal, J. J., & Pepper, M. P. (1974). Social class, life events, and psychiatric symptoms: A longitudinal study. In B. S. Dohrenwend & B. P. Dohrenwend (Eds.), *Stressful life events: Their nature and effects.* New York: Wiley.
Nash, J. (1970). *Developmental psychology: A psychobiological approach.* Englewood Cliffs, NJ: Prentice-Hall.
Nathanson, C. A., & Lorenz, G. (1982). Women and health: The social dimensions of biomedical data. In J. Z. Giele (Ed.), *Women in the middle years.* New York: Wiley.
Neimark, E. (1975). Intellectual development during adolescence. In F. D. Horowitz (Ed.), *Review of child development research* (Vol. 4). Chicago: University of Chicago Press.
Nelson, K. (1973). Structure and strategy in learning how to talk. *Monographs of the Society for Research in Child Development, 38,* (Serial No. 149).
Neugarten, B. L. (1968). Adult personality: Toward a psychology of the life cycle. In B. L. Neugarten (Ed.), *Middle age and aging.* Chicago: University of Chicago Press.
Neugarten, B. L. (1968). The awareness of middle age. In B. L. Neugarten (Ed.), *Middle age and aging.* Chicago: University of Chicago Press.
Neugarten, B. L. (1975). The future and the young-old. *The Gerontologist, 15,* 4–9.
Neugarten, B. L. (1979). Policy for the 1980's: Age or need entitlement? In National Journal Issues Book, *Aging: Agenda for the 80's.* Washington: Government Research Corporation.
Neugarten, B. L. (1979). Time, age and the life cycle. *American Journal of Psychiatry, 136,* 887–895.
Neugarten, B. L., & Datan, N. (1973). Sociological perspectives on the lifespan. In P. B. Baltes & K. W. Schaie (Eds.), *Life span developmental psychology: Personality and socialization.* New York: Academic Press.
Neugarten, B. L., Moore, J. W., & Lowe, J. C. (1965). Age norms, age constraints, and adult socialization. *American Journal of Sociology, 70,* 229–236.
Neugarten, B. L., Wood, V., Kraines, R. J., & Loomis, B. (1968). Women's attitude toward the menopause. In B. L. Neugarten (Ed.), *Middle age and aging.* Chicago: University of Chicago Press.
Newson, J. (1979). The growth of shared understanding between infant and caregiver. In M. Bullowa (Ed.), *Before speech.* Cambridge, England: University of Cambridge Press.
Noam, G. G., Kohlberg, L., & Snarey, J. (1983). Steps toward a model of the self. In B. Lee & G. G. Noam (Eds.), *Developmental approaches to the self.* New York: Plenum.
Norton, A. J. (1983). Family life cycle 1980. *Journal of Marriage and the Family, 45,* 267–275.
Norton, A. J., & Glick, P. C. (1976). Marital instability, past, present and future. *Journal of Social Issues, 32,* 5–20.
Notman, M. T. (1980). Changing roles for women at mid-life. In W. H. Norman & T. J. Scarmella (Eds.), *Mid-life developmental and clinical issues.* New York: Brunner/Mazel.

Nowak, C. A. (1977). Socialization to become an old hag. Paper presented at the 85th annual convention of the American Psychological Association, San Francisco.

Oden, S. (1982). Peer relationship development in childhood. In L. G. Katz (Ed.), *Current topics in early childhood education* (Vol. 4). Norwood, NJ: Ablex Publishing Co.

Offer, D., Ostrov, E., & Howard, K. (1981). *The adolescent: A psychological self-portrait*. New York: Basic Books.

Opie, I., & Opie, P. (1959). *The lore and language of schoolchildren*. Oxford: Clarendon Press.

Opie, I., & Opie, P. (1969). *Children's games in street and playground*. Oxford: Clarendon Press.

O' Rand, A. M., & Landerman, R. (1984). Women's and men's retirement income status: Early family role effects. *Research on Aging, 6,* 25–45.

Osherson, S. D. (1980). *Holding on or letting go: Men and career change at midlife*. New York: The Free Press.

Osterman, P. (1980). *Getting started: The youth labor market*. Cambridge, MA: The M. I. T. Press.

Ostfeld, A. M. (1983). Depression, disability and demise in older people. In L. D. Breslau & M. R. Haug (Eds.), *Depression and aging*. New York: Springer Publishing Co.

Otto, L. B. (1979). Antecedents and consequences of marital timing. In W. R. Burr (Ed.), *Contemporary theories about the family* (Vol. 1). New York: The Free Press.

Palmore, E. (1977). Facts on aging: A short quiz. *The Gerontologist, 17,* 315–320.

Palmore, E. B. (1981). The facts on aging quiz: Part two. *The Gerontologist, 21,* 431–437.

Palmore, E., Cleveland, W. P., Nowlin, J. B., Ramm, D., & Siegler, I. C. (1979). Stress and adaption in later life. *Journal of Gerontology, 34,* 841–851.

Parelius, A. P. (1975). Emerging sex role attitudes, expectations and strains among college women. *Journal of Marriage and the Family, 37,* 146–153.

Paris, S. G., & Lindauer, B. K. (1977). Constructive aspects of children's comprehension and memory. In R. V. Kail & J. W. Hagand (Eds.), *Perspectives on development and memory*. Hillsdale, NJ: Lawrence Erlbaum Associates.

Parke, R. D., & Lewis, N. G. (1981). The family in context: A multilevel interactional analysis of child abuse. In R. W. Henderson (Ed.), *Parent-child interaction*. New York: Academic Press.

Parke, R. D., Power, T. G., Tinsley, B. R., & Hymel, S. (1980). The father's role in the family system. In P. M. Taylor (Ed.), *Parent-infant relations*. New York: Grune & Stratton.

Parkes, C. M. (1972). *Bereavement: Studies of grief in adult life*. New York: International University Press.

Parr, J. (1980). The interaction of persons and living environments. In L. W. Poon (Ed.), *Aging in the 1980's*. Washington: American Psychological Association.

Patterson, G. R. (1980). Mothers: The unacknowledged victims. *Monographs of the Society for Research in Child Development, 45* (5, Serial No. 186).

Pattison, E. M. (1977). *The experience of dying*. Englewood Cliffs, NJ: Prentice-Hall.

Pearlin, L. I. (1975). Sex role and depression. In N. Datan & L. H. Ginsberg (Eds.), *Life span developmental psychology: Normative life events*. New York: Academic Press.

Pearlin, L. I. (1980). Life strains and psychological distress among adults. In N. J. Smelser & E. H. Erikson (Eds.), *Themes of work and love in adulthood*. Cambridge, MA: Harvard University Press.

Peck, R. C. (1968). Psychological developments in the second half of life. In B. L. Neugarten (Ed.), *Middle age and aging*. Chicago: University of Chicago Press.

Peplau, L. A., Rubin, Z., & Hill, C. T. (1977). Sexual intimacy in dating relationships. *Journal of Social Issues, 33,* 86–109.

Perlmutter, J. F. (1978). A gynecological approach to menopause. In M. T. Notman & C. C. Nadelson (Eds.), *The woman patient* (Vol. 1). New York: Plenum.

Perlmutter, M. (1983). Learning and memory through adulthood. In M. W. Riley, B. B. Hess, & K. Bond (Eds.), *Aging and society*. Hillsdale, NJ: Lawrence Erlbaum Associates.

Perry, W. G. (1970). *Forms of intellectual and ethical development in the college years: A scheme*. New York: Holt, Rinehart & Winston.

Peshkin, H. (1967). Pubertal onset and ego functioning. *Journal of Abnormal Psychology, 72,* 1–15.

Peterson, A. C., & Taylor, B. (1980). The biological approach to adolescence. In J. Adelson (Ed.), *Handbook of adolescent psychology*. New York: Wiley.

Petranek, C. F. (1975). Post parental period: An opportunity for redefinition. Paper presented at the 1975 meeting of the National Council on Family Relations, Salt Lake City.

Piaget, J. (1927/1969). *The child's conception of time*. New York: Basic Books.

Piaget, J. (1952). *The origins of intelligence in children*. New York: International University Press.

Piaget, J. (1960). *The child's conception of physical causality*. Paterson, NJ: Littlefield, Adams & Co.

Piaget, J. (1967). *Six psychological studies*. New York: Vintage Books.

Piaget, J. (1970). *Genetic epistemology*. New York: Columbia University Press.

Piaget, J. (1972). Intellectual evolution from adolescence to adulthood. *Human Development, 15,* 1–12.

Piaget, J. (1974). *The language and thought of the child.* New York: New American Library.

Piaget, J. (1975). *The development of thought.* New York: Viking Press.

Piaget, J. (1980). *Adaptation and intelligence.* Chicago: University of Chicago Press.

Piaget, J., & Inhelder, B. (1969). *The psychology of the child.* New York: Basic Books.

Piaget, J., & Szeminska, A. (1952). *The child's conception of number.* New York: Humanities Press.

Piotrkowski, C. S. (1979). *Work and the family system.* New York: The Free Press.

Piotrkowski, C. S., & Katz, M. H. (1982). Indirect socialization of children: The effects of mother job on academic behaviors. *Child Development, 53,* 1520–1529.

Pleck, J. H. (1983). Husband's paid work and family roles: Current research issues. In H. Z. Lopata & J. H. Pleck (Eds.), *Research in the interweaves of social roles: Family and jobs* (Vol. 3). Greenwich, CT: Jai Press.

Plomin, R., DeFries, J. C., & McClearn, G. E. (1980). *Behavioral genetics: A primer.* San Francisco: Freeman.

Polonko, K. A., Scanzoni, J., & Teachman, J. D. (1982). Childlessness and marital satisfaction. *Journal of Family Issues, 3,* 545–575.

Price-Bonham, S., & Balswick, J. O. (1980). The noninstitutions: Divorce, desertion, and remarriage. *Journal of Marriage and the Family, 42,* 959–972.

Raphael, B. (1983). *The anatomy of bereavement.* New York: Basic Books.

Rappaport, A. F., Payne, D., Steinmann, A. (1970). Perceptual differences between married and single college women for the concept of self, ideal woman and man's ideal woman. *Journal of Marriage and the Family, 32,* 441–442.

Raush, H. L., Barry, W. A., Hertel, R. K., & Swain, M. A. (1974). *Communication, conflict and marriage.* San Francisco: Jossey-Bass.

Reinhard, D. W. (1977). The reaction of adolescent boys and girls to the divorce of their parents. *Journal of Clinical Child Psychology, 6,* 21–23.

Reiss, I. L. (1980). *Family systems in America* (3rd ed.). New York: Holt, Rinehart & Winston.

Reiss, I. L., Banwart, A., & Foreman, H. (1975). Premarital contraceptive usage. *Journal of Marriage and the Family, 37,* 619–630.

Rhyne, D. (1981). Bases of marital satisfaction among men and women. *Journal of Marriage and the Family, 43,* 941–955.

Riegel, K. F. (1972). Influence of economic and political ideologies on the development of developmental psychology. *Psychological Bulletin, 78,* 129–141.

Reigel, K. F. (1973). An epitaph for a paradigm. *Human Development, 16,* 1–7.

Riegel, K. F. (1975). Toward a dialectic theory of development. *Human Development, 18,* 50–64.

Riegel, K. F., & Riegel, R. M. (1972). Development, drop, and death. *Developmental Psychology, 6,* 306–319.

Ritchie, O. W., & Koller, R. (1964). *Sociology of childhood.* New York: Appleton-Century-Crofts.

Roberts, P., Papalia-Finlay, D., Davis, E. S., Blackburn, J., & Dellman, M. (1982). "No two fields ever grow grass the same way": An assessment of conservation abilities in the elderly. *International Journal of Aging and Human Development, 15,* 185–195.

Robertson, J. F. (1977). Grandmotherhood: A study of role conception. *Journal of Marriage and the Family, 39,* 165–177.

Robertson, J. F. (1978). Women in midlife: Crises, reverberations, and support networks. *The Family Coordinator, 27,* 375–382.

Robinson, J. P. (1979). *How Americans use time: A social-psychological analysis of everyday behavior.* New York: Praeger.

Roby, P. (Ed.). (1973). *Child care—who cares.* New York: Basic Books.

Roche, A. F. (Ed.). (1979). Secular trends in human growth, maturation, and development. *Monographs of the Society for Research in Child Development, 44* (Serial No. 179).

Rockstein, M., & Sussman, M. (1979). *The biology of aging.* Belmont, CA: Wadsworth.

Rodgers, D. (1981). *Adolescents and youth* (4th ed.). Englewood Cliffs, NJ: Prentice-Hall.

Rodgers, R. H. (1973). *Family interaction and transaction.* Englewood Cliffs, NJ: Prentice-Hall.

Rogel, M. J., Zuehlke, M. E., Petersen, A. C., Tobin-Richards, M., & Shelton, M. (1980). Contraceptive behavior in adolescence: A decision making perspective. *Journal of Youth and Adolescence, 9,* 491–507.

Rohwer, W. D. (1971). Prime time for education: Early childhood or adolescence? *Harvard Educational Review, 41,* 316–346.

Rollins, B. C., & Feldman, H. (1970). Marital satisfaction over the life cycle. *Journal of Marriage and the Family, 32,* 20–28.

Rollins, B. C., & Galligan, R. (1978). The developing child and marital satisfaction of parents. In R. M. Lerner & G. B. Spanier (Eds.), *Child influences on marital and family interaction.* New York: Academic Press.

Rosenbaum, J. E. (1976). *Making inequality.* New York: Wiley.

Rosenberg, M. (1979). *Conceiving the self.* New York: Basic Books.

Rosenman, R. H. (1978). The interview method of assessment of the coronary-prone behavior pattern. In T. M. Dembroski (Ed.), *Coronary prone behavior.* New York: Springer-Verlag.

Rosenman, R. H., Brand, R. J., Jenkins, D., Friedman, M., Straus, R., & Wurm, M. (1975). Coronary heart disease in the Western Collaborative Group study. *Journal of the American Medical Association, 233,* 872–877.

Rosow, I. (1973). The social context of the aging self. *The Gerontologist, 3,* 82–87.

Ross, H. L., & Sawhill, I. V. (1975). *Time of transition: The growth of families headed by women.* Washington, DC: The Urban Institute.

Rossi, A. S. (1968). Transition to parenthood. *Journal of Marriage and the Family, 30,* 26–39.

Rossi, A. S. (1980). Aging and parenthood in the middle years. In P. B. Baltes & O. G. Brim, Jr. (Eds.), *Life span development and behavior* (Vol. 3). New York: Academic Press.

Rotham, S. M. (1973). Other people's children: The day care experience in America. *The Public Interest, 30,* 11–27.

Rowles, G. D. (1980). Growing old "inside": Aging and attachment to place in an Appalachian community. In N. Datan & N. Lohmann (Eds.), *Transitions of aging.* New York: Academic Press.

Rubin, K. H., & Pepler, D. J. (1980). The relationship of child's play to social-cognitive growth and development. In H. C. Foot, A. J. Chapman, & J. R. Smith (Eds.), *Friendship and social relations in children.* New York: Wiley.

Rubin, L. B. (1976). *Worlds of pain.* New York: Basic Books.

Rubin, L. B. (1979). *Women of a certain age: The midlife search for self.* New York: Harper & Row.

Rubin, V. (1983). Family work patterns and community resources: An analysis of children's access to support and services outside school. In C. D. Hayes & S. B. Kamerman (Eds.), *Children of working parents.* Washington, DC: National Academy Press.

Rubin, Z. (1980). *Children's friendships.* Cambridge, MA: Harvard University Press.

Ruble, D. N., Boggiano, A. K., Feldman, N. S., & Loebl, J. H. (1980). Developmental analysis of the role of social comparison in self-evaluation. *Developmental Psychology, 16,* 105–115.

Rutter, M. (1979). Maternal deprivation 1972–1978. New findings, new concepts, new approaches. *Child Development, 50,* 283–305.

Rutter, M. (1979). Protective factors in children's responses to stress and disadvantage. In M. W. Kent & J. E. Rolf (Eds.), *Primary prevention of psychopathology* (Vol. 3). Hanover, NH: University Press of New England.

Rutter, M. (1980). *Changing youth in a changing society.* Cambridge, MA: Harvard University Press.

Rutter, M. (1983). School effects on pupil progress: Research findings policy implications. *Child Development, 54,* 1–29.

Rutter, M., Graham, P., Chadwick, O. F. D., & Yule, W. (1976). Adolescent turmoil: Fact or fiction? *Journal of Child Psychology and Psychiatry, 17,* 35–56.

Rutter, M., Maughan, B., Mortimore, P., Ouston, J., & Smith, A. (1979). *Fifteen thousand hours.* Cambridge, MA: Harvard University Press.

Ryder, R. G., Kafka, J. S., & Olsen, D. H. (1971). Separating and joining influences in courtship and early marriage. *American Journal of Orthopsychiatry, 42,* 450–463.

Saltzstein, H. D. (1976). Social influence and moral development: A perspective on the role of parents and peers. In T. Lickona (Ed.), *Moral development and behavior.* New York: Holt, Rinehart & Winston.

Sameroff, A. (1975). Transactional models in early social relations. *Human Development, 18,* 65–79b.

Sameroff, A. J., & Chandler, M. J. (1975). Reproductive risk and the continuum of caretaking causality. In F. D. Horowitz (Ed.), *Review of child development research* (Vol. 4). Chicago: University of Chicago Press.

Sarason, S. B. (1977). *Work, aging and social change.* New York: The Free Press.

Satariano, W. A., & Syme, S. L. (1981). Life changes and disease in elderly populations. In J. L. McGaugh & S. B. Kiesler (Eds.), *Aging.* New York: Academic Press.

Saunders, L. E. (1975). Husband-wife consensus in their perception of life satisfaction in the post parental years. Paper presented at the annual meeting of the National Council on Family Relations, Salt Lake City.

Scanzoni, J., & Fox, G. L. (1980). Sex roles, family and society: The seventies and beyond. *Journal of Marriage and the Family, 42,* 743–756.

Scanzoni, L. D., & Scanzoni, J. (1981). *Men, women and change* (2nd ed.). New York: McGraw-Hill.

Scarr, S., (1984). *Mother care, other care.* New York: Basic Books.

Scarr-Salapatek, S. (1976). An evolutionary perspective on infant intelligence: Species patterns and individual variations. In M. Lewis (Ed.), *Origins of intelligence.* New York: Plenum.

Schaeffer, R. (1977). *Mothering.* Cambridge, MA: Harvard University Press.

Schafer, R. B., & Keith, P. M. (1981). Equity in marital role across the family life cycle. *Journal of Marriage and the Family, 43,* 359–367.

Schaie, K. W. (1965). A general model for the study of developmental problems. *Psychological Bulletin, 64,* 92–107.

Schaie, K. W. (1979). The primary mental abilities in adulthood: An exploration in the development of psychometric intelligence. In P. B. Baltes & O. G. Brim (Eds.), *Life span development and behavior* (Vol. 2). New York: Academic Press.

Schaie, K. W. (Ed.). (1983). *Longitudinal studies of*

adult psychological development. New York: Guilford Press.
Schaie, K. W. (1983). The Seattle longitudinal study: A twenty-one year exploration of psychometric intelligence in adulthood. In K. W. Schaie (Ed.), *Longitudinal studies of adult psychological development*. New York: Guilford Press.
Schenkel, S., & Marcia, J. E. (1972). Attitudes toward premarital intercourse in determining identity status in college women. *Journal of Personality, 40*, 472–482.
Schiavo, R. S. (1976). Finding time for companionship: Couples with young children. Paper presented at the 1976 meeting of the National Council on Family Relations, New York.
Schofield, J. W. (1981). Complementary and conflicting identities: Images and interaction in an interracial school. In S. R. Asher & J. M. Gottman (Eds.), *The development of children's friendships*. Cambridge, England: Cambridge University Press.
Selman, R. (1976). Social-cognitive understanding: A guide to educational and clinical practice. In T. Lickona (Ed.), *Moral development and behavior*. New York: Holt, Rinehart & Winston.
Selman, R. L. (1980). *The growth of interpersonal understanding*. New York: Academic Press.
Seltzer, V. C. (1982). *Adolescent social development: Dynamic functional interaction*. Lexington, MA: Heath.
Selye, H. (1978). *The stresses of life*. New York: McGraw-Hill.
Severne, L. (1982). Psychosocial aspects of the menopause. In A. M. Voda, M. Dinnerstein, & S. R. O'Donnell (Eds.), *Changing perspectives on menopause*. Austin, TX: University of Texas Press.
Shaffer, D. R., & Brody, G. H. (1981). Parental and peer influences on moral development. In R. W. Henderson (Ed.), *Parent-child interaction*. New York: Academic Press.
Shanas, E., & Sussman, M. B. (1981). The family in later life: Social structure and social policy. In R. W. Fogel, E. Hatfield, S. B. Kiesler, & E. Shanas (Eds.), *Aging: Stability and change in the family*. New York: Academic Press.
Sheehy, G. (1976). *Passages: Predictable crises of adult life*. New York: Dutton.
Sheppard, H. L. (1979). Work and retirement. In R. H. Binstock & E. Shanas (Eds.), *Handbook of aging and the social sciences*. New York: Van Nostrand Reinhold Co.
Shipman, V. C. (1976). Disadvantaged children and their first school experience. *ETS-Headstart longitudinal study*. Princeton, NJ Educational Testing Service Report #PR 76-21.
Shirley, M. M. (1931). *The first two years: A study of twenty-five babies*. Minneapolis: University of Minnesota Press.
Shock, N. W. (1977). Biological theories of aging. In J. E. Birren & K. W. Schaie (Eds.), *Handbook of the psychology of aging*. New York: Van Nostrand Reinhold Co.
Shostak, A. B. (1980). *Blue-collar stress*. Reading, MA: Addison-Wesley.
Siegler, I. C. (1983). Psychological aspects of the Duke longitudinal studies. In K. W. Schaie (Ed.), *Longitudinal studies of adult psychological development*. New York: Guilford Press.
Siegler, R. S. (1976). Three aspects of cognitive development. *Cognitive psychology, 8*, 481–520.
Silverman, P. R. (1972). Widowhood and preventive intervention. *Family Coordinator, 21*, 95–102.
Silverman, P. R. (1978). Mutual help: An alternate network. In *Women in midlife—Security and fulfillment*. Select Committee on Aging, U.S. House of Representatives. Washington: U.S. Government Printing Office.
Simon, W., Gagnon, J. H., & Buff, S. A. (1972). Son of Joe: Continuity and change among white working class adolescents. *Journal of Youth and Adolescence, 1*, 13–25.
Siqueland, E. R., & DeLucia, C. A. (1969). Visual reinforcements of nonnutritive sucking in human infants. *Science, 165*, 1144–1146.
Skolnick, A. (1976). *Rethinking childhood*. Boston: Little, Brown.
Skolnick, A. (1978). *The intimate environment: Exploring marriage and the family* (2nd ed.). Boston: Little, Brown.
Skolnick, A. (1981). Married lives: Longitudinal perspectives on marriage. In D. E. Eichorn, J. A. Clausen, N. Haan, M. P. Honzik, & P. H. Mussen (Eds.), *Past and present in middle life*. New York: Academic Press.
Smith, A. D. (1980). Age differences in encoding, storage and retrieval. In L. W. Poon, J. L. Fozard, L. S. Cermak, D. Arenberg, & L. W. Thompson (Eds.), *New directions in memory and aging*. Hillsdale, NJ: Lawrence Erlbaum Associates.
Smith, D. H., & Macaulay, J. (1980). *Participation in social and political activities*. San Francisco: Jossey-Bass.
Smith, D. W., & Bierman, E. L. (1973). *The biologic ages of man*. Philadelphia: W. B. Saunders.
Smith, D. W., Bierman, E. L., & Robinson, N. M. (1978). *The biologic ages of man* (2nd ed.). Philadelphia: W. B. Saunders.
Smith, M. S., & Bissel, J. S. (1970). Report analysis: The impact of Head Start. *Harvard Educational Review, 40*, 51–104.
Sofer, C. (1970). *Men in mid-career: A study of British managers and technical specialists*. Cambridge, England: Cambridge University Press.
Sokol, R. J. (1981). Alcohol and abnormal outcomes of pregnancy. *Canadian Medical Association Journal, 125*, 143–148.

Sommer, B. B. (1978). *Puberty and adolescence.* New York: Oxford University Press.

Spanier, G. B., & Lewis, R. A. (1980). Marital quality: A review of the seventies. *Journal of Marriage and the Family, 42,* 825–841.

Sprey, J., & Matthews, S. H. (1982). Contemporary grandparent: A systematic transition. *Annals of the American Academy of Political and Social Science, 464,* 91–104.

Sroufe, L. A. (1977). Early experience: Evidence and myth. (Review of *Early experience: Myth and evidence* by A. M. Clarke & A. D. B. Clarke.) *Contemporary Psychology, 22,* 878–880.

Sroufe, L. A. (1979). Socioemotional development. In J. D. Osofsky (Ed.), *Handbook of Infant Development.* New York: Wiley.

Stefanko, M. (1984). Trends in adolescent research. *Adolescence, 19,* 1–14.

Steinberg, L. D., Greenberger, E., Jacobi, M., & Garduque, L. (1981). Early work experience: A partial antidote for adolescent egocentrism. *Journal of Youth and Adolescence, 10,* 141–159.

Sternglanz, S. H., & Serbin, L. A. (1974). Sex role stereotyping in children's television programs. *Developmental Psychology, 10,* 710–715.

Steuer, J., & Jarvik. L. F. (1981). Cognitive functioning in the elderly: Influence of physical health. In J. L. McGaugh & S. B. Kiesler (Eds.), *Aging.* New York: Academic Press.

Stevens-Long, J. (1979). *Adult life.* Palo Alto, CA: Mayfield Publishing Co.

Stipek, D. J. (1981). Adolescents—too young to earn, too old to learn? *Journal of Youth and Adolescence, 10,* 113–141.

Stott, L. H. (1967). *Child development.* New York: Holt, Rinehart & Winston.

Stott, L. H. (1974). *The psychology of human development.* New York: Holt, Rinehart & Winston.

Strange, C. C., & King, P. M. (1981). Intellectual development and its relationship to maturation during the college years. *Journal of Applied Developmental Psychology, 2,* 281–295.

Streib, G. F. (1976). Social stratification and aging. In R. H. Binstock & E. Shanas (Eds.), *Handbook of aging and the social sciences.* New York: Van Nostrand Reinhold Co.

Streib, G. F., & Beck, R. W. (1980). Older families: A decade review. *Journal of Marriage and the Family, 42,* 937–956.

Streib, G. F., & Schneider, C. J. (1971). *Retirement in American society.* Ithaca, NY: Cornell University Press.

Stroud, J. G. (1981). Women's careers: Work, family and personality. In D. H. Eichorn, J. A. Clausen, N. Haan, M. P. Honzik, & P. H. Mussen (Eds.), *Present and past in middle life.* New York: Academic Press.

Sullivan, H. S. (1953). *The interpersonal theory of psychiatry.* New York: Norton.

Sullivan, K., & Sullivan, A. (1980). Adolescent-parent separation. *Developmental Psychology, 16,* 93–99.

Swensen, C. H., Eskew, R. W., & Kohlhepp, K. A. (1981). Stages of family life cycle, ego development and the marriage relationship. *Journal of Marriage and the Family, 43,* 841–853.

Sylva, K., Bruner, J. S., & Genova, P. (1976). The relationship between play and problem solving behavior in children three to five years old. In J. S. Bruner, A. Jolly, & K. Sylva (Eds.), *Play: Its role in development and evolution.* New York: Basic Books.

Szinovacz, M. E. (1980). Female retirement. *Journal of Family Issues, 1,* 423–440.

Szinovacz, M. E. (1982). Personal problems and adjustment to retirement. In M. Szinovacz (Ed.), *Women's retirement.* Beverly Hills, CA: Sage Publications.

Taft, L. T., & Cohen, H. J. (1967). Neonatal and infant reflexology. In J. Hellmuth (Ed.), *Exceptional infant* (Vol. I). New York: Brunner/Mazel.

Takanishi, R. (1978). Childhood as a social issue: Historical roots of contemporary child advocacy movements. *Journal of Social Issues, 34,* 8–28.

Tamir, L. M. (1982). *Men in their forties: The transition to middle age.* New York: Springer Publishing Co.

Tanner, J. M. (1961). *Education and physical growth.* New York: International University Press.

Tanner, J. M. (1972). Sequence, tempo and individual variation in growth and development of boys and girls aged twelve to sixteen. In J. Kagan & R. Coles (Eds.), *Early adolescence.* New York: Norton.

Tanner, J. M., Whitehouse, R. H., & Takaishi, M. (1966). Standards from birth to maturity for height, weight, height velocity and weight velocity in British children, 1965. *Archives of the Diseases of Childhood, 41,* 455–471.

Tanzer, D. T., & Block, J. L. (1972). *Why natural childbirth?* New York: Doubleday.

Targ, D. P. (1979). Toward a reassessment of women's experience at middle age. *The Family Coordinator, 28,* 377–381.

Terkel, S. (1974). *Working: People talk about what they do all day and how they feel about what they do.* New York: Pantheon Books.

Thiessen, D. B. (1972). *Gene organization and behavior.* New York: Random House.

Thomas, A., & Chess, S. (1977). *Temperament and development.* New York: Brunner/Mazel.

Thomas, A., & Chess, S. (1980). *The dynamics of psychological development.* New York: Brunner/Mazel.

Thomas, A., Chess, S., & Birch, H. G. (1968). *Temperament and behavior disorders in children.* New York: New York University Press.

Thompson, S. K. (1975). Gender labels and early sex-role development. *Child development, 46,* 339–347.

Timiras, P. S. (1978). Biological perspectives on aging. *American Scientist, 66,* 605–613.

Tokuno, K. A. (1983). Family and friends as support during the early adult transition years. Paper presented at the 1983 meeting of the Society for Research in Child Development, Detroit, MI.

Torrey, B. B. (1982). The lengthening of retirement. In M. W. Riley, R. P. Abeles, & M. S. Tietelbaum (Eds.), *Aging from birth to death.* Vol. II. *Sociotemporal perspectives.* Washington, DC: American Association on the Advancement of Science.

Traupmann, J., & Hatfield, E. (1981). Love and its effect on mental and physical health. In R. W. Fogel, E. Hatfield, S. B. Kiesler, & E. Shanas (Eds.), *Aging: Stability and change in the family.* New York: Academic Press.

Treas, J., & Bengston, V. L. (1982). The demography of mid- and late-life transitions. *The Annals of the American Academy of Political and Social Science, 464,* 11–22.

Trent, J. W., & Medsker, L. L. (1968). *Beyond high school.* San Francisco: Jossey-Bass.

Troll, L. E. (1982). *Continuations: Adult development and aging.* Monterey, CA: Brooks/Cole.

Troll, L. E., & Nowak, C. (1976). "How old are you?"—The question of age bias in the counseling of adults. *The Counseling Psychologist, 6,* 41–44.

Tulkin, S. R., & Kagan, J. (1972). Mother-child interaction in the first year of life. *Child Development, 43,* 31–41.

Turner, B. F. (1977). Sex-roles among wives in middle and late life. Paper presented at the annual meeting of the American Psychological Association, San Francisco.

Ullian, D. Z. (1981). The child's construction of gender: Anatomy as destiny. In E. K. Shapiro & E. Weber (Eds.), *Cognitive and effective growth.* Hillsdale, NJ: Lawrence Erlbaum Associates.

Uzgiris, I. C. (1973). Patterns of cognitive development in infancy. *Merrill-Palmer Quarterly, 19,* 181–205.

Uzgiris, I. C. (1976). Organization of sensorimotor intelligence. In M. Lewis (Ed.), *Origins of intelligence.* New York: Plenum.

Uzgiris, I. C. (1977). Some observations on early cognitive development. In M. H. Appel & L. S. Goldberg (Eds.), *Topics in cognitive development* (Vol. I). New York: Plenum.

Vachon, M. L. S., Lyall, W. A. L., Rogers, J., Freedman-Letofsky, K., & Freeman, S. J. J. (1980). A controlled study of self-help interventions for widows. *American Journal of Psychiatry, 137,* 1380–1384.

Vaillant, G. E. (1977). *Adaption to life.* Boston: Little, Brown.

Vale, J. R. (1980). *Genes, environment and behavior: An interactionist approach.* New York: Harper & Row.

Vandenberg, B. (1980). Play, problem solving and creativity. In K. H. Rubin (Ed.), *Children's play: New directions for child development, 9,* 49–69.

Veroff, J., Douvan, E., & Kulka, R. A. (1981). *The inner American.* New York: Basic Books.

Volpe, J. S. (1981). The development of concepts of self: An interpersonal perspective. *Contributions to Human Development, 5,* 131–144.

Vygotsky, L. (1962). *Thought and language.* Cambridge, MA: The M.I.T. Press.

Vygotsky, L. (1978). *Mind and society.* Cambridge, MA: Harvard University Press.

Waddington, C. H. (1957). *The strategy of the genes.* London: Allen and Son.

Waddington, C. H. (1962). *New patterns in genetics and development.* New York: Columbia University Press.

Waddington, C. H. (1966). *Principles of development and differentiation.* New York: MacMillan.

Waldron, I. (1978). The coronary-prone behavior pattern, blood pressure, employment, and socioeconomic status in women. *Journal of Psychosomatic Research, 22,* 79–87.

Waldrop, M. F., & Halverson, C. F. (1975). Intensive and extensive peer behavior: Longitudinal and cross sectional analysis. *Child Development, 46,* 19–27.

Walker, A. J., & Thompson, L. (1983). Intimacy and intergenerational aid and contact among mothers and daughters. *Journal of Marriage and the Family, 45,* 841–849.

Walker, L. J. (1984). Sex differences in the development of moral reasoning: A critical review. *Child Development, 55,* 677–692.

Wallerstein, J. S. (1983). Children of divorce: The psychological tasks of the child. *American Journal of Orthopsychiatry, 53,* 230–243.

Wallerstein, J., & Kelly, J. (1974). The effects of parental divorce: The adolescent experience. In E. Anthony & C. Koupernik (Eds.), *The child in his family.* New York: Wiley.

Walters, J., & Walters, L. H. (1980). Trends affecting adolescent views of sexuality, employment, marriage and childrearing. *Family Relations, 29,* 191–198.

Wandersman, L. P. (1978). Longitudinal changes in the adjustment to parenthood. Paper presented at the 1978 meeting of the American Psychological Association, Toronto.

Ward, R. A. (1977). The impact of subjective age and stigma on older persons. *Journal of Gerontology, 32,* 227–232.

Ward, R. A. (1979). The meaning of voluntary association participation to older people. *Journal of Gerontology, 34,* 438–445.

Wartella, E., Wackman, D. B., Ward, S., Shamir, J., & Alexander, A. The young child as a consumer.

In E. Wartella (Ed.), *Children communicating.* Beverly Hills, CA: Sage Publications.

Waterman, C. K., & Nevin, J. S. (1977). Sex differences in the resolution of the identity crisis. *Journal of Youth and Adolescence, 6,* 337–343.

Weg, R. B. (1975). Changing physiology of aging: Normal and pathological. In D. S. Woodruff & J. E. Birren (Eds.), *Aging.* New York: D. Van Nostrand Co.

Weisler, A., & McCall, R. B. (1976). Exploration and play. *American Psychologist, 31,* 492–508.

Weisman, A. D. (1972). *On dying and denying.* New York: Behavioral Publications Inc.

Weiss, R. S. (1975). *Marital separation.* New York: Basic Books.

Weiss, R. S. (1979). Growing up a little faster: The experience of growing up in a single parent household. *Journal of Social Issues, 35,* 97–112.

Weitz, S. (1977). *Sex roles.* New York: Oxford University Press.

Werner, E. E., & Smith, R. S. (1977). *Kauai's children come of age.* Honolulu: University of Hawaii Press.

Werner, H. (1957). The concept of development from a comparative and organismic point of view. In D. B. Harris (Ed.), *The concept of development.* Minneapolis, MN: University of Minnesota Press.

West, M., & Newton, P. (1983). *The transition from school to work.* London: Croom Helm.

Whitbourne, S. K., & Weinstock, C. S. (1979). *Adult development: The differentiation of experience.* New York: Holt, Rinehart & Winston.

White, B. L. (1971). *Human infants.* Englewood Cliffs, NJ: Prentice-Hall.

White, G. (1977). *Socialisation.* London: Longman.

White, L. K. (1983). Determinants of spousal interaction: Marital structure on marital happiness. *Journal of Marriage and the Family, 45,* 511–519.

White, R. W. (1959). Motivation reconsidered: The concept of competence. *Psychological Review, 66,* 297–333.

White, R. W. (1972). *The enterprise of living.* New York: Holt, Rinehart & Winston.

White, S. (1970). Some general outlines of the matrix of developmental changes between five and seven years. *Bulletin of the Orton Society, 20,* 41–57.

White, S. H. (1965). Evidence for a hierarchical arrangement of learning processes. In L. P. Lipsitt & C. C. Spiker (Eds.), *Advances in child development and behavior* (Vol. 2). New York: Academic Press.

Wilkie, J. R. (1981). The trend toward delayed parenthood. *Journal of Marriage and the Family, 43,* 583–591.

Williams, J. W., & Stith, M. (1980). *Middle childhood* (2nd ed.). New York: MacMillan.

Williams, R. H., & Wirths, C. G. (1965). *Lives through the years.* New York: Atherton.

Wilson, J. G. (1977). Embryotoxicity of drugs in man. In J. G. Wilson & F. C. Fraser (Eds.), *Handbook of teratology: Vol. VI. General principles and etiology.* New York: Plenum.

Windley, P. G., & Scheidt, R. J. (1980). Person-environment dialects: Implications for competent functioning in old age. In L. W. Poon (Ed.), *Aging in the 80's.* Washington, DC: American Psychological Association.

Winship, E. C. (1972). *Ask Beth: You can't ask your mother.* Boston: Houghton Mifflin.

Wohlwill, J. (1973). *The study of behavioral development.* New York: Academic Press.

Wolff, M., & Stein, A. (1966). Six months late: A comparison of children who had Head Start summer 1964 with their classmates in kindergarten. (ERIC Document ED 015025).

Woodruff, D. S. (1975). A physiological perspective of the psychology of aging. In D. S. Woodruff & J. E. Birren (Eds.), *Aging.* New York: D. Van Nostrand Co.

Woodruff, D. S. (1983). A review of aging and cognitive processes. *Research on Aging, 5,* 139–155.

Woodruff, D. S., & Birren, J. E. (1975). *Aging: Scientific perspectives and social issues.* New York: D. Van Nostrand Co.

Woods, N. F. (1982). Menopausal distress: A model for epidemiologic investigation. In A. M. Voda, M. Dinnerstein, & S. R. O'Donnell (Eds.), *Changing perspectives on menopause.* Austin, TX: University of Texas Press.

Yankelovich, D. (1974). *The new morality.* New York: McGraw-Hill.

Yankelovich, D. (1981). *New rules.* New York: Random House.

Yogev, S. (1981). Do professional women have egalitarian marital relations? *Journal of Marriage and the Family, 43,* 865–871.

Youniss, J. (1980). *Parents and peers in social development.* Chicago: University of Chicago Press.

Zelnick, M., & Kanter, J. F. (1977). Adolescent sexual behavior. *Family Planning Perspective, 9,* 55–73.

Zeskind, P. S., & Ramey, C. T. (1981). Preventing intellectual and interactional sequelae of fetal malnutrition: A longitudinal, transactional and synergistic approach to development. *Child Development, 52,* 213–218.

Ziegler, M., & Reid, D. W. (1983). Correlates of changes in desired control scores and in life satisfaction scores among elderly persons. *International Journal of Aging and Human Development, 16,* 135–146.

ILLUSTRATION CREDITS

PHOTOGRAPHS

Chapter 1

2, Joan Liftin/Archive Pictures, Inc.; 9, Danuta Otfinowski, 1977/Jeroboam, Inc.; 12, © Charles Harbutt/Archive Pictures, Inc.; 16, © Jane Scherr/Jeroboam, Inc.; 18, FPG; 21, © Suzanne Wu, 1981/Jeroboam, Inc.; 25, © Ethan Hoffman, 1981/Archive Pictures, Inc.; 26, © Joel Gordon, 1975; 32, Suzanne Arms, 1977/Jeroboam, Inc.; 35 (top), © Erika Stone, 1980/Peter Arnold, Inc.; 35 (bottom), © Marcia Weinstein, 1979; 44, Sybil Shelton/© Peter Arnold, Inc.; 46, James Pozarik/Gamma-Liaison; 47, © Jeffry W. Myers/FPG; 53, Bill Aron, 1985/Jeroboam, Inc.; 55, © Jane Scherr, 1982/Jeroboam, Inc.

Chapter 2

58, Peter Gridley/FPG; 63 (A, B, C, and D), Dr. Sundstroem/Gamma-Liaison; 78, © Suzanne Arms, 1977/Jeroboam, Inc.; 80 (left), Tom Marotta-UNICEF/FPG; 80 (right), © Marcia Weinstein, 1979; 91 (left), © M. Rastellini/Gamma-Liaison; 91 (right), © M. Rastellini/Gamma-Liaison.

Chapter 3

96, R. Nowitz/FPG; 102, © Charles Harbutt/Archive Pictures, Inc.; 109 (Figure 3–2), Enrico Ferorelli/Dot; 117, © Steve Malone, 1980/Jeroboam, Inc.; 120, © Mariette Pathy Allen/Peter Arnold, Inc.; 122, Erika Stone/Peter Arnold, Inc.; 125, Eileen Christelow, 1979/Jeroboam, Inc.; 131, George B. Gibbons III, 1983/FPG; 132, © Erika Stone, 1979/Peter Arnold, Inc.; 147, James M. Mejuto/FPG; 151, © Suzanne E. Wu, 1979/Jeroboam, Inc.

Chapter 4

158, © Ginger Chih, 1980/Peter Arnold, Inc.; 161, David Frankenstein/Jeroboam, Inc.; 163, Michael R. Stoklos/FPG; 167 (top), © Marcia Weinstein, 1977; 167 (bottom), © Marcia Weinstein, 1977; 170, Paul L. Mathews/FPG; 175 (top left), G. Giblom/FPG; 175 (top right), © Erika Stone, 1982/Peter Arnold, Inc.; 175 (bottom right), Robert Wm. Simpson/FPG; 178, © Peter L. Gould, 1975/FPG; 182, Jeffry W. Myers/FPG; 184, Forman and Hill, 1984; 186, © Peter L. Gould, 1975/FPG; 195, Bill Owens/Archive Pictures, Inc.; 199, Suzanne Arms/Jeroboam, Inc.; 205, © Mary Ellen Mark/Archive Pictures, Inc.; 208, U. S. Department of Housing and Urban Development.

Chapter 5

212, © Cheryl A. Traendly, 1981/Jeroboam, Inc.; 215, © Marcia Weinstein, 1981; 218, © Jay Lurie/FPG; 225, © Jeffry W. Myers/FPG; 229, © Erika Stone, 1984/Peter Arnold, Inc.; 235, Erika Stone/Peter Arnold, Inc.; 239, © Irene Kane, 1979/Jeroboam, Inc.; 242, © Erika Stone, 1980/© Peter Arnold, Inc.; 245, Sybil Shelton/© Peter Arnold, Inc.; 249, Culver Pictures, Inc.; 250, © Sylvia Plachy/Archive Pictures, Inc.; 254, © Joel Gordon, 1975; 258, Sybil Shelton/© Peter Arnold, Inc.

Chapter 6

266, © Susan Lapides, 1981/Design Conceptions; 271, © Marcia Weinstein, 1982; 272, © T. Simon, 1984/Gamma-Liaison; 278, © Terry Parke/Gamma-Liaison; 289, © Mary Ellen Mark/Archive Pictures, Inc.; 291, © Joel Gordon, 1983; 293, © Terry Parke/Gamma-Liaison; 296, © Sybil Shelton/Peter Arnold, Inc.; 297, © Carolyn A. McKeone/FPG; 300, © Jonathan A. Meyers, 1984/FPG; 303, © Mary Ellen Mark/Archive Pictures, Inc.; 304, © Joe Traver, 1984/Gamma-Liaison; 306, Richard Laird/FPG; 314, E. Adams/Gamma-Liaison.

Chapter 7

318, Tom Griffin Studio; 325, © Robert Alexander, 1982/Gamma-Liaison; 328, © William Thompson, 1980/Jeroboam, Inc.; 331, © Jim Howard, 1981/FPG; 341, Steve Liss/Gamma-Liaison; 349, © Jane Scherr/Jeroboam, Inc.; 353, © Michelle Vignes/Jeroboam,

Inc.; 354, Carolyn A. McKeone/FPG; 356, King Photography/FPG; 359, © Susan Lapides, 1980/Design Conceptions; 360, © Frank D. Smith, 1982/Jeroboam, Inc.; 364, © Ed Kashi, 1984/Gamma-Liaison; 367, © Bill Burke/Archive Pictures, Inc.; 369, © Mary Ellen Mark/Archive Pictures, Inc.; 373, Richard A. Verdi, 1978/FPG; 376, © Peter L. Gould/FPG.

Chapter 8

380, FPG; 384, © Kent Reno, 1981/Jeroboam, Inc.; 389, Sybil Shelton/© Peter Arnold, Inc.; 395, Abigail Heyman/Archive Pictures, Inc.; 396, © Jeffry W. Myers/FPG; 399, Robert George Gaylord/Jeroboam, Inc.; 402, Marc E. Burkhart/FPG; 404 (left), © Jeffry W. Myers/FPG; 404 (right), © Marcia Weinstein, 1984; 407 (top), Mark Godfrey/Archive Pictures, Inc.; 407 (bottom), © Marcia Weinstein, 1984; 415, © Robin Moyer/Gamma-Liaison; 417, Sylvia Plachy/Archive Pictures, Inc.; 419, © Carolyn A. McKeone/FPG; 421, Terry Parke/Gamma-Liaison; 423, © Rebecca Choa, 1983/Archive Pictures, Inc.; 427 (left), © Norman Mosallem/FPG; 427 (right), © Bettina Cirone/Photo Researchers, Inc.; 432, Steve Liss/Gamma-Liaison.

Chapter 9

436, Rose Skytta, 1977/Jeroboam, Inc.; 441, Chas. Schneider/FPG; 445, HBJ Photo; 446, Carolyn A. McKeone/FPG; 449, Carolyn A. McKeone/FPG; 452, © Earl Dotter/Archive Pictures, Inc.; 457, © Jim Howard, 1981/FPG; 462, H. Gris/FPG; 464, Paul M. Schrock/FPG; 468, David Burnett/Gamma-Liaison; 474, © George B. Gibbons III, 1985/FPG; 477, © Mary Ellen Mark/Archive Pictures, Inc.; 481, © Bruce Roberts/Photo Researchers, Inc.; 484, © Jane Scherr/Jeroboam, Inc.

Chapter 10

488, Martin Parr/Archive Pictures, Inc.; 493, FPG; 496, © Marcia Weinstein, 1984; 498, © Marcia Weinstein, 1984; 500, © Mike Valeri, 1982/FPG; 502, Charles Schneider/FPG; 506, © Suzanne Arms, 1983/Jeroboam, Inc.; 508, Sybil Shelton/© Peter Arnold, Inc.; 511, Paul M. Schrock/FPG; 517, Michael Abramson/Gamma-Liaison; 518, © Mike Valeri, 1983/FPG; 519, Susan Greenwood/Gamma-Liaison; 520, © Joel Gordon, 1976; 524, © John Neubauer, 1983/FPG; 526, Evan Johnson/Jeroboam, Inc.; 531, Carole Graham/FPG; 535, © Jay Lurie, 1979/FPG; 539, Catherine Leroy, 1982/Gamma-Liaison.

Chapter 11

544, © Horst Schäfer/Peter Arnold, Inc.; 549, Nedra Westwater/FPG; 551, © F. B. Grunzweig, 1978/Photo Researchers, Inc.; 553, © Suzanne Arms, 1980/Jeroboam, Inc.; 560, Terry Parke/Gamma-Liaison; 563, James M. Mejuto/FPG; 566, © David Bellak/Jeroboam, Inc.; 571, © Erika Stone/Peter Arnold, Inc.; 573, Jacob Sutton/Gamma-Liaison.

FIGURE CREDITS

1–1 From Duvall, E. M. *Marriage and Family Development,* 5th ed. Philadelphia: J. B. Lippincott, 1977.

1–2 From Schaie, K. W., & Strother. *Multivariate Behavioral Research.* Figure 4, p. 81. Fort Worth, TX: Texas Christian University Press, 1968. Reprinted by permission of publisher.

2–2 From Moore, K. L. *Before We Are Born.* Figure 6–2, p. 45. Philadelphia: W. B. Saunders, 1974. Reprinted by permission of publisher.

2–3 From Moore, K. L. *Before We Are Born.* Figure 1–2, p. 5. Philadelphia: W. B. Saunders, 1974. Reprinted by permission of publisher.

2–5 From Lerner, I. Michael, & Libby, William J. *Heredity, Environment, and Society.* Figure 3–6, p. 97. New York: W. H. Freeman and Company, 1976. Reprinted by permission of publisher.

2–6 From Moore, K. L. *Before We Are Born.* Figure 9–11, p. 96. Philadelphia: W. B. Saunders, 1974. Reprinted by permission of publisher.

2–7 From Birch, H. G., & Gussow, J. D. *Disadvantaged Children: Health, Nutrition, and School Failure.* Figure on p. 268. New York: Harcourt, Brace, & World, Inc., 1970. Reprinted by permission of publisher.

Box 2–1 Reprinted with permission of Macmillan Publishing Company from *Becoming a Sexual Person* by J. T. Francoeur. Figure 5.11, p. 221. Copyright © 1982 by John Wiley and Sons.

2–9 From Moore, K. L. *Before We Are Born.* Figure 8–9, p. 76. Philadelphia: W. B. Saunders, 1974. Reprinted by permission of publisher.

2–10 From Smith, D. W., Bierman, E. L., & Robinson, N. M. *The Biological Ages of Man*, 2d ed. Figure 5–4, p. 84. Philadelphia: W. B. Saunders, 1978. Reprinted by permission of publisher.

Box 3–1 From Montagu, A. *Growing Young*. Figure 11, p. 38. New York: McGraw-Hill Book Company, 1981. Reprinted by permission of publisher.

3–1 The Diagram Group. *Child's Body: An Owner's Manual*. Paddington Press, 1977, Section D–03.

Box 3–2 From Frantz, R. L., & Nevis, S. "Pattern Preference and Perceptual Cognitive Development in Infancy." *Merrill-Palmer Quarterly*, 1967, 13, pp. 77–108, Figure 4–12. Reprinted by permission of publisher.

3–6 From Shirley, M. M. *The First Two Years*. Figure 4.13, p. 162. Institute of Child Welfare Monograph No. 7. Minneapolis: University of Minnesota Press. © Copyright 1933, 1961. Preprinted by permission of publisher.

3–7 From Brazelton, T. B. "Behavioral Competence of Newborn Infants." In P. M. Taylor (Ed.), *Parent-Infant Relationships*. Figure 4–3. New York: Grune & Stratton, 1980. Reprinted by permission of publisher.

3–8 From Brazelton, T. B. "Behavioral Competence of Newborn Infants." In P. M. Taylor (Ed.), *Parent-Infant Relationships*. Figure 4–4. New York: Grune & Stratton, 1980. Reprinted by permission of publisher.

Box 4–1 Photograph from Forman, G. E., & Hill, F. *Construction Play: Applying Piaget in the Classroom*, Rev. ed. Menlo Park, CA: Addison-Wesley Publishing Company, Inc., 1984. Reprinted by permission of publisher.

4–1 From Parke, R. D., & Lewis, N. G. "The Family in Context: A Multilevel Interactional Analysis of Child Abuse." Figure on p. 171. In R. W. Henderson (Ed.), *Parent-Child Interaction*. New York: Academic Press, 1981. Reprinted by permission of publisher.

6–1 Reproduced from Tanner, J. M., "Sequence, Tempo, and Individual Variation," in *Twelve to Sixteen: Early Adolescence*, by Jerome Kagan and Robert Cole, by permission of W. W. Norton & Company, Inc. Copyright © 1972 by the American Academy of Arts and Sciences.

6–3 From Marshall, W. A., & Tanner, J. M. "Variations in the Pattern of Pubertal Change in Boys." *Archives of the Diseases of Childhood*, 1970, 45, 13, Figure 4, p. 915.

6–4 From Tanner, J. M., Whitehouse, R. H., & Takaishi, M. "Standards from Birth to Maturity for Height, Weight, Height Velocity, and Weight Velocity in British Children, 1965." *Archives of the Diseases of Childhood*, 1966, 41, pp. 455–471.

6–5 From Shuttleworth, K. The Physical and Mental Growth of Boys and Girls Age Six through Nineteen in Relation to Age of Maximum Growth." *Monographs of the Society for Research in Child Development*, 1939, 4, 245–247. Copyright © The Society for Research in Child Development, Inc.

6–6 Adapted from J. H. Tanner, *Education & Physical Growth*, 1961, by permission of Hodder & Stoughton Educational Ltd.

8–1 From Spanier, G. B., & Lewis, R. A., "Marital Quality: A Review of the Seventies," *Journal of Marriage and the Family*, 1980, 42, pp. 825–841. Copyright 1980 by the National Council on Family Relations, 1910 West County Road B, Suite 147, St. Paul, Minnesota 55113. Reprinted by permission.

9–1 From Smith, D. W., Bierman, E. L., & Robinson, N. M. *The Biological Ages of Man*, 2d ed. Figure 10–12, p. 226. Philadelphia: W. B. Saunders, 1978. Reprinted by permission of publisher.

9–2 From Smith, D. W., Bierman, E. L., & Robinson, N. M. *The Biological Ages of Man*, 2d ed. Figure 10–11, p. 226. Philadelphia: W. B. Saunders, 1978. Reprinted by permission of publisher.

10–1 From Kuypers, J. A., & Bengtson, V. I. "Social Breakdown and Competence." *Human Development*, 1973, 16, 181–201, figure 1. Reprinted by permission.

Box 10–2 Living Will reprinted by permission of Concern for Dying, 250 West 57th Street, New York, NY 10107.

11–1 From Kegan, R. *The Evolving Self*, Cambridge, MA: Harvard University Press, 1982, figure 4. Reprinted by permission of publisher.

INDEX

A 6
B 7
C 8
D 9
E 0
F 1
G 2
H 3
I 4
J 5